France

Vintage	Red Bordeaux				White Bordeaux				Alsace	
	Médoc/Graves		Pom/St-Ém		Sauternes & sw		Graves & dry			
2021	5–7	♈	5–7	♈	7–8	♈	7–9	♈	6–8	♈
2020	7–8	♈	7–9	♈	7–8	♈	7–8	♈	8–9	♈
2019	7–9	♈	6–9	♈	7–8	♈	7–9	♈	6–8	♈
2018	8–9	♈	8–9	♈	7–9	♈	7–8	♈	7–9	♈
2017	6–8	♈	6–7	♈	8–9	♈	7–8	♈	9–10	♈
2016	8–9	♈	8–9	♈	8–10	♈	7–9	♈	7–8	♈
2015	7–9	♈	8–10	♈	8–10	♈	7–9	♈	7–9	♈
2014	7–8	♈	6–8	♈	8–9	♈	8–9	♈	7–8	♈
2013	4–7	◇	4–7	◇	8–9	♈	7–8	♈	8–9	♈
2012	6–8	♈	7–8	♈	5–6	♈	7–9	♈	8–9	♈
2011	7–8	♈	7–8	♈	8–10	♈	7–8	♈	5–7	♈
2010	8–10	♈	7–10	♈	7–8	♈	7–9	♈	8–9	♈
2009	7–10	♈	7–10	♈	8–10	♈	7–9	♈	8–9	♈
2008	6–8	♈	6–9	♈	6–7	♈	7–8	♈	7–8	♈
2007	5–7	♈	6–7	◇	8–9	♈	8–9	♈	6–8	◇
2006	7–8	♈	7–8	♈	7–8	♈	8–9	♈	6–8	♈
2005	9–10	♈	8–9	♈	7–9	♈	8–10	♈	8–9	♈
2004	7–8	♈	7–9	♈	5–7	◇	6–7	◇	6–8	♈
2003	5–9	◇	5–8	◇	7–8	♈	6–7	◇	6–7	◇

France, continued

Vintage	Burgundy						Rhône			
	Côte d'Or red		Côte d'Or white		Chablis		North		South	
2021	6–7	♈	6–7	♈	7–8	♈	6–7	♈	6–8	♈
2020	6–8	♈	7–9	♈	6–8	♈	7–8	♈	6–8	♈
2019	7–9	♈	7–10	♈	7–9	♈	8–9	♈	7–9	♈
2018	6–9	♈	7–8	♈	7–9	♈	7–9	♈	6–8	♈
2017	6–9	♈	8–9	♈	8–9	♈	7–9	♈	7–9	♈
2016	7–8	♈	6–8	♈	5–7	◇	7–9	♈	8–9	♈
2015	7–9	♈	6–8	♈	7–8	♈	8–9	♈	8–9	♈
2014	6–8	♈	8–9	♈	8–9	♈	7–8	♈	7–9	♈
2013	5–7	♈	6–7	◇	6–8	◇	7–9	♈	7–8	♈
2012	8–9	♈	7–8	♈	7–8	♈	7–9	♈	7–9	♈
2011	7–8	♈	7–8	◇	7–8	◇	7–8	♈	6–8	◇
2010	8–10	♈	8–10	◇	8–10	♈	8–10	♈	8–9	♈
2009	7–10	♈	7–8	◇	7–8	♈	7–9	♈	7–8	♈
2008	7–9	♈	7–9	◇	7–9	♈	6–7	◇	5–7	◇

Beaujolais 21 20 19 18 17 15 14; crus will keep. **Mâcon-Villages** (w) drink: 21 20 19 18 17 15. **Loire** (sw Anjou and Touraine) best recent vintages: 21 20 19 18 15 10 09 07 05 02 97 96 93 90; Bourgueil, Chinon, Saumur-Champigny: 21 20 19 18 17 15 14 10 09 06 05 04. **Upper Loire** (Sancerre, Pouilly-Fumé): 21 20 19 18 17 15 14. **Muscadet**: DYA.

HUGH
JOHNSON'S
POCKET
WINE BOOK
2023

Hugh Johnson's Pocket Wine Book 2023

Edited and designed by Mitchell Beazley,
an imprint of Octopus Publishing Group Limited,
Carmelite House, 50 Victoria Embankment
London EC4Y 0DZ
www.octopusbooks.co.uk

An Hachette UK Company
www.hachette.co.uk

Distributed in the US by Hachette Book Group
1290 Avenue of the Americas
4th and 5th Floors
New York, NY 10020
www.octopusbooksusa.com

First edition published 1977

Revised editions published 1978, 1979, 1980, 1981, 1982,
1983, 1984, 1985, 1986, 1987, 1988, 1989, 1990, 1991, 1992,
1993, 1994, 1995, 1996, 1997, 1998, 1999, 2000, 2001,
2002 (twice), 2003, 2004, 2005, 2006 (twice), 2007, 2008,
2009, 2010, 2011, 2012, 2013, 2014, 2015, 2016, 2017, 2018,
2019, 2020, 2021, 2022

A CIP record for this book is available from
the British Library.

ISBN (UK): 978-1-78472-814-4
ISBN (US): 978-1-78472-843-4

The author and publishers will be grateful for any
information that will assist them in keeping future editions
up to date. Although all reasonable care has been taken in
preparing this book, neither the publishers nor the author
can accept any liability for any consequences arising from
the use thereof, or from the information contained herein.

General Editor **Margaret Rand**
Commissioning Editor **Hilary Lumsden**
Senior Editor **Pauline Bache**
Proofreader **David Tombesi-Walton**
Art Director **Yasia Williams-Leedham**
Designer **Jeremy Tilston**
Picture Researchers **Giulia Hetherington and Jennifer Veall**
Senior Production Manager **Katherine Hockley**

Printed and bound in China

Mitchell Beazley would like to acknowledge and thank the following
for supplying photographs for use in this book:

123RF natika 330; **Alamy Stock Photo** Album 321; Jeremy Pembrey 324;
Montager Productions 325; Photo Alto/Isabelle Rozenbaum 329; Vitalii
Marchenko 11; **Cephas Picture Library** Herbert Lehmann 323, 326; **Dom
Fleming** 8; **Getty Images** Stockfood 336; Topical press 331; **iStock** kcline 12;
love_life 6; zoranm 7; **Shutterstock** hxdbzxy 335; Marian Weyo 334; Unni
Karin Henning 14; Valentyn Volkov 10; Verca 328; **Unsplash** Svetlana
Gumerova 332

HUGH
JOHNSON'S
POCKET
WINE BOOK
2023

MARGARET RAND

MITCHELL BEAZLEY

Contributors

To demonstrate that no one person can be up-to-date with the vinous affairs of every wine-producing country in the world, this is the book's team of contributors:

Key: ❶ Facebook; ❷ Instagram; ❸ Twitter; ❹ website; ❺ book/press; ❻ blog/other media

Helena Baker, Czechia, Slovakia: ❶ helena.baker.73; ❷ @bakerwine837; ❸ @HelenaB62554469; ❻ bakerwine.cz

Kristel Balcaen, Belgium: ❷ @kristel_balcaen; ❹ wineandwords.be

Amanda Barnes, S America: ❶ SouthAmericaWineGuide; ❷ @southamericawineguide; ❸ @amanda_tweeter; ❺ *South America Wine Guide*

Lana Bortolot, US SW States, Mexico: ❷ @pourLana; ❻ Forbes

Juliet Bruce Jones MW, Midi, Provence, Corsica: ❶ latasque; ❷ @domainelatasque; ❹ domainelatasque.com

Jim Budd, Loire: ❶ jim.budd.94; ❷ @jymbudd; ❹ les5duvin.wordpress.com; ❻ jimsloire.blogspot.com, investdrinks-blog.blogspot.com

Ch'ng Poh Tiong, Asia: ❶+❷ @ChngPohtiong, 100TopChineseRestaurants; ❹ chngpohtiong.com, wineguru.com.sg, 100chineserestaurants.com

Samantha Cole-Johnson, Oregon: ❷ @samanthacolejohnson; ❸ @S_ColeJohnson; ❹ jancisrobinson.com

Michael Cooper ONZM, NZ: ❶ MichaelCooperNZWine; ❹ michaelcooper.co.nz; ❺ *New Zealand Wines: Michael Cooper's Buyer's Guide, Wine Atlas of New Zealand*

Peter Csizmadia-Honigh, India: ❹ www.thewinesofindia.com

Ian D'Agata, Alsace, Italy: ❷ @iandagata_vino; ❹ iandagata.com, terroirsense.com/en; ❺ *Italy's Native Wine Grape Terroirs, Native Wine Grapes of Italy*

Michael Edwards, Champagne: ❶ Michael's Bulles de Champagne; ❷ @edwardsmichaelfrank; ❸ @michaelfrankedw; ❺ *Finest Wines of Champagne*

Sarah Jane Evans MW, Spain: ❶ Sarah Jane Evans MW; ❷+❸ @sjevansmw; ❹ sarahjaneevans.com; ❺ *The Wines of Northern Spain*

Jane Faulkner, Australia: ❶+❷+❸ @winematters; ❺ *Halliday Wine Companion*

Caroline Gilby MW, E Europe, Cyprus: ❶ Caroline Gilby MW; ❷ @Caroline Gilby; ❸ @CarolineGilbyMW; ❺ *The Wines of Bulgaria, Romania and Moldova*

Anthony Gismondi, Canada: ❶+❷ @Anthony.Gismondi; ❸ @thespitter; ❹ Gismondionwine.com; ❺ *The Vancouver Sun*; ❻ BC Food & Wine Radio Podcast

Susan H Gordon MFA, US E States: ❷ @susanhillaryg; ❺ *ForbesLife, Gastronomica*

Michael Karam, Lebanon: ❺ *Wines of Lebanon, Arak and Mezze, Tears of Bacchus, Lebanese Wine: A Complete Guide*; ❻ Wine and War (Amazon Prime)

Konstantinos Lazarakis MW, Greece: ❶ Konstantinos Lazarakis II; ❷ @Konstantinos Lazarakis MW; ❸ @Lazarakis; ❹ Lazarakis.gr; ❺ *The Wines of Greece*

James Lawther, B'x: ❺ *The Finest Wines of Bordeaux, The Heart of Bordeaux, On Bordeaux*

John Livingstone-Learmonth, Rhône: ❸ @drinkrhone; ❹ sdrinkrhone.com; ❺ *The Wines of the Northern Rhône*

Michele Longo, Italy: ❷ @michele_nebbiolo; ❹ iandagata.com; ❺ *The Grapes and Wines of Italy, Barolo&Co*

Wink Lorch, Bugey, Jura, Savoie: ❶+❷+❸ @WinkLorch; ❹ winetravelmedia.com; ❺ *Jura Wine, Wines of the French Alps*; ❻ jurawine.co.uk

Adam Sebag Montefiore, Israel, Turkey, N Africa: ❹ adammontefiore.com, wines-israel.com; ❺ *Jerusalem Post, The Wine Route of Israel, Wines of Israel*

Jasper Morris MW, Burgundy: ❷ @jaspermorris_insideburgundy; ❸ @justjasper; ❹ insideburgundy.com; ❺ *Inside Burgundy*

Marcel Orford Williams, SW France: buyer, The Wine Society; ❸ @owmarcel

Margaret Rand, General Editor, England: *see back flap of book*; ❹ margaretrand.com

André Ribeirinho, Portugal: ❶+❷+❸ @andrerib; ❹ andre.wine

Ulrich Sautter, Germany, Switzerland, Luxembourg: ❹ falstaff.com, weinverstand.com

Luzia Schrampf, Austria: ❶+❷ @Luzia Schrampf; ❻ *111 Austrian Wines You Must Not Miss, 111 Sparkling Wines Worldwide You Must Not Miss* (both with D Dejnega)

Eleonora Scholes, Black Sea & Caucasus: ❷ @spaziovino; ❹ spaziovino.com

Sean P Sullivan, Washington State & Idaho: ❶+❷ @wawinereport; ❸ @wawinereport; ❹ wawinereport.com; ❺ *Wine Enthusiast*

Tim Teichgraeber, California: ❷ @timskyscraper; ❻ modernwine.blogspot.com

Philip van Zyl, S Africa: ❺ *Platter's by Diners Club South African Wine Guide*

Contents

The top line of most entries consists of the following information:

1. Aglianico del Vulture Bas

2. ★★★

3. 12' 13 14 15 16' 17 (18)

1. Aglianico del Vulture Bas

Wine name and region. Abbreviations of regions are listed in each section.

2. ★★★

Indication of quality – a necessarily rough-and-ready guide:

★	plain, everyday quality
★★	above average
★★★	excellent
★★★★	outstanding, compelling

★ etc. **Stars are coloured for any wine that, in our experience, is usually especially good within its price range. There are good everyday wines as well as good luxury wines. This system helps you find them.**

We try to be objective, but no wine rating can ever be wholly objective.

3. 12' 13 14 15 16' 17 (18)

Vintage information: those recent vintages that are outstanding, and of these, which are ready to drink this year, and which will probably improve with keeping. Your choice for current drinking should be one of the vintage years printed in **bold** type.
Buy light-type years for further maturing.

17 etc. recommended years that may be currently available

16' etc. vintage regarded as particularly successful for the property in question

13 etc. years in bold should be ready for drinking (those not in bold will benefit from keeping)

15 etc. vintages in colour are those recommended as first choice for drinking in 2023. (*See also* Bordeaux introduction, p.100)

(18) etc. provisional rating

The German vintages work on a different principle again: *see* p.170.

Abbreviations

Style references appear in brackets where required:

r	red
w	white
dr	dry
sw	sweet
s/sw	semi-sweet
sp	sparkling

DYA	drink the youngest available
NV	Non-Vintage; in Champagne this means a blend of several vintages for continuity
CHABLIS	properties, areas or terms cross-referred within the section; all grapes cross-ref to Grape Varieties chapter on pp.14–24
Foradori	entries styled this way indicate wine especially enjoyed by Margaret Rand and/or the author of that section (mid-2021–22).

For help in sourcing wines in this book we recommend winesearcher.com

If you have any feedback, please contact us on pocketwine@octopusbooks.co.uk

Agenda 2023

Wine styles are changing. This is not some apocalyptic warning – just fact. Wine does not taste the way it used to.

Is this good or bad? It's certainly riper. New drinkers, coming to red Bordeaux or burgundy without preconceptions, might prefer this new opulent style. Older drinkers might regret a certain loss of tension. But it is what it is. Good growers are adapting, and whereas in the past they focused their viticulture on producing maximum ripeness, now high levels of ripeness are almost a given. The next stage, and sometimes the current stage, is overripeness: too much opulence, not enough acidity. So, growers have changed their focus to acidity, or freshness, or extending the ripening period. It can be done – for a while, anyway. Perhaps in the future the great wines will not be, as they have been for a few lifetimes now, the ripest years; perhaps the greatest terroirs will not be those that deliver more reliable ripeness. Ripeness can be too much of a good thing. Different terroirs are already coming into their own.

We've started with France, so let's have the good news. Loire Chenin, for a long time a challenging proposition in youth, has moved into its comfort zone. It's in a sweet spot now for ripeness, acidity and balance. How long will it stay there? Maybe a decade, according to some estimates, if the world keeps getting warmer.

Cote d'Or burgundy, in most years now, tastes far more opulent than it used to. St-Émilion can sometimes be reminiscent of Tuscany's Maremma, with floral and balsamic notes. Barolo is more approachable than ever before, partly because of better tannin management, but also because those tannins are riper. German Riesling? Making classic light Kabinett is a problem now, but dry wines are better than ever. NV Champagne, blended to consistency year in year out, is gradually giving way to NV blends that are more reflective of change, with ups and downs of style being promoted as a virtue rather than hidden away in the blending room. English wines are having the time of their lives – though not so much in frost-hit, rainy 2021.

Greater ripeness has implications for when we drink wine, and for how long we cellar it. The most obvious change is that we don't need to keep wine as long. The 2018 St-Émilions – not the top Grands Crus Classés, but the good Grands Crus – were often drinking well at three years old. The 2017 Left Bank Crus Classés, at four years old, however, were sleek but closed. In the Supplement this year we explore the whys and wherefores of how wines age – and indeed whether we should care as much about such things as we used to.

Scattered throughout the book you'll find boxes on The New Fine Wines, in which the book's contributors put together their own lists of outstanding examples that break with tradition and will make you sit up and take notice – for all the right reasons.

It was sommelier Raj Parr who came up with the idea of The New Fine Wines (and we'll drop the capital initials soon, promise). They encompass non-classic regions, indigenous grape varieties we hadn't heard of until a few years ago, unconventional winemaking techniques, minimal intervention in the cellar – all sorts of things that go against the ideal of super-polished safe perfection that has become necessary for wines at the highest price points in an international market.

Most of us don't want to buy wines at the highest price points, and can't afford to. Moves away from a uniform international style have opened up the world to us. There are superb wines – original, subtle, thoughtful – from everywhere. Lift your eyes from the classic regions and you'll find them.

But these new wines have also multiplied the confusion. Instead of just remembering a few grape varieties and a few regions, we need to be more aware of the names of producers, and it helps if we have some awareness of different techniques, to help us remember styles we might like and styles we might not. There are pitfalls: not all winemakers are brilliant, and just being minimum-intervention and right-on doesn't automatically mean good. The aim of this book is to guide you through – with pointers, definitions and opinions.

We are increasingly focusing on producers rather than regions because we believe that that is what matters most now. Regional styles have an influence, of course, but it's the name of the producer that distinguishes the best from the mediocre. There is still mediocrity to be found in wine; this book aims to help you bypass it in favour of the interesting, the fun, the distinguished.

And talking of the distinguished: non-fungible tokens (NFTs) have arrived in wine. They have appeared in the form of a tasting note attached to every barrel sold at Barolo En Primeur – a new charity auction, which is what the name suggests. No doubt by the time this book appears there will be others: artists' labels? A version of this book? (Maybe not.)

It reminds us that there are a great many ways to spend money in wine, and a great many different definitions of value. Presumably those buyers of Barolo barrels will regard their NFT as added value. At the opposite end of the scale, "value" is a synonym for cheap: wine bought and sold at the lowest possible price. It reminds me of a supermarket wine buyer who described one of her "value" purchases thus: "Well, it won't kill you".

We hope to do better than that.

Well, you wouldn't call it a great year, on the whole.

Emotionally draining, yes. Much of Europe suffered spring frosts and a chilly, damp summer – considerably more than damp, in fact, in parts of **Germany**, where the Ahr had catastrophic floods. We talk more about these in the Germany chapter, p.154.

Overall, though, Germany was an exception in reporting a more generous harvest than usual; most European countries were down, which is helping to push wine prices up. Frost was the main reason: a few **English** vineyards will make no wine in 2021. For those that managed to hang on to their crop, the autumn was nerve-racking. When I visited a couple of Kent vineyards at the end of October, they needed two more weeks of decent weather: November was mild, but even so, it was a bit of a near thing.

Italy had a decent summer, even a good one, but the spring frosts, especially in the north, took their toll, and most areas are down by anything from 10–50%. But quality is good. In Valpolicella they're very happy, and no mildew equals a good year for Amarone: you can't successfully dry mildewed grapes. Tuscany had a summer drought, but balance is good in the wines, and lovely colour too. Summer was also dry in Piedmont, and acidity seems to have been concentrated as well as flavour and everything else. Campania and Sicily are also reporting remarkable quality, in spite of some frost and in spite of a scorching summer with some wildfires.

The centre of **Spain** had frosts, and Ribera del Duero had localized hail: production will be on the low side, but the country didn't suffer the extremes of climate that France found so punishing. Quality seems good.

Portugal's Douro Valley had its share of hail and storms but actually had a calmer year than 2020 with its record-breaking heat. "Freshness and elegance" is how Fladgate Partnership technical director David Guimaraens characterizes the Port vintage. Roughly 10% of the vineyards in the region are estimated to have suffered real hail damage, but as ever, it was patchy.

France really caught it, though. The 21 harvest was the smallest for several decades, with the Loire, Bordeaux, Champagne, Provence and the Rhône all hit by April frosts on vines encouraged to grow by a warm March. In Champagne the frosts were the worst since the 43% damage of 2003, and they were compounded in 2021 by hail, lots of rain and the inevitable accompaniment, mildew. It will be a Chardonnay year, Chardonnay always being a better bet in a cool wet year than Pinot, and the best parts of the Côte des Blancs made some very good wines. But quantities are well down. A good autumn saved the Loire, or what was left; the Rhône made good whites, and reds that are mostly on the light side and will drink young. Burgundy made light reds and probably better whites. In Bordeaux the lack of hype about the vintage tells its own tale. There will be some good wines, but taste before you buy. The whites are terrific: Pessac-Léognan made whites of beautiful balance and concentration, and Sauternes had plenty of botrytis, providing it had grapes left after frost, rain and mildew.

The low yields that the spring imposed did have an upside, in that only a smaller crop could have ripened in such a year. Most red producers had a race to the finish: did they dare leave the grapes a few more days before picking in order to get full phenolic ripeness, or would the weather turn on them? Sometimes there had to be a compromise, and dodgy berries had to be removed via one of those very expensive bells-and-whistles sorting tables. It was a year when a fair bit of intervention was needed in the winery, including some chaptalization.

Outside Europe, quality in **New Zealand** is good, though quantity less so; and in **Australia** the word "bumper" is being used. Plenty of grapes, plenty of rain at the right time, plenty of nice calm warmth without terrifying heat: conditions seem to have been as good as you can get. Growers there are pretty happy.

As they are in **California**, mostly. Napa and Sonoma had drought, which cut the crop but concentrated the flavours. In Russian River Valley, water allowances were cut, forcing some growers to make uncomfortable choices. But August brought cooler conditions, which slowed ripening a little and gave the Cabernet better balance. And wildfires were thankfully less of a problem than they were in 2020, when smoke taint was a widespread issue.

On the other side of the country, **New York State** had a wet, warm summer, with good whites and decent reds. Berries were notably big, which will translate into bigger yields, but the opposite of California's extra concentration. Grape prices are up. What isn't?

Ten wines to try in 2023

These have all made me go "Wow!" this year.
And I don't wow easily.

White and orange wines

Viña Corrales, Pago Balbaína, Jerez, Spain
Lovely petrichor nose, very delicate, tense and salty – intense
delicacy. Huge length. Peter Sisseck reckons that Fino is the
greatest white wine of Spain, and he bought a bodega here in
2017. The vineyard is chalky and overlooks the sea; the blend
has an average age of nine years. There are plenty of good
Finos around, but this is exceptional.

Exton Park RB45 Blanc de Blancs, Hampshire, England
You can smell the sea (the English Channel this time)
from this Hampshire vineyard, and unpredictable English
summers mean that every year is very different. Hence the
decision to blend 45 different reserve wines into a single
multi-vintage expression. It's smoky, tense, rich and powerful,
but crisp and saline too. The RB (Reserve Blend) wines put
Exton Park right at the top level of English wine.

Quarzit Oestrich, Peter Jakob Kühn, Rheingau, Germany
German wine has never been better than it is now, or more
suited to contemporary food. If you don't know what to drink
with something, try this bio number from a family of growers
who've been in their vineyards for 230 years. It's a blend of
Riesling grapes from various sites with, as the name suggests,
quartzite soils, and that crystalline quality is there in the wine
too. Pear and tarragon flavours; pure, light, firm, very long.

Dragon Langhe Bianco, Luigi Baudana, Piedmont, Italy
A white from the coolest corners of commune Serralunga in
Barolo territory, a semi-field-blend of Chardonnay/Nascetta/
Sauvignon Blanc/Riesling: complex and unusual, winey and
salty, very long. Superb value. A semi-field-blend because the
Nascetta, a local speciality, is planted separately; but all arc
co-fermented. Luigi and Fiorina Baudana originally had just
2.6 ha, which is about as tiny as you can get, and in 2009 the Vajra
family took over the estate while maintaining the Baudana name.

Ni Malagouzia, Chatzivaritis Estate, Macedonia, Greece
Foot-trodden, spontaneous fermentation, unfined, unfiltered, low
sulphites: this orange wine had one month in clay amphorae on skins,
and couldn't be trendier if it tried. Malagouzia is one of those amazingly
aromatic Greek grapes that have been brought back from near-extinction.
And it's lovely: very delicate and poised, with a nice bit of grip, savoury,
fresh and long. Sourdough, fresh plums. Pretty pale orange.

Red wines

Pleiades XXVIII, Sean Thackrey, California, US

An unlikely red blend of Sangiovese, Viognier, Pinot Noir, Syrah and Mourvèdre (the precise blend changes each vintage) sourced from old vines in different vineyards and made in Marin County. Thackrey is largely self-taught and works by intuition; he lets the fermenting grapes "rest" outside under the stars, which is not an idea you will find embraced at multinationals. Big, black, figgy fruit, plenty of acidity and huge depth; a tense balance of lush ripeness and dancing freshness. Drink now.

Los Peros Tinto, Ca' di Mat, Gredos, Spain

From 75-year-old vines, and a brilliant advertisement for vine age. Lavender and rosemary fruit, pure and complex, in a taut frame of depth and precision. Mountainous Gredos is a poster region for new-wave Garnacha, often from old vines that had been abandoned until rediscovered by a new tribe of winemakers. Expect Pinot-like lightness and aroma, and a delicacy you never knew Garnacha could have – not least because Los Peros is a village at 850m (2789ft) altitude.

Grifalco, Basilicata, Italy

We can thank volcanoes and the magically enlivening effect their soil has on wine for this one, although its particular volcano is not technically extinct. The name is a mix of the griffin, the symbol of Montepulciano, where the family came from, and the falcon, the symbol of Monte Vulture. (Yes, I know it should be a vulture. I don't control these things.) The 45-year-old Aglianico vines give great depth and detail, plenty of acidity and perfect balance. It's both powerful and dancing. Very long.

Signé Syrah/Viognier, Yves Cuilleron, Rhône, France

Here is a Côtie-Rôtie blend of Syrah with 15% Viognier, from outside the AOC, made to drink earlier – from a superb grower, whose wines are always complex and precise, and always noteworthy in any lineup. "Signé" = Syrah + Viognier: get it? Blackberries, violets and cream combine in a deep and seductive wine for early drinking. Serious and aromatic.

Sercial 1964, Henriques & Henriques, Madeira

H&H always has old, very old and very, very old vintages of Madeira available, and you should try one at least once in your life. They seem to be immortal: you might even find 18th-century Madeiras at auction, if you want to splash the cash. I picked the 1964 because I happened to taste it; just try whatever you find. Old Sercial is powerful and delicate, searing and rich, reminiscent (if you'll forgive me for lowering the tone) of salt and vinegar crisps. Well, I think so.

Grape varieties

In the past two decades a radical change has come about in all except the most long-established wine countries: the names of a handful of grape varieties have become the ready-reference to wine. In senior wine countries, above all France and Italy, more complex traditions prevail. All wine of old prestige is known by its origin, more or less narrowly defined – not just by the particular fruit juice that fermented. For the present the two notions are in rivalry. Eventually the primacy of place over fruit will become obvious, at least for wines of quality. But for now, for most people, grape tastes are the easy reference point – despite the fact that they are often confused by the added taste of oak. If grape flavours were really all that mattered, this would be a very short book. But of course they *do* matter, and a knowledge of them both guides you to flavours you enjoy and helps comparisons between regions. Hence the originally Californian term "varietal wine", meaning, in principle, made from one grape variety. At least seven varieties – Cabernet Sauvignon, Pinot Noir, Riesling, Sauvignon Blanc, Chardonnay, Gewurztraminer and Muscat – taste and smell distinct and memorable enough to form international wine categories. To these add Merlot, Malbec, Syrah, Sémillon, Chenin Blanc, Pinots Blanc and Gris, Sylvaner, Viognier, Nebbiolo, Sangiovese, Tempranillo. The following are the best and/or most popular wine grapes.

All grapes and synonyms are cross-referenced in SMALL CAPITALS throughout every section of this book.

Grapes for red wine

Agiorgitiko Greek; the grape of Nemea, now planted all over Greece. Versatile and delicious, from soft and charming to dense and age-worthy. A must-try.

Aglianico S Italy's best red, the grape of Taurasi; dark, deep and fashionable.

Alicante Bouschet Used to be shunned, now stylish in Alentejo, Chile, esp old vines.

Aragonêz *See* TEMPRANILLO.

Auxerrois *See* MALBEC, if red. White Auxerrois has its own entry in White Grapes.

Băbească Neagră Traditional "black grandmother grape" of Moldova; light body and ruby-red colour.

Babić Dark grape from Dalmatia, grown in stony seaside vyds around Šibenik. Exceptional quality potential.

Baga Portugal. Bairrada grape. Dark, tannic, fashionable. Needs a gd grower.

Barbera Widely grown in Italy, best in Piedmont: high acidity, low tannin, cherry fruit. Ranges from serious and age-worthy to semi-sweet and frothy. Fashionable in California and Australia; promising in Argentina.

Blauburger Austrian cross of BLAUER PORTUGIESER, BLAUFRÄNKISCH. Simple wines.

Blauburgunder *See* PINOT N.

Blauer Portugieser Central European, esp Germany (Rheinhessen, Pfalz, mostly for rosé), Austria, Hungary. Light, fruity reds: drink young, slightly chilled.

Blaufränkisch (Kékfrankos, Lemberger, Modra Frankinja) Widely planted in Austria's Mittelburgenland: medium-bodied, peppery acidity, fresh, berry aromas, eucalyptus. Can be top quality: Austria's star red. Often blended with CAB SAUV or ZWEIGELT. Lemberger in Germany (esp Württemberg), Kékfrankos in Hungary, Modra Frankinja in Slovenia.

Bobal Spain. Can be rustic; best at high altitude. Gd acidity.

Bogaskere Tannic and Turkish. Produces full-bodied wines.

Bonarda Ambiguous name. In Oltrepò Pavese, an alias for Croatina, soft fresh *frizzante* and still red. In Lombardy and Emilia-Romagna an alias for Uva Rara. Different in Piedmont. Argentina's Bonarda can be any of these, or something else. None is great.

Bouchet St-Émilion alias for CAB FR.

Brunello SANGIOVESE, splendid at Montalcino.

Cabernet Franc [Cab Fr] In B'x: more important than CAB SAUV in St-Émilion. Outperforms Cab Sauv in Loire (Chinon, Saumur-Champigny, rosé), in Hungary (depth and complexity in Villány and Szekszárd) and often in Italy. Much of NE Italy's Cab Fr turned out to be CARMENÈRE. Used in B'x blends of Cab Sauv/MERLOT across the world.

Cabernet Sauvignon [Cab Sauv] Characterful: slow-ripening, spicy, herby, tannic, with blackcurrant aroma. Main grape of the Médoc; also makes some of the best California, S American, E European reds. Vies with SHIRAZ in Australia. Grown almost everywhere, but few places make great varietal Cab Sauv: usually benefits from blending with eg. MERLOT, CAB FR, SYRAH, TEMPRANILLO, SANGIOVESE, etc. Makes aromatic rosé. Top wines need ageing.

Cannonau GRENACHE in its Sardinian manifestation; can be v. fine, potent.

Carignan (Carignane, Carignano, Cariñena) Low-yielding old vines now fashionable everywhere from S France to Chile, via S Africa. Lots of depth, vibrancy, but must never be overcropped. Common in N Africa, Spain (as Cariñena), California.

Carignano *See* CARIGNAN.

Cariñena *See* CARIGNAN.

Carmenère An old B'x variety now a star, rich and deep, in Chile (where it's pronounced "carmeneary"); B'x is looking at it again.

Castelão *See* PERIQUITA.

Cencibel *See* TEMPRANILLO.

Chiavennasca See NEBBIOLO.

Cinsault (Cinsaut) A staple of S France, v.gd if low-yielding, hopeless if not. Makes gd rosé. One of parents of PINOTAGE.

Cornalin du Valais Swiss speciality with high potential, esp in Valais.

Corvina Dark and spicy; one of best grapes in Valpolicella blend. Corvinone, even darker, is a separate variety.

Côt See MALBEC.

Dolcetto Source of relatively light red in Piedmont. Now high fashion.

Dornfelder Gives deliciously light reds, straightforward, often rustic, and well coloured in Germany, parts of the US, even England.

Duras Spicy, peppery, structured; exclusive to Gaillac and parts of Tarn V, SW France.

Fer Servadou Exclusive to SW France, aka Mansois in Marcillac, Braucol in Gaillac and Pinenc in St-Mont. Redolent of red summer fruits and spice.

Fetească Neagră Romania: "black maiden grape" with potential as showpiece variety; can give deep, full-bodied wines with character.

Frühburgunder An ancient German mutation of PINOT N, mostly in Ahr but also in Franken and Württemberg, where it is confusingly known as Clevner. Lower acidity than Pinot N.

Gamay The Beaujolais grape: light, fragrant wines, best young, except in Beaujolais crus (see France) where quality can be high, wines for 2–10 yrs. Grown in Loire Valley, Central France, Switzerland, Savoie, Canada. California's Napa Gamay is Valdiguié.

Gamza See KADARKA.

Garnacha (Cannonau, Garnatxa, Grenache) Important pale, potent grape for warm climates, fashionable with *terroiristes* because it expresses its site. Can be beefy (Priorat) or delicate (Gredos), but usually quite high alc. Oak, extraction now less. The base of Châteauneuf-du-Pape. Rosé and *vin doux naturel* in S France, Spain, California. Old-vine versions prized in S Australia. Often blended. Cannonau in Sardinia, Grenache in France.

Garnatxa See GARNACHA.

Graciano Spanish; part of Rioja blend. Aroma of violets, tannic, lean structure, a bit like PETIT VERDOT. Difficult to grow but increasingly fashionable.

Grenache See GARNACHA. **GSM**: Grenache/SHIRAZ/MOURVÈDRE blend.

Grignolino Italy: gd everyday table wine in Piedmont.

Kadarka (Gamza) Makes spicy, light reds in E Europe. In Hungary revived, esp for Bikavér.

Kalecik Karasi Turkish: sour-cherry fruit, fresh, supple. Bit like GAMAY. Drink young.

Kékfrankos Hungarian BLAUFRÄNKISCH.

Lagrein N Italian, dark, bitter finish, rich, plummy. DOC in Alto Adige (see Italy).

Lambrusco Productive grape of lower Po V; cheerful, sweet, fizzy, can be v.gd.

Lefkada In Cyprus, higher quality than MAVRO. Usually blended, as tannins can be aggressive. Called Vertzami in its Greek homeland.

Lemberger See BLAUFRÄNKISCH.

Malbec (Auxerrois, Côt) Minor in B'x, major in Cahors (alias Auxerrois) and the star in Argentina. Dark, dense, tannic but fleshy wine capable of real quality. High-altitude versions in Argentina best. Bringing Cahors back into fashion.

Maratheftiko Deep-coloured Cypriot grape with quality potential.

Marselan CAB SAUV X GRENACHE, 1961; gd colour, structure, supple tannins, ages well. A success in China.

Mataro See MOURVÈDRE.

Mavro Most planted black grape of Cyprus but only moderate quality. Best for rosé.

Mavrodaphne Greek; means "black laurel". Sweet fortifieds, speciality of Patras, also in Cephalonia. Dry versions too, great promise.

Mavrotragano Greek; almost extinct; now revived; found on Santorini. Top quality.

Mavrud Probably Bulgaria's best. Spicy, dark, plummy late-ripener native to Thrace. Ages well.

Melnik Bulgarian; from region of same name. There are two Melniks: Shiroka (Broadleafed) M and its offspring, Early M. Both have dark colour and nice dense, tart-cherry character, and age well.

Mencía Making waves in Bierzo, N Spain. Aromatic, with steely tannins and lots of acidity.

Merlot The grape behind the great fragrant and plummy wines of Pomerol and (with CAB FR) St-Émilion, a vital element in the Médoc, soft and strong in California, Washington, Chile, Australia. Lighter, often gd in N Italy (can be world-class in Tuscany), Italian Switzerland, Slovenia, Argentina, S Africa, NZ, etc. More often dull than great. Much planted in E Europe, esp Romania.

Meunier See PINOT MEUNIER.

Modra Frankinja See BLAUFRÄNKISCH.

Modri Pinot See PINOT N.

Monastrell See MOURVÈDRE.

Mondeuse In SAVOIE; the skier's red; deep-coloured, gd acidity. Related to SYRAH.

Montepulciano Deep-coloured, dominant in Italy's Abruzzo and important along Adriatic coast from Marches to S Puglia. Also name of a Tuscan town, unrelated.

Morellino SANGIOVESE in Maremma, S Tuscany. Esp Scansano.

Mourvèdre (Mataro, Monastrell) Star of S France (eg. Bandol, growing influence Châteauneuf-du-Pape), Australia (aka Mataro), Spain (aka Monastrell). Excellent dark, aromatic, tannic; gd for blending. Also S Australia, California, S Africa.

Napa Gamay Identical to Valdiguié (S France). Nothing to get excited about.

Nebbiolo (Chiavennasca, Spanna) One of Italy's best; makes Barolo, Barbaresco, Gattinara and Valtellina. Intense, nobly fruity, perfumed wine; tannins now better managed, still improves for yrs.

Negroamaro Puglian "black bitter" red grape with potential for either high quality or high volume.

Nerello Mascalese Sicilian red grape, esp Etna; characterful, best v. elegant, fine.

Nero d'Avola Dark-red grape of Sicily, quality levels from sublime to industrial.

Nielluccio Corsican; plenty of acidity and tannin. Gd for rosé.

Öküzgözü Soft, fruity Turkish grape, usually blended with BOĞASKERE, rather as MERLOT in B'x is blended with CAB SAUV.

País (Listán Prieto, Mission) Traditional/trendy in Chile, rustic. Listán Prieto in Canaries, Mission in California.

Pamid Bulgarian: light, soft, everyday red

Periquita (Castelão) Common in Portugal, esp around Setúbal. Originally nicknamed Periquita after Fonseca's popular (trademarked) brand. Firm-flavoured, raspberryish reds develop a figgish, tar-like quality.

Petite Sirah Nothing to do with SYRAH; gives rustic, tannic, dark wine. Brilliant blended with ZIN in California; also found in S America, Mexico, Australia.

Petit Verdot Excellent but awkward Médoc grape, now increasingly planted in CAB areas worldwide for extra fragrance. Mostly blended but some gd varietals, esp in Virginia.

Pinotage Singular S African cross (PINOT N x CINSAULT). Has had a rocky ride; getting better from top producers. Gd rosé too. "Coffee Pinotage" is espresso-flavoured, sweetish, aimed at youth.

Pinot Crni See PINOT N.

Pinot Meunier (Schwarzriesling) [Pinot M] The 3rd grape of Champagne, better known as Meunier, great for blending but occasionally fine in its own right. Best on chalky sites (Damery, Leuvigny, Festigny) nr Épernay.

Pinot Noir (Blauburgunder, Modri Pinot, Pinot Crni, Spätburgunder) [Pinot N]
Glory of Burgundy's Côte d'Or. Fine in Alsace and Germany; v.gd in Austria, esp in Kamptal, Burgenland, Thermenregion. Light in Hungary; mainstream, light to weightier in Switzerland (aka Clevner). Splendid in Sonoma, Carneros, Central Coast, also Oregon, Ontario, Yarra V, Adelaide Hills, Tasmania, NZ's Central Otago, S Africa's Walker Bay. Some v. pretty Chileans. New French clones promise improvement in Romania. Modri Pinot in Slovenia; probably country's best red. In Italy, best in ne, gets worse as you go s. PINOTS BL and GR mutations of Pinot N.

Plavac Mali (Crljenak) Croatian, and offspring of ZIN, aka PRIMITIVO, Crljenak, Kratosija. Lots of quality potential, can age well, but can also be alcoholic, dull.

Primitivo S Italian grape, originally from Croatia, making big, dark, rustic wines, now fashionable because genetically identical to ZIN. Early ripening, hence the name. The original name for both seems to be Tribidrag.

Refosco (Refošk) Various DOCs in Italy, esp Colli Orientali. Deep, flavoursome, age-worthy, esp in warmer climates. Dark, high acidity. Refošk in Slovenia and points e, genetically different, tastes similar. On limestone karst in Slovenia takes PDO of Teran, which otherwise is a grape. Got it?

Refošk See REFOSCO.

Roter Veltliner Austrian; unrelated to GRÜNER V. There is also a Frühroter and a Brauner Veltliner.

Rubin Bulgarian cross, NEBBIOLO X SYRAH. Peppery, full-bodied.

Sagrantino Italian grape grown in Umbria for powerful, cherry-flavoured wines.

St-Laurent Dark, smooth, full-flavoured Austrian speciality, tricky to grow and make. Can be light and juicy or deep and structured. Also in Pfalz.

Sangiovese (Brunello, Morellino, Sangioveto) Principal red grape of Tuscany and central Italy. Its characteristic astringency is hard to get right, but sublime and long-lasting when it is. Dominant in Chianti, Vino Nobile, Brunello di Montalcino, Morellino di Scansano and various fine IGT offerings. Also in Umbria (eg. Montefalco and Torgiano) and across the Apennines in Romagna and Marches. Not so clever in the warmer, lower-altitude vyds of the Tuscan coast, nor in other parts of Italy despite its nr-ubiquity. Interesting in Australia.

Sangioveto See SANGIOVESE.

Saperavi The main red of Georgia, Ukraine, etc. Blends well with CAB SAUV (eg. in Moldova). Huge potential, seldom gd winemaking.

Schiava See TROLLINGER.

Schioppettino NE Italian, high acidity, high quality. Elegant, refined, can age.

Schwarzriesling PINOT M in Württemberg.

Sciacarello Corsican, herby and peppery. Not v. tannic.

Shiraz See SYRAH.

Spanna See NEBBIOLO.

Spätburgunder German for PINOT N.

Syrah (Shiraz) The great Rhône red grape: tannic, purple, peppery, matures superbly. Important as Shiraz in Australia, increasingly gd under either name in Chile, S Africa, terrific in NZ (esp Hawke's Bay). Widely grown, gd traveller.

Tannat Raspberry-perfumed, highly tannic force behind Madiran, Tursan and other firm reds from SW France. Also rosé. The star of Uruguay.

Tempranillo (Aragonêz, Cencibel, Tinto Fino, Tinta del País, Tinta Roriz, Ull de Llebre) Aromatic, fine Rioja grape, called Ull de Llebre in Catalonia, Cencibel in La Mancha, Tinto Fino in Ribera del Duero, Tinta Roriz in Douro, Tinta del País in Castile, Aragonêz in S Portugal, now in Australia too; v. fashionable; elegant in cool climates, beefy in warm. Early ripening, long maturing.

Teran (Terrano) Close cousin of REFOSCO.

Teroldego Rotaliano Trentino's best indigenous variety; serious, full-flavoured, esp on the flat Campo Rotaliano.

Tinta Amarela *See* TRINCADEIRA.

Tinta del País *See* TEMPRANILLO.

Tinta Negra (Negramoll) Until recently called Tinta Negra Mole. Easily Madeira's most planted grape and the mainstay of cheaper Madeira. Now coming into its own in colheita wines (*see* Portugal).

Tinta Roriz *See* TEMPRANILLO.

Tinto Fino *See* TEMPRANILLO.

Touriga Nacional [Touriga N] The top Port grape, now widely used in the Douro for floral, stylish table wines. Australian Touriga is usually this; California's Touriga can be either this or Touriga Franca.

Trincadeira (Tinta Amarela) Portuguese; v.gd in Alentejo for spicy wines. Tinta Amarela in the Douro.

Trollinger (Schiava, Vernatsch) Popular pale red in Württemberg; aka Vernatsch and Schiava. Covers group of vines, not necessarily related. In Italy, snappy, brisk.

Ull de Llebre *See* TEMPRANILLO.

Vernatsch *See* TROLLINGER.

Vranac W Balkans, old and varied. Related to ZIN – they're all connected around there. Soft, high tannin, colour, alc, can age.

Xinomavro Greece's answer to NEBBIOLO. "Sharp-black"; the basis for Naoussa, Rapsani, Goumenissa, Amindeo. Some rosé, still or sparkling. Top quality, can age for decades. Being tried in China.

Zinfandel [Zin] California. Blackberry-like, sometimes metallic, flavour. Love it or hate it. Can be serious, structured, long-ageing, or simple. Pink too. Same as S Italian PRIMITIVO.

Zweigelt (Blauer Zweigelt) BLAUFRÄNKISCH X ST-LAURENT, popular in Austria for aromatic, dark, supple wines. Underrated. Also in Hungary, Germany.

Grapes for white wine

Airén Bland workhorse of La Mancha, Spain: fresh if made well. Old-vine versions can surprise.

Albariño (Alvarinho) Fashionable, expensive in Spain: apricot-scented, gd acidity. Superb in Rías Baixas; shaping up elsewhere, but not all live up to the hype. Alvarinho in Portugal just as gd: aromatic Vinho Verde, esp in Monção, Melgaço.

Aligoté Burgundy's 2nd white grape, now trendy and often serious. Widely planted in E Europe, Russia.

Alvarinho *See* ALBARIÑO.

Amigne One of Switzerland's speciality grapes, traditional in Valais, esp Vétroz. Full-bodied, tasty, often sweet but also bone-dry.

Ansonica *See* INSOLIA.

Arinto Portuguese, rather gd; mainstay of aromatic, citrus Bucelas; also adds welcome zip to blends, esp in Alentejo.

Arneis Fine, aromatic, appley-peachy, high-priced NW Italian grape, DOCG in Roero, DOC in Langhe, Piedmont.

Arvine Rare but excellent Swiss *spécialité*, from Valais. Also Petite Arvine. Dry or sweet, fresh, long-lasting wines with salty finish.

Assyrtiko From Santorini; one of the best grapes of the Mediterranean, balancing power, minerality, extract and high acid. Built to age. Could conquer the world…

Auxerrois Red Auxerrois is a synonym for MALBEC, but white Auxerrois is like a fatter, spicier PINOT BL. Found in Alsace and much used in Crémant; also Germany.

Bacchus German-bred crossing, England's answer to NZ SAUV BL; v. aromatic, can be coarse.

Beli Pinot *See* PINOT BL.

Blanc Fumé *See* SAUV BL.

Boal *See* BUAL.

Bourboulenc This and the rare Rolle make some of the Midi's best wines.

Bouvier Indigenous aromatic Austrian grape, esp gd for Beerenauslese and Trockenbeerenauslese, rarely for dry wines.

Bual (Boal) Makes top-quality sweet Madeira wines, not quite so rich as MALMSEY.

Cabernet Blanc (Cab Bl) German PiWi (*see* Germany) with SAUV BL-like flavours, no need to spray.

Carricante Italian. Principal grape of Etna Bianco, regaining ground.

Catarratto Prolific white grape found all over Sicily, esp in w in DOC Alcamo.

Cerceal *See* SERCIAL.

Chardonnay (Morillon) [Chard] Grape of Burgundy and Champagne, ubiquitous worldwide, easy to grow and vinify. Reflects terroir but also winemaker's intentions: often a fashion victim. Can be steely or fat. Also name of a Mâcon-Villages commune. Morillon in Styria, Austria.

Chasselas (Fendant, Gutedel) Swiss (originated in Vaud). Neutral flavour, can be elegant (Geneva); refined, full (Vaud); exotic, racy (Valais). Fendant in Valais. Makes almost 3rd of Swiss wines but giving way, esp to red. Also in France, esp Savoie. Gutedel in Germany, esp S Baden. Elsewhere usually a table grape.

Chenin Blanc [Chenin Bl] Wonderful white grape of the middle Loire (Vouvray, Layon, etc). Wine can be dry or sweet (or v. sweet), but with plenty of acidity. Superb old-vine versions in S Africa, esp Swartland.

Cirfandl *See* ZIERFANDLER.

Clairette Important Midi grape, low-acid, part of many blends. Improved winemaking helps.

Colombard Slightly fruity, nicely sharp grape, makes everyday wine in S Africa, California and SW France. Often blended.

Dimiat Perfumed Bulgarian grape, made dry or off-dry, or distilled. Far more synonyms than any grape needs.

Encruzado Portuguese, serious; fresh, versatile, ages well. Esp gd in Dão.

Ermitage Swiss for MARSANNE.

Ezerjó Hungarian, with sharp acidity. Name means "thousand blessings".

Falanghina Italian: ancient grape of Campanian hills; gd dense, aromatic dry whites.

Fendant *See* CHASSELAS.

Fernão Pires *See* MARIA GOMES.

Fetească Albă / Regală (Királyleanyka, Leanyka) Romania has two Fetească grapes, both with slight MUSCAT aroma. F Regală is a cross of F Albă and Frâncușă; more finesse, gd for late-harvest wines. F NEAGRĂ (unrelated) is dark-skinned.

Fiano High quality, giving peachy, spicy wine in Campania, S Italy.

Folle Blanche (Gros Plant) High acid/little flavour make this ideal for brandy. Gros Plant in Brittany, Picpoul in Armagnac, but unrelated to true PICPOUL. Also respectable in California.

Friulano (Sauvignonasse, Sauvignon Vert) N Italian: fresh, pungent, subtly floral. Used to be called Tocai Friulano. Best in Collio, Isonzo, Colli Orientali. Found in nearby Slovenia as Sauvignonasse; also in Chile, where it was long confused with SAUV BL. Ex-Tocai in Veneto now known as Tai.

Fumé Blanc *See* SAUV BL.

Furmint (Šipon) Superb, characterful. The trademark of Hungary, both as principal grape in Tokaji and as vivid, vigorous dry wine, sometimes mineral, sometimes apricot-flavoured, sometimes both. Šipon in Slovenia. Some grown in Rust, Austria for sweet and dry.

Garganega Best grape in Soave blend; also in Gambellara. Top, esp sweet, age well.

Garnacha Blanca (Grenache Blanc) White version of GARNACHA/Grenache, much used in Spain and S France. Low acidity. Can be innocuous, or surprisingly gd.

Gewurztraminer (Traminac, Traminec, Traminer, Tramini) [Gewurz] One of the most pungent grapes, spicy with aromas of rose petals, face cream, lychees, grapefruit. Often rich and soft, even when fully dry. Best in Alsace; also gd in Germany (Baden, Pfalz, Sachsen), E Europe, Australia, California, Pacific Northwest and NZ. Can be relatively unaromatic if just labelled Traminer (or variants). Italy uses the name Traminer Aromatico for its (dry) "Gewurz" versions. (Name takes umlaut in German.) Non-aromatic version is SAVAGNIN.

Glera Uncharismatic new name for Prosecco vine: Prosecco is now wine only in the EU, but still a grape name in Australia.

Godello Top quality (intense, mineral) in nw Spain. Called Verdelho in Dão, Portugal, but unrelated to true VERDELHO.

Grasă (Kövérszőlő) Romanian; name means "fat". Prone to botrytis; important in Cotnari, potentially superb sweet wines. Kövérszőlő in Hungary's Tokaj region.

Graševina See WELSCHRIESLING.

Grauburgunder See PINOT GR.

Grechetto Ancient grape of central and S Italy noted for vitality, stylishness of wine. Blended, or used solo in Orvieto.

Greco S Italian: there are various Grecos, probably unrelated, perhaps of Greek origin. Brisk, peachy flavour, most famous as Greco di Tufo. Greco di Bianco is from semi-dried grapes. Greco Nero is a black version.

Grenache Blanc See GARNACHA BLANCA.

Grillo Italy: main grape of Marsala. Also v.gd full-bodied dry table wine.

Gros Plant See FOLLE BLANCHE.

Grüner Veltliner [Grüner V] Austria's fashionable flagship white; v. diverse – from simple, peppery, everyday, to great complexity, ageing potential. Useful because gd at all levels. Found elsewhere in Central Europe and outside.

Gutedel See CHASSELAS.

Hárslevelű Other main grape of Tokaji, but softer, peachier than FURMINT. Name means "linden-leaved"; gd in Somló, Eger as well.

Heida Swiss for SAVAGNIN.

Humagne Swiss speciality, older than CHASSELAS. Fresh, plump, not v. aromatic. Humagne Rouge is not related but increasingly popular: same as Cornalin du Aosta. Cornalin du Valais is different. (Keep up at the back, there.)

Insolia (Ansonica, Inzolia) Sicilian; Ansonica on Tuscan coast. Fresh, racy wine at best. May be semi-dried for sweet wine.

Irsai Olivér Hungarian cross; aromatic, MUSCAT-like wine for drinking young.

Johannisberg Swiss for SILVANER.

Kéknyelű Low-yielding, flavourful grape giving one of Hungary's best whites. Has the potential for fieriness and spice. To be watched.

Kerner Quite successful German cross. Early ripening, flowery (but often too blatant) wine with gd acidity.

Királyleanyka See FETEASCĂ ALBĂ/REGALĂ.

Koshu More or less indigenous Japanese table-turned-wine grape, much hyped. Fresh, tannic. Orange versions gd.

Kövérszőlő See GRASĂ.

Laški Rizling See WELSCHRIESLING.

Leányka See FETEASCĂ ALBĂ.

Listán See PALOMINO.

Longyan (Dragon Eye) Chinese original; gd substantial, aromatic wine.

Loureiro Best Vinho Verde grape after ALVARINHO: delicate, floral. Also in Spain.

Macabeo See VIURA.

Maccabeu *See* VIURA.

Malagousia Rediscovered Greek grape for gloriously perfumed wines.

Malmsey *See* MALVASIA. The sweetest style of Madeira.

Malvasia (Malmsey, Malvazija, Malvoisie, Marastina) Italy, France and Iberia. Not a single variety but a whole stable, not necessarily related or even alike. Can be white or red, sparkling or still, strong or mild, sweet or dry, aromatic or neutral. Slovenia's and Croatia's version is Malvazija Istarka, crisp and light, or rich, oak-aged. Sometimes called Maŕastina in Croatia. Malmsey (as in the sweetest style of Madeira) is a corruption of Malvasia.

Malvoisie *See* MALVASIA. Name used for several varieties in France, incl BOURBOULENC, Torbato, VERMENTINO. Also PINOT GR in Switzerland's Valais.

Manseng, Gros / Petit Gloriously spicy, floral whites from SW France. The key to Jurançon. Superb late-harvest and sweet wines too.

Maria Gomes (Fernão Pires) Portuguese; aromatic, ripe-flavoured, slightly spicy whites in Bairrada and Tejo.

Marsanne (Ermitage) Principal white grape (with ROUSSANNE) of the N Rhône (Hermitage, St-Joseph, St-Péray). Also gd in Australia, California and (as Ermitage Blanc) the Valais. Soft, full wines that age v. well.

Melon de Bourgogne *See* MUSCADET.

Misket, Red Bulgarian. Pink-skinned, mildly aromatic; the basis of most country whites. There are many other Miskets, all recent crosses and less planted.

Morillon CHARD in parts of Austria.

Moscatel *See* MUSCAT.

Moscato *See* MUSCAT.

Moschofilero Pink-skinned, rose-scented, high-quality, high-acid, low-alc Greek grape. Makes white, some pink, some sparkling.

Müller-Thurgau [Müller-T] Aromatic wines to drink young. Makes gd sweet wines but usually dull, often coarse, dry ones. In Germany, most common in Pfalz, Rheinhessen, Nahe, Baden, Franken. Has some merit in Italy's Trentino-Alto Adige, Friuli. Sometimes called RIES x SYLVANER (incorrectly) in Switzerland.

Muscadelle Adds aroma to white B'x, esp Sauternes. In Victoria used (with MUSCAT, to which it is unrelated) for Rutherglen Muscat.

Muscadet (Melon de Bourgogne) Light, refreshing, dry wines with seaside tang to complex ones around Nantes. Also found (as Melon) in parts of Burgundy.

Muscat (Moscatel, Moscato, Muskateller) Many varieties; the best is Muscat Blanc à Petits Grains (alias Gelber Muskateller, Rumeni Muškat, Sarga Muskotály, Yellow Muscat, Tămâioasă Românească). Widely grown, easily recognized, pungent grapes, mostly made into perfumed sweet wines, often fortified, as in France's *vin doux naturel*. Superb, dark and sweet in Australia. Sweet, sometimes v.gd in Spain. Most Hungarian Muskotály is Muscat Ottonel except in Tokaj, where Sarga Muskotály rules, adding perfume (in small amounts) to blends. Occasionally (eg. Alsace, Austria, parts of S Germany) made dry. Sweet Cap Corse Muscats often superb. Light Moscato fizz in N Italy.

Muskateller *See* MUSCAT.

Narince Turkish; fresh and fruity wines.

Neuburger Austrian, rather neglected; mainly in the Wachau (elegant, flowery), Thermenregion (mellow, ample-bodied) and n Burgenland (strong, full).

Olaszrizling *See* WELSCHRIESLING.

Païen *See* SAVAGNIN.

Palomino (Listán) Great grape of Sherry; little intrinsic character, but gains all from production method. Now table wine too. As Listán, makes dry white in Canaries.

Pansa Blanca *See* XAREL·LO.

Pecorino Italian: not a cheese but alluring dry white from a revived variety.

Pedro Ximénez [PX] Makes sweet brown Sherry under its own name, also in Montilla, Málaga. In Argentina, the Canaries, Australia, California and S Africa.

Picpoul (Piquepoul) Southern French, best known in Picpoul de Pinet. Should have high acidity. Picpoul Noir is black-skinned.

Pinela Local to Slovenia. Subtle, lowish acidity; drink young.

Pinot Bianco *See* PINOT BL.

Pinot Blanc (Beli Pinot, Pinot Bianco, Weissburgunder) [Pinot Bl] Mutation of PINOT N, similar to but milder than CHARD. Light, fresh, fruity, not aromatic, to drink young. Gd for Italian *spumante*, and potentially excellent in the ne, esp high sites in Alto Adige. Widely grown. Weissburgunder in Germany and best in s: often racier than Chard.

Pinot Gris (Pinot Grigio, Grauburgunder, Ruländer, Sivi Pinot, Szürkebarát) [Pinot Gr] Popular as Pinot Grigio in N Italy, even for rosé, but top, characterful versions can be excellent (from Alto Adige, Friuli). Cheap versions are just that. Terrific in Alsace for full-bodied, spicy whites. Once important in Champagne. In Germany can be alias Ruländer (sw) or Grauburgunder (dr): best in Baden (esp Kaiserstuhl) and S Pfalz. Szürkebarát in Hungary, Sivi P in Slovenia (characterful, aromatic).

Pošip Croatian; mostly on Korčula. Quite characterful and citrus; high-yielding.

Prosecco Old name for grape that makes Prosecco. Now you have to call it GLERA.

Renski Rizling Rhine RIES.

Rèze Super-rare ancestral Valais grape used for *vin de glacier*.

Ribolla Gialla / Rebula Acidic but characterful. In Italy, best in Collio. In Slovenia, traditional in Brda. Can be v.gd. Favourite of amphora users.

Rieslaner Rare German cross (SILVANER x RIES); fine Auslesen in Franken and Pfalz.

Riesling Italico *See* WELSCHRIESLING.

Riesling (Renski Rizling, Rhine Riesling) [Ries] The greatest, most versatile white grape, diametrically opposite in style to CHARD. From steely to voluptuous, always perfumed, far more ageing potential than Chard. Great in all styles in Germany; forceful and steely in Austria; lime cordial-and-toast fruit in S Australia; rich and spicy in Alsace; Germanic and promising in NZ, NY State, Pacific Northwest; has potential in Ontario, S Africa.

Rkatsiteli Found widely in E Europe, Russia, Georgia. Can stand cold winters and has high acidity; protects to a degree from poor winemaking. Also in NE US.

Robola In Greece (Cephalonia), top-quality, floral; unrelated to RIBOLLA GIALLA but related to Rebula.

Roditis Pink grape, all over Greece, usually making whites; gd when yields low.

Roter Veltliner Austrian; unrelated to GRÜNER V. There is also a Frühroter and an (unrelated) Brauner Veltliner.

Rotgipfler Austrian; indigenous to Thermenregion. With ZIERFANDLER, makes lively, lush, aromatic blends.

Roussanne (Bergeron) Rhône grape of real finesse, called Bergeron in Savoie. Now popping up in California and Australia. Can age many yrs.

Ruländer *See* PINOT GR.

Sauvignonasse *See* FRIULANO.

Sauvignon Blanc [Sauv Bl] Distinctive aromatic, grassy to tropical wines, pungent in NZ, often mineral in Sancerre, riper in Australia; v.gd in Rueda, Austria, N Italy (Isonzo, Piedmont, Alto Adige), Chile's Casablanca Valley and S Africa. Blended with SÉM in B'x. Can be austere or full (or indeed nauseating). Sauv Gr is a pink-skinned, less aromatic version of Sauv Bl with untapped potential.

Sauvignon Vert *See* FRIULANO.

Savagnin (Heida, Païen) Grape for VIN JAUNE from Jura: aromatic form is GEWURZ. In Switzerland known as Heida, Païen or Traminer. Full-bodied, high acidity.

Scheurebe (Sämling) Grapefruit-scented German RIES x SILVANER (possibly), in Pfalz v. successful, esp Auslese and up. Can be weedy: must be v. ripe to be gd.

Sémillon [Sém] Contributes lusciousness to Sauternes but decreasingly important for Graves and other dry white B'x. Grassy if not fully ripe, but can make soft dry wine of great ageing potential. Superb in Australia; NZ and S Africa promising.

Sercial (Cerceal) Portuguese: makes the driest Madeira. Cerceal, also Portuguese, seems to be this plus any of several others.

Seyval Blanc [Seyval Bl] French hybrid of French and American vines; v. hardy, attractively fruity. Popular and reasonably successful in E US and England.

Silvaner (Johannisberg, Sylvaner) Can be excellent in Germany's Rheinhessen, Pfalz, esp Franken, with plant/earth flavours and mineral notes; v.gd (and powerful) as Johannisberg in Valais, Switzerland. Lightest of Alsace grapes.

Šipon *See* FURMINT.

Sivi Pinot *See* PINOT GR.

Spätrot *See* ZIERFANDLER.

Sylvaner *See* SILVANER.

Tămâioasă Românească *See* MUSCAT.

Torrontés Name given to a number of grapes, mostly with an aromatic, floral character, sometimes soapy. A speciality of Argentina; also in Spain. DYA.

Traminac Or Traminec. *See* GEWURZ.

Traminer Or Tramini (Hungary). *See* GEWURZ.

Trebbiano (Ugni Blanc) Principal white grape of Tuscany, found all over Italy in many different guises. Rarely rises above the plebeian except in Tuscany's Vin Santo. Some gd dry whites under DOCs Romagna or Abruzzo. Trebbiano di Soave, aka VERDICCHIO, only distantly related. T di Lugana now called Turbiana. Grown in southern France as Ugni Blanc, and Cognac as St-Émilion. Mostly thin, bland wine; needs blending (and more careful growing).

Ugni Blanc [Ugni Bl] *See* TREBBIANO.

Verdejo The grape of Rueda in Castile, potentially fine and long-lived.

Verdelho Great quality in Australia (pungent, full-bodied); rare but gd (and medium-sweet) in Madeira.

Verdicchio Potentially gd, muscular, dry; central-E Italy. Wine of same name.

Vermentino Italian, sprightly, satisfying texture; ageing capacity. Potential here.

Vernaccia Name given to many unrelated grapes in Italy. V di San Gimignano is crisp, lively; V di Oristano is Sherry-like.

Vidal French hybrid much grown in Canada for Icewine.

Vidiano Most Cretan producers love this. Powerful, stylish. Lime/apricot, gd acidity.

Viognier Ultra-fashionable Rhône grape, finest in Condrieu, less fine but still aromatic in the Midi. California, Virginia, Uruguay, Australia gd examples.

Viura (Macabeo, Maccabéo, Maccabeu) Workhorse of N Spain, widespread in Rioja, Catalan Cava country, over border in SW France; gd quality potential.

Weissburgunder PINOT BL in Germany.

Welschriesling (Graševina, Laški Rizling, Olaszrizling, Riesling Italico) Not related to RIES. Light and fresh to sweet and rich in Austria; ubiquitous in Central Europe, where it can be remarkably gd for dry and sweet wines.

Xarel·lo (Pansa Blanca) Traditional Catalan grape, used for Cava (with Parellada, MACABEO). Tannic, can age, can be top quality. Lime-cordial character in Alella, as Pansa Blanca.

Xynisteri Cyprus's most planted white grape. Can be simple and is usually DYA; but when grown at altitude makes appealing, minerally whites.

Zéta Hungarian; BOUVIER x FURMINT used by some in Tokaji Aszú production.

Zierfandler (Spätrot, Cirfandl) Found in Austria's Thermenregion; often blended with ROTGIPFLER for aromatic, orange-peel-scented, weighty wines.

Wine & food

Wine and food evolve all the time, and attitudes evolve too. Every year there are new ingredients to try in the kitchen, and not all cuisines are easy to match with wine. The most important thing is matching weight: hefty wines = hefty food, light wines = light food. Then, match acidity. If the dish has acidity (tomatoes, lime juice, tamarind...), the wine will need acidity too. The suggestions that follow, which vary from the very general to the precise, are just that: suggestions, to point you towards flavours you might not have thought of, wines that you might like to try.

On p.37 you'll find a box of can't-go-wrong favourites with food; the entries below give lots of specific matches enjoyed over the years.

Before the meal – apéritifs
The obvious apéritifs are gd fizz, Fino, Ries and gd rosé, and you can't beat them. But don't stint: nothing lowers the mood faster than cut-price fizz tasting of dishwater. Magnums are fun, and everybody likes Ries – everybody.

First courses
Aïoli More about mood than matching. Cold Provence rosé, PECORINO, ALIGOTÉ. Beer, marc or grappa... you'll hardly notice.

Antipasti / tapas / mezze You can be in Italy, Spain, Greece or Edinburgh: a selection of savoury, salty, meaty, cheesy, fishy, veggie bits and pieces works perfectly with Fino Sherry, XYNISTERI, orange wines. Roasted peppers, aubergines suit fruity reds: CAB FR, KADARKA. Or gd Prosecco in emergencies.

Burrata Forget mozzarella; this is the crème de la crème. So a top Italian white: FIANO or Cusumano's GRILLO. The veg with it will probably need some acidity.

Carpaccio, beef or fish Beef version works well with most wines, incl reds. Tuscan is appropriate, but fine CHARDS are gd. So are pink and Vintage Champagnes. Give Amontillado a try. **Salmon** Chard or Champagne. **Tuna** VIOGNIER, California Chard, Marlborough SAUV BL. Or sake.

Charcuterie / prosciutto / salami High-acid, unoaked red works better than white. Simple Beaujolais, Valpolicella, REFOSCO, SCHIOPPETTINO, TEROLDEGO, BARBERA. If you must have white, it needs acidity. Chorizo makes wines taste metallic. Prosciutto with melon or figs needs full dry or medium white: CHENIN BL, FIANO, MUSCAT, VIOGNIER.

Dim sum Classically, China tea. PINOT GR or classic German dry RIES; light PINOT N. For reds, soft tannins are key. Bardolino, GARNACHA, Rioja; Côtes du Rhône. Also NV Champagne or English fizz.

Eggs See also SOUFFLÉS. Not easy: eggs have a way of coating your palate. Omelettes: follow the other ingredients; mushrooms suggest red; Côtes du Rhone is a safe bet. With a truffle omelette, Vintage Champagne. As a last resort, I can bring myself to drink Champagne with scrambled eggs or eggs Benedict. Florentine, with spinach, is not a winey dish.

    gulls' eggs Push the luxury: mature white burgundy or Vintage Champagne.

    oeufs en meurette Burgundian genius: eggs in red wine with a glass of the same.

    quails' eggs Blanc de Blancs Champagne; VIOGNIER.

Mozzarella with tomatoes, basil Fresh Italian white, eg. Soave, Alto Adige. VERMENTINO from Liguria or Rolle from the Midi. See also VEGETABLE/AVOCADO.

Oysters , raw NV Champagne, Chablis, MUSCADET, white Graves, Sancerre, English Bacchus, or Guinness. Experiment with Sauternes. Manzanilla is gd. Flat oysters worth gd wine; Pacific ones drown it in brine.

   stewed, grilled or otherwise cooked Puligny-Montrachet or gd NZ CHARD. Champagne is gd with either.

Pasta Red or white according to the sauce:

   cream sauce (eg. carbonara) Orvieto, GRECO di Tufo. Young SANGIOVESE.

   meat sauce MONTEPULCIANO d'Abruzzo, Salice Salentino, MALBEC.

   pesto (basil) sauce BARBERA, VERMENTINO, NZ SAUV BL, Hungarian FURMINT.

   seafood sauce (eg. vongole) VERDICCHIO, Lugana, Soave, GRILLO, unoaked CHARD.

   tomato sauce Chianti, Barbera, Sicilian red, ZIN, S Australian GRENACHE.

Risotto Follow the flavour:

   with vegetables (eg. Primavera) PINOT GR from Friuli, Gavi, youngish SÉM, DOLCETTO or BARBERA d'Alba.

   with funghi porcini Finest mature Barolo or Barbaresco.

   seafood A favourite dry white.

   Nero A rich dry white: VIOGNIER or even Corton-Charlemagne.

Soufflés As show dishes, these deserve ★★★ wines:

   with cheese Mature red burgundy or B'x, CAB SAUV (not Chilean or Australian), etc. Or fine mature white burgundy.

   with fish (esp smoked haddock with chive cream sauce) Dry white: ★★★ burgundy, B'x, Alsace, CHARD, etc.

   with spinach (tough on wine) Mâcon-Villages, St-Véran or Valpolicella. Champagne (esp Vintage) can also spark things with the texture of a soufflé.

Fish

Abalone Dry or medium white: SAUV BL, unoaked CHARD. A touch of oak works with soy sauce, oyster sauce, etc. In Hong Kong: Dom Pérignon (at least).

Anchovies Fino, obviously. Franciacorta or Cava.

   salade niçoise Provence rosé.

   bocquerones VERDEJO, unoaked SÉM.

Bacalão Salt cod needs acidity: young Portuguese red or white. Or Italian ditto. Orange can be gd.

Bass, sea Fine white, eg. Clare RIES, Chablis, white Châteauneuf, VERMENTINO from Sardinia, WEISSBURGUNDER from Baden or Pfalz. Rev the wine up for more seasoning, eg. ginger, spring onions; more powerful Ries, not necessarily dry.

Beurre blanc, fish with Deserves gd unoaked white with maturity: Hunter SEM, Premier Cru Chablis, CHENIN BL, RIES, Swartland white blend. Applies to most veg with beurre blanc too.

Caviar Iced vodka (and) full-bodied Champagne (eg. Bollinger, Krug). Don't (ever) add raw onion.

Ceviche Can be applied to anything now, but here, it's fish. Australian RIES or VERDELHO, Chilean SAUV BL, TORRONTÉS. Manzanilla.

Crab (esp Dungeness) and RIES together are part of the Creator's plan. But He also created Champagne.

   Chinese, with ginger & onion German RIES Kabinett or Spätlese Halbtrocken. Tokaji FURMINT, GEWURZ.

   cioppino SAUV BL; but West Coast friends say ZIN. Also California sparkling.

   cold, dressed Top Mosel Ries, dry Alsace or Australian Ries or Assyrtiko.

   crab cakes Any of the above.

   softshell Unoaked CHARD, ALBARIÑO or top-quality German Ries Spätlese.

Cured fish Salmon can have a whisky cure, a beetroot cure; all have sweetness and pungency. With gravadlax, sweet mustard sauce is a complication. SERCIAL

Madeira (eg. 10-yr-old Henriques), Amontillado, Tokaji Szamarodni, orange wine. Or NV Champagne.

Curry S African CHENIN BL, Alsace PINOT BL, Franciacorta, fruity rosé, not too pale and anodyne; look at other flavours. Prawn and mango need more sweetness, tomato needs acidity. Fino can handle heat. So can IPA or Pilsner.

Fish pie (with creamy sauce) ALBARIÑO, Soave Classico, RIES Erstes Gewächs, Mâcon Bl, Spanish GODELLO.

Grilled, roast or fried fish Also applies to fish & chips, tempura, fritto misto, baked...

  cod, haddock CHARD, PINOT BL, MALVAZIJA.

  Dover sole Perfect with fine wines: white burgundy or equivalent.

  halibut, turbot, brill Best rich, dry white; top Chard, mature RIES Spätlese.

  monkfish Meaty but neutral; full-flavoured white or red, according to sauce.

  mullet, grey VERDICCHIO, unoaked Chard, rosé.

  oily fish like herrings, mackerel, sardines More acidity, weight: ASSYRTIKO, VERDELHO, FURMINT, orange, rosé.

  perch, sandre Top white burgundy, Mosel, Grand Cru Alsace, mature top fizz.

  plaice, flounder Light, fresh whites.

  red fish like salmon, red mullet PINOT N. For salmon, also best Chard, Grand Cru Chablis, top Ries.

  skate, ray Delicate, but brown butter, capers need oomph: Alsace, ROUSSANNE, CHENIN BL.

  swordfish Full-bodied dry white (or why not red?) of the country. Nothing grand.

  trout Gd Chard, Furmint, Ries, Malvazija, Pinot N.

  tuna Best served rare (or raw) with light red: young Loire CAB FR or red burgundy. Young Rioja is a possibility.

  whitebait Crisp dry whites, eg. Furmint, Greek, Touraine SAUV BL, Verdicchio, white Dão, Fino Sherry, rosé. Or beer.

Kedgeree Full white, still or sparkling: Mâcon-Villages, S African CHARD, GRÜNER V, German Grosses Gewächs or (at breakfast) Champagne.

Lobster with a rich sauce Eg. Thermidor: Vintage Champagne, fine white burgundy, Cru Classé Graves, Roussanne, top Australian CHARD. Alternatively, for its inherent sweetness, Sauternes, Pfalz Spätlese, even Auslese.

  plain grilled, or cold with mayonnaise NV Champagne, Alsace RIES, Premier Cru Chablis, Condrieu, Mosel Spätlese, GRÜNER V, Hunter SEM, or local fizz.

Mussels marinière MUSCADET *sur lie*, Premier Cru Chablis, unoaked CHARD.

  curried Alsace RIES or PINOT BL.

Paella, shellfish Full-bodied white or rosé, unoaked CHARD, ALBARIÑO, or GODELLO. Or local Spanish red.

Prawns, crayfish with garlic Keep the wine light, white, or rosé, and dry.

  with mayonnaise Menetou-Salon or Reuilly.

  with spices Up to and incl chilli, go for a bit more body, but not oak: dry RIES or Italian, eg. FIANO, Grillo. *See also* CURRY.

Sardines in saor Try top Prosecco, and I mean top.

Sashimi Koshu comes into its own here, either as orange or white, and can deal with wasabi and soy, within reason. Otherwise, try white with body (Chablis Premier Cru, Alsace RIES) with white fish, PINOT N with red. Both need acidity. Simple Chablis can be too thin. If soy is involved, then low-tannin red (again, Pinot). Remember sake (or Fino).

Scallops An inherently slightly sweet dish, best with medium-dry whites.

  in cream sauces German Spätlese, Montrachets or top Australian CHARD.

  grilled or seared Hermitage Bl, GRÜNER V, Pessac-Léognan Bl, Vintage Champagne or PINOT N.

  with Asian seasoning NZ Chard, CHENIN BL, GODELLO, Grüner V, GEWURZ.

Scandi fish dishes Scandinavian dishes often have flavours of dill, caraway and cardamom and combine sweet and sharp notes. Go for acidity and some weight: FALANGHINA, GODELLO, VERDELHO, Australian, Alsace or Austrian RIES.

Shellfish Dry white with plain boiled shellfish, richer wines with richer sauces. RIES is the grape.

    plateaux de fruits de mer Chablis, MUSCADET de Sèvre et Maine, PICPOUL de Pinet, Alto Adige PINOT BL.

Smoked fish All need freshness and some pungency; Fino Sherry works with all.

    eel Often with beetroot, crème fraiche: Fino again, or Mosel RIES.

    haddock Gd Chablis, MARSANNE, GRÜNER V. *See also* SOUFFLÉS.

    kippers Try Oloroso Sherry or Speyside malt.

    mackerel Not wine-friendly. Try Fino.

    salmon Condrieu, Alsace PINOT GR, Grand Cru Chablis, German Ries Spätlese, Vintage Champagne, vodka, schnapps, or akvavit.

    trout More delicate: Mosel Ries.

Squid / octopus Fresh white: ALBARIÑO, MUSCADET, sparkling, esp with salt-and-pepper squid. Squid ink (risotto, pasta) needs Soave.

Sushi Hot wasabi is usually hidden in every piece. Koshu is 1st choice. Failing that, German QbA Trocken, simple Chablis, ALVARINHO or NV Brut Champagne or KOSHU. Obvious fruit doesn't work. Or, of course, sake or beer.

Tagine N African flavours need substantial whites to balance – Austrian, Rhône – or crisp, neutral whites that won't compete. Go easy on the oak. VIOGNIER or ALBARIÑO can work well.

Taramasalata A Med white with personality, Greek if possible. Fino Sherry works well. Try Rhône MARSANNE.

Teriyaki A way of cooking, and a sauce, used for meat as well as fish. Germans favour off-dry RIES with weight: Kabinett can be too light.

Meat / poultry / game

Barbecues The local wine: Australian, S African, Chilean, Argentine are right in spirit. Reds need tannin and vigour. Or the freshness of cru Beaujolais.

Beef (*see also* **Steak**) **boiled** Red: B'x (eg. Fronsac), Roussillon, Gevrey-Chambertin or Côte-Rôtie. Medium-ranking white burgundy is gd, eg. Auxey-Duresses. In Austria you may be offered skin-fermented TRAMINER. Mustard softens tannic reds, horseradish kills your taste; can be worth the sacrifice.

    roast An ideal partner for fine red of any kind. Even Amarone. *See* above for mustard. The silkier the texture of the beef (wagyu, Galician, eg.), the silkier the wine. Wagyu, remember, is about texture; has v. delicate flavour.

    stew, daube Sturdy red: Pomerol or St-Émilion, Hermitage, Cornas, BARBERA, SHIRAZ, Napa CAB SAUV, Ribera del Duero, or Douro red.

    stroganoff Dramatic red: Barolo, Valpolicella Amarone, Priorat, Hermitage, late-harvest ZIN. Georgian SAPERAVI or Moldovan Negru de Purkar.

Boudin blanc Loire CHENIN BL, esp when served with apples: dry Vouvray, Saumur, Savennières; mature red Côte de Beaune if without.

Boudin noir / morcilla Local SAUV BL or CHENIN BL (esp in Loire). Or Beaujolais cru, esp Morgon. Or light TEMPRANILLO. Or Fino.

Brazilian dishes Pungent flavours that blend several culinary traditions. Rhônish grapes work for red, or white with weight: VERDICCHIO, California CHARD. Or a Caipirinha (better not have two).

Cajun food Gutsy reds, preferably New World: ZIN, CARMENÈRE, SHIRAZ. Fish or white meat: off-dry RIES, MARSANNE, ROUSSANNE. Or, of course, cold beer.

Cassoulet Red from SW France (Gaillac, Minervois, Corbières, St-Chinian or Fitou) or SHIRAZ. But best of all, Fronton, Beaujolais cru or young TEMPRANILLO.

Chicken / turkey / guinea fowl, roast Virtually any wine, incl v. best bottles of dry to medium white and finest old reds (esp burgundy). Sauces can make it match almost any fine wine (eg. coq au vin; the burgundy can be red or white, or *vin jaune* for that matter).

chicken Kyiv Alsace RIES, Collio, CHARD, Bergerac rouge.

fried Sparkling works well.

Chilli con carne Young red: Beaujolais, TEMPRANILLO, ZIN, Argentine MALBEC, Chilean CARMENÈRE. Or beer.

Chinese dishes Food in China is regional – like Italian, only more confusing. It's easiest to have both white and red; no one wine goes with all. Peking duck is pretty forgiving. Champagne becomes a thirst-quencher. Beer too.

Cantonese Big, slightly sweet flavours work with slightly oaky CHARD, PINOT N, off-dry RIES. GEWURZ is often suggested but rarely works; GRÜNER V is a better bet. You need wine with acidity. Dry sparkling (esp Cava) works with textures.

Shanghai Richer and oilier than Cantonese, not one of wine's natural partners. Shanghai tends to be low on chilli but high on vinegar of various sorts. German and Alsace whites can be a bit sweeter than for Cantonese. For reds, try MERLOT – goes with the salt. Or mature Pinot N, but a bit of a waste.

Szechuan VERDICCHIO, Alsace PINOT BL, or v. cold beer. Mature Pinot N can also work; but *see* above. The Creator intended tea.

Taiwanese LAMBRUSCO works with traditional Taiwan dishes if you're tired of beer.

Choucroute garni Alsace PINOT BL, PINOT GR, RIES, or lager.

Cold roast meat Generally better with full-flavoured white than red. Mosel Spätlese or Hochheimer and Côte Chalonnaise are v.gd, as is Beaujolais. Leftover Champagne too.

Confit d'oie / de canard Young, brisk red B'x, California CAB SAUV or MERLOT, Priorat cut richness. Alsace PINOT GR or GEWURZ match it.

Coq au vin Red burgundy. Ideal: one bottle of Chambertin in the dish, two on the table. *See also* CHICKEN.

Dirty (Creole) rice Rich, supple red: NZ PINOT N, GARNACHA, Bairrada, MALBEC.

Duck / goose PINOT N is tops. Also other red in the Pinot idiom, like GARNACHA; also BLAUFRÄNKISCH. Or rich white, esp for goose: Pfalz Spätlese or off-dry Grand Cru Alsace. With oranges or peaches, the Sauternais propose drinking Sauternes, others Monbazillac or RIES Auslese. Mature and weighty Vintage Champagne handles accompanying red cabbage surprisingly well. So does decent Chianti.

Peking *See* CHINESE DISHES.

roast breast & confit leg with Puy lentils Madiran (best), St-Émilion, Fronsac.

wild duck Worth opening gd Pinot N. Austrian or Tuscan red (easy on the oak) is also gd.

with olives Top-notch Chianti or other Tuscans.

Filipino dishes Spanish-influenced flavours, lots of garlic, bell peppers, adobo, not necessarily super-spicy. Fashionable in San Francisco and soon nr you. Straightforward unoaked white with acidity (adobo can have a burst of vinegar) or fizz, Côtes de Gascogne, Rueda, RIES, rosé, even light red.

Foie gras Sweet white: Sauternes, Tokaji Aszú 5 Puttonyos, late-harvest PINOT GR or RIES, Vouvray, Montlouis, Jurançon *moelleux*, GEWURZ. Old dry Amontillado can be sublime.

hot Mature Vintage Champagne. But never CHARD, SAUV BL, or (shudder) red.

Game birds, young, roast The best red wine you can afford, but not too heavy. Partridge is more delicate than pheasant, which is more delicate than grouse. Up the weight of wine accordingly, starting with youngish PINOT N, BLAUFRÄNKISCH, SYRAH, GARNACHA, and moving up.

cold game Best German RIES or mature Vintage Champagne.

older birds in casseroles Gevrey-Chambertin, Pommard, Châteauneuf, Dão, or Grand Cru Classé St-Émilion, Rhône.

well-hung game Vega Sicilia, great red Rhône, Ch Musar.

Game pie, hot Red: Oregon PINOT N, St-Émilion Grand Cru Classé.

cold Gd-quality white burgundy or German Erstes Gewächs, cru Beaujolais, or Champagne.

Goat (hopefully kid). As for lamb.

Jamaican curry goat See INDIAN DISHES.

Goulash Flavoursome young red: Hungarian Kékoportó, ZIN, Uruguayan TANNAT, Douro red, MENCÍA, young Australian SHIRAZ, SAPERAVI; or dry Tokaji Szamarodni.

Haggis Fruity red, eg. young claret, young Portuguese red, New World CAB SAUV or MALBEC or Châteauneuf. Or, of course, malt whisky.

Ham, cooked Softer red burgundies: Volnay, Savigny, Beaune; Chinon or Bourgueil; sweetish German white (RIES Spätlese); lightish CAB SAUV (eg. Chilean), or New World PINOT N. And don't forget the heaven-made match of ham and Sherry.

Hare Jugged hare calls for flavourful red: not-too-old burgundy or B'x, Rhône (eg. Gigondas), Bandol, Barbaresco, Ribera del Duero, Rioja Res. The same for saddle or for hare sauce with pappardelle.

Indian dishes Various options: dry Sherry is brilliant. Choose a fairly weighty Fino with fish, and Palo Cortado, Amontillado, or Oloroso with meat, according to weight of dish; heat's not a problem. The texture works too. Otherwise, medium-sweet white, v. cold, no oak: Orvieto *abboccato*, S African CHENIN BL, Alsace PINOT BL, TORRONTÉS, Indian sparkling, Cava or NV Champagne. Rosé is gd all-rounder. For tannic impact Barolo or Barbaresco, or deep-flavoured reds, ie. Châteauneuf, Cornas, Australian GRENACHE or MOURVÈDRE, or Valpolicella Amarone – will emphasize the heat. Hot-and-sour flavours need acidity.

Sri Lankan More extreme flavours, coconut. Sherry, rich red, rosé, mild white.

Japanese dishes A different set of senses come into play. Texture and balance are key; flavours are subtle. A gd mature fizz works well, as does mature dry RIES; you need acidity, a bit of body, and complexity. Dry FURMINT can work well. Umami-filled meat dishes favour light, supple, bright reds: Beaujolais perhaps, or mature PINOT N. Full-flavoured yakitori needs lively, fruity, younger versions of the same reds. KOSHU with raw fish – but why not sake? Orange Koshu with wagyu beef. *See also* FISH/SASHIMI, SUSHI, TERIYAKI.

Korean dishes Fruit-forward wines seem to work best with strong, pungent Korean flavours. PINOT N, Beaujolais, Valpolicella can all work: acidity is needed. Non-aromatic whites: GRÜNER V, SILVANER, VERNACCIA. Beer too.

Lamb, roast One of the traditional and best partners for v.gd red B'x, or its CAB SAUV equivalents from the New World. In Spain, finest old Rioja and Ribera del Duero Res or Priorat, in Italy ditto SANGIOVESE. Fresh mint is gd, but mint sauce should be banned.

milk-fed Is delicate and deserves top, delicate: B'x, burgundy, Spanish.

slow-cooked roast Flatters top reds, but needs less tannin than pink lamb. *See also* TAGINES.

Liver Young red: Beaujolais-Villages, St-Joseph, Médoc, Italian MERLOT, Breganze CAB SAUV, ZIN, Priorat, Bairrada.

calf's Red Rioja Crianza, Fleurie. Or a big Pfalz RIES Spätlese.

Mexican food Californians favour RIES, esp German. Or beer.

Moussaka Red or rosé: Naoussa, SANGIOVESE, Corbières, Côtes de Provence, Ajaccio, young ZIN, TEMPRANILLO.

Mutton A stronger flavour than lamb, and not usually served pink. Needs a strong sauce. Robust red; top-notch, mature CAB SAUV, SYRAH. Sweetness of fruit (eg. Barossa) suits it.

'Nduja Calabria's spicy, fiery spreadable salumi needs a big, juicy red: young Rioja, Valpolicella, CAB FR, AGLIANICO, CARIGNAN, NERELLO MASCALESE.

Osso bucco Low-tannin, supple red such as DOLCETTO d'Alba or PINOT N. Or dry Italian white such as Soave.

Ox cheek, braised Superbly tender and flavoursome, this flatters the best reds: Vega Sicilia, St-Émilion. Best with substantial wines.

Oxtail Rather rich red: St-Émilion, Pomerol, Pommard, Nuits-St-Georges, Barolo, or Rioja Res, Priorat or Ribera del Duero, California or Coonawarra CAB SAUV, Châteauneuf, mid-weight SHIRAZ, Amarone.

Paella Young Spanish wines: red, dry white, or rosé: Penedès, Somontano, Navarra, or Rioja.

Pastrami Alsace RIES, young SANGIOVESE, or St-Émilion.

Pâté Chicken liver calls for pungent white (Alsace PINOT GR or MARSANNE), a smooth red eg. light Pomerol, Volnay, or NZ PINOT N. More strongly flavoured (duck, etc.) needs Gigondas, Moulin-à-Vent, Chianti Classico, or gd white Graves. Amontillado can be marvellous match.

Pigeon or squab PINOT N perfect; young Rhône, Argentine MALBEC, young SANGIOVESE. Try Franken SILVANER Spätlese. With luxurious squab, top quite tannic red.

    pastilla Depends on sweetness of dish. As above, or if authentically sweet, try RIES Spätlese, Alsace PINOT GR with some sweetness.

Pork A perfect rich background to a fairly light red or rich white.

    belly Slow-cooked and meltingly tender, needs red with some tannin or acidity. Italian would be gd: Barolo, DOLCETTO, or BARBERA. Or Loire red, or lightish Argentine MALBEC. With Chinese spices, VIOGNIER, CHENIN BL.

    Mangalica pork Fashionable, fatty. KÉKFRANKOS, or other brisk red.

    pulled Often with spicy sauce: juicy New World reds.

    roast Deserves ★★★ treatment: Médoc is fine. Portugal's suckling pig is eaten with Bairrada; S America's with CARIGNAN; Chinese is gd with PINOT N.

    with prunes or apricots Something sweeter: eg. Vouvray.

Pot au feu, bollito misto, cocido Rustic red wines from region of origin; SANGIOVESE di Romagna, Chusclan, Lirac, Rasteau, Portuguese Alentejo or Spain's Yecla, or Jumilla.

Quail Succulent, delicate: try red or white. Rioja Res, mature claret, PINOT N. Or mellow white: Vouvray, St-Péray.

Quiche Egg and bacon are not great wine matches, but one must drink something. Alsace RIES or PINOT GR, even GEWURZ, is classical. Beaujolais could be gd too.

Rabbit Lively, medium-bodied young Italian red, eg. AGLIANICO del Vulture, REFOSCO; Chiroubles, Chinon, Saumur-Champigny or Rhône rosé.

    as ragu Medium-bodied red with acidity: Aglianico, NEBBIOLO.

    with mustard Cahors.

    with prunes Bigger, richer, fruitier red.

Satay McLaren Vale SHIRAZ, Alsace or NZ GEWURZ. Peanut sauce: problem for wine.

Sauerkraut (German) German RIES, lager or Pils. (But see also CHOUCROUTE GARNI.)

Singaporean dishes Part Indian, part Malay and part Chinese, Singaporean food has big, bold flavours that don't match easily with wine – not that that bothers the country's many wine-lovers. Off-dry RIES is as gd as anything. With meat dishes, ripe, supple reds: Valpolicella, PINOT N, DORNFELDER, unoaked MERLOT, or CARMENÈRE.

Steak Rare steak needs brisker, more tannic reds; well done needs juicy, fruity reds, eg. young Argentine MALBEC. Fattier cuts need acidity, tannin.

    au poivre A fairly young Rhône red or CAB SAUV. Nothing too sweetly fruity.

    fillet Silky red: Pomerol or PINOT N.

fiorentina (bistecca) Chianti Classico Riserva or BRUNELLO.

from older cattle Has deep, rich savouriness. Top Italian, Spanish red.

Korean yuk whe (world's best steak tartare) Sake.

ribeye, tomahawk, tournedos Big, pungent red: Barolo, Cahors, SHIRAZ, Rioja.

sirloin Suits most gd reds. B'x blends, Tuscans.

tartare Vodka or light young red: Beaujolais, Bergerac, Valpolicella. Aussies drink GAMAY with kangaroo tartare, charred plums, Szechuan pepper.

T-bone Reds of similar bone structure: Barolo, Hermitage, Australian CAB SAUV or Shiraz, Chilean SYRAH, Douro.

wagyu Delicate, silky red, or orange wine.

Steak-&-kidney pie or pudding Red Rioja Res or mature B'x. Pudding (with suet) wants vigorous young wine. Madiran with its tannin is gd.

Stews & casseroles Burgundy such as Nuits-St-Georges or Pommard if fairly simple; otherwise lusty, full-flavoured red, eg. young Côtes du Rhône, BLAUFRÄNKISCH, Corbières, BARBERA, SHIRAZ, ZIN, etc.

Sweetbreads A rich dish, so grand white wine: Rheingau RIES or Franken SILVANER Spätlese, Grand Cru Alsace PINOT GR, or Condrieu, depending on sauce.

Tagines Depends on what's under the lid, but fruity young reds are a gd bet: Beaujolais, TEMPRANILLO, SANGIOVESE, MERLOT, SHIRAZ. Amontillado is great. Amarone is fashionable.

chicken with preserved lemon, olives VIOGNIER.

Tandoori chicken RIES or SAUV BL, young red B'x or light N Italian red served cool. Also Cava, NV Champagne or, of course, Palo Cortado or Amontillado Sherry.

Thai dishes Ginger and lemon grass call for pungent SAUV BL (Loire, Australia, NZ, S Africa) or RIES (Spätlese or Australian). Most curries suit aromatic whites with a touch of sweetness: GEWURZ also gd.

Tongue Any red or white of abundant character. Alsace PINOT GR or GEWURZ, gd GRÜNER V. Also Beaujolais, Loire reds, BLAUFRÄNKISCH, TEMPRANILLO and full, dry rosés.

Veal A friend of fine wine. Rioja Res, CAB blends, PINOT N, NEBBIOLO, German or Austrian RIES, Vouvray, Alsace PINOT GR, Italian GRECO di Tufo.

Venison Big-scale reds, incl MOURVÈDRE, solo as in Bandol or in blends. Rhône, Languedoc, B'x, NZ Gimblett Gravels or California CAB SAUV of a mature vintage; or rather rich white (Pfalz Spätlese or Alsace PINOT GR). Top Italian, too.

with sweet & sharp berry sauce Try a German Grosses Gewächs RIES, or Chilean CARMENÈRE, or SYRAH.

Vietnamese food RIES, dry or up to Spätlese, German, Austrian, NZ, also GRÜNER V, SÉM. For reds, PINOT N, CAB FR, BLAUFRÄNKISCH.

Vitello tonnato Full-bodied whites: CHARD. Light reds (eg. Valpolicella) served cool. Or a southern rosé.

Wild boar Serious red: top Tuscan or Priorat. NZ SYRAH. I've even drunk Port; Amarone would be a compromise.

Vegetable dishes

With few tannins to help or hinder, matching wine to veg is about sweetness, acidity, weight and texture. *See also* FIRST COURSES.

Agrodolce Italian sweet-and-sour, with pine kernels, sultanas, capers, vinegar and perhaps anchovies. Go to fresh white: VERDICCHIO, unoaked CHARD. Or orange.

Artichokes Not great for wine. Incisive dry white: NZ SAUV BL; Côtes de Gascogne or Greek (precisely, 4-yr-old MALAGOUSIA, but easy on the vinaigrette); VERMENTINO. Orange wine, yes; red no, unless you absolutely have to, in which case go for acidity: LAGREIN, DOLCETTO.

Asparagus Is lightly bitter, and needs acidity, if it needs wine at all. Asparagus solo: skip the wine. With other ingredients: go by those.

Aubergine Comes in a multitude of guises, usually strongly flavoured. Sturdy reds with acidity are a gd bet: SHIRAZ, Greek, Lebanese, Bulgarian, Hungarian, Turkish. Structured white, eg. VERDICCHIO, or go further and have orange wine.

Avocado Not a wine natural. Dry to slightly sweet RIES Kabinett will suit the dressing. Otherwise, light and fresh: ALIGOTÉ, TREBBIANO, PINOT GRIGIO.

Beetroot Mimics a flavour found in red burgundy. You could return the compliment. New-wave (ie. light) GARNACHA/Grenache is gd, as well, as is Gamay.

Bitter leaves: radicchio etc. Bone-dry, aged Palo Cortado. But easy on the dressing. Fab combo. Or NEBBIOLO, LAGREIN, white VERMENTINO, orange.

    roast radicchio, chicory etc. Valpolicella, SANGIOVESE, BLAUFRÄNKISCH.

Cauliflower roast, etc. Go by the other (usually bold) flavours. Try Austrian GRÜNER V, Valpolicella, NZ PINOT N.

    cauliflower cheese Crisp, aromatic white: Sancerre, RIES Spätlese, MUSCAT, ALBARIÑO, GODELLO. CHARD too, and Beaujolais-Villages.

    with caviar – yes, really. Vintage Champagne.

Celeriac, slow-roast or purée Won't interfere with rest of dish. Acidity works well, so classic CAB blends, Beaujolais, PINOT N, according to dish.

    remoulade with smoked ham Needs bright red: DOLCETTO, simple GAMAY, Valpolicella. Or white GRÜNER V.

Chestnuts, slow-cooked daube or as purée Earthy, rich red: Tuscan or S Rhône.

Chickpeas Look at other flavours. Casserole works with TEMPRANILLO, S French reds.

    hummus Any simple red, pink or white, or, of course, Fino.

Chilli Some like it hot, but not with your best bottles. Tannic wines become more tannic; if you like that, go for it. Light, fruity reds and whites are refreshing: TEMPRANILLO, Chilean MERLOT, NZ SAUV BL. Same for **harissa**. *See also* MEAT/CHILLI CON CARNE, CHINESE DISHES, INDIAN DISHES.

Couscous with vegetables Young red with a bite: SHIRAZ, Corbières, Minervois; rosé; orange wine; Italian REFOSCO or SCHIOPPETTINO.

Dhal Comes with many variations, but all share aromatic earthiness. Simple, warm-climate reds work best: CARMENÈRE, Dão, S Italian.

Fennel-based dishes SAUV BL: Pouilly-Fumé or NZ; SYLVANER or English SEYVAL BL; or young TEMPRANILLO. **Deep-flavoured braised fennel** Light GAMAY, PINOT N.

Fermented foods *See also* SAUERKRAUT, CHOUCROUTE GARNI, KOREAN DISHES. Kimchi and miso are being worked into many dishes. Fruit and acidity are generally needed. If in sweetish veg dishes, try Alsace.

Grilled Mediterranean vegetables Italian whites, or for reds Brouilly, BARBERA, TEMPRANILLO or SHIRAZ.

Lentil dishes Sturdy reds such as Corbières, ZIN or SHIRAZ. *See also* DHAL.

Macaroni cheese As for CAULIFLOWER CHEESE.

Mushrooms (in most contexts) A boon to most reds and some whites. Context matters as much as species. Pomerol, California MERLOT, Rioja Res, top burgundy or Vega Sicilia.

    button or Paris with cream Fine whites, even Vintage Champagne.

    ceps / porcini Ribera del Duero, Barolo, Chianti Rùfina, Pauillac or St-Estèphe, NZ Gimblett Gravels.

    on toast Best claret, even Port.

Onion / leek tart / flamiche Fruity, off-dry or dry white: Alsace PINOT GR or GEWURZ is classic; Canadian, Australian or NZ RIES; Jurançon. Or Loire CAB FR.

Peppers, cooked Mid-weight Rhône grapes, CARMENÈRE, Rioja; or ripe SAUV BL (esp with green Hungarian wax peppers).

    stuffed Full-flavoured red, white, or pink: Languedoc, Greek, Spanish. Or orange.

Pickled foods & vinegar Vinegar and wine don't go, it's true, but pickled foods are everywhere. Try Alsace, German RIES with CHOUCROUTE/SAUERKRAUT (*see* MEAT). With pickled veg as part of a dish, just downgrade the wine a bit (no point in opening best bottles) and make sure it has some acidity. (Or have beer.) In dressings, experiment with vinegars: Sherry vinegar can work with Amontillado, etc., big reds; Austrian apricot vinegar is delicate; balsamic can work with rich Italian reds. Wine just has to work harder than it used to.

Pumpkin / squash ravioli or risotto Full-bodied, fruity dry or off-dry white: VIOGNIER or MARSANNE, demi-sec Vouvray, Gavi or S African CHENIN. If you want red, MERLOT, ZIN.

Ratatouille (or piperade) Vigorous young red: Chianti, NZ CAB SAUV, MERLOT, MALBEC, TEMPRANILLO, Languedoc. Or gd rosé.

Roasted veg Can be root veg or more Med, but all have plenty of sweetness. Rosé, esp with some weight, or orange wine. Lightish reds with acidity to match the dressing. Pesto will tilt it towards white with weight.

Saffron Found in sweet and savoury dishes, and wine-friendly. Rich white: ROUSSANNE, VIOGNIER, PINOT GR. Orange wines can be gd too. With desserts, Sauternes or Tokaji. *See also* MEAT/TAGINES.

Salsa verde Whatever it's with, it points to more acidity, less lushness in the wine.

Seaweed (nori) Depends on the context. *See also* SUSHI. Iodine notes go well with Austrian GRÜNER V, RIES.

Sweetcorn fritters Often served with a hot, spicy sauce. Rosé, orange or neutral white all safe.

Tahini Doesn't really affect wine choice. Go by rest of dish.

Tapenade Manzanilla or Fino Sherry, or any sharpish dry white or rosé. Definitely not Champagne.

Truffles Black truffles are a match for finest Right Bank B'x, but even better with mature white Hermitage or Châteauneuf. White truffles call for best Barolo or Barbaresco of their native Piedmont. With buttery pasta, Lugana. Or at breakfast, on fried eggs, BARBERA.

Watercress, raw Makes every wine on earth taste revolting.

Wild garlic leaves, wilted Tricky: a fairly neutral white with acidity will cope best.

Desserts

Apple pie, strudel, or tarts Sweet German, Austrian or Loire white, Tokaji Aszú, or Canadian Icewine.

Apples Cox's Orange Pippins with Cheddar cheese Vintage Port.
    Russets with Caerphilly Old Tawny, or Amontillado.

Bread-&-butter pudding Fine 10-yr-old Barsac, Tokaji Aszú, Australian botrytized SEM.

Cakes *See also* CHOCOLATE, COFFEE, RUM. BUAL or MALMSEY Madeira, Oloroso or Cream Sherry. Asti, sweet Prosecco.

Cheesecake Sweet white: Vouvray, Anjou, or Vin Santo – nothing too special.

Chocolate Don't try to be too clever. Texture matters. BUAL, California Orange MUSCAT, Tokaji Aszú, Australian Liqueur Muscat, 10-yr-old Tawny or even young Vintage Port; Asti for light, fluffy mousses. Or *vins doux naturels* Banyuls, Maury or Rivesaltes. Some like Médoc with bitter black chocolate, though it's a bit of a waste of both. A trial of Syrah with bitter chocolate showed that you shouldn't, ever. Armagnac, or a tot of gd rum.

Christmas pudding, mince pies Tawny Port, Cream Sherry or that liquid Christmas pudding itself, PEDRO XIMÉNEZ Sherry. Tokaji Aszú. Asti, or Banyuls.

Coffee desserts Sweet MUSCAT, Australia Liqueur Muscats, or Tokaji Aszú.

Creams, custards, fools, syllabubs *See also* CHOCOLATE, COFFEE, RUM. Sauternes, Loupiac, Ste-Croix-du-Mont or Monbazillac.

Crème brûlée Sauternes or Rhine Beerenauslese, best Madeira, or Tokaji Aszú.

Ice cream & sorbets PX with vanilla, or Australian Liqueur MUSCAT. Sorbets: give wine a break.

Lemon flavours For dishes like tarte au citron, try sweet RIES from Germany or Austria or Tokaji Aszú; v. sweet if lemon is v. tart.

Meringues (eg. Eton mess) Recioto di Soave, Asti, mature Vintage Champagne.

Nuts (incl praline) Finest Oloroso Sherry, Madeira, Vintage or Tawny Port (nature's match for walnuts), Tokaji Aszú, Vin Santo, or Setúbal MOSCATEL. Cashews and Champagne. Pistachios with Fino. **Salted nut parfait** Tokaji Aszú, Vin Santo.

Orange flavours Experiment with old Sauternes, Tokaji Aszú, or California Orange MUSCAT.

Panettone Vin Santo. Jurançon *moelleux*, late-harvest RIES, Barsac, Tokaji Aszú.

Pears in red wine Rivesaltes, Banyuls, or RIES Beerenauslese.

Pecan pie Orange MUSCAT or Liqueur Muscat.

Raspberries (no cream, little sugar) Excellent with fine reds that themselves taste of raspberries: young Juliénas, Regnié. Even better with cream and something in the Sauternes spectrum.

Rum flavours (baba, mousses) MUSCAT – from Asti to Australian Liqueur, according to weight of dish.

Strawberries, wild no cream With red B'x (most exquisitely Margaux) poured over. **with cream** Sauternes or similar sweet B'x, Vouvray *moelleux*, or Vendange Tardive Jurançon.

Summer pudding Fairly young Sauternes of a gd vintage.

Sweet soufflés Sauternes or Vouvray *moelleux*. Sweet (or rich) Champagne.

Tiramisù Vin Santo, young Tawny Port, MUSCAT de Beaumes-de-Venise, Sauternes, or Australian Liqueur Muscat. Better idea: skip the wine.

Trifle Should be sufficiently vibrant with its internal Sherry (Oloroso for choice).

Zabaglione Light-gold Marsala or Australian botrytized SEM, or Asti.

Wine & cheese

Counterintuitively, white is a safer option than red. Fine red wines are slaughtered by strong cheeses. Principles to remember (despite exceptions): 1st, the harder the cheese, the more tannin the wine can have; 2nd, the creamier the cheese, the more acidity is needed in the wine – and don't be shy of sweetness. Cheese is classified by its texture and the nature of its rind, so its appearance is a guide to the type of wine to match it.

Bloomy-rind soft cheeses: Brie, Camembert, Chaource Full, dry white burgundy or Rhône. Not tannic red.

Blue cheeses The extreme saltiness of Roquefort or most blue cheeses needs sweetness: Sauternes, Tokaji, youngish Vintage or Tawny Port, esp with Stilton. Intensely flavoured old Oloroso, Amontillado, Madeira, Marsala and other fortifieds go with most blues. Dry red does not. Trust me.

Cooked cheese dishes fondue Trendy again. Light, fresh white as above.
frico Traditional in Friuli. Cheese baked or fried with potatoes or onions; high-acid local REFOSCO (r), or RIBOLLA GIALLA (w).
macaroni or cauliflower cheese See VEGETABLE DISHES/CAULIFLOWER.
Mont d'Or Delicious baked, and served with potatoes. Fairly neutral white with freshness: GRÜNER V, Savoie.

Fresh cream cheese, fromage frais, mozzarella Light crisp white: Chablis, Bergerac, Entre-Deux-Mers; juicy rosé can work too.

Hard cheeses – Gruyère, Manchego, Parmesan, Cantal, Comté, old Gouda, Cheddar Hard to generalize, relatively easy to match. Gouda, Gruyère, some

Spanish, and a few English cheeses complement fine claret or CAB SAUV and great SHIRAZ/SYRAH. But strong cheeses need less refined wines, preferably local ones. Granular old Dutch red Mimolette, Comté or Beaufort gd for finest mature B'x. Also for Tokaji Aszú. But try tasty whites too.

Natural rind (mostly goats' cheese) – St-Marcellin Sancerre, light SAUV BL, Jurançon, Savoie, Soave, Italian CHARD; or young Vintage Port.

Semi-soft cheeses – Livarot, Pont l'Evêque, Reblochon, St-Nectaire, Tomme de Savoie Powerful white B'x, even Sauternes, CHARD, Alsace PINOT GR, dryish RIES, S Italian and Sicilian whites, aged white Rioja, dry Oloroso Sherry. The strongest of these cheeses kill almost any wines. Try marc or Calvados.

Washed-rind soft cheeses – Carré de l'Est, mature Époisses, Langres, Maroilles, Milleens, Münster Local reds, esp for Burgundian cheeses; vigorous Languedoc, Cahors, Côtes du Frontonnais, Corsican, S Italian, Sicilian, Bairrada. Also powerful whites, esp Alsace GEWURZ, MUSCAT. Gewurz with Münster, always.

Food & your finest wines

With v. special bottles, the wine guides the choice of food rather than vice versa. The following is based largely on gastronomic conventions, some bold experiments and much diligent and ongoing research.

Red wines

Amarone Classically, in Verona, risotto all'Amarone or pastissada. But if your butcher doesn't run to horse, then shin of beef, slow-cooked in more Amarone.

Barolo, Barbaresco Risotto with white truffles; pasta with game sauce (eg. pappardelle alla lepre); porcini mushrooms; Parmesan.

Great Syrahs: Hermitage, Côte-Rôtie, Grange; Vega Sicilia Beef, venison, well-hung game; bone marrow on toast; English cheese (Lincolnshire Poacher) but also hard goats'-milk and ewes'-milk cheeses such as England's Lord of the Hundreds. I treat Côte-Rôtie like top red burgundy.

Great Vintage Port or Madeira Walnuts or pecans. A Cox's Orange Pippin and a digestive biscuit is a classic English accompaniment.

Red Bordeaux v. old, light, delicate wines, (eg. pre-75) Leg or rack of young lamb, roast with a hint of herbs (not garlic); entrecôte; simply roasted partridge; roast chicken never fails.

fully mature great vintages (eg. 82 85 89 90) Shoulder or saddle of lamb, roast with a touch of garlic; roast ribs or grilled rump of beef.

mature but still vigorous (eg. 00 05) Shoulder or saddle of lamb (incl kidneys) with rich sauce. Fillet of beef marchand de vin (with wine and bone marrow). Grouse. Avoid beef Wellington: pastry dulls the palate.

Merlot-based Beef (fillet is richest) or well-hung venison. In St-Émilion, lampreys.

Red burgundy Consider the weight and texture, which grow lighter/more velvety with age. Also the character of the wine: Nuits is earthy, Musigny flowery, great Romanées can be exotic, Pommard is renowned for its foursquareness. Roast chicken or (better) capon is a safe standard with red burgundy; guinea fowl for slightly stronger wines, then partridge, grouse, or woodcock for those progressively more rich and pungent. Hare and venison are alternatives.

great old burgundy The Burgundian formula is cheese: Époisses (unfermented); a fine cheese but a terrible waste of fine old wines. *See* above.

vigorous younger burgundy Duck or goose roasted to minimize fat. Or faisinjan (pheasant cooked in pomegranate juice). Coq au vin, or lightly smoked gammon.

Rioja Gran Reserva, top Duero reds Richly flavoured roasts: wild boar, mutton, saddle of hare, whole suckling pig.

White wines

Beerenauslese / Trockenbeerenauslese Biscuits, peaches, greengages. But TBAs don't need or want food.

Condrieu, Ch-Grillet, Hermitage Bl Pasta, v. light, scented with herbs and tiny peas or broad beans. Or v. mild tender ham. Old white Hermitage loves truffles.

Grand Cru Alsace Gewurz Cheese soufflé (Münster cheese).

  Pinot Gr Roast or grilled veal. Or truffle sandwich. (Slice a whole truffle, make a sandwich with salted butter and gd country bread – not sourdough or rye – wrap and refrigerate overnight. Then toast it in the oven. Thanks, Dom Weinbach.)

  Ries Truite au bleu, smoked salmon, or choucroute garni.

  Vendange Tardive Foie gras or tarte tatin.

Old vintage Champagne (not Blanc de Blancs) As an apéritif, or with cold partridge, grouse, woodcock. The evolved flavours of old Champagne make it far easier to match with food than the tightness of young wine. Hot foie gras can be sensational. Don't be afraid of garlic or even Indian spices, but omit the chilli.

  late-disgorged old wines These have extra freshness plus tertiary flavours. Try with truffles, lobster, scallops, crab, sweetbreads, pork belly, roast veal, chicken. Saffron is flattering to old Champagne.

  old Vintage Rosé Pigeon, veal.

Sauternes Simple crisp buttery biscuits (eg. langues de chat), white peaches, nectarines, strawberries (without cream). Not tropical fruit. Pan-seared foie gras. Lobster or chicken with Sauternes sauce. Ch d'Yquem recommends oysters (and indeed lobster). Experiment with blue cheeses. Rocquefort is classic but needs one of the big Sauternes. Savoury food, apart from cheese, seldom works. Cantonese dishes can be gd. A chilled glass as an apéritif can be even better.

Sherry VOS or VORS Just some almonds or walnuts, or gd cheese. And time to appreciate them.

Tokaji Aszú (5–6 Puttonyos) Foie gras recommended. Fruit desserts, cream desserts, even chocolate can be wonderful. Roquefort. It even works with some Chinese, though not with chilli – the spice has to be adjusted to meet the sweetness. Szechuan pepper is gd. Havana cigars are splendid. So is the naked sip.

Top Chablis White fish simply grilled or *meunière*. Dover sole, turbot, halibut are best; brill, drenched in butter, can be excellent. (Sea bass is too delicate; salmon passes but does little for the finest wine.)

Top white burgundy, top Graves, top aged Ries Roast veal, farm chicken stuffed with truffles or herbs under the skin, or sweetbreads; richly sauced white fish (turbot for choice) or scallops, white fish as above. Lobster, wild salmon.

Vouvray moelleux, etc. Buttery biscuits, apples, apple tart.

Fail-safe face-savers

Some wines are more useful than others – more versatile, more forgiving. If you're choosing restaurant wine to please several people, or just stocking the cellar with basics, these are the wines: **Red** Alentejo, BARBERA d'Asti/d'Alba, BLAUFRÄNKISCH, Beaujolais, Chianti, GARNACHA/ Grenache, young MALBEC (easy on the oak), PINOT N, SYRAH (more versatile than Shiraz), Valpolicella. **White** Alsace PINOT BL, ASSYRTIKO, cool-climate CHARD, CHENIN BL from the Loire or S Africa, Fino Sherry, GRÜNER v, dry RIES, Sancerre, gd Soave, VERDICCHIO. And the greatest of these is Ries. Always go for the best producer you can afford, and don't get too hung up on appellations or even vintages, within reason. If you can't afford gd Chablis, buy Assyrtiko, not cheap Chablis.

France

More heavily shaded areas are
the wine-growing regions.

Abbreviations used in the text:

Al	Alsace
Beauj	Beaujolais
Burg	Burgundy
B'x	Bordeaux
Cas	Castillon-Côtes de Bordeaux
Chab	Chablis
Champ	Champagne
Cors	Corsica
C d'O	Côte d'Or
L'doc	Languedoc
Lo	Loire
Mass C	Massif Central
Prov	Provence
N/S Rh	Northern/Southern Rhône
Rouss	Roussillon
Sav	Savoie
SW	Southwest
AC	appellation contrôlée
ch, chx	château(x)
dom, doms	domaine(s)

Le Havre
Caen
Brest
LOIRE
Loire
Nantes
Muscadet
Anjou-Saumur
La Rochelle
BORDEAU
Médoc
Pomero
St-Emil
Bordeaux
Entre
Deu
Graves
Sauternes
Côtes du
Marmanc
Buzet
Côtes
Tursan
St-Mc
Madir
Biarritz
Jurançon

At a time when wine quality across the world is better than it has
ever been, what is the purpose of the classic wines of France? The
demand for top Bordeaux and top burgundy is as strong as ever, which
keeps prices up: the days when burgundy looked good value compared to
Bordeaux are over. If you want good value from France now, look to the
Rhône: 2020 is a lovely vintage, and so far prices have not rocketed. The
south has plenty of gems too, and so do Alsace and the Loire. Climate
change is obvious in the glass. New drinkers may love the new opulence
of classic wines, but uniqueness is being lost, and we will have to find
other ways of distinguishing these wines from others: France has a task
on its hands. For many drinkers, particularly the less adventurous, France
is still the touchstone of quality. A glance through this book will reveal
many insights into the new fine wines from all sorts of countries. Ideally
these are not copies of French styles, but inventions that suit their terroirs
and the vines that grow there. They build on the traditions of France,
very often, and the traditions of Spain, Italy and many other countries –
Georgia, for example. Wine is evolving in many different directions, and
one idea sparks off another. All are building on giants' shoulders.

France entries also cross-reference to Châteaux of Bordeaux

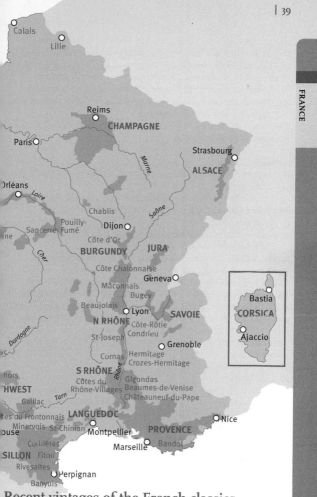

Recent vintages of the French classics

Red Bordeaux

Médoc / Red Graves For many wines, bottle-age is optional: for these it is indispensable. Minor chx from light vintages may need only 1 or 2 yrs these days, but even modest wines of gd yrs can improve for 10 or so, and the great chx of these yrs can profit from double that time.

2021 Complicated: frost, rain, mildew. Cab Sauv variable but classic. Be selective.

2020 Cab Sauv rich, dark, expressive. High alc but elegant. Merlot, Petit Verdot gd too. Low yields.

2019 Great balance; concentration but freshness as well. High percentage of Cab Sauv in blends. Potential to age.

2018 Pure, aromatically intense Cab Sauv. Rich, powerful (alc high), but balance there. Long-term potential.

2017 Attractive wines: gd balance, fairly early drinking. Volumes often small.

2016 Cab Sauv with colour, depth, structure. Vintage to look forward to.
2015 Excellent Cab Sauv yr, but not structure of 05 10. Some variation. Keep.
2014 Cab Sauv bright, resonant. Gd to v.gd; classic style, beginning to open.
2013 Worst since 92. Patchy at Classed Growth level. For early drinking.
2012 Difficulties ripening Cab Sauv but early drinking charm. Don't reject.
2011 Mixed quality, better than its reputation. Classic freshness, moderate alc.
        After an awkward phase, opening up.
Earlier fine vintages: 10 09 08 06 05 00 98 96 95 90 89 88 86 85 82 75 70 66
62 61 59 55 53 49 48 47 45 29 28.

St-Émilion / Pomerol

2021 Merlot hit hard by frost, mildew. Low yields, variable quality. Cab Fr
        faired better.
2020 Merlot rich and gourmand. Cabs Fr/Sauv also v.gd. Great potential.
2019 Drought. Merlot excellent on limestone, clay soils. Suffered on
        sandier reaches.
2018 Powerful but pure. Best from limestone, clay soils. Yields affected
        by mildew.
2017 Gd balance, classic fruit-cake flavours. Will be quite early drinking.
2016 Conditions as Méd. Some young vines suffered in drought, overall v.gd.
2015 Great yr for Merlot. Perfect conditions. Colour, concentration, balance.
2014 More rain than the Méd so Merlot variable; v.gd Cab Fr. Drinking now.
2013 Difficult flowering (so tiny crop), rot in Merlot. Modest yr, early drinking.
2012 Conditions as Méd. Merlot marginally more successful. Drinking now.
2011 Complicated, as Méd. Gd Cab Fr. Pomerol best overall? Don't shun it.
Earlier fine vintages: 10 09 05 01 00 98 95 90 89 88 85 82 71 70 67 66 64 61
59 53 52 49 47 45.

Red Burgundy

Côte d'Or Côte de Beaune reds generally mature sooner than grander wines
of Côte de Nuits. Earliest drinking dates are for lighter commune wines,
eg. Volnay, Beaune; latest for GCs, eg. Chambertin, Musigny. Even the best
burgundies are more attractive young than equivalent red B'x. It can seem
magical when they really blossom yrs later.

2021 Best will be elegant, lighter wines for medium term. Others may
        be fragile.
2020 Concentrated: some will be legends; others overcooked. Avoid most
        later-picked wines.
2019 Lush and luxurious, many great, a few too high-octane. Best will
        age effortlessly.
2018 Many brilliant sumptuous reds, should drink well young or old. Some
        overripe or flawed.
2017 Attractive, mostly ripe enough, stylish lighter wines enjoyable already.
        Slight preference for Côte de Nuits.
2016 Some spectacular reds with great energy, fresh acidity. Keep them locked
        away, though.
2015 Dense, concentrated, as 05 but with juiciness of 10. Earning stellar
        reputation; don't miss. Lesser AOPs gd now.
2014 Attractive fresh reds, some lack depth. Take a look soon.
2013 Overlooked. Côte de Beaune tricky but some stars in Côte de Nuits,
        esp GCs. Approachable now.
2012 Côte de Beaune gd if no hail. Fine Côte de Nuits, exuberant yet classy.
        Overlooked – time to dig these out.

2010 Great classic: pure, fine-boned, yet impressive density. All wines
     already delicious.
Earlier fine vintages: (drink or keep) 09 05 02 99 96 93 (mature) 90 89 85 78
76 71 69 66 64 62 59.

White Burgundy

Côte d'Or White wines are now rarely made for ageing as long as they were
in the past, but top wines should still improve for 10 yrs or more. Most
Mâconnais and Chalonnais (St-Véran, Mâcon-Villages, Montagny) usually best
drunk early (2–3 yrs).
2021 All but wiped out by frost but survivors are promising in perfumed style.
2020 Gd crop, unlike reds. Powerful full-bodied, surprising freshness.
     Extremely promising.
2019 Great if you are comfortable with richer style; most have balance.
     Structure great, aromatics riper.
2018 Saved by size of crop in a hot yr. More potential than 1st thought.
     Delicious from start yet with power to age.
2017 Modern classic, becoming v. interesting. Ripe, balanced, consistent,
     enough acidity. Most can be approached now.
2016 Small, frosted crops, inconsistent results. Most ready now.
2015 Rich, concentrated. Most picked early have done well, later wines may
     be too heavy. Similar to 09 but more successes. Keep the winners.
2014 Finest, most consistent yr for generation. Whites elegant, balanced.
     Top wines seem to be getting younger, keep.
2013 Rot affected, should be drunk up.
2012 Clean, classic; crisp finish. Holding up nicely but mostly approachable.
Earlier fine vintages (all ready): 10 09 08 07 05 02 99 96 93 92 85 79 73 59.

Chablis Chablis GC of vintages with both strength and acidity really need at
least 5 yrs, can age superbly for 15 or more; PCs proportionately less, but give
them 3 yrs at least. Then, for the full effect, serve them at cellar temperature,
not iced, and decant them. Yes, really.
2021 Frost calamity. Survivors should be fine.
2020 Hot and dry but wines have turned out well, esp at the higher end.
2019 Small crop of concentrated wine: typical marine Chab, overlaid by
     exotic notes.
2018 Vintage of century for volume: attractive quality, will be ready soon.
     But 17 even better.
2017 Despite frost again, classical style though with additional richness.
     Start drinking.
2016 Hail, frost, more hail, hardly any wine, often unbalanced. Move along.
Earlier fine vintages: 12 10 08 02 00.

Beaujolais

21 Frost and a dreadfully wet summer. Challenging! 20 Frighteningly early
harvest but v. promising quality. 19 Hot dry summer, gorgeously juicy and
satisfying wines. 18 Hot summer, but some exciting and well-balanced wines.
17 Large crop but hideous hail in Fleurie, Moulin-à-Vent.

Languedoc-Roussillon

2021 Frost, drought, pre-harvest rain in L'doc, variable. Rouss fared better, v.gd.
2020 Summer not too hot, lovely balanced wines with gd acidity, freshness.
     Rouss small vintage, gd quality.

2019 Hot dry summer; some sunburn; gd to v.gd, but small vintage.

2018 Mildew, but best are vibrant (r/w). Pic St-Loup: powerful, rich, v.gd reds.

2017 Small, spring frost, v.gd: Rouss possibly better than 15; Cabardes (r) balanced, will age, (w) acidity, fruit.

SW France

2021 Relentless succession of frost, rain, mildew and more rain. Keep to older vintages for reds.

2020 Summer heatwave, drought. A little uneven maybe, but in general, a gd and occasionally exceptional yr.

2019 Spring storms brought havoc, esp Bergerac. Heatwave summer rescued what was left. Small crop, fine quality.

2018 Big yr. Dry white mostly ready. Most reds (esp oaked) and sweet whites too. Heftier styles will need longer.

2017 Subtler yr than most. Excellent drinking now. Madiran, Cahors will keep.

2015 Marvellous full, fruity yr. Sweet whites, big reds will keep for many yrs.

Northern Rhône

Region of small plots, specific terroirs, can live as long as burgundy. White Hermitage can age as red. Hard to generalize, but don't be in a hurry (r or w).

2021 Rainswept yr, v. tricky. Much lower alc, fluid, aromatic reds, purity over power. Hermitage, Cornas more full than Côte-Rôtie. Not long-keeping. Beautiful, balanced whites.

2020 Stylish reds, charm, v.gd. Côte-Rôtie aromatic, convincing, major Hermitage, well-filled Cornas. Top whites, esp if Marsanne-based, v.gd.

2019 Tremendous reds, concentrated, highish alc, flair at Côte-Rôtie; like 16 with more stuffing, acidity slightly better than 18. No hurry. Whites concentrated, sunswept; note Hermitage, St-Joseph.

2018 Scaled-up reds, v.gd; yr for top terroirs; v. rich Hermitage (r), similar Côte-Rôtie. Crozes variable. Whites v. successful: depth, surprising freshness. Note Condrieu, Hermitage, St-Péray.

2017 V.gd, esp Côte-Rôtie. Full reds, deeper than 16, show sunshine, packed-in, firm tannins, so need time. Whites for hearty food, Condrieu variable.

2016 Reds harmonious, classic at Côte-Rôtie, gd to v.gd. Note Cornas, Crozes-Hermitage reds. Marvellous Hermitage whites, other whites gd, clean.

2015 Excellent, v. concentrated, full-tannin reds; long-lived Hermitage, Côte-Rôtie. Time essential. Full whites, can be heady.

2014 Juicy reds, gained depth, give ready pleasure. Excellent whites: content, style, freshness.

2013 Reds v.gd, just starting to open. 20–25 yrs life. Exceptional whites (Hermitage, St-Joseph, St-Péray).

2012 Hermitage v.gd, open-book Côte-Rôtie. Fresh reds will last 15 yrs+. Whites have style, freshness.

Southern Rhône

2021 April frost, uneven summer, late harvest. Vacqueyras -50% crop. Standard reds gd, probably underestimated. Whites v.gd indeed, expressive, charming.

2020 Open reds, notable fruit in best. Can be variable. Top (esp gd terroirs): rich, full, gd acidity. Whites tricky from less gd zones, stick to best.

2019 Splendid reds; v.gd across board. Deep Grenache. Châteauneuf back on form, Gigondas excellent. Top names demand patience. Whites v.gd, full-bodied.

2018 Mixed, can be gd. Note Valréas, Visan, Vinsobres: higher, later vyds. Also Lirac, Rasteau. Whites v. full.

2017 V.gd, but can be variable. Full, bold reds, but tannins can be chewy. Top doms best, will be stylish. Rasteau, Visan gd. Full whites.

2016 Excellent for all (Châteauneuf, old-vines Grenache triumph). Sensuous reds: fruit bonanza, masses of life. Sun-filled whites; keep some to mature.

2015 Rich, dark, body, firm tannins, often enticing flair; v.gd. Quality high across board (Gigondas). Full whites v.gd.

2014 Aromatic finesse returns to Châteauneuf. Stick to best names. Note Gigondas, Rasteau, Cairanne. Fresh whites.

2013 Sparky, vibrant, slow-burn reds, v. low Grenache yields. Châteauneuf best from old vines. Long-lived whites.

2012 Dashing reds, lively, gd tannins. Open-book vintage. Food-friendly whites.

Champagne

2021 Cold spring, April frost, mildew = v. low yields. Sad in Marne V. Chard will be winner.

2020 Miraculous trio: 18 19 20. Beautiful ripe wines.

2019 Excellent quality. Fresh, pure fruit, tension, better acidity than much-lauded 18.

2018 Best in Pinot GCs of N Montagne. Chard more mixed: heat stress, some lowish acidity.

2017 Rain: Chard on Côte des Blancs the saviour grape. Athletic wines of grace, energy.

2016 Underrated. Gently expressive Pinot Ns give much pleasure.

2015 Hot yr. Some wines sweaty, blurred aromas; others are little triumphs; classic finely sculpted wines from a late season

2014 Suzuki fruit fly plagued black grapes. Gd ripe yr for rich Chard.

Earlier fine vintages: 12 10 09 08 07 06 04 02 00 98 96 95 92 90.

The Loire

2021 Difficult, complicated yr: frost, mildew, hail in places. Small crop saved by fine autumn. Fresher style than recently.

2020 Incredible 7th successive v.gd yr. Early vintage: gd quality, quantity.

2019 Well-balanced wines but volume down, esp Muscadet, Anjou (April frost).

2018 Exceptional quality/quantity, esp reds; gd sweet, some dry whites lack freshness; as gd as 47.

2017 Quality gd, April frosts again: Bourgueil, Savennières; Sancerre spared.

Alsace

2021 Tumultuous rain and disease. Pinot family suffered, but sunny Sept gave fine Ries.

2020 May be most refined, elegant, scented across board, as growers master climate change.

2019 Hot; e-facing hills, esp GCs, may be v.gd, ripe, dry (Ries, Pinot Gr, Gewurz). Wines from plains could be a problem.

2018 Warm, but fresh wines. Gewurz, Pinot Gr, Ries tops in high-altitude GCs.

2017 One of best since World War Two, but small crop. Recalls 71 08.

2016 Poised, finely balanced vintage and plenty of it, unlike 13 14 15.

2015 Rich vintage, one of driest ever. Great Pinot Gr; little volume but ripe, still fresh Sylvaner.

Earlier fine vintages: 12 10 08 07 04 02 96 95 92 90.

Abelé, Henri Champ ★★★ New name for Abel Lepitre, oldest house, now focusing on exports. Best CUVÉE Sourire de Reims 12 15 19', new Sourire Rosé 15 voluptuous, from Les Riceys (AUBE), expansive burg style. Gd value.

Abymes Sav On limestone rubble of Mont Granier by APREMONT. DYA. Vin de SAV AOP CRU. Jacquère grape. Try A&M QUENARD, ★★ Giachino, ★ Labbé, L'Epervière, P et S Ravier.

AC or AOC (Appellation Contrôlée) / AOP Government control of origin and production (but not quality) of most top French wines; around 45% of total. Now being converted to AOP (appellation d'origine protégée – which is much nearer the truth than contrôlée).

Agenais SW Fr ★ DYA IGP of Lot-et-Garonne. Beautiful country, best for Vieille Prune spirit. Simple wines from wide choice of SW grape varieties. A few DOMS (eg. Boiron, Campet, Lou Gaillot).

Agrapart Champ ★★★★ Pascal A makes fine-drawn CHAMP from scrupulously tended vyds. Focus on terroir. Fine top Avizoise 13' 17 19' 20. Best place for CHARD in 21.

Allemand, Thierry N Rh ★★★★ 01' 05' 06' 07' 08' 09' 10' 12' 13' 15' 16' 17' 18' 19' 20' 21 Offbeat, talented owner, started with nothing, magisterial CORNAS from organic 5-ha DOM, low sulphur. Two v. deep, smoky wines, dizzy prices, hard to find. Top is Reynard (profound, complex; 20 yrs+), Chaillot (bursting fruit, floral) drinks earlier.

Alliet, Philippe Lo ★★★ 17' 18' 19' 20' (21) Top CHINON, low yields. Tradition, VIEILLES VIGNES and two steep s-facing vyds e of Chinon: Coteau de Noiré (top CUVÉE), L'Huisserie. Age-worthy.

Aloxe-Corton Burg ★★→★★★ 05' 09' 10' 12' 15' 17 18' 19' 20' The n end of CÔTE DE BEAUNE, famous for GC CORTON, CORTON-CHARLEMAGNE, but less interesting at village or PC level. Reds attractive if not overextracted. Recent warm vintages have softened tannins. Best DOMS: Capitain-Gagnerot, Follin-Arbelet, Rapet, Senard, TOLLOT-BEAUT.

Alsace ★★→★★★★ Vosges rain-shadow effect means driest French wine region; 1800 sun hrs do the rest: aromatic, fruity, full whites, improving reds (PINOT N); now back to (mostly) classically dry styles. Huge geological diversity means terroir-lovers' bonanza; magnificent age-worthiness (40 yrs not a stretch), esp 61' 67' 71' 83' 90' 10' 17' 18 19.

Alsace Grand Cru Al ★★★→★★★★ 08' 10' 12 13 14 (esp RIES) 15 17' 18 19' 20' Only 51 best vyd sites (approx 1600 ha, 800 in production), but some too large; and only four "noble" grapes (PINOT GR, Ries, GEWURZ, MUSCAT) allowed, but three now named for PINOT N too. PC in gd sites in the works. Among famous houses, TRIMBACH now embracing GC status too.

Amirault, Yannick Lo ★★★→★★★★ 16' 17' 18' 19' 20' 21 Yannick and son Benoît. Organic. Top range of carefully crafted BOURGUEIL/ST NICOLAS DE BOURGUEIL. Small parcel of CHENIN BL.

Angerville, Marquis d' C d'O ★★★★ Discreet, classical, bio superstar in VOLNAY, with great range of PCs topped by legendary CLOS des Ducs (MONOPOLE). *See also* DOM DU PÉLICAN for Jura interests.

Anjou Lo ★→★★★★ Region and AC: ANJOU, SAUMUR. Mainly CHENIN BL dry whites, juicy reds, incl GAMAY; fruity CAB FR-based Anjou Rouge; robust tannic ANJOU-VILLAGES, CAB FR/CAB SAUV. Mainly dry SAVENNIÈRES; lightly sweet to rich COTEAUX DU LAYON Chenin Bl; rosé (dr, s/sw), (esp CABERNET D'ANJOU), fizz (esp CRÉMANT). Natural wine big here, often VDF.

Anjou Blanc Lo ★→★★★★ 18' 19' 20' 21 Increasingly important, exciting AOP for dry white CHENIN BL in starring role, from fresh young to complex, v.gd, long-lived. Increasingly made in Layon's best sites as sweet sadly hard to sell. Move to create white crus. Best: Angeli (VDF), Bablut, Baudouin, BELARGUS, Bergerie, Ch Bonnezeaux (VdF), CADY, Drost (formerly Delesvaux), Forges, Juchepie, Leroy (VdF), Ogereau, Passavant, PIERRE-BISE, Plaisance, Terra Vita Vinum.

Anjou-Villages Lo ★→★★★ 16' 17' 18' 19' 20' (21) Structured red AOP (CAB FR/CAB SAUV, a few pure Cab Sauv). Can be tannic; needs ageing. Best: Bergerie, Branchereau, Brizé, CADY, CH PIERRE-BISE, CLOS de Coulaine, Drost (formerly Delesvaux), Ogereau, Sauveroy, Soucherie. Sub-AOP Anjou-Villages-Brissac same area as COTEAUX DE L'AUBANCE; Try Bablut, Fontaines, Hardières, Haute Perche, Montigilet, Princé, Rochelles, Terra Vita Vinum, Varière.

Aphillanthes, Dom Les S Rh ★★→★★★ 16' 18' 19' 20' Organic, bio DOM in PLAN DE DIEU, progressive vyd work. Punchy, terroir wines, value. Two CUVÉES: des Galets, VIEILLES VIGNES (50S GRENACHE). CÔTES DU RH (r); GIGONDAS PROMESSE; RASTEAU 1921.

Apremont Sav Largest cru of SAV, thus quality varies. Steely, light Jacquère on limestone rubble. Keep up to 4 yrs. Try ★ 13 Lunes, Apffel, Blard, ★ Dupraz, ★★ Giachino, ★★ *Musson*, Perrier, Richel.

Arbin Sav ★★ A SAV cru. Dark, spicy MONDEUSE on steep slopes. Drink to 8 yrs+. Try ★ A&M QUENARD, Genoux, Jacquet, L'Idylle, ★★ F Trosset, *Magnin*.

Arbois Jura AOP of N Jura, great spot for wine, cheese, walking and Louis Pasteur museum. CHARD and/or SAVAGNIN whites, VIN JAUNE, reds from Poulsard/Ploussard, Trousseau or PINOT N. Try terroir-true ★★★ Stéphane TISSOT; burgundian styles from ★★★ *Pélican*, ★ Rijckaert, sulphur-free from MAISON OVERNOY, plus all-rounders ★★ Aviet, ★ Pinte, ★ Renardière, Rolet, ★★ *Tournelle*.

Recap on Al: world's best Gewurz (dr/sw); underrated Sylvaner (dr/sw).

Ardèche S Rh ★ →★★ 19' IGP Rocky granite hills w of S Rh, can use looser VDF status. Also base for Rhône Vin Nature movement. Quality up, often gd value. Fresh, clear reds; VIOGNIER (eg. CHAPOUTIER, Mas de Libian), MARSANNE. Best from SYRAH, also GAMAY (often old vines), CAB SAUV (Serret). Restrained, burgundy-style Ardèche CHARD by LOUIS LATOUR (Grand Ardèche too oaky). CH de la Selve; DOMS de Vigier, du Grangeon, Flacher, JF Jacouton; Mas d'Intras (organic).

Ardoisières, Dom des Sav IGP Vin des Allobroges organic estate with vyds outside AOP. Superb, eclectic range incl ★★★ Améthyste (r) blend, Quartz Altesse, Schist (w) blend.

Arlaud C d'O ★★★→★★★★ Leading MOREY-ST-DENIS estate energized by Cyprien A, reached new heights with 2019 vintage. Beautifully poised, modern wines with depth, class from exceptional BOURGOGNE Roncevie up to GCS. Fine range of Morey PCS, esp Ruchots.

Arlot, Dom de l' C d'O ★★→★★★ AXA-owned estate; stylish, fragrant wines across range from HAUTES-CÔTES to GC ROMANÉE-ST-VIVANT. Star buy is NUITS CLOS des Forêts St-Georges. Interesting whites too.

Armand, Comte C d'O ★★★★ MONOPOLE CLOS des Epeneaux may be POMMARD's most graceful wine, effortless and ageless. Gd value from AUXEY, VOLNAY too.

Arnoux-Lachaux C d'O ★★★★ New star in VOSNE-ROMANÉE. All change under Charles Lachaux: no-till, high-trellised, unhedged vines, v.-light-hand vinification with whole bunches. Breathtaking ethereal quality, and pricing for top wines, PCS Grand Suchots, Reignots and GC ROMANÉE-ST-VIVANT.

Aube CHAMP's s vyds, aka Côte des Bar – v.gd PINOT N by great Reims houses eg. KRUG, VEUVE CLICQUOT. Aube 11' excels. Nightmare of torrential rain and heat spikes in 2021; a few gd producers picked later in September.

Aubert et Matthieu L'doc ★★★ New, go-ahead NÉGOCIANT started by two friends. From best L'DOC terroirs eg. La Livinière, TERRASSES DU LARZAC. Hors Piste IGP. CHARD, PINOT N from cooler areas. Sustainable, eco-packaging, striking labels.

Aupilhac, Dom d' L'doc ★★★ Sylvain Fadat, pioneer of MONTPEYROUX, cultivates s-facing old-vine MOURVÈDRE, CARIGNAN as well as n-facing, higher-altitude SYRAH and whites on Mt Baudile for Les Cocalières.

Auxey-Duresses C d'O ★★→★★★ (r) 15′ 16 17 18′ 19′ 20′ (w) 14′ 15′ 17′ 18 19′ 20′ CÔTE DE BEAUNE village in valley behind MEURSAULT, enjoying climate change. Similar *whites offer value*, reds now ripen properly. Best: (r) COCHE-DURY, COMTE ARMAND, d'Auvenay (Boutonniers), Gras, Jessiaume, Paquet, Prunier; (w) Diconne, Lafouge, LEROUX, Paquet, Vincent.

Aveyron SW Fr ★ IGP DYA. Quaffing wines from Roquefort country. Try DOMS Bertau, Bias.

Avize Champ ★★★★ Côte des Blancs GC CHARD, home to finest growers AGRAPART, Bonville, SELOSSE, Thienot. Huge co-op Union CHAMP provides base wines to biggest houses. Its *Pierre Vaudon* brand is worth seeking out.

Aÿ Champ Revered PINOT N village, home of BOLLINGER, DEUTZ. Mix of merchants' and growers' wines, some in barrel (eg. Claude Giraud, master of Argonne oak, son-in-law Sébastien now in charge), some in tanks. *Gosset-Brabant* Noirs d'Aÿ excels. Aÿ Rouge (AOP COTEAUX CHAMPENOIS) now excellent in riper yrs, esp 15′ 18′ 19 ★★★(★). Wines typically have weight, power, finesse.

Ayala Champ Revitalized AÿЀ house, owned by BOLLINGER. Precision, energy, complexity under Caroline Latrive, chef de CAVE. Two stand-outs: v. fine BRUT Zéro, savoury but never rasping; lovely BLANC DE BLANCS 13 ★★★★ 14 19 20′.

Ries in Al grown on granite has a slight flavour of bergamot. Try it.

Ayze Sav ★★ Cru closest to Mont Blanc, sometimes spelt Ayse. Rare Gringet grape for fruity fizz and mineral-laden (dr) whites. Made famous by Dominique Belluard, RIP 2021.

Bachelet Burg ★★→★★★★ Widespread family name in S C D'O. Look out for: B-Monnot (esp BÂTARD-MONTRACHET, PULIGNY), Bernard B (Maranges), Jean-Claude B (ST-AUBIN, CHASSAGNE etc.). Note no relation to Denis B (great GEVREY-CHAMBERTIN).

Bandol Prov ★★★ Noble reds for ageing, majority MOURVÈDRE plus GRENACHE, CINSAULT. Rosé now main production; can be v. pale, Prov-style, or gutsier, gastronomic from Mourvèdre. Some white: CLAIRETTE, UGNI BL, occasionally SAUV BL. Tops: DOMS DE LA BÉGUDE, du Gros'Noré, La Bastide Blanche, Lafran Veyrolles, La Suffrène, Mas de la Rouvière, Pibarnon, Pradeaux, Ray-Jane, TEMPIER, Terrebrune, Val d'Arenc, Vannières.

Banyuls Rouss ★★★ Undervalued, sometimes brilliant VDN, oxidized or not, from GRENACHE of all colours. Young, fresh style is *rimage*, but stars are RANCIOS, long-aged, pungent, intense. Try with chocolate. Best: DOMS de la Rectorie, du Mas Blanc, la Tour Vieille, Madeloc, Vial Magnères; Coume del Mas, Les Clos de Paulilles. *See also* MAURY.

Barbier, Christopher L'doc ★★★ Works magic with underrated BOURBOULENC: maritime vyds nr LA CLAPE. DOM de Simonet IGP, unoaked, gd value. Les Terres Salées, oaked, keep 5 yrs for full glory. Terres Salées red succeeds with MERLOT where most L'DOC growers struggle. Competent La Clape under CH Bouïsset label.

Barmes-Buècher, Dom Al ★★→★★★ Geneviève Barmès (née Buecher) est Wettolsheim estate in 1995; son Maxime now makes wines: gd CRÉMANT and SGNS, bevy of GCs; lieu-dit CLOS Sand RIES often best buy.

Barrique Term for oak barrel holding 225 litres in B'x (and Cognac). Fashion now dictates less new oak and more subtle toasting. Average price €750/barrel.

Barroubio, Dom L'doc ★★★ Raymond Miquel is Monsieur Muscat with five VDN Muscat de St Jean de MINERVOIS; white chalky soils at 300m (984ft) give purity, freshness. Black label v.gd value. New Les Cresses aged in amphorae.

FRANCE

Barsac Saut ★★→★★★★ 09' 11' 13 14 15' 16' 18 19 20 Neighbour of SAUT with similar botrytized wines from lower-lying limestone soil; fresher, less powerful; vintage 20 gd but scarce, 21 frost. Top: *Climens*, COUTET, DOISY-DAËNE, *Doisy-Védrines*. Value: Cantegril, La Clotte Cazalis, Liot.

Barthod, Ghislaine C d'O ★★★→★★★★ A reason to fall in love with CHAMBOLLE-MUSIGNY, if you haven't already. Son Clément in charge from 2019, further refining style. Wines of perfume, delicacy yet depth, concentration. Unbeatable range of 11 different PCS, incl Baudes, Beaux Bruns, Cras, Fuées.

Bâtard-Montrachet C d'O ★★★★ 08' 09' 10 12 14' 15 17' 18 19' 20' 12-ha GC downslope from LE MONTRACHET. Grand, hefty wines that should need time; more power than neighbours Bienvenues-B-M (more graceful) and CRIOTS B-M. Seek out: BACHELET-Monnot, BOILLOT (both H and JM), FAIVELEY, GAGNARD, LATOUR, LEFLAIVE, LEROUX, MOREY, OLIVIER LEFLAIVE, PERNOT, Ramonet, SAUZET, VOUGERAIE. Also J-C BACHELET, J CARILLON for Bienvenues version.

Baudry, Dom Bernard Lo ★★→★★★ 17' 18' 19' 20' (21) Organic. Diverse sites and soils: sand/clay, gravel, limestone. Fine CHINON, CAB FR (r/rosé), CHENIN BL too; fruity Les Granges; CLOS Guillot (r/w – limestone), Croix Boissée, Les Grézcaux.

Baudry-Dutour Lo ★★→★★★ 18' 19' 20' 21 CHINON's largest producer. Reliable range: light to age-worthy (r/w). CHX de la Grille, de la Perrière, de St Louans (r/w), du Roncée; 3 Coteaux (w). Bought 60 ha in E TOURAINE (2020), 55 ha SAUV BL. Justine Baudry winemaker.

Baumard, Dom des Lo ★★→★★★ 17' 18' 19' 20' 21 ANJOU DOM with well-made range incl CLOS de Ste Catherine, QUARTS DE CHAUME (use of cryoextraction now banned), SAVENNIÈRES.

Baux-de-Provence, Les Prov ★★→★★★ Stunning, touristy village, in the centre of the AC, almost all organic/bio. Mostly red: CAB SAUV, GRENACHE, SYRAH. White: CLAIRETTE, GRENACHE BL, Rolle, ROUSSANNE. TRÉVALLON best, prefers IGP Alpilles. Others: Dalmeran, d'Estoublon, DOM Hauvette, Gourgonnier, Lauzieres, Mas de Carita, Mas de la Dame, Mas Ste Berthe, ROMANIN, Terres Blanches, Valdition; atypical Milan.

Béarn SW Fr AOP (r) ★→★★ 18 19 (20) (w/rosé) DYA Reds from ★ JURANÇON co-op, DOMS Lapeyre Guilhémas. Same grapes as MADIRAN and Jurançon.

Beaucastel, Ch de S Rh ★★★★ 01' 05' 06' 07' 09' 10' 12' 13' 15' 16' 17' 18 19' 20 Big, long-time organic CHÂTEAUNEUF estate, classic large galet stone soils: old MOURVÈDRE, 100-yr-old ROUSSANNE. Dark-fruited, recently more polished wines, upfront, suave elegance, drink at 2 yrs or from 7–8. Breathtaking, top-quality 60% Mourvèdre Hommage à Jacques Perrin (r). Wonderful, complex old-vine Roussanne: enjoy over 5–25 yrs. Genuine, stylish own-vines CÔTES DU RH Coudoulet de Beaucastel (r), lives 10 yrs+. Famille Perrin GIGONDAS (v.gd), RASTEAU, VINSOBRES (best) all gd, authentic. Note organic Perrin Nature Côtes du Rh (r/w). Growing N Rh merchant venture, Maison Les Alexandrins (elegant). (*See also* Tablas Creek, California.)

Beaujolais ★ DYA Basic appellation of huge Beauj region. Time to fall in love again with the simple fresh fruit, and more from growers in the hills. Avoid overcropped industrial examples. Also sold as COTEAUX BOURGUIGNONS.

Beaujolais Primeur / Nouveau Beauj More of an event than a drink. The BEAUJ of the new vintage, hurriedly made for release at midnight on the 3rd Wednesday in Nov. Enjoy juicy fruit but don't let it put you off real thing.

Beaujolais-Villages Beauj ★★ 17 18' 19' 20' Challenger vyds to the ten named crus, eg. MOULIN-À-VENT. May specify best village, eg. Lantigné. Try CH de Basty, Ch des Vergers, F Berne, F Forest, JM BURGAUD, N Chemarin.

Beaumes-de-Venise S Rh ★★ (r) 10' 15' 16' 17' 18 19' Village nr GIGONDAS, mix high ochre-stony and plain vyds, noted for VDN MUSCAT apéritif/dessert. Serve v. cold:

grapey, honeyed, can age well. eg. DOMS Beaumalric, Bernardins (complex), Coyeux, Durban (rich, long life), Fenouillet (brisk), JABOULET, Perséphone (stylish), Pigeade (fresh, v.gd), VIDAL-FLEURY, co-op Rhonéa. Also robust, smoky, grainy reds. CH Redortier; Doms Cassan, de Fenouillet, Durban, la Ferme St-Martin (organic), Les Baies Gouts, Mathiflo, St-Amant (gd w). Leave for 2–3 yrs. Simple whites (some dry Muscat, VIOGNIER).

Beaumont des Crayères Champ ★★★ Bijou co-op nr Épernay for top MEUNIER; on ideal chalk. Fleur de Prestige top value 18' 19' 20'; CHARD-led CUVÉE Nostalgie 16'; new Fleur de Meunier BRUT Nature 18' 19' 20'. A question mark over future quality with departure of key wine director (2021).

Beaune C d'O ★★★ 05' 09' 10' 12 15' 16 17 18' 19' 20 Centre of Burgundy wine trade, classic merchants: BOUCHARD, CHANSON, DROUHIN, JADOT, LATOUR; and more recent contenders Bernstein, Lemoine, LEROUX, Pacalet. Top DOMS: Bellène, Besancenot, Croix, DE MONTILLE, Dominique LAFON, Morot, plus iconic HOSPICES DE BEAUNE. Graceful, perfumed PC reds offer value, eg. Bressandes, Cras, VIGNES Franches; more power from Grèves. Try Aigrots, CLOS St-Landry and esp *Clos des Mouches (Drouhin)* for whites.

Beauregard, Ch de Burg ★★→★★★ Exceptional range of unmissable POUILLY-FUISSÉ from Frédéric Burrier. Try Les Reisses, Ménétrières, Vers Cras. Fine BEAUJ too: FLEURIE, MOULIN-À-VENT.

Bee, Dom of the Rouss ★★★ Justin Howard-Sneyd MW farms "4 ha of nectar" nr MAURY. Bee-side GRENACHE gd value. Les Genoux single-vyd, 100-yr-old vines the bee's knees.

Begude, Dom L'doc ★★→★★★ AOP LIMOUX, 300m (984ft) altitude gives freshness, pristine fruit. Flagship CHARD Arcturus v. classy. IGP L'DOC, GEWURZ. PINOT N, VIOGNIER gd value.

Bégude, Dom de la Prov ★★★★ Guillaume Tari makes astonishing, long-lived BANDOL. La Brulade (r), powerful, complex; characterful rosé, rewards ageing.

Belargus, Dom Lo ★★★→★★★★ 18' 19 20' (21) Mainly dry, v. pure CHENIN BL from single vyds: high-quality, bio DOM owned by Ivan Massonnat. ANJOU, QUARTS DE CHAUME, SAVENNIÈRES. Bought Dom de Beauséjour, CHINON (2021)

Bellivière, Dom de Lo ★★→★★★ 19' 20' (21) COTEAUX DU LOIR, JASNIÈRES, Pineau d'Aunis, all bio; v. precise CHENIN BL incl Les Rosiers. Les Arches de Bellivière (NÉGOCIANT).

Bergerac SW Fr ★→★★★ 15' 17 18 19 20 AOP More continental climate produces fuller-bodied wines. Côtes de Bergerac for sweet whites or fuller reds from lower yields. Huge variations in styles, quality. Best ★★★ *Tour des Gendres*, DOM de l'Ancienne Cure, ★★ CH de la Jaubertie, Moulin Caresse, Thénac, Tirecul La Gravière. *See* sub-AOPs MONBAZILLAC, MONTRAVEL, PÉCHARMANT, ROSETTE, SAUSSIGNAC.

Berthet-Bondet, Dom Jura ★★ Biggest producer (still small) of CH-CHALON VIN JAUNE but reliably covers all bases for CÔTES DU JURA red, white and CRÉMANT; gd-value organic.

Besserat de Bellefon Champ ★★ Épernay house specializing in gently sparkling CHAMP (old CRÉMANT style). Part of LANSON-BCC group. Respectable quality, gd value, esp 13 14 17' (tiny but excellent). Opulent flavours in trio 18 19 20.

Beyer, Léon Al ★★★★ Top wines labelled Comtes d'Eguisheim: GEWURZ best and underrated, but RIES Comtes d'Eguisheim the calling card. Arch-traditionalist, v. famous family house, delivers intense, dry gastronomic wines listed by many Michelin-starred restaurants; many old vintages still in stock. SGNS grand.

Bichot, Maison Albert Burg ★★→★★★★ Major BEAUNE merchant/grower with bio DOMS in BEAUJOLAIS (Rochegrès), CHABLIS (LONG-DEPAQUIT), MERCUREY (Adélie), NUITS (CLOS Frantin), POMMARD (Pavillon); v. sound source with increasing flair.

Bienvenues-Bâtard-Montrachet C d'O *See* BÂTARD-MONTRACHET.

Billaud Chab ★★★→★★★★ Difficult CHAB choice between DOM Billaud-Simon back on form under FAIVELEY ownership and Samuel B's sensational wine under his own label. Both brilliant.

Billecart-Salmon Champ ★★★★ Revered family house, 7th-generation, gifted young chef de CAVE, top quality. New innovative longer lees-aged Rendezvous de Billecart by variety: No 1 = MEUNIER; No 2 = PINOT N; No 3 = CHARD. CUVÉE Louis BLANC DE BLANCS 07 ready, will keep. Superb CLOS St-Hilaire 02. NF Billecart, summit of range, perhaps greatest 02; excellent Elisabeth Salmon Rosé 02 06'.

Bize, Simon C d'O ★★★ Chisa B makes sensational bio whole-bunch-style reds at all levels from BOURGOGNE to GC LATRICIÈRES-CHAMBERTIN. Look out for SAVIGNY Grands Liards, PCS Guettes, Vergelesses and tasty whites too.

Blagny C d'O ★★→★★★ 05' 09' 10' 12 15' 16' 17 18' 19' 20' Hamlet on hillside above MEURSAULT and PULIGNY. Own AOP for austere yet fragrant reds, diminishing volumes. Whites sold as Meursault-Blagny PC. Best vyds: La Jeunelotte, Pièce Sous le Bois, Sous le Dos d'Ane. Best growers: (r) Lamy-Pillot, LEROUX, MATROT; (w) de Cherisey, JOBARD, LATOUR, LEFLAIVE, Matrot.

Blanc de Blancs Any white wine made from white grapes only, esp CHAMP. Description of style, not quality.

Blanc de Noirs White (or slightly pink or "blush", or "gris") wine from red grapes, esp CHAMP: can be solid in style, but many now more refined; better PINOT N and new techniques. With climate change, often escapes the harsh tannins of old.

Blanck, Paul & Fils Al ★★★→★★★★ Kientzheim grower among best in Al. Finest from 6-ha GC Furstentum (GEWURZ, PINOT GR, RIES), sumptuous GC SCHLOSSBERG (great Ries 17' 18'). Top-value entry-level wines.

Climate change in Burg: can't replace Pinot, but add Aligoté in white villages?

Blanquette de Limoux L'doc ★★ Crisp, appley fizz, pretty, fun; 90% Mauzac plus CHARD, CHENIN BL. AOP CRÉMANT de LIMOUX, more classic with Chard, Chenin Bl, PINOT N, and less Mauzac. Sieur d'Arques co-op is biggest, gd enough. Also *Antech*, Delmas, Jo Riu, La Coume-Lumet, LAURENS, Robert, Les Hautes Terres. *Monsieur S.*

Blaye B'x ★ →★★ 14 15 16 17 19 20 Designation for better reds (lower yields, higher vyd density, longer maturation) from AOP BLAYE-CÔTES de B'x.

Blaye-Côtes de Bordeaux B'x ★→★★ 15 16' 17 19 20 Mainly MERLOT-led red AOP on right bank of Gironde. A little dry white (mainly SAUV BL). Best CHX: Bel-Air la Royère, Cantinot, de la Chapelle, des Tourtes, Gigault (CUVÉE Viva), Haut-Bertinerie, Haut Grelot, Jonqueyres, Monconseil-Gazin, Mondésir-Gazin, Montfollet, Peybonhomme Les Tours, Roland la Garde, Ste-Luce Bellevue. Also go-ahead VIGNERONS de Tutiac co-op for whites and reds.

Boeckel, Dom Al ★★★ An AL vigneron from C16; DOM est 1853, 23 ha, organic. RIES CLOS Eugénie rich, rounded 18 19' 20. Outstanding *Sylvaner* from Zotzenberg GC; Ries great in 17', generous yet refined; for shellfish, truffles.

Boillot C d'O Leading Burgundy family. Look for ★★★ Jean-Marc (POMMARD), esp fine, long-lived whites; ★★→★★★★ Henri (MEURSAULT), potent, stylish (r/w); ★★★ Louis (CHAMBOLLE) for great reds from both Côtes, and his brother Pierre (DOM Lucien B) ★★→★★★ (GEVREY). Marthe Henry B (Meursault), no close relation, makes interesting post-modern wines.

Boisset, Jean-Claude Burg Ultra-successful merchant/grower group created over last 50 yrs. Boisset label from amazing new winery in NUITS and esp own vyds *Dom de la Vougeraie* excellent. Recent additions to empire: (Burg) brands Alex Gambal, VINCENT GIRARDIN. Also projects in BEAUJ, Jura and overseas: California (Gallo connection), Canada, Chile, Uruguay.

Boizel Champ ★★★ Exceptional value, rigorous quality, family-run. Well-aged

BLANC DE BLANCS NV, esp on base of 13' 17'. CUVÉE Sous Bois purity without overwoodiness. Prestige Cuvée Joyau de France, esp Rosé 12 drinking sublimely. Florent Rocques-Boizel taking reins. Old yrs available in rationed quantities back to 80s for collectors.

Bollinger Champ ★★★★ Great classic house, ever better, more tension. BRUT Special NV singing since 2012; RD 04; Grande Année 08 12 14, innovative Vintage Rosé 12 powerful, incl high 30% of Côte aux Enfants rouge (14). New special small CUVÉES showing new faces of PINOT N villages; fine PN VZ 15 fairly priced. *See also* LANGLOIS-CH.

Bonneau du Martray, Dom C d'O (r) ★★★ (w) ★★★★ Reference producer for CORTON-CHARLEMAGNE, bought 2016 by Stanley Kroenke, owner of Screaming Eagle (California) and Arsenal FC (UK). Intense wines designed for long (c.10 yrs) ageing, glorious mix of intense fruit, underlying minerals. Small amount of fine red CORTON.

Bonnes-Mares C d'O ★★★★ 90' 93 96' 99' 02' 05' 09' 10' 12' 15' 16' 18 19 20 A GC between CHAMBOLLE-MUSIGNY and MOREY-ST-DENIS with some of latter's wilder character. Sturdy, long-lived wines, less fragrant than MUSIGNY. Best: ARLAUD, d'Auvenay, Bart, BRUNO CLAIR, DE VOGÜÉ, Drouhin-Laroze, DUJAC, Groffier, H BOILLOT, JADOT, MORTET, ROUMIER, VOUGERAIE.

Bonnezeaux Lo ★★★→★★★★ 16 17 18' 19 20 (21) Sweet CHENIN BL, v. age-worthy; three sw-facing slopes (schist) in COTEAUX DU LAYON. Best: Deux Arcs, Fesles, Fontaines, les Grandes VIGNES, Mihoudy, Petit Val, Petite Croix, Varière. Potential GC.

Mme Bollinger's famous bicycle was standard bike – with Hermès saddle.

Bordeaux ★→★★ 18 19 20 Catch-all AOP for generic B'x (represents c. half region's production). Most brands (*Dourthe*, Michel Lynch, MOUTON CADET, *Sichel*) are in this category. E-2-M zone principal source. Look for CHX Bauduc, Beauregard-Ducourt, BONNET, Bouillerot, La Freynelle (CAB SAUV), Lamothe-Vincent, Reignac.

Bordeaux Supérieur B'x ★→★★ 16 17 18 19 20 Higher min alc, lower yield, longer ageing than the last. Mainly bottled at property. Consistent CHX: Argadens, Bellevue Peycharneau, Camarsac, Grand Village, Grée-Laroque, Jean Faux, Landereau, Méaume, *Parenchère* (CUVÉE Raphaël), Penin, *Pey la Tour* (Rés), Pierrail, Reignac, *Thieuley*, Turcaud.

Borie-Manoux B'x Admirable B'x shipper, CH-owner: BATAILLEY, BEAU-SITE, DOM DE L'EGLISE, LYNCH-MOUSSAS, TROTTEVIEILLE. Also owns NÉGOCIANT Mähler-Besse.

Bouchard Père & Fils Burg ★★→★★★★ Dynamic BEAUNE merchant, all-round, robust style, age-worthy. Whites best in MEURSAULT and GC, esp CHEVALIER-MONTRACHET. Flagship reds: Beaune VIGNE de L'Enfant Jésus, CORTON, *Volnay Caillerets Ancienne Cuvée Carnot. See also* WILLIAM FÈVRE (CHAB).

Boulay, Gérard Lo ★★★→★★★★ 16' 17' 18 19' 20' (21) Top SANCERRE producer who for yrs flew under radar. Wonderfully pure, age-worthy wines, esp CLOS de Beaujeu, La Côte, Monts Damnés and rare Comtesse. Now joined by son Thibault.

Bouley Burg ★★→★★★ Developing stars in VOLNAY, at van of cutting-edge higher-trained viticulture. Smart reds, esp Volnay PCS, from cousins Thomas and Pierrick at their respective DOMS.

Bourgeois, Famille Lo ★★→★★★ 17' 18' 19' 20' (21) SANCERRE grower-merchant: v.gd, best v. age-worthy. Organic from 2023. Patriarch Jean-Marie stepping back at 79. All Central Lo AOPS, Petit Bourgeois (IGP). Best: Etienne Henri, Jadis, La Bourgeoise (r/w), Le Graveron (r Monts Damnés), Les Côtes aux Valets, Les Ruchons (w Silex), Sancerre d'Antan. Also v.gd CLOS HENRI (NZ).

Bourgogne Burg ★→★★ (r) 15' 17 18' 19' 20' (w) 14' 17' 18 19' 20' Ground-floor AOP for burg, ranging from mass-produced to bargain beauties. Sometimes comes with subregion attached, eg. CÔTE CHALONNAISE, HAUTES-CÔTES and now, C D'O. Whites from CHARD unless B ALIGOTÉ. Reds from PINOT N unless declassified BEAUJ crus (sold as B GAMAY) or B Passetoutgrains (Pinot/Gamay mix, must have 30%+ of former).

Bourgueil Lo ★★→★★★ 16 17′ 18′ 19′ 20′ (21) AOP Full-bodied, age-worthy, mainly CAB FR. *Amirault*, Ansodelles, Audebert, Chevalerie, Courant, *La Butte*, Gambier, LAMÉ DELISLE BOUCARD, Ménard, Minière, Nau Frères, Omasson, Petit Bondieu, Revillot, Rochouard.

Bouscassé, Dom SW Fr ★★★ 15′ 17 18 19 (20) MADIRAN Palace. BRUMONT's Napa V-style home. Reds a shade quicker to mature than oaky flagship Montus. VIEILLES VIGNES (100% TANNAT) is star.

Bouvet-Ladubay Lo ★★→★★★ Big sparkling range: Brut Zéro, Rubis, Saphir, Trésor (w/rosé). Also SAUMUR-CHAMPIGNY, esp Les Nonpareils.

Bouzereau C D'O ★★→★★★ The B family infests MEURSAULT, in a gd way. DOM Michel B is leader but try also Jean-Marie B, Philippe B (CH de Cîteaux), Vincent B or B-Gruère & Filles for gd-value whites.

Bouzeron Burg ★★ 17′ 18 19 20′ CÔTE CHALONNAISE village with unique AOP for ALIGOTÉ, esp golden version. Stricter rules and greater potential than straight BOURGOGNE Aligoté. Chanzy and esp *A & P de Villaine* outstanding. Same village makes gd CHARD, PINOT N as Bourgogne CÔTE CHALONNAISE.

Bouzy Rouge Champ ★★★ 09 12 15′ 18′ Still red of famous PINOT N village. Formerly like v. light burg, now with more intensity (climate change, better viticulture), also refinement. VEUVE CLICQUOT, Colin and Paul Bara best producers.

Boxler, Albert Al ★★★★ Top few, classic AL DOM, age-worthy wines as complex as great burg, not just GC but GC subzones too: RIES Brand (Kirchthal, Kirchberg), Ries Sommerberg (Dudenstein, Eckberg). Outstanding VTS, esp PINOT GR Wibtal 17. Exceptional quality/price ratio PINOT BL, SYLVANER 18.

Brice Champ ★★★(★) Exciting, gastronomy-orientated Bouzy DOM recruited brilliant Christophe Constant from JL Vergnon as chef de CAVE: 1st-rate Poteau CHARD 19′ vinified in large 350-litre oak tuns, magnums only.

Brocard, J-M Chab ★★→★★★ Quality and commercial acumen under one roof: CHAB Ste Claire and fine range of PC and GC. Son Julien B has impressive bio range under 7 Lieux label.

Brochet, Emmanuel Champ ★★★ Bijou producer, exceptional, from steep Mont Bernard. Extra BRUT pure, exhilarating, winemaking slow and patient, in barrel 9 mths. Certified organic, no fining or filtration. Excelled in sumptuous 18′; more classic, racy 19.

Brouilly Beauj ★★ 17 18′ 19′ 20′ Largest of ten BEAUJ crus: solid, rounded, with some depth of fruit, approachable early but can age 3–5 yrs. Top growers: CH de la Chaize, DOMS Chermette, JC Lapalu, L&R Dufaitre, Piron. Even better is adjacent Côte de Brouilly, esp Ch THIVIN.

Brumont, Alain SW Fr ★★★★ MADIRAN's pioneer and living icon, creator of BOUSCASSÉ, La Tyre, MONTUS, but also quaffing Gascogne ★ Torus and range of quaffable IGPS. ★★★ PACHERENCS (dr/sw) outstanding, named after mth of picking, using revolutionary calender: Vendemiaire = Oct, Frimaire = Dec.

Brut Champ Term for dry classic wines of CHAMP. Most houses/growers have reduced dosage (adjustment of sweetness) in recent yrs. But great Champ can still be made at 8–9g residual sugar.

Brut Ultra / Zéro Term for bone-dry wines (no dosage) in CHAMP (also known as Brut NATURE); fashionable, esp with sommeliers, quality better with warmer summers: needs ripe yr, old vines, max care, eg. *Pol Roger Pure*, ROEDERER Brut Nature Philippe Starck 12′, Veuve Fourny Nature.

Bugey ★★ Small AOP w of SAV. Light, fresh sparkling and all colours of still. Crus incl Cerdon (pink GAMAY/Poulsard *méthode ancestrale*), Montagnieu (Altesse, MONDEUSE, sp). Whites mainly Altesse (Roussette du Bugey AOP), CHARD. Red/rosé: GAMAY, Mondeuse, PINOT N. Try Angelot, Balivet P (Cerdon), ★ Bonnard, Bugiste, ★ Grangeons de l'Albarine, ★★ Renardat-Fache (Cerdon), ★★ Peillot, Tissot, Trichon.

Burgaud Beauj ★★★ Jean-Marc from MORGON, top bottlings of Charmes, Côte du Py, Grands Cras, etc., with fine ageing potential, but juicily attractive from day one. Nephew Alexandre promising too.

Burn, Ernest / Clos Saint-Imer Al ★★→★★★★ Francis B loves to pick late and makes some of AL's richest, most opulent wines. Owns MONOPOLE CLOS Saint-Imer in GC GOLDERT; top wines labelled La Chapelle. Incredibly rich SYLVANER and MUSCAT (Al best?) 10 13 15 16 17 18 19. Excellent PINOT GR 10 12 15′ 18.

Bursin, Agathe Al ★★★→★★★★ Leading Westhalten family estate revitalized by Agathe. Stellar, modern SYLVANER (Lutzental and esp Eminence) of depth and class from v. old vines (and top sites Bollenberg and GC Zinnkoepflé). Also v.gd RIES, SGN.

Buxy, Caves de Burg ★→★★ Leading CÔTE CHALONNAISE CO-op for decent CHARD, PINOT N, source of many merchants' own-label ranges. Easily largest supplier of AOP MONTAGNY.

Buzet SW Fr ★★ 18 19 20 AOP Plummy cousin of B'x. Dominated by exemplary co-op, incl CHX de Guèyze, Padère. Ch Pierron is worthy indie producer.

Cahors black wine (juice concentrated by heat) famous in Middle Ages: travelled well.

Cabernet d'Anjou Lo ★→★★ ANJOU's largest AOP. CABS FR/SAUV: soft red-fruits flavour. Try Bablut, Bergerie, Chauvin, Clau de Nell, de Sauveroy, Grandes VIGNES, Montgilet, Ogereau, Plessis-Duval.

Cabidos SW Fr ★★★ 15 16 17 Proper CH in BÉARN, but outside JURANÇON. Similar wines. ★★ Gaston Phoebus is dry PETIT MANSENG. ★★★ St Clément is gorgeous, golden, sweet version, worth seeking. L'Or de Cabidos is heavenly, concentrated.

Cabrol, Dom de L'doc ★★★ Longstanding star of AOP Cabardès, nr Carcassonne where Med and B'x varieties allowed. SYRAH-dominant Vent d'Est, fragrant, fiercely Med style. Vent d'Ouest, more CAB SAUV, quieter, no less fine.

Cadillac-Côtes de Bordeaux B'x ★→★★ 16 18 19 20 Long, narrow, hilly zone on right bank of Garonne. Mainly MERLOT with CABS SAUV/FR. Medium-bodied, fresh reds; quality v. varied. Best: Alios de Ste-Marie, Biac, Brethous, *Carsin*, CH Carignan, CLOS Chaumont, Clos Ste-Anne, de Ricaud, Grand-Mouëys, Lamothe de Haux, Le Doyenné, Mont-Pérat, Plaisance, Réaut (Carat), *Reynon*, Suau.

Cady, Dom Lo ★★→★★★ 19′ 20 (21) Organic family dom in ANJOU, v.gd, esp CHENIN, incl Cheninsolite (dr), COTEAUX DU LAYON. Winery burned down April 2021, getting back on feet.

Cahors SW Fr ★★★ 15′ 16 18′ 19 20 Historical AOP on River Lot. Now wrestling back its MALBEC birthright. Badly affected by frosts in 2021. All red (some IGP w). Styles continue to change with, mercifully, less extraction. Easy-drinking ★★ CH de Hauterive, CLOS Coutale. More substance from ★★★ CH DU CÈDRE, Clos d'Un Jour, *Clos Triguedina*, Clos Troteligotte, de la Bérengeraie, DOM Cosse-Maisonneuve, Haut-Monplaisir; ★★★ Chx Gaudou, Hautes-Serres, La Coustarelle, Lamartine, Les Croisille, Mas La Périé, Ponzac.

Cailloux, Les S Rh ★★★ 90′ 98′ 03′ 05′ 09′ 10′ 16′ 18 19′ 20′ Family CHÂTEAUNEUF DOM; elegant, profound, spiced, handmade reds, local ID, fab value. Special wine Centenaire, oldest GRENACHE 1889 noble, dear, max elegant 16′ 19′. Also DOM André Brunel (esp gd-value CÔTES DU RH red Est-Ouest).

Cairanne S Rh ★★→★★★ 10′ 15′ 16′ 17′ 18 19′ 20′ Classy S Rh area, many options from *garrigue* soils, wines of character, dark fruits, mixed herbs, refinement, texture, esp CLOS des Mourres (organic), Clos Romane (punchy), DOMS Alary (style, organic), Amadieu (pure, bio), Boisson (hearty), Brusset (deep), Cros de Romet, Escaravailles (flair), Grands Bois (organic), Grosset, Hautes Cances (sleeker since 19), Jubain, *Oratoire St Martin* (classy, bio), Rabasse-Charavin (spark), Richaud (great fruit), Roche. Food-friendly, full whites.

Canard-Duchêne Champ House owned by ALAIN THIÉNOT. Now run by next generation. ★★★ CUVÉE Léonie. Improved Authentique Cuvée (organic) 12' 13 15 17' 18' 20. Single-vyd Avize Gamin 12 13 17' 19' 20'.

Canon-Fronsac B'x ★★→★★★ 14 15 16 18 19' 20' Tiny enclave within FRON, otherwise same wines. Environmental action. Best: rich, full and finely structured. Try Barrabaque, Canon Pécresse, Cassagne Haut-Canon la Truffière, GABY, Grand-Renouil, La Fleur Cailleau, MOULIN PEY-LABRIE, Pavillon, Toumalin.

Carillon C d'O ★★★ Contrasting PULIGNY brothers: Jacques unchangingly classical; try PC Referts. François for exciting modern approach. Try Combettes, Folatières. Village Puligny great from both.

Castelmaure L'doc ★★→★★★ Excellent co-op in wilds of CORBIÈRES. Lots of CARIGNAN. Easy-drinking La Pompadour. Grand CUVÉE screams terroir: dark fruit, wild herbs, hot stone. No3 flagship, brooding, big. Keep 5 yrs.

Castelnau, De Champ ★★★ Rising co-op. Cellarmaster Carine Bailleul insists on longer lees-ageing in excellent BLANC DE BLANCS and Vintage 02. Innovative Prestige Collection Hors d'Age blended from best wines, different each yr: current release CCF2067 led by fine MEUNIER.

Castigno, Ch L'doc ★★ Belgian-owned ST-CHINIAN estate turning sleepy village of Assignan into tourist spot with restaurants, hotel, sustainable, cork-clad, bottle-shaped winery. Secrets des Dieux oaky, aged SYRAH/CARIGNAN/GRENACHE. Stylish Grace des Anges ROUSSANNE-based white.

Castillon-Côtes de Bordeaux B'x ★★ →★★★ 15 16 18 19 20 Appealing e neighbour of ST-ÉM; similar wines, usually gd value. Some ageing potential. 230 growers; 25% organic. Top: Alcée, Ampélia, Cap de FAUGÈRES, CLOS Les Lunelles, Clos Louie, *Clos Puy Arnaud*, Côte Montpezat, *d'Aiguilhe, de l'A*, de Pitray, Joanin Bécot, *La Clarière-Laithwaite*, l'Aurage, *Le Rey*, *l'Hêtre*, Montlandrie, Poupille, Roquevieille, Veyry.

Cathiard, Dom Sylvain C d'O ★★★★ Sébastien C makes wines of astonishing quality from VOSNE-ROMANÉE esp Malconsorts, Orveaux, Reignots, plus NUITS-ST-GEORGES Aux Thorey, Murgers. New range of generic Bourgognes. Style to pick late and destem.

Cave Cellar, or any wine establishment.

Cave coopérative Growers' co-op winery; over half of all French production. Wines often well priced, probably but not most exciting. Many co-ops closing down.

Cazes, Dom Rouss ★★★ Big, bio, impressive. VDN eg. RIVESALTES Ambré, Tuilé, Grenat and sensational aged Aimé Cazes. MAURY SEC. Ambre (SW GRENACHE BL), Le Canon du Maréchal GRENACHE/SYRAH; Top red Crédo CÔTES DU ROUSS-VILLAGES with Ego, Alter. CLOS de Paulilles BANYULS, COLLIOURE.

Cébène, Dom de L'doc ★★★★ Brigitte Chevalier makes thrilling FAUGÈRES from high-altitude, organic vyds. Flagship Felgaria, mostly MOURVÈDRE, is elegance, power. Ages well. Les Bancels SYRAH et al earlier drinking, no less fine. Belle Lurette shows what CARIGNAN can do.

Cédre, Ch du SW Fr ★★→★★★ 15' 16 17 18 19 20 Verhaeghe brothers make best-known modern-style CAHORS. Choice of styles, prices. Delicious ★★ MALBEC IGP for everyday drinking.

Cellier aux Moines Burg ★★★ Now top DOM in GIVRY based in vyd of same name, reached new heights after investment in new winery and winemaker Guillaume Marko. Interesting options in CÔTE DE BEAUNE, but star is home vyd.

Cendrillon, Dom de la L'doc ★★★ Joyeux family makes polished, terroir-driven wines nr Narbonne. Organic. Nuance, creamy, oaky blend of eight varieties, incl PETIT MANSENG/GRENACHES BL/GR/ALBARIÑO. Inédite (r) classy CORBIÈRES.

Cépage Grape variety. *See* pp.14–24 for all.

Cérons B'x ★★ 15' 16 18 19 Tiny AOP (20 ha). Sweet wines next to SAUT, less intense. Best: CHX de Cérons, DE CHANTEGRIVE, du Seuil, Grand Enclos; CLOS Bourgelat.

Chablis ★★→★★★ 14' 15 17' 18' 19' 20 Such an evocative name; such beautiful wines when they stay true to their marine mineral origins; CHARD grape with crushed oyster shells. Also PETIT CHAB, DYA lighter version.

Chablis Grand Cru Chab ★★★→★★★★ 10' 12' 14' 15 17' 18' 19' 20 Contiguous s-facing block overlooking River Serein, most concentrated CHAB, needs 5–15 yrs to show detail. Seven vyds: Blanchots (floral), Bougros (incl Côte Bouguerots), CLOS (usually best), Grenouilles (spicy), Preuses (cashmere), Valmur (structure), Vaudésir (plus brand La Moutonne). Many gd growers.

Chablisienne, La Chab ★★→★★★ Exemplary co-op responsible for huge slice of CHAB production, esp supermarket own-labels. Trade up to bio CUVÉES of PETIT CHAB and Chab. Top wine is GC CH Grenouilles.

Chablis Premier Cru Chab ★★★ 14' 15 17' 18' 19' 20 Well worth premium over straight CHAB: better sites on rolling hillsides. Mineral favourites: Montmains, Vaillons, Vaucoupin, softer-style Côte de Léchet, Fourchaume; greater opulence from Mont de Milieu, *Montée de Tonnerre*, Vaulorent.

Chambertin C d'O ★★★★ 90' 93 96' 99' 02' 05' 09' 10' 12' 14' 15' 16 17 18' 19' 20' Called the King of Burg, but in touch with its feminine side. Imperious wine; amazingly dense, sumptuous, long-lived, expensive. Producers who match potential incl Bernstein, BOUCHARD PÈRE & FILS, Charlopin, Damoy, DOM LEROY, DUGAT-Py, DROUHIN, MORTET, ROSSIGNOL-TRAPET, ROUSSEAU, TRAPET. Clos de Bèze next door is slightly more accessible in youth, velvet texture, deeply graceful: Bart, B CLAIR, Damoy, Drouhin, Drouhin-Laroze, Duroché, FAIVELEY, Groffier, JADOT, Prieuré-Roch, Rousseau.

Chambolle-Musigny C d'O ★★★→★★★★ 93 99' 02' 05' 09' 10' 12' 15' 16 17 19' 20 Silky, velvety wines from CÔTE DE NUITS: Charmes, Combe d'Orveau for substance, more chiselled from Cras, Fuées, seduction from Amoureuses, plus GCS BONNES-MARES, MUSIGNY. Superstars: BARTHOD, MUGNIER, ROUMIER, VOGÜÉ. Try Amiot-Servelle, DROUHIN, Felettig, Groffier, HUDELOT-Baillet, Pousse d'Or, RION, Sigaut.

Champagne Sparkling wines of PINOT N, MEUNIER and CHARD: 33,805 ha, heartland c.145 km (90 miles) e of Paris. Sales 300 million+ bottles/yr. Some PINOT BL in AUBE adds freshness. Other sparkling wines, however ace, cannot be called Champ.

Champagne le Mesnil Champ ★★★→★★★★ Top-flight co-op, centred in Le Mesnil-

---

Chablis

There is no better expression of the all-conquering CHARD than the full but tense, limpid but stony wines it makes on the rolling limestone hills of CHAB. Most use little or no new oak to avoid masking the precision of their terroirs. **Top:** BILLAUD, DAUVISSAT (V), Droin, FÈVRE (W), LAROCHE, Michel (L), MOREAU (C), Pinson, RAVENEAU. **Challengers:** Bessin, BROCARD (J-M), CHABLISIENNE, Collet, Dampt (D), Davenne, Defaix (B), DROUHIN-Vaudon, Duplessis, Fèvre (N&G), Grossot, LAROCHE, LONG-DEPAQUIT, Malandes, MOREAU-Naudet, Picq, Piuze, Pommier, Oudin, Tribut. **Up-and-coming:** Dauvissat (J & Fils), d'Henri, Gautheron, Lavantureux, Vocoret (E&E), Vrignaud. **Organic/natural:** Brocard (J), CH de Béru, de Moor, Goulley, Pattes Loup.

sur-Oger ; exceptional GC CHARD village. CUVÉE Sublime 08' 09' 13 15 17 19 20 from finest sites. Majestic Cuvée Prestige 05 triumphs against odds. Real value.

Chandon de Briailles, Dom C d'O ★★★ Defined by bio farming, min sulphur, whole bunches, no new oak. Brilliantly pure perfumed reds reaching new heights, esp CORTON-Bressandes and PERNAND-VERGELESSES PC Île de Vergelesses.

Chanson Père & Fils Burg ★→★★★ BEAUNE merchant, quality whites (CLOS DES MOUCHES, CORTON-Vergennes) and idiosyncratic reds (whole-cluster aromatics), esp CLOS des Fèves.

Chapelle-Chambertin C d'O ★★★ 99' 02' 05' 09' 10' 12' 15' 16 18' 19' 20' Lighter neighbour of CHAMBERTIN; thin soil does better in cooler, damper yrs. Fine-boned wine, less meaty. Top: Damoy, DROUHIN-Laroze, JADOT, PONSOT, ROSSIGNOL-TRAPET, TRAPET, Tremblay.

Chapoutier N Rh ★★·¹★★★★ Talkative grower-merchant at HERMITAGE, 1st to put braille on labels. Stylish reds via low-yield, plot-specific, expensive CUVÉES. Intense GRENACHE CHÂTEAUNEUF – Barbe Rac, Croix de Bois (r); CÔTE-RÔTIE La Mordorée; Hermitage – L'Ermite (outstanding r/w), Le Pavillon (granite, deep r), Cuvée de l'Orée (w), Le Méal (w). Also ST-JOSEPH Les Granits (r/w). *Hermitage whites* outstanding, all old-vine MARSANNE. Meysonniers Crozes gd value. Also vyds in COTEAUX D'AIX-EN-PROV, CÔTES DU ROUSS-VILLAGES (gd DOM Bila-Haut), RIVESALTES. And owns FERRATON at Hermitage, BEAUJ house Trenel, CH des Ferrages (Prov), has AL vyds, also past Australian joint ventures, esp Doms Tournon and Terlato & Chapoutier (fragrant); hotel, wine bars in Tain.

Charbonnière, Dom de la S Rh ★★★ 05' 09' 10' 16' **17 18** 19' 20' CHÂTEAUNEUF estate run by sisters; punchy, herbal wines Consistent Tradition (r), deep, special: VIEILLES VIGNES (vigour, best), authentic Mourre des Perdrix, also Hautes Brusquières; v. stylish, pure white. Also peppery VACQUEYRAS red.

Chardonnay As well as a white wine grape, also the name of a MÂCON-VILLAGES commune, hence Mâcon-Chardonnay.

Charlemagne C d'O ★★★★ 14' 15' 17' 18 19' 20' Almost extinct sister appellation to CORTON-C, revived by DOM DE LA VOUGERAIE from 2013. Same rules as sibling.

Charlopin C d'O ★★→★★★ Philippe C makes impressive range of reds from GEVREY base. BOURGOGNE C D'O, MARSANNAY for value, gd range GC for top of line. Son Yann C, DOM C-Tissier also exciting.

Charmes-Chambertin C d'O ★★★★ 99' 02' 03 05' 09' 10' 12' 15' 16 **17** 18' 19' 20' GEVREY GC, 31 ha, incl neighbour MAZOYÈRES-CHAMBERTIN. Raspberries and cream, dark-cherry fruit, sumptuous texture, fragrant finish. So many gd names: ARLAUD, BACHELET, Castagnier, Coquard-Loison-Fleurot, DUGAT, DUJAC, Duroché, LEROY, MORTET, Perrot-Minot, Roty, ROUSSEAU, Taupenot-Merme, VOUGERAIE.

Chartogne-Taillet Champ A disciple of SELOSSE, Alexandre Chartogne is a star. Ideal BRUT Ste Anne NV and single-vyds Le Chemin de Reims and Les Barres; striking energy. High hopes for magic trio 18 19 20.

Charvin, Dom S Rh ★★★ 06' 07' 09' 10' 12' 15' 16' 17' 18' 19' 20' Close-up terroir truth at 8-ha CHÂTEAUNEUF estate, one of best; 85% GRENACHE, no oak, only one handmade CUVÉE. Spiced, mineral, high-energy red, vintage accuracy. Recent gd white. Top-value, genuine, mineral, long-lived CÔTES DU RH (r).

Chassagne-Montrachet C d'O ★★→★★★★ (w) 08' 09' 12' **14'** 15 17' 18' 19' 20' Large village at s end of CÔTE DE BEAUNE. Great white vyds Blanchot, Cailleret, LA ROMANÉE, Ruchottes and GCS. Try COLIN, GAGNARD, MOREY, PILLOT families plus DOMS Heitz, MOREAU, Niellon, Ramonet. Undersung reds from eg Boudriotte, CLOS St-Jean, Morgeot.

Château (Ch) Means an estate, big or small, gd or indifferent, particularly in B'X (*see* pp 100 121). Literally, castle or great house. In Burg, DOM is usual term.

Château-Chalon Jura ★★★→★★★★ 99' 05' 09 10' 14 Not a CH but AOP and village, the summit of VIN JAUNE style from SAVAGNIN grape. An unfortified, winey version of Sherry. Min 6 yrs barrel-age under veil of yeast, not cheap but gd value. Drink with Comté cheese (or with and in a chicken dish). Ages for decades. Search out ★★ BERTHET-BONDET, ★★★★ *Macle*, ★★★ Stéphane TISSOT or Bourdy for old vintages.

Château-Grillet N Rh ★★★★ 07' 09' 10' 12' 14' 15' 16' 17' 18 19' 20' 21' France's smallest AC, 3.7-ha, picturesque amphitheatre s of CONDRIEU, sandy-granite terraces, detailed vyd care. Owned by F Pinault of CH LATOUR, prices v. high, wine *en finesse*, less rich now. Can be great at 20 yrs. Scented, oily, precise VIOGNIER: drink cool, decanted, with refined dishes.

Châteaumeillant Lo ★→★★ 20' 21 Isolated AOP undergoing renaissance. Mostly light red GAMAY, PINOT N, plus VIN GRIS with 10% PINOT GR allowed. BOURGEOIS, Chaillot, Gabrielle, Goyer, Joffre, Joseph MELLOT, Lecomte, Nairaud-Suberville, Roux, Rouzé, Siret-Courtaud.

Châteauneuf-du-Pape S Rh ★★★→★★★★ 01' 07' 09' 10' 12 15 16' 17 19' 20 Large, 3200 ha+, nr Avignon; multiple soils incl clay, sand, limestone, so different styles. About 50 gd DOMS (remaining 85 fair, uneven to poor). Up to 13 grapes (r/w), led by GRENACHE, also SYRAH, MOURVÈDRE (increasing), Counoise. Warm, spiced, textured, long-lived; should be fine, pure, magical, but until mid-2010s too many heavy, sip-only wines. Small, traditional names often gd value. Prestige old-vine wines (v.gd Grenache 16' 19'). To avoid: late-harvest, new oak, 16% alc, too pricey. Grand whites: fresh, fruity, or replete, smooth, best can age 15 yrs+. For top names, *see* box, below.

Chave, Dom Jean-Louis N Rh ★★★★ 00 01' 03' 04 05' 07' 09' 10' 11' 12' 13' 15' 16' 17' 18' 19' 20' Excellent, world-famous family DOM at heart of HERMITAGE, gd mix of soils, so blending artful. Classy, silken, long-lived SYRAH reds, incl v. occasional Cathelin. Stylish, v.gd, complex, long-lived white (mainly MARSANNE). ST-JOSEPH reds dark, tangy, great fruit, plot-specific CLOS Florentin (since 2015) v. stylish; also lively J-L Chave brand St-Joseph Offerus, jolly CÔTES DU RH Mon Coeur, sound-value merchant Hermitage Farconnet (r), Blanche (w).

Chavignol Lo SANCERRE village; v. steep vyds Cul de Beaujeu/Les Monts Damnés dominate picturesque, vibrant village. Clay-limestone soil gives excellent full-bodied, mineral whites, reds ageing 15 yrs+. Young producers v. fine: Pierre Martin, Vincent Delaporte. Also: Alphonse Mellot, Anthony Girard, *Boulay*, Bourgeois, Cotat, *Dagueneau*, Paul Thomas, Thomas Laballe.

Chénas Beauj ★★★ 15' 16 18' 19' 20' Smallest BEAUJ cru, between MOULIN-À-VENT and JULIÉNAS, gd value, meaty, age-worthy, merits more interest. Thillardon is reference DOM, but try also Janodet, LAPIERRE, Pacalet, Piron, Trichard, co-op.

Chevalier-Montrachet C d'O ★★★★ 08 09' 10 12 14' 15 17' 18' 19' 20' Just above

---

Châteauneuf: kings of the castle

Top names in this enormous and varied (soils, blends, styles) appellation: CHX DE BEAUCASTEL, Fortia, Gardine (also w), Mont-Redon, Nalys, Nerthe, RAYAS, Sixtine, Vaudieu; DOMS Barroche, Beaurenard (bio, classy), Bosquet des Papes, Chante Cigale, Chante Perdrix, CHARBONNIÈRE, CHARVIN, CLOS DES PAPES, CLOS du Caillou, Clos du Mont-Olivet, Clos St-Jean (scale), Cristia, de la Biscarelle, de la Janasse, de la Vieille Julienne (bio), du Banneret (traditional), Fontavin, Font-de-Michelle, Galet des Papes, Grand Tinel, Grand Veneur, Henri Bonneau, LES CAILLOUX (value), Isabel Ferrando, Marcoux (esp VIEILLES VIGNES), Mas du Boislauzon, Pegaü, Pères de l'Eglise, Pierre André (bio, classic), Porte Rouge, P Usseglio, R Usseglio (bio, style), Roger Sabon, Sénéchaux, Vieux Donjon (classic), VIEUX TÉLÉGRAPHE.

MONTRACHET on hill, just below in quality; brilliant crystalline wines, dancing white fruit and flowers. Long-lived but can be accessible early. Top grower is LEFLAIVE, special CUVÉES Les Demoiselles from JADOT, LOUIS LATOUR and La Cabotte from BOUCHARD. Also: Chartron, COLIN (P), Dancer, DE MONTILLE, Niellon, VOUGERAIE.

Cheverny Lo ★→★★ 19′ 20′ 21 AOP white from SAUV BL (mostly)/CHARD blend. Light reds mainly GAMAY, PINOT N (CAB FR, CÔT). *Cour-Cheverny* 100% Romorantin: ages v. well, best up to ★★★. Try Bellier, Cazin, Clos Tue-Boeuf (+ VDF), *Huards* (bio), Montcy, Moulin, Sauger, Tessier, Veilloux, Villemade. Frost-prone, bad 21.

Chevillon, R C d'O ★★★ With GOUGES, reference DOM for NUITS-ST-GEORGES, dark fruited but not overtannic PCS: Bousselots, Chaignots, Pruliers more accessible; Cailles, Les St-Georges, Vaucrains for long term.

Chevrot C d'O ★★ Pablo and Vincent C great source for juicy MARANGES, esp Sur les Chênes, PC Croix Moines. Decent whites, esp ALIGOTÉ Tilleul, Maranges Fussière. CRÉMANT too. Lively wines, fair prices.

Chidaine, François Lo ★★★ 17′ 18′ 19′ 20′ (21) Bio MONTLOUIS, v.gd, single vyds. Accent on precise SEC, DEMI SEC that ages well. His VOUVRAY, incl CLOS Baudoin, stupidly has to be VDF, TOURAINE (Cher V). Try Clos du Breuil, Clos Habert, Les Truffeaux. Project in Spain.

Chignin Sav ★→★★ AC: Jacquère, MONDEUSE. Chignin-Bergeron is ROUSSANNE. Try ★ A&M QUENARD, J-F Quenard, ★★ *Berthollier*, ★★★ Partagé.

Chinon Lo ★★ →★★★ 16′ 17′ 18′ 19′ 20′ (21) Light to rich top TOURAINE CAB FR from sand, gravel, limestone. Best age 30 yrs+. A little dry CHENIN BL. Best: *Alliet*, Baudry, BAUDRY-DUTOUR, Beauséjour, Couly Dutheil, Grosbois, JM Raffault, Jourdan-Pichard, Landry, L'R, *Noblaie*, P&B Couly, Pain, Pallus, Petit Thouars, Pierre Sourdais, Saut au Loup.

Chiroubles Beauj ★★ 15′ 18′ 19′ 20′ BEAUJ cru in hills above FLEURIE: fresh, fruity, savoury wines. Growers: Berne, CH de Javernand, Cheysson, LAFARGE-Vial, Métrat, Passot, Raousset, or merchants DUBOEUF, Trenel.

Chorey-lès-Beaune C d'O ★★ 15′ 17 18 19′ 20 Village just s of BEAUNE. TOLLOT-BEAUT remains reference for this affordable, fruit-forward AC. Try also Arnoux, DROUHIN, Gay, Guyon, JADOT, Rapet, ROUGET.

Clair, Bruno C d'O ★★★→★★★★ Top-class CÔTE DE NUITS estate for supple, subtle, savoury wines. Could go even further with new generation; gd-value MARSANNAY, old-vine SAVIGNY La Dominode, GEVREY-CHAMBERTIN (CLOS ST JACQUES, Cazetiers), standout CHAMBERTIN-Clos de Bèze. Best whites from CORTON-CHARLEMAGNE, MOREY-ST-DENIS.

Clairet B'x Between rosé/red. B'x Clairet is AC. CHX Penin, Ste-Catherine, Turcaud.

Clairette de Die N Rh ★★ NV Rh/low-Alpine bubbly – flinty or (better) MUSCAT sparkling (s/sw), v. underrated, muskily fruited, gd value, low degree. Or dry CLAIRETTE, can age 3 yrs. Note Achard-Vincent, Carod, David Bautin (organic), Jaillance (value), J-C Raspail (organic, IGP SYRAH), Poulet & Fils (terroir, Chatillon-en-Diois r). Beautiful apéritif, must try.

Clape, Dom Pierre, Olivier N Rh ★★★→★★★★ 01′ 03′ 05′ 06′ 07′ 09′ 10′ 12′ 14′ 15′ 16′ 17′ 18′ 19′ 20′ 21 *Les rois de* CORNAS. Top location SYRAH vyds, old vines, gd granite soil work. Profound, complex reds, vintage truth, need 6 yrs+, live 25+. Bright fruit in youngish-vines label Renaissance. Superior CÔTES DU RH, VDF (r), ST-PÉRAY (gd style, improved).

Clape, La L'doc ★★★→★★★★ Limestone massif on Med nr Narbonne. MOURVÈDRE hotspot with SYRAH, GRENACHE for characterful herb-scented reds plus salty, herbal whites from BOURBOULENC, will age. CHX Anglès, Camplazens, *La Combe St-Paul*, LA NÉGLY, Laquirou, l'Hospitalet, Mire l'Etang, Pech-Céleyran, Pech-Redon, Ricardelle, ROUQUETTE-SUR-MER, SARRAT DE GOUNDY.

A trio of Champagne growers to watch
Didier Doué, Montueux: exemplary, organic. Note BRUT NATURE. **Lancelot-Pienne**, Cramant: model NV, plus Arthurian CUVÉE names like Table Ronde. **Veuve Fourny**, Vertus: finest maker of Brut Nature from old vines; ace CLOS Faubourg Notre Dame de Vertus single-vyd CHARD, old vines.

Climat Burg Individual named vyd at any level, esp in C D'O, eg. MEURSAULT Tesson, MAZOYÈRES-CHAMBERTIN. UNESCO World Heritage status.

Clos Distinct (walled) vyd, often in one ownership (esp AL, Burg, CHAMP). Often prestigious.

Clos Alivu Cors ★★→★★★ AOP Patrimonio. Stylish, fresh wines from talented Eric Poli, also owner of DOM Poli, source of gd-value AOP CORSE (r/w/rosé), IGP Île de Beauté.

Clos Canarelli Cors ★★★ Revered bio estate in S. AOP CORSE Figari. Reviving indigenous grapes plus more mainstream NIELLUCCIO, SCIACARELLO. Core range aged in and named after amphorae, worth seeking out. Excellent Tarra di Sognu (r/w), rare Tarra d'Orasi (r/w) from ungrafted vines.

Clos Cibonne Prov ★★★ Small estate nr Toulon, making red/rosé from rare local Tibouren as AOP CÔTES DU PROV. Wonderful traditional labels, gastronomic, ethereal wines. Tentations range is fruitier, early drinking.

Clos de Tart C d'O ★★★★ 02' 05' 08' 10' 13' 14 15' 16' 17 18' 19' 20' Expensive MOREY-ST-DENIS GC. MONOPOLE of Pinault/Artemis empire (CHX GRILLET, Latour, etc.). Becoming more refined with earlier picking and rebuilt winery, while retaining natural intensity.

Clos de Vougeot C d'O ★★★→★★★★ 90' 99' 02' 03' 05' 09' 10' 12' 13' 15' 16 17 18' 19' 20' CÔTE DE NUITS GC with many owners. Occasionally sublime, needs 10 yrs+ to show real class. Loving recent warm vintages. Style, quality depend on producer's philosophy, technique, position. Top: ARNOUX-LACHAUX, CH de la Tour, EUGÉNIE, *Faiveley*, GRIVOT, *Gros (Anne)*, HUDELOT-Noëllat, LEROY, LIGER-BELAIR (both), MÉO-CAMUZET, MORTET, *Vougeraie*. Also v.gd BOUCHARD, Castagnier, Clerget (Y), Coquard-Loison-Fleurot, DROUHIN, Forey, MONTILLE, MUGNERET-Gibourg.

Clos des Fées Rouss ★★ Organic philosophy, wines with character from range of terroirs. Les Sorcières (r/w) consistently gd CÔTES DU ROUSS. Les CLOS des Fées Herve Bizeul (r) top-notch. IGP COTES DE CATALANES (old-vine GRENACHE BL).

Clos des Lambrays C d'O ★★★ 05' 09' 10' 15' 16' 18' 19' 20' All-but-MONOPOLE GC vyd at MOREY-ST-DENIS, now belongs to LVMH. Big investmest and new winemaker from 2019 to bring leap in quality. Previously attractive, early picked, spicy, stemmy style.

Clos des Mouches C d'O ★★★ (w) 02 05' 09' 10' 14' 15 17' 18 19' 20' PC vyd in several Burg AOPs. Mostly reds (PINOT N), but most famous for glorious Beaune white. *Mouches* = honeybees; *see* label of DROUHIN's iconic BEAUNE bottling. Also BICHOT, CHANSON (Beaune); plus CLAIR, MOREAU, Muzard (SANTENAY); Germain (MEURSAULT).

Clos des Papes S Rh ★★★★ 01' 03' 04' 05' 07' 09' 10' 12' 13' 14' 15' 16' 17' 18' 19' 20' 21 Outstanding CHÂTEAUNEUF DOM of Avril family, v. small yields, burg elegance, long life; v. stylish, provocative red (GRENACHE, high MOURVÈDRE, drink at 2–3 yrs or from 8+); top white (six varieties, intricate, allow time, merits noble cuisine; 2–3 yrs, then 10–20).

Clos du Mesnil Champ ★★★★ KRUG's famous walled vyd in GC Le Mesnil. Long-lived, pure CHARD vintage, great mature yrs like 95 *à point* till 2024+; 02 and remarkable 03 08' 13' will be classics, as will 17' 19'.

Clos de la Roche C d'O ★★★★ 90' 93' 96' 99' 02' 05' 08 09' 10' 15' 16' 17 18' 19' 20' Underrated though not underpriced: maybe finest GC of MOREY-ST-DENIS, as much grace as power, more savoury than sumptuous, blueberries. Needs time. ARLAUD, *Dujac*, H LIGNIER, PONSOT references but try Amiot, Bernstein, Castagnier, Coquard, LEROY, LIGNIER-Michelot, Pousse d'Or, Remy, ROUSSEAU.

Clos du Roi C d'O ★★→★★★ Frequent Burg vyd name, sometimes as Clos du Roy. The king usually chose well. Best vyd in GC CORTON (DE MONTILLE, Pousse d'Or, VOUGERAIE); top PC vyd in MERCUREY, future PC (still waiting) in MARSANNAY. Less classy in BEAUNE.

Clos Rougeard Lo ★★★★ 09 10 11 12 14 15 Was iconic, almost invisible small DOM, bought by Bouygues (CH MONTROSE). Finesse, long-lived: SAUMUR Bl, SAUMUR-CHAMPIGNY. Big winery under construction.

Clos St-Denis C d'O ★★★ 90' 93' 96' 99' 02' 05' 09' 10' 12' 15' 16' 17 18' 19' 20' MOREY-ST-DENIS GC. Sumptuous in youth, silky with age. Outstanding from DUJAC, PONSOT (Laurent P from 2016). Try also Amiot-Servelle, ARLAUD, Bertagna, Castagnier, Coquard-Loison-Fleurot, Heresztyn-Mazzini, JADOT, Jouan, LEROUX.

Clos Ste-Hune Al ★★★★ Legendary TRIMBACH single site from GC ROSACKER. World's greatest dry RIES? Super 71' 75' 13' 16' 17' 18', needs a decade ageing; complex wine for gastronomy, scallops, crayfish, salmon.

Clos St-Jacques C d'O ★★★★ 90' 93' 96' 99' 02' 05' 09' 10' 12' 15' 16 17 18' 19' 20' Hillside PC in GEVREY-CHAMBERTIN with perfect se exposure. Shared by five excellent producers: CLAIR, ESMONIN, FOURRIER, JADOT, ROUSSEAU; powerful, age-worthy, velvety reds ranked (and often priced) above many GCs.

Clusel-Roch, Dom N Rh ★★★ 01 05' 09' 10' 11 12' 13' 14 15' 16' 17' 18' 19' 20' Organic CÔTE-RÔTIE DOM (higher costs, rare), gd range vyds, mostly Serine (pre-clone SYRAH). Tight, smoky wines, patience rewarded. Les Schistes gd entry point, high-quality La Viallière, Les Grandes Places (schist, iron). Son Guillaume C makes fab v. drinkable COTEAUX DU LYONNAIS (GAMAY, w).

Coche-Dury C d'O ★★★★ Top MEURSAULT DOM led by Raphaël C in succession to legend Jean-François. Exceptional whites from ALIGOTÉ to CORTON-CHARLEMAGNE; v. pretty reds too. Stratospheric prices. Cousin Coche-Bizouard (eg. Meursault Goutte d'Or) sound, gd-value, but not same style.

Colin C d'O ★★★→★★★★ Leading CHASSAGNE and ST-AUBIN family; current generation turning heads with brilliant whites, esp Pierre-Yves C-MOREY, DOM Marc C, Joseph C and their cousins Bruno C, Philippe C, Simon C.

Colin-Morey Burg ★★★ Pierre-Yves C M has made his name with vibrant tingling whites, esp from *St-Aubin* and CHASSAGNE PC, with their characteristic gunflint bouquets. Fine wines at all levels incl NÉGOCIANT CUVÉES.

Collin, Ulysse Champ Cerebral grower on Coteaux du Petit Morin, sw of Vertus. Single-vyd wines only, all subtly oaked, low but not zero dosage. ★★★ Les Pierrières BLANC DE BLANCS, base of 15.

Collines Rhodaniennes N Rh ★★ IGP major value, character, quality, incl v.gd Seyssuel (v. nr Vienne, schist, steep), live, crisp granite hillside, plateau reds, often from top estates. ("Rhodanienne" = "of the Rhône.") Mostly SYRAH (best), plus GAMAY, mini-CONDRIEU VIOGNIER (best). Reds: A Paret, A PERRET, Bonnefond, CLOS de la Bonnette (organic), E Barou, Hameau Touche Boeuf, *Jamet*, Jasmin, J-M Gérin, L Chèze, Monier-Pérreol (bio), N Champagneux, s OGIER, S Pichat, ROSTAING, Y CUILLERON. Whites: Alexandrins, Amphores, A Perret (v.gd), Barou, F Merlin, *G Vernay*, P-J Villa, P Marthouret, X Gérard, Y Cuilleron.

Collioure Rouss ★★→★★★ Same vyds as BANYULS, stunning views from steep terraces overlooking Med. Mainly GRENACHE of all colours. Rosé can be serious. Top: DOMS Augustin, Bila-Haut, de la Rectorie, du Mas Blanc, du Traginer, La Tour Vieille, Madeloc, Vial-Magnères; Coume del Mas, Les CLOS de Paulilles. Co-ops Cellier des Templiers, l'Étoile. Pedres Blanques new, natural VDF.

Combe Blanche, Dom L'doc ★★→★★★ Eclectic mix of of fine, long-lived La

Livinière (Chandelière, La Galine), PINOT N, TEMPRANILLO from n-facing slopes. AOP MINERVOIS and Misunderstood CINSAULT gd value. CLOS du Causses gd, fragrant GRENACHE.

Comte Abbatucci Cors ★★★ Saviour of indigenous varieties, bio DOM. Humble VDF but quality top. CUVÉE Fustine (r/rosé), Valle de Nero (r/rosé). Collection range recalls military connections of ancestor, pal of Napoleon.

Comté Tolosan SW Fr ★ Usually DYA. Catch-all IGP covering most of SW. Kaleidoscope of styles, mostly entry level; ★★ DOM de Ribonnet stands out among a heap of mostly moderate wines.

Condrieu N Rh ★★★→★★★★ 18' 19 20' 21' Ancestral home of VIOGNIER; floral, perfumed airs, pear, apricot flavours from sandy-granite slopes. Best: pure, precise (20 and 21 over 19); but beware excess oak, sweetness, alc. Growers adapting to v. hot yrs; 80 growers, not all gd. Rare white companion for asparagus. Best: A Paret, *A Perret* (all three wines gd), Boissonnet, CHAPOUTIER, CLOS de la Bonnette (organic), C Pichon, DELAS, Faury (esp La Berne), F Merlin, F Villard (lighter recently), Gangloff, GUIGAL, *G Vernay* (fine, incl wonderful, long-lived Coteau de Vernon), Monteillet, Niéro, ROSTAING, St Cosme, Semaska, X Gérard (value), Y CUILLERON.

Corbières L'doc ★→★★★ Huge AOP, varied but some real gems. Characterful reds, styles reflect contrasts of terroir from coastal lagoons to dry foothills of Pyrénées. Plenty of CARIGNAN, particularly Boutenac. Some v.gd white. Try CHX Aiguilloux, Aussières, Borde-Rouge, CARAGUILHES, de Sérame, Grand Moulin, LA BARONNE, Lastours, La Voulte-Gasparets, Les Palais, *Ollieux Romanis*, Pech-Latt, VAUGELAS; DOMS de Fontsainte, DE LA CENDRILLON, de Villemajou, du Grand Crès, du Vieux Parc, Trillol; CLOS de l'Anhel, FAMILLE FABRE, Grand Arc, Les Clos Perdus, MAXIME MAGNON, Sainte-Croix, Serres Mazard. CASTELMAURE CO-op. Phew!

Over half of Corsican wine is pink. Not a political statement.

Cornas N Rh ★★★ 01' 05' 09' 10' 12' 15' 16' 17' 18' 19' 20' 21 Top-quality N Rh granite SYRAH, v. fashionable. Dark, strongly fruited, always mineral-lined. Some made for early fruit, ready 5 yrs+. Top: A&E Verset, *Allemand* (top two), Balthazar (traditional, incl zero-sulphur), *Clape* (benchmark), Colombo (modern), Courbis (modern), DELAS, DOM du Tunnel, Dumien Serrette, G Gilles (style), J&E Durand (racy fruit), Lemenicier, Lionnet (organic), M Barret (bio), M Bourg, P&V Jaboulet, Tardieu-Laurent (full, oak), Voge (swish, oak), V Paris.

Corsica / Corse ★→★★★ "Île de Beauté" aptly named IGP for whole island. Plenty of variety; altitude, sea winds give freshness. Reds elegant, spicy from SCIACARELLO, structured from rarer NIELLUCCIO aka SANGIOVESE. *Gd rosés.* Tangy VERMENTINO whites. Also VDN sweet MUSCATS. Local varieties championed by top producers CLOS CANARELLI, COMTE ABBATUCCI et al. Nine AOPS incl crus PATRIMONIO in n, Ajaccio to w. AOP Corse plus villages Calvi, Coteaux du Cap Corse, Sartène. Top: Alzipratu, CLOS Alivu, Clos Calviani, Clos Capitoro, Clos d'Alzeto, Clos Nicrosi, Clos Poggiale, Clos Venturi; Doms Clos Columbo, *de Grenajolo*, Fiumicicoli, Peraldi, PIERETTI, Saperale, Torraccia, *Yves Leccia*, Vaccelli.

Corton C d'O ★★★→★★★★ 99' 02' 03' 05' 09' 10' 12' 15' 17 18' 19' 20' Largely overpromoted GC, but can be underrated from best vyds CLOS DU ROI, Bressandes, Renardes, Rognet. Wines can be fine, elegant, not all blockbusters. References: BOUCHARD, CHANDON DE BRIAILLES, DRC, Dubreuil-Fontaine, FAIVELEY (CLOS des Cortons), Follin-Arbelet, MÉO-CAMUZET, Rapet, TOLLOT-BEAUT. Under the radar and can be gd value: Bichot, Camille Giroud, Capitain-Gagnerot, Clavelier, DOM des Croix, H&G Buisson, Mallard, Pousse d'Or, Terregelesses. Best whites from Vergennes vyd, eg. CH de MEURSAULT, CHANSON, HOSPICES DE BEAUNE.

Corton-Charlemagne C d'O ★★★→★★★★ 05' 09' 10' 14' 15' 17' 18 19' 20' Potentially scintillating GC, invites mineral descriptors, should age well; sw- and w-facing limestone slopes, plus band around top of hill. Top: BIZE, *Bonneau du Martray*, BOUCHARD, CLAIR, *Coche-Dury*, FAIVELEY, HOSPICES DE BEAUNE, JADOT, Javillier, LATOUR, Mallard, MONTILLE, Rapet, Rollin. DRC from 2019. *Dom Vougeraie* uses rarely seen sister AC, CHARLEMAGNE.

Costières de Nîmes S Rh ★→★★ Located n of Rhône delta, sw of CHÂTEAUNEUF; similar v. stony soils, Mistral-blown; gd quality, value. Red (GRENACHE, SYRAH) full, spiced, up to 10 yrs. Best: CHX de Grande Cassagne, de Valcombe, d'Or et de Gueules (full), L'Ermitage, Mas Carlot (gd fruit), Mas des Bressades (top fruit), Mas Neuf, Montfrin (organic), Mourgues-du-Grès (organic), Nages, Roubaud, Tour de Béraud, Vessière (w); DOMS de la Patience (organic), du Vieux Relais, Galus, M Gassier, M KREYDENWEISS (bio), Petit Romain, Terres des Chardons (bio). Lively, table-friendly rosés; some stylish whites (gd ROUSSANNE).

Coteaux Bourguignons Burg ★ DYA Mostly reds, GAMAY, PINOT N. New AOP since 2011 to replace BOURGOGNE Grand Ordinaire and to sex up basic BEAUJ. Market accepting the change. Rare whites ALIGOTÉ, CHARD, MELON, PINOTS BL/GR.

Coteaux Champenois Champ ★★★ (w) DYA AOP for still wines of CHAMP, eg. BOUZY. Vintages as for Champ. Better reds with climate change (12'). Impressive range of Coteaux Champenois Grands Blancs based on 17' by CHARLES HEIDSIECK, as gd as fine white burg.

Coteaux d'Aix-en-Provence Prov ★★ Myriad styles from big AOP centred on Aix: Lots of pale, fruity rosé based on GRENACHE, CINSAULT. Reds: CAB SAUV in cooler n: Pigoudet, Revelette, VIGNELAURE. Med grapes often more interesting in warmer spots. CHX Beaupré, Calissanne, La Realtière, Les Bastides, Les Béates; DOM d'Eole (on Alpilles), du Ch Bas; Villa Baulieu. *See also* LES BAUX-DE-PROV.

Coteaux d'Ancenis Lo ★→★★ 19 20' (21) Small AOP, slopes, both sides of Loire e of Nantes. Age-worthy sweet *Malvoisie* (PINOT GR); light red, rosé GAMAY, (CAB FR, 10% max), esp Galloires, Guindon, Landron-Chartier, Merceron-Martin, Paonnerie (natural).

Coteaux de l'Aubance Lo ★★→★★★ 15' 16 17 18' 19 20 (21) Small AC. Age-worthy sweet CHENIN BL; nervier, usually less rich than COTEAUX DU LAYON; less steep slopes. Esp *Bablut*, CH Princé, Dittiere, Montgilet, Rochelles, Ste-Anne, Terra Vita Vinum, Varière.

Coteaux du Giennois Lo ★ ★★ 20' 21 Citrus SAUV BL can be gd value. Light reds blend GAMAY/PINOT N. Frost-prone scattered vyds. Try Balland, Berthier, Bourgeois, Charrier, Langlois, Treuillet, Terres Blanches, Villargeau.

Coteaux du Layon Lo ★★→★★★★ 16 17 18' 19 20' 21 Long-lived CHENIN BL v.gd but v. difficult to sell. Seven villages may add name to AC. Chaume now Layon PC. Top AOPs: BONNEZEAUX, QUARTS DE CHAUME. Growers: Baudouin, Bellevue, Breuil, *Ch Pierre-Bise*, Chauvin, Drost (formerly Delesvaux), Fesles, Forges, Gueguniard, Juchepie, *Ogereau*, Soucherie. Also ANJOU-Coteaux de la Loire (sw).

Coteaux du Loir Lo ★→★★★ 19' 20' (21) Exciting, dynamic AOP incl *Jasnières*. Le Loir is a tributary, not a typo. Steely, fine, long-lived CHENIN BL; GAMAY reds; peppery Pineau d'Aunis; fizz; plus Grolleau (rosé), CAB, CÔT. Best: Ange Vin, BELLIVIÈRE, Breton, Briseau, Cezin, Gigou, Janvier, Maisons Rouges, *Roche Bleue*, Ryke.

Coteaux du Lyonnais Beauj ★ DYA Jnr BEAUJ. Most best en PRIMEUR. CLUSEL-ROCH best, incl 120-yr GAMAY Les Pessiaux.

Coteaux du Quercy SW Fr ★ 16 18 20 AOP between CAHORS and GAILLAC. Hearty country wines based on CAB FR plus TANNAT or MALBEC. Active co-op challenged by independents: ★★ DOMS du Guillau, Lacoste, Revel.

Coteaux du Vendômois Lo ★→★★ 20' 21 VIN GRIS – Pineau d'Aunis plus peppery reds, also blends of CAB FR, GAMAY, PINOT N. Whites: CHARD, CHENIN BL. Best: Brazilier, Cave du Vendômois (gd), Patrice Colin (gd), Four à Chaux.

Coteaux Varois-en-Provence Prov ★→★★ Higher, inland "Provence Vert" is cooler. Home of "Hollywood corner": Miraval by Brad Pitt and Angelina Jolie, and neighbouring Margui by George Lucas. Otherwise, SYRAH and VIOGNIER in cooler n, interesting IGP Coteaux du Verdon. Try CHX Duvivier, Trians; DOMS des Aspras, du Deffends, du Loou, La Grand'vigne, Les Terres Promises, Routas, St Mitre.

Côte Chalonnaise Burg ★★ Region immediately s of C D'O; always threatening to be rediscovered. Lighter wines, lower prices. BOUZERON for ALIGOTÉ, *Rully* for accessible, juicy wines in both colours; *Mercurey* and GIVRY have more structure and can age; MONTAGNY for leaner CHARD.

Côte d'Or Burg Département name applied to central and principal Burg vyd slopes: CÔTE DE BEAUNE and CÔTE DE NUITS. Not used on labels except for BOURGOGNE C D'O AC, since 17 vintage.

Côte de Beaune C D'O ★★→★★★★ The s half of C D'O. Also a little-seen AOP in its own right applying to top of hill above BEAUNE itself. DROUHIN's versions (r/w) also incl declassified Beaune PC. Try DOM VOUGERAIE too. Confusingly C de B-Villages is different – often a blend from lesser villages of s half of C D'O.

Côte de Brouilly Beauj ★★ 15′ 17 18′ 19′ 20′ Range of styles as soils vary on different flanks of Mont Brouilly. Merits a premium over straight BROUILLY. Reference is CH THIVIN, but try also Brun, Dufaitre, LAFARGE-Vial, Martray, Pacalet.

Côte de Nuits C D'O ★★→★★★★ The n half of C D'O. Nearly all red, from CHAMBOLLE-MUSIGNY, MARSANNAY, FIXIN, GEVREY-CHAMBERTIN, MOREY-ST DENIS, NUITS-ST GEORGES, VOSNE-ROMANÉE, VOUGEOT.

Côte de Nuits-Villages C D'O ★★ 10′ 12′ 15′ 16 17 18′ 19′ 20′ A jnr AOP for extreme n/s ends of CÔTE DE NUITS; can be bargains. Chopin, Gachot-Monot, Jourdan specialists. Top single-vyds CLOS du Chapeau (Arlot), Croix Violette (FOURNIER, Pernot), Faulques (Millot), Leurey (JJ Confuron), Meix Fringuet (TRAPET), Montagne (many), Robignotte (Jourdan), Vaucrains (JADOT). Monts de Boncourt best for whites.

Côte Roannaise Lo ★★→★★★ 19′ 20′ 21 Makes v.gd GAMAY: Bonneton, Désormière, Giraudon, Paroisse, Plasse, Pothiers, Rochette, SÉROL, Vial Fine (w) IGP Urfé: ALIGOTÉ, CHARD, CHENIN BL, ROUSSANNE, VIOGNIER.

Côte-Rôtie N Rh ★★★→★★★★ 01′ 05′ 09′ 10′ 12′ 15′ 16′ 17′ 18′ 19′ 20′ Most refined Rh red, mainly SYRAH, some VIOGNIER, granite, schist soils, style links to Burg. Violet aromas, pure (esp 16 20), complex, v. fine with age (5–10 yrs+). Exceptional, v. long-lived 10 15 19, silken 20. Top: *Barge*, B Chambeyron, Billon (energy), Bonnefond (oak), Bonserine (esp La Garde), Burgaud, CHAPOUTIER, *Clusel-Roch* (organic), DELAS, DOM de Rosiers, Duclaux, Gaillard (oak), Garon, GUIGAL (long oaking), *Jamet*, Jean-Luc Jamet, Jasmin, J-M Gérin, J-M Stéphan (organic), Lafoy, Levet (traditional), *Rostaing* (fine), S OGIER (oak), Semaska, VIDAL-FLEURY (La Chatillonne), Xavier Gérard, Y CUILLERON.

Côtes Catalanes Rouss ★★★ Arguably most thrilling IGP in France. From fruity gd value from big Vignerons Catalans CO-OP, to serious age-worthy wines from some of ROUSS's finest growers. Danjou-Banessy, GAUBY, CH DE L'OU among others use it for top wines.

Côtes d'Auvergne Lo, Mass C ★★ 19′ 20′ 21′ Well worth a detour to Clermont-Ferrand – yes, really. GAMAY, PINOT N, CHARD. Villages: Boudes, Chanturgue, Châteaugay, Corent (rosé), Madargues (r). Producers: Cave St-Verny, *Les Chemins de l'Arkose*, Miolanne, *Montel*, Pelissier, Sauvat. Also IGP Puy de Dôme (incl SYRAH).

Côtes de Bordeaux B'x ★ AOP launched in 2008 for reds. Embraces and permits cross-blending between CAS, FRANCS, BLAYE, CADILLAC and Ste-Foy. Growers who want to maintain *the identity of a single terroir* have stiffer controls (note) but can put Cas, Cadillac, etc. before Côtes de B'x. BLAYE-CÔTES DE B'X, FRANCS-CÔTES

DE B'x and Ste-Foy Côtes de B'x also produce a little dry white. 950 growers in group; represents 10% B'x production (65 million bottles/yr). Try CHX Dudon, Lamothe de Haux, Malagar.

Côtes de Bourg B'x ★→★★ 16 18 19 20 Solid, savoury reds, a little white from e bank of Gironde. Environmental certicate obligatory from 2025. Top CHX: Brûlesécaille, Bujan, *Falfas*, Fougas-Maldoror, Grand-Maison, Grave (Nectar VIEILLES VIGNES), Haut-Guiraud, Haut-Macô, Haut-Mondésir, Macay, Mercier, Nodoz, *Roc de Cambes*, Rousset, Sociondo, Tour des Graves.

Côtes de Duras SW Fr ★→★★ 17' 18' 19 20 Affordable AOP s of BERGERAC; Bergerac lookalike. Best known for crisp white. Berticot co-op is sound. ★★ DOMS de Laulan, Grand Mayne.

Côtes de Gascogne SW Fr ★→★★ DYA IGP. Mostly in Gers; largest producer of IGP white in France. Largely contiguous with Armagnac. Grassa family of ★★ DOM TARIQUET and La Hitaire helped introduce countless grape varieties. Typically, clean, aromatic, light. ★★ Doms Joÿ, Ménard, Miselle, Pellehaut; Combebelle from ever-impressive PLAIMONT co-op a well-known brand.

Côtes de Millau SW Fr IGP ★ DYA From nearby Gorges du Tarn. Foster's Millau viaduct celebrated in wines from popular co-op. ★ DOMS du Vieux Noyer, La Tour-St-Martin best of independents.

Côtes de Provence Prov ★→★★★ DYA Fashionably pale-pink rosé, often more style over substance, from swathe of vyds from Marseilles to Cannes and inland. Whites (increasingly 100% Rolle) and reds (SYRAH and GRENACHE with MOURVÈDRE nearer coast) can be more interesting. Fréjus, La Londe, Pierrefeu, STE-VICTOIRE, Notre Dame des Anges are subzones. Leaders: CLOS CIBONNE (primarily Tibouren), *Gavoty* (superb); CHX d'ESCLANS, de Selle (Ott), Gasqui (bio), La Gordonne, La Mascaronne; Estandon VIGNERONS, Mirabeau, Rimauresq. *See* BANDOL, COTEAUX D'AIX, COTEAUX VAROIS.

Côtes du Brulhois SW Fr ★ 16 17 18 19 (20) Small AOP nr Agen producing a softer version of TANNAT (obligatory) with CAB SAUV, MERLOT, MALBEC in support. Local co-op unusually supportive of a few independents.

Côtes du Forez Lo ★→★★ 19' 20' 21 Most s Lo AC, ripe, vibrant GAMAY (r/rosé). Try Bonnefoy, CLOS de Chozieux, Guillot, Mondon & Demeure, Poyet, Verdier/ Logel. Exciting IGP (w): CHARD, CHENIN BL, PINOT GR, RIES, ROUSSANNE, VIOGNIER.

Côtes du Jura AOP covering all Jura, but mainly s. Organic, natural encouraged exports and can be pricey. Whites dominate from mineral CHARD to deliberately oxidative SAVAGNIN (incl VIN JAUNE). Light perfumey reds from PINOT N, Poulsard, Trousseau. Great food wines. Try classic Badoz or CH d'Arlay; organic bankers from ★★ BERTHET-BONDET, ★ Buronfosse, ★★ *Labet*, ★ Marnes Blanches and ★★ *Pignier*, and natural cults ★★★ GANEVAT or Miroirs. *See also* ARBOIS, CH-CHALON, L'ÉTOILE AOPs.

Côtes du Marmandais SW Fr ★→★★ (r) 15' 16 18 19 20 AOP. Neighbour of B'x, but increasingly eccentric in style thanks to local Abouriou grape and SYRAH. ★★★ cult winemaker Elian da Ros, CH Beaulieu (ages v. well); ★★ DOMS Beyssac, Bonnet, Cavenac and Ch Lassolle blend it with usual B'x grapes. ★★ Ch de Beaulieu Syrah-based. Co-ops (95% total production) still dull.

Côtes du Rhône S Rh ★→★★ 19' 20 The base of S Rh, 170 communes, incl gd SYRAH of Brézème, St-Julien-St-Alban (N Rh). Ranges between enjoyable, handmade, high quality (esp CHÂTEAUNEUF estates, numbers rising, top value) and dull, mass produced. Lively fruit now common. Mainly GRENACHE, also Syrah, CARIGNAN. Most best drunk young. Vaucluse best, then Gard (Syrah). Whites improving fast, rocking value, note 21.

Côtes du Rhône-Villages S Rh ★→★★★ 19' 20 Filled, spiced reds from 7700 ha, incl 22 named S Rh villages (Nyons new in 2020), numbers on the up, some

> **Village life: the best of Côtes du Rhône-Villages**
> For best of this gd-value appellation, try Gadagne, MASSIF D'UCHAUX,
> Ste-Cécile, Signargues, VISAN. CHX Bois de la Garde, Fontségune, Signac;
> DOMS Aure, Bastide, Bastide St Dominique, *Biscarelle*, Bois de St Jean,
> Cabotte (bio), Coulange, Coste Chaude (organic), Crève Coeur (bio),
> Echevin (gd w), Florane (bio), Grand Veneur, Grands Bois (organic),
> Gravennes, Janasse (top class), Jérome, *Les Aphillanthes* (bio), Mas
> de Libian (bio, flair), Montbayon, Montmartel (organic), Mourchon,
> Pasquiers (organic), Pique-Basse (organic, gd w), *Rabasse-Charavin*,
> Réméjeanne (v.gd), Renjarde, Romarins, Saladin (organic), St-Siffrein,
> STE-ANNE, Valériane, Viret (cosmo); CAVE de RASTEAU, Les VIGNERONS
> d'Estézargues (gd range).

obscure. Best are generous, deep, lively, gd value. Red heart is GRENACHE, plus
SYRAH, MOURVÈDRE. Improving whites, often incl VIOGNIER, ROUSSANNE added
to rich base CLAIRETTE, GRENACHE BL – gd with food, note 21. *See* LAUDUN,
PLAN DE DIEU (gd choice), SABLET, St-Gervais, SÉGURET (quality), VALRÉAS, VISAN
(improving, many organic). (*See* box, above, for best growers.)

Côtes du Roussillon (Villages) Rouss ★→★★★ AOP for ROUSS, varied styles, often
v.gd: lots of CARIGNAN, also old-vine GRENACHES BL/Gr, Maccabeo for whites.
AOP Côtes du Rouss-Villages smaller area, just reds, some excellent. 32 villages:
Caramany, Latour de France, Les Aspres, Lesquerde, Tautavel singled out on
label. Brial co-op gd, plus v.gd individual estates: Boucabeille, CAZES, Charles
Perez, CLOS DES FÉES, Clot de l'Oum, des Chênes, GAUBY, Les VIGNES de Bila-Haut
from CHAPOUTIER, Mas Becha, Mas Crémat, Modat, Piquemal, Rancy, Roc des
Anges, Thunevin-Calvet, Venus. *See also* CÔTES CATALANES.

Coulée de Serrant Lo ★★ 14 15 16 18 19 (20) Variable wines that really ought to
be better for the site, price. Historic, remarkable CHENIN BL AOP 7-ha site
overlooking the Loire in heart of SAVENNIÈRES. Nicolas Joly bio-pope, daughter
Virginie in charge. Frost-prone.

Coupe-Roses, Ch L'doc ★★→★★★ MINERVOIS DOM: high-altitude vyds, bio, Med with
freshness. Granaxa mostly GRENACHE, gd. No-added-sulphite Naturamania,
brimming with crunchy fruit.

Courcel, Dom de C d'O ★★★ Idiosyncratic POMMARD estate, late-picking specialist
of whole-bunch techniques; fine floral wines when it works. Top: age-worthy PCS
Rugiens and Épenots, plus interesting Croix Noires.

Crémant AOP for quality classic-method sparkling from AL, B'X, BOURGOGNE, Die,
Jura, LIMOUX, LO, Luxembourg, SAV. Many gd examples.

Crémant de Loire Lo ★★→★★★ Big volumes of gd fizz, wide range of styles from
CAB FR, CAB SAUV, CHARD, CHENIN BL, Grolleau, Orbois, Pineau d'Aunis, PINOT N.
Try Ackerman, Arnaud Lambert, *Aulée*, Bouvet-Ladubay, Langlois-Chateau,
Monmousseau, Nerleux.

Criots-Bâtard-Montrachet C d'O ★★★ 09 10 12 14' 15 17' 18 19' Tiny and much
morsellated MONTRACHET satellite, 1.57 ha. Anybody had a great one, apart from
d'Auvenay if you're v. rich? Maybe from Blain- or Fontaine-GAGNARD, LAMY and
now Caroline MOREY.

Cros, Pierre L'doc ★★★ Maverick producer; mainstream wines incl MINERVOIS
VIEILLES VIGNES CARIGNAN; Les Aspres, 100% SYRAH, butch and brilliant. Age-
worthy. Mal Amiès light, fruity from forgotten L'DOC varieties. NEBBIOLO and
TOURIGA N here too.

Cros Parantoux Burg ★★★★ Cult PC in VOSNE-ROMANÉE made famous by the late
Henri Jayer. Now made to great acclaim and greater price by DOMS ROUGET and
MÉO-CAMUZET. But Brûlées better in cool vintages?

FRANCE

Crozes-Hermitage N Rh ★★→★★★ 17′ 18 19′ 20 SYRAH from mostly flat, alluvial vyds nr River Isère, hot summer challenges: dark berry, licorice, tar; most early drinking (2–5 yrs). Reserved, complex, cooler from granite hills nr HERMITAGE: fine, red-fruited, take time. Best (simple CUVÉES) ideal for grills, parties. Some oaked, older-vine wines cost more, can age. Top: *A Graillot*, Aléofane (r/w), Belle (organic), Chapoutier, *Dard & Ribo* (organic), Delas (Le CLOS v.gd), DOM des Grands Chemins, G Robin; DOMS Combier (organic), de Thalabert of JABOULET, des Entrefaux, des Hauts-Châssis, des Lises (fine), *du Colombier* (*Gaby* great), Dumaine (organic), Habrard (organic), *Laurent Fayolle* (v. stylish, v.gd Clos Cornirets), Les Bruyères (bio, big fruit), Machon, Martinelles (traditional), Melody, Michelas St Jemms, Mucyn (fine), Remizières (oak), Rousset, Ville Rouge, Vins de Vienne, Y Chave. Drink *white* (MARSANNE) early, v.gd vintages recently, esp 21. Value.

Cuilleron, Yves N Rh ★★★ Important name at CONDRIEU, N Rh, always deft hand with whites. Top CONDRIEU: Les Chaillets. Note old-vine ST-JOSEPH (r) Les Serines. CÔTE-RÔTIE okay, oaked, best La Viaillière. MARSANNE, VIOGNIER VDP gd. Much recent vyd expansion.

Cuve close Quicker method of making sparkling in tank. Bubbles die away in glass much quicker than with *méthode traditionnelle*.

Cuvée Usually indicates a particular blend. In CHAMP, means 1st and best wines off the press.

Dagueneau, Didier / Louis-Benjamin Lo ★★★ →★★★★ 14′ 15′ 16′ 17 18′ 19 (20) (21) Bio SAUV BL benchmark. Stunningly precise, age-worthy. Louis-Benjamin (winemaker) and sister Charlotte D. Try Buisson Renard, Pur Sang, Silex. Also SANCERRE (Le Mont Damné, CHAVIGNOL), Les Jardins de Babylone (JURANÇON).

Danjou-Banessy, Dom Rouss ★★★★ Danjou bros make extraordinary complex, fresh wines in Agly V. Bio. IGP CÔTES CATALANES. Les Myrs 100% CARIGNAN, La Truffière (r/w) v.gd.

Dard & Ribo N Rh ★★★ Quirky duo, rightly loved by natural-wine aficionados. Old-vine CROZES-HERMITAGE, ST-JOSEPH, relaxed winemaking, character, charm. Note: whites incl ROUSSANNE; Crozes Les Bâties (r/w); St-Joseph Pitrou (r/w).

Dauvissat, Vincent Chab ★★★★ Imperturbable bio producer of CHAB using old barrels and local 132-litre *feuillettes*. Grand, age-worthy wines similar to RAVENEAU cousins. Best: La Forest, Les Clos, Preuses, Séchet. Try also DOM Jean D & Fils (no relation).

Deiss, Dom Marcel Al ★★★ Famous Bergheim estate now run mostly by Jean-Michel's son Mathieu. Terroir-proponent favouring co-plantation and blends. Best wine Schoenenbourg 13′ 17′ outstanding.

Delamotte Champ Fine, small, CHARD-dominated house. Managed with SALON by LAURENT-PERRIER. BRUT, BLANC DE BLANCS 07′ 08 13 ★★★; great trio 18 19 20.

Delas Frères N Rh ★★★ Owner/merchant in N Rh, with CONDRIEU, CROZES-HERMITAGE, CÔTE-RÔTIE, HERMITAGE vyds. Polished reds, quality high, chic new cellars. Best: Côte-Rôtie Landonne, Hermitage DOM des Tourettes (r/w), *Les Bessards* (r, granite terroir, sublime finesse 15 yrs+, smoky), ST-JOSEPH Ste-Épine (r, tight, interesting); S Rh: esp CÔTES DU RH St-Esprit (r), Grignan-les-Adhémar (r, value). Whites lighter recently. Owned by ROEDERER.

Delaunay, Édouard C d'O ★★→★★★ Old Burg name revived in NUITS and l'Étang-Vergy by Laurent D; gd range NÉGOCIANT CUVÉES all price points. Gaining widespread acclaim already.

Demi-sec Half-dry: but in practice more like half-sweet (eg. CHAMP typically 45g/l dosage)

Derenoncourt, Stéphane B'x Leading international consultant (team of six); self-taught, focused on terroir. Own property, *Dom de l'A* in CAS.

Deutz Champ One of best mid-sized houses, owned by Roederer family. Quality ever upwards. Top-flight CHARD CUVÉE Amour de Deutz 08, Deutz Rosé 12 ★★★★. *Superb Cuvée William Deutz 08'*, BLANC DE BLANCS 13, classic style. Leader of non-oaked school. Ace quality/price ratio.

Dirler-Cadé, Dom Al ★★→★★★★ Substance and finesse. Marvellous old-vines MUSCAT GC Saering and Spiegel and exceptional SYLVANER VIEILLES VIGNES 16 17' 18' 19 20; v.gd Saering RIES 10 14' 16 17' 18 19 20. Jean Dirler/Ludivine Cadé own vines in Heisse Wanne, best part of GC Kessler.

Domaine (Dom) Property, except next entry. *See* under name, eg. TEMPIER, DOM.

Dom Pérignon Champ Vincent Chaperon, now chef de CAVE, brings own style to luxury CUVÉE of MOËT & CHANDON. Ultra-consistent quality, huge quantities. Seductive creamy allure, esp after 10–15 yrs. Plénitude releases: long bottle-age, recent disgorgement, huge price, at 7, 16, 30 yrs+ (P1, P2, P3); superb P2 98; still-vibrant P3 70. More PINOT N focus in DP since 2000, esp underrated exquisite 06 Blanc, Rosé. New release: excellent 12.

Dopff au Moulin Al ★★★→★★★★ Famous name in pretty Riquewihr, today less concentration, quality variable: best GEWURZ GCS Brand, Sporen 12 16 18' 20' and RIES SCHOENENBOURG 13 17' and *Sylvaner de Riquewihr 19'*. Pioneer of AL CRÉMANT.

Doué, Didier Champ ★★★★ Fine, sculptured CHARD from high sunny hill above Troyes. Shuns oak, wants finesse, elegance. A little rich in 18', better in 19', fresh, terroir-dominant.

Dourthe B'x Sizeable merchant/grower with nine properties; quality emphasis; incl CHX BELGRAVE, LA GARDE, LE BOSCQ, Grand Barrail Lamarzelle Figeac, PEY LA TOUR, RAHOUL, REYSSON. *Dourthe No 1* (esp w) well-made generic B'x.

Drappier, Michel Champ Great family-run AUBE house, children of Michel D now in charge. Fine PINOT N, 60 ha+, certified bio; *Pinot-led NV*, BRUT ZÉRO, Brut *sans souffre*, Millésime d'Exception 12, Prestige CUVÉE Grande Sendrée 08' 09 12 ★★★★; 15 18 both rich crowd-pleasing yrs, 19 20 even better in a balanced fresh style, all finesse. CUVÉE Quatuor (four CÉPAGES). Superb 95' 82 (magnums). Constant research into early C17 vines resistant to climate change. More use of large oak *foudres*.

DRC (Dom de la Romanée-Conti) C d'O ★★★★ Grandest estate in Burg (or world). This emperor is definitely wearing clothes. MONOPOLES ROMANÉE-CONTI and LA TÂCHE, major parts of ÉCHÉZEAUX, GRANDS-ÉCHÉZEAUX, RICHEBOURG, ROMANÉE-ST-VIVANT and a tiny part of MONTRACHET. Also CORTON from 2009, CORTON-CHARLEMAGNE from 2019. Crown-jewel prices. Keep top vintages for decades.

Drouhin, Joseph & Cie Burg ★★★→★★★★ Fine family-owned grower/NÉGOCIANT in BEAUNE; vyds (all bio) incl (w) Beaune *Clos des Mouches*, MONTRACHET (Marquis de LAGUICHE) and large CHAB holdings. Stylish, fragrant reds from pretty CHOREY-LÈS-BEAUNE through great ranges in Beaune, CHAMBOLLE, VOSNE (note Petits Monts) and now GEVREY. Also Dom Drouhin Oregon (*see* US).

Duboeuf, Georges Beauj ★★→★★★★ From hero (saviour of BEAUJ) to less so (too much BEAUJ NOUVEAU). Always major player, still v. sound source for multiple Beauj crus, MÂCON bottlings. Georges D RIP 2020; son continues gd work.

Dubosc, André SW Fr Pioneering grower/agronomist; man behind PLAIMONT, AOP SAINT MONT and renaissance of Arrufiac, Manseng N, TANNAT vines. Host of ancient varieties uncovered, now subject of study. Heart and soul of Gascony.

Dugat C d'O ★★★ Cousins Claude and Bernard (Dugat-Py) made excellent, deep-coloured GEVREY-CHAMBERTIN, respective labels. Both flourishing with new generation. Tiny volumes, esp GCS, huge prices, esp Dugat-Py. Collector territory. But try village Gevrey from either for a (just) affordable thrill.

Dujac, Dom C d'O ★★★ →★★★★ MOREY-ST-DENIS grower originally noted for sensual, smoky reds, from unbeatable village Morey to outstanding GCS, esp CLOS DE LA

ROCHE, CLOS ST-DENIS, ÉCHÉZEAUX. Slightly more mainstream these days; gd *whites* from Morey and PULIGNY. Lighter merchant wines as D Fils & Père and DOM Triennes in COTEAUX VAROIS.

Dureuil-Janthial Burg ★★ Vincent D-J runs outstanding DOM in RULLY, with *fresh, punchy whites* and cheerful, juicy reds. All recommended but esp Maizières (r/w) or PC Meix Cadot (w).

Duval-Leroy Champ Family-owned house, 200 ha of mainly fine CHARD crus, focus on DOM wines. Fleur de CHAMP NV v.gd. Top Blanc de Prestige *Femme 13* ★★★★ one of Champ's top prestige CUVÉES. Also great in 18' 19.

Échézeaux C d'O ★★★ 99' 02' 05' 09' 10' 12' 15' 16 17 18' 19' 20' GC next to CLOS DE VOUGEOT, but totally different style: lacy, ethereal, scintillating. Can vary depending on exact location. Outstanding in quality and price: ARNOUX-LACHAUX, Bizot, Coquard-Loison-Fleurot, DRC, DUJAC, EUGÉNIE, G NOËLLAT, GRIVOT, GROS, LIGER-BELAIR, MÉO-CAMUZET, *Mugneret-Gibourg*, ROUGET, Tremblay. Other fine choices: Berthaut-Gerbet, *Guyon*, Lamarche, Millot, MUGNERET, Naudin-Ferrand, Tardy.

Edelzwicker Al ★ DYA Blended light white; CH d'Ittenwiller, HUGEL Gentil gd.

Entraygues et du Fel and Estaing SW Fr ★ ›★★ DYA Two tiny AOP neighbours in almost vertical terraces above Lot V. Bone-dry CHENIN BL for whites, esp ★★ DOMS Laurent Mousset (gd reds, esp La Pauca, excellent rosé), Méjanassère. ★★ Nicolas Carmarans making wines in and out of AOP.

Entre-Deux-Mers B'x ★→★★ DYA Often gd-value dry white B'x (drink *entre deux huitres*) from between the rivers Garonne and Dordogne. Best: CHX BONNET, Fontenille, Haut-Rian, La Freynelle, Landereau, La Mothe du Barry (French Kiss), l'Aubrade, Les Arromans, Lestrille, Marjosse, Martinon, Nardique-la-Gravière, Sainte-Marie, *Tour de Mirambeau*, Turcaud, Vignol.

Esclans, Ch D' Prov ★→★★★ Sacha Lichine founder of canny estate marketing rosé as lifestyle choice for export. Part owned by LVMH. Garrus GRENACHE/ROLLE, oaked is top-notch, expensive, will age. Rock Angel gd quality, value. Whispering Angel hugely successful NÉGOCIANT brand.

Esmonin, Dom Sylvie C d'O ★★★ Rich, dark wines from fully ripe grapes, whole-bunch vinification and new oak. Best: GEVREY-CHAMBERTIN VIEILLES VIGNES, CLOS ST-JACQUES. Fairly priced for top burg.

Etoile, L' Jura AOP for elegant CHARD and SAVAGNIN on limestone and marl, usually oxidative. VIN JAUNE and VIN DE PAILLE also allowed but not reds (sold as AOP CÔTES DU JURA). Try ★★ *Montbourgeau*, P Vandelle, Rolet.

Eugénie, Dom C d'O ★★★→★★★★ Artemis Estates' 1st foray into burg. Intense, dark wines now enlivened by more whole-bunch vinification. CLOS VOUGEOT, GRANDS-ÉCHÉZEAUX outstanding, but try village CLOS d'Eugénie too.

Fabre, Famille L'doc ★★★ Go-ahead CORBIÈRES family with five CHX. Ch de Luc Boutenac (r), masterful, rugged. Ch Fabre Gasparets Les Amouries red v. smart, gd. MOURVÈDRE-based rosé too.

Faiveley, Dom Burg ★★→★★★★ More grower than merchant, revitalized by Erwan F since 2005, incl major overhaul of facilities. Now making high-class reds and sound whites. Leading light in CÔTE CHALONNAISE, but save up for top wines from CHAMBERTIN-Clos de Bèze, CHAMBOLLE-MUSIGNY, CORTON *Clos des Cortons*, NUITS. Also owns classy DOM Billaud-Simon (CHAB).

Faugères L'doc ★★★ Rare instance of single-soil AC: schist on s-facing foothills of Cevennes; signature style is freshness to balance rich fruit. Will age. Elegant whites from GRENACHE BL, MARSANNE, ROUSSANNE, VERMENTINO. Drink CH DE LA LIQUIÈRE, DOMS Ancienne Mercerie, Bardi-Alquier, CÉBÈNE, Chenaie, DES TRINITÉS, Estanilles, Grézan, Léon Barral, Mas d'Alezon, Mas Gabinèle, Météore, Ollier-Taillefer, St Antonin, Sarabande.

Félines-Jourdan, Dom L'doc ★★→★★★ Classic, tangy PICPOUL from Claude Jourdan nr Mèze. Top white Féline, low yields, *bâttonage*, for contemplation, not quaffing.

Ferraton Père & Fils N Rh ★★★ Grower-merchant in HERMITAGE, CHAPOUTIER-owned. Bio; wide range, gd to v.gd: CROZES-HERMITAGE Grand Courtil (r/w), Le Méal HERMITAGE (r), Le Reverdy (w), Les Dionnières (r), ST-JOSEPH Bonneveau (r).

Fèvre, William Chab ★★★→★★★★ Biggest owner of CHAB GCS; Bougros Côte Bougerots and Les CLOS outstanding. Small yields, no expense spared, priced accordingly, top source for rich, age-worthy wines and some more humble. Look out also for cousins N&G Fèvre, esp PC Vaulorent and GC Preuses.

Fiefs Vendéens Lo ★→★★★ 19' 20' (21) Serious producers, characterful age-worthy wines. CHARD, CHENIN BL, MELON, SAUV BL, CAB FR, CAB SAUV, GAMAY, Grolleau Gr, Négrette, PINOT N (r/rosé). Try DOM St-Nicolas (bio), Jumeaux (bio), Mourat, Prieuré-la-Chaume (bio).

Fitou L'doc ★★→★★★ Mostly CARIGNAN, so expect rugged richness, spice. Wines that taste of the sun. Two parts: schist on inland hills s of Narbonne; and chalk, limestone nr coast. Seek out: CHX de Nouvelles, Grand Guilhem, Champs des Soeurs; DOMS Astruc, Bertrand-Bergé, JONES, Lérys.

Fixin C d'O ★★★ 05' 09' 10' 12' 14 15' 16 17 18' 19' 20' Worthy, undervalued n neighbour of GEVREY-CHAMBERTIN. Sturdy, sometimes splendid reds, can be rustic but enjoying warmer vintages. Best vyds: Arvelets, CLOS de la Perrière, Clos du Chapitre, Clos Napoléon. Top locals: Berthaut-Gerbet, Gelin, Joliet, Naddef. Try also Bart, CLAIR, FAIVELEY, MORTET.

Fleurie Beauj ★★★ 15' 18' 19' 20' Top BEAUJ cru for perfumed, strawberry fruit, silky texture. Racier from La Madone hillside, richer below. Classic: CHX BEAUREGARD, Chatelard, de Poncié; DOMS Brun, Chignard; CLOS de la Roilette, Depardon, DUBOEUF, Métrat, co-op. Naturalists: Balagny, Dutraive, Métras, Pacalet, Sunier. Newcomers: Chapel, Clos de Mez, Dom de Fa, Hoppenot, Lafarge-Vial.

Fourrier, Dom C d'O ★★★★ Jean-Marie F taken sound GEVREY-CHAMBERTIN DOM to top, with cult prices to match. Sensual vibrant reds at all levels, made from ancient vines. Best: CLOS ST-JACQUES, Combe aux Moines, GRIOTTE-CHAMBERTIN. *See also* Bass Phillip (Australia).

Francs-Côtes de Bordeaux B'x ★★ 15 16 18 19 20 Tiny B'X AOP next to CAS. Fief of Thienpont (PAVIE MACQUIN) family. Mainly red; MERLOT-led (60%). Some gd white (eg. Charmes-Godard, Puyanché). Top CHX: Cru Godard, Francs, Godard Bellevue, La Prade, Marsau, Puyfromage, *Puygueraud*.

Fronsac B'x ★★→★★★ 14 15 16 18 19' 20' Great-value, hilly AOP w of POM. MERLOT-led on clay-limestone; some ageing potential. Top CHX: Arnauton, DALEM, Fontenil, *La Dauphine*, La Rivière, La Rousselle, LA VIEILLE CURE, LES TROIS CROIX, Haut-Carles, Mayne-Vieil (CUVÉE Alienor), *Moulin Haut-Laroque*, Puy Guilhem, Tour du Moulin, Villars. *See also* CANON-FRON.

Fronton SW Fr ★★ 16 18 19 (20) AOP n of Toulouse. Must be based on rare (sometimes unblended) Négrette grape (flavours of violets, cherries, licorice). Often blended with SYRAH. ★★★ CHX Baudare, Bouissel, Caze, du Roc, Laurou, Plaisance; ★★ *Ch Bellevue-la-Forêt* best known. Also ★ Boujac, Clamens, La Colombière, Viguerie de Belaygues. No AOP for whites as yet.

Fuissé, Ch Burg ★★→★★★ Smart operation in POUILLY-FUISSÉ with long track record. Concentrated oaky whites. Top terroirs Le CLOS, Combettes. Also BEAUJ crus, eg. JULIÉNAS.

Gagnard C d'O ★★→★★★ Respected clan in CHASSAGNE. Long-lasting wines, esp Caillerets, BÂTARD from Jean-Noël G; while Blain-G, Fontaine-G have full range incl rare CRIOTS-BÂTARD, MONTRACHET itself. Gd value offered by all Gagnards. Decent Chassagne reds all round.

Gaillac SW Fr ★→★★ Ramshackle vyds ne of Toulouse. Cornucopia of grapes incl

FRANCE

Braucol (Fer), Duras, SYRAH (r), Len de l'El, Mauzac (w). Prunelard is gaining
ground for red, while Ondenc (w) has legendary status at PLAGEOLES. Quality
variable, but also look for ★★★ DOMS Brin, Causse-Marines, d'Escausses, La
Ramaye, La Vignereuse, Le Champ d'Orphée, L'Enclos des Roses, Peyres-Roses,
Plageoles, Rotier. Perlé is refreshing, summery white with slight prickle; can be
delicious. Co-ops do it well and cheaply.

Ganevat Jura ★★★→★★★★ CÔTES DU JURA superstar bought by A Pumpyanskiy
(2021), but Ganevat family remains. Single-vyd CHARD (eg. Chalasses, Grand
Teppes). Also innovative reds. Cult pricing.

Gauby, Dom Gérard Rouss ★★★ Pioneer nr village of Calce. Bio, increasingly
natural; lots of innovation; son taking over. High-altitude vyds up to 550m
(1804ft), chalk for fresh acidity. Try Les Calcinaires VIEILLES VIGNES (r/w),
Muntada. Associated with DOM Le Soula.

Gaure, Ch de L'doc ★★★★ Sublime AOP LIMOUX CHARD Oppidum from Pierre Fabre.
Pour Mon Père CARIGNAN/GRENACHE/SYRAH, AOP L'DOC. Vyds in Agly V, ROUSS.

GC (Grand Cru) Official term meaning different things in different areas. One of top
Burg vyds with its own AC. In AL, one of 51 top vyds, each now with its own rules.
In ST-ÉM, 60% of production is St Ém GC, often run of the mill. In MÉD: five tiers
of GC CLASSÉS. In CHAMP, top 17 villages are GCs. Since 2011 in Lo for QUARTS DE
CHAUME, and emerging system in L'DOC. Take with pinch of salt in Prov.

Gevrey-Chambertin C d'O ★★★ 05' 09' 10' 12' 15' 16 17 18' 19' 20' Major AOP for
fine savoury reds at all levels up to great CHAMBERTIN and GC cousins. Top PCS
Cazetiers, Combe aux Moines, Combottes, CLOS ST-JACQUES. Value single-vyd
village wines (En Champs, La Justice), VIEILLES VIGNES bottlings. Top: BACHELET,
BOILLOT, Burguet, Damoy, Drouhin-Laroze, DUGAT, Dugat-Py, Duroché,
ESMONIN, FAIVELEY, FOURRIER, Harmand-Geoffroy, Heresztyn-Mazzini, LEROY,
Magnien (H), Marchand-Grillot, MORTET, ROSSIGNOL-TRAPET, Roty, ROUSSEAU,
Roy, SÉRAFIN, TRAPET, and all gd merchants. Rebourseau gd from 18.

Syrah gets its black-pepper flavour from rotundone, also found in – black pepper.

Gigondas S Rh ★★→★★★★ 05' 06' 09' 10' 12' 13' 15' 16' 17' 18 19' 20 Top S Rh red.
Breathtaking vyds on stony clay-sand *garrigue* plain rise to alpine limestone hills e
of Avignon; GRENACHE, plus SYRAH, MOURVÈDRE. Full, clear, menthol fresh wines;
best give fine dark red fruit. Top 10 15 16 19. More oak recently, higher prices,
but genuine local feel in many. Try Boissan, Bosquets (gd modern), Bouïssière
(punchy), Brusset, Cayron (gd traditional), CH de Montmirail, CH DE ST COSME
(flair), *Clos des Cazaux* (value), CLOS du Joncuas (organic), DOM *Famille Perrin*,
Goubert, Gour de Chaulé (fine), Grapillon d'Or, Les Pallières, Longue Toque,
Moulin de la Gardette (stylish), Notre Dame des Pallières, P Amadieu (consistent),
Pesquier, Pourra (robust), *Raspail-Ay* (ages), Roubine (hearty), *St Gayan* (ages),
Santa Duc (now stylish), Semelles de Vent, Teyssonnières. Heady rosés.

Gimonnet, Pierre Champ ★★★★★ Didier G makes beautifully consistent CHARD
on N Côte des Blancs, 28 GCS, PCS. Not a fan of single-vyd CHAMP. Special Club
is unmatched complex expression of great Chard 13 17' 19 20 for long ageing.
Vintage 12' 1st rate.

Ginglinger, Dom Paul Al Eguisheim house making pure, racy RIES, PINOT BL, gd 18
19. Excellent CRÉMANT.

Girardin, Vincent C d'O ★★→★★★ White-specialist MEURSAULT-based NÉGOCIANT,
part of BOISSET group. Pierre-Vincent G, son of the original, is installed afresh in
Meursault, showing promise, no expense spared.

Givry Burg ★★ 15' 17 18' 19' 20' Top tip in CÔTE CHALONNAISE for tasty reds that can
age. Better value than MERCUREY. Rare whites nutty in style. Best (r): CELLIER AUX
MOINES, CLOS Salomon, *Faiveley*, F Lumpp, Joblot, Masse, Thénard.

Goisot Burg ★★★ Guilhem and J-H G, outstanding bio producers of single-vyd bottlings of ST-BRIS (SAUV BL) and Côtes d'Auxerre for CHARD, PINOT N. Nobody else comes close.

Goldert Al Exceptional S AL GC, marl-limestone, v. aromatic. Exceptional MUSCAT (Al's best?), SYLVANER and GEWURZ. Best: ERNEST BURN, ZIND HUMBRECHT.

Gonon, Dom N Rh ★★★ 09' 10' 12 13' 14 15' 16' 17 18' 19' 20' Top estate at ST-JOSEPH, in v. high demand; bros Pierre and Jean work organically, hand-graft cuttings on 10 ha prime old vyds. Mainly whole-bunch, old 600-litre casks, aromatic, peppered, iron-toned red, most savoury, suave, compelling *Les Oliviers* (w, great *à table*), both live 20 yrs.

Gosset Champ Oldest house, based in AŸ, owned by Cointreau. Chef de CAVE Odilon de Varine is passionate about terroir. Grand Blanc de MEUNIER a 1st for house; mainly 07, elegant, aged on CHARD lees. Prestige Celebris Extra BRUT one of best, a classic in 04. Sublime double-aged Gosset Célébrissimes 95' in same spirit; long-aged Gosset 12 Ans de Cave a Minima.

Gosset-Brabant Champ AŸ grower, great PINOT N vyds. Noirs d'Aÿ 12 16 18' 19' 20' cathedral of PINOT flavours.

Gouges, Henri C d'O ★★★ Reference point over several generations for meaty, long-lasting NUITS-ST-GEORGES with Grégory and Antoine G now at helm. Great range of PC vyds, eg. CLOS des Porrets, Vaucrains and esp Les St-Georges. Also rare, excellent *white Nuits*, from PINOT BL.

Graillot, Dom Alain N Rh ★★★ 15' 16' 17 18' 19' 20' Organic vyds on stony plain, whole-bunch ferments. Dashing CROZES (r): La Guiraude special selection, serious, long life. Crozes (w), ST-JOSEPH (r). Son Maxime: Crozes DOM des Lises, gd merchant range Equis (CORNAS, St-Joseph).

Gramenon, Dom S Rh ★★→★★★ 19' 20' At height in lower Drôme, bio since 2007. Beautiful fruit purity, v. low sulphur, gd range. CÔTES DU RH: La Papesse (GRENACHE), La Sagesse, Poignée des Raisins (fun), Sierra du Sud (SYRAH).

Grande Rue, La C d'O ★★★ 05' 06 09' 10' 12' 15' 16 17 18' 19' 20' MONOPOLE of DOM Lamarche, GC between LA TÂCHE, ROMANÉE-CONTI. Quality, consistency improved under Nicole Lamarche. Fascinating blood-orange hallmark across vintages. Also special bottling dubbed 1959 – not better, just different

Grands-Échézeaux C d'O ★★★★ 90' 93 96' 99' 02' 05' 09' 10' 12' 15' 17 18' 19' 20' Superlative GC next to CLOS DE VOUGEOT, but with a MUSIGNY silkiness. More weight than most ÉCHÉZEAUX. Top: BICHOT (CLOS Frantin), Coquard-Loison-Fleurot, DRC, DROUHIN, EUGÉNIE, Lamarche (lost vyd in 2021), Millot, NOËLLAT G.

Grange des Pères, Dom de la H'doc ★★★ Next door to MAS DE DAUMAS GASSAC, and IGP. Red from CAB SAUV, SYRAH, MOURVÈDRE; white 80% ROUSSANNE, plus CHARD, MARSANNE. Difficult to get hold of but pretty special if you can.

Gratien & Meyer / Alfred Gratien Champ ★★★ BRUT 93 12 13 15' 18' Small but beautiful CHAMP house, owned by Henkell Freixenet. BRUT NV. CHARD-led Prestige CUVÉE Paradis Brut, Rosé (NV). Fine, v. dry, lasting, oak-fermented wines incl *The Wine Society's house Champagne*. Careful buyer of top crus, esp Chard from favourite growers. Also Gratien & Meyer in SAUMUR.

Graves B'x ★→★★ 15 16 18 19 20 Appetizing grainy reds from MERLOT, CAB SAUV, fresh SAUV/SÉM (dr w). Graves Supérieures denotes *moelleux*. "Ambassadeur des Graves" special selection. Some of best values in B'x today. Top CHX: ARCHAMBEAU, Brondelle, CHANTEGRIVE, CLOS Bourgelat, *Clos Floridène*, CRABITEY, de Cérons, Ferrande, Fougères, Grand Enclos du Ch de Cérons, Haura, Magneau, Pont de Brion, RAHOUL, *Respide Medeville*, Roquetaillade La Grange, Saint-Robert (CUVÉE Poncet Deville), Seuil, *Vieux Ch Gaubert*, Villa Bel-Air.

Graves de Vayres B'x ★ DYA. Tiny AOP within E-2-M zone. Red, white, *moelleux*.

Grignan-les-Adhémar S Rh ★→★★ AOP, stony soils; best reds spiced, herbal, drink within 5 yrs. Leaders: Baron d'Escalin, DELAS (value), La Suzienne (value); CHX Bizard, La Décelle (incl CÔTES DU RH w); DOMS de Bonetto-Fabrol, de Montine (stylish r, gd w/rosé, also Côtes du Rh r), Grangeneuve best (esp VIEILLES VIGNES), St-Luc.

Griotte-Chambertin C d'O ★★★★ 90' 96' 99' 02' 05' 09' 10' 12' 15' 16 17 18' 19' 20' Small GC next to CHAMBERTIN; nobody has much volume. Brisk red fruit, depth and ageing potential, from DROUHIN, DUGAT, Duroché, FOURRIER, *Ponsot (L)*.

Gripa, Dom N Rh ★★★ 13' 15' 16' 17' 18' 19' 20' Top ST-JOSEPH, ST-PÉRAY DOM, v. refined whites. St-Joseph Le Berceau (w) 100% 60-yr+ MARSANNE; *St-Péray Les Figuiers*, mainly ROUSSANNE, v. classy. Both St-Joseph reds gd, top Le Berceau tracks vintage: deep 15, pure-fruit 16, dense 17, bold 18, rich 19, stylish 20.

Grivot, Jean C d'O ★★★ ·★★★★ VOSNE-ROMANÉE DOM that may improve even further as Mathilde G takes over. Superb range of PCS (Note Beaux Monts and NUITS Boudots) topped by GCS CLOS DE VOUGEOT, ÉCHÉZEAUX, RICHEBOURG. Higher prices these days.

Gros, Doms C d'O ★★★→★★★★ Family of vignerons in VOSNE-ROMANÉE with stylish wines from Anne (sumptuous RICHEBOURG), succulent reds from Michel (CLOS de Réas), Anne-Françoise (now in BEAUNE) and Gros Frère & Soeur (CLOS VOUGEOT En Musigni). Not just GC; try value HAUTES-CÔTES DE NUITS. Also Anne's DOM Gros-Tollot in MINERVOIS. Change of generation at all four doms, but no changes in style evident yet. From 2022, some vyds changing hands within family.

Gros Plant du Pays Nantais Lo ★→★★ AOP DYA Gros Plant (FOLLE BLANCHE); much better than reputation. Fresh, saline, married to oysters/shellfish. Try Basse Ville, Haut-Bourg, LUNEAU-PAPIN, Poiron Dabin, Preuille. Sparkling: pure or blended.

Guigal, Ets E N Rh ★★→★★★★ Justly famous, always-expanding grower-merchant: CÔTE-RÔTIE mainly, plus CONDRIEU, CROZES-HERMITAGE, HERMITAGE, ST-JOSEPH, 52-ha CHÂTEAUNEUF CH de Nalys, plus two lots of 7-ha and 18-ha vyds there, top TAVEL Ch d'Aquéria. Merchant: Condrieu, Côte-Rôtie, Crozes-Hermitage, Hermitage, S Rh. Owns DOM de Bonserine (Côte-Rôtie), VIDAL-FLEURY (quality on up). Top, v. expensive Côte-Rôties La Mouline, La Landonne, La Turque (rich, 42 mths new oak, so atypical), also v.gd Hermitage, St-Joseph VIGNES de l'Hospice. Standard also gd: *brilliant-value Côtes du Rh* (r/w/rosé). Best whites: Condrieu, Condrieu La Doriane (oaky), Hermitage, St-Joseph Lieu-dit St-Joseph.

Hautes-Côtes de Beaune/Nuits C d'O ★★ (r) 15' 18' 19' 20' (w) 17' 18 19' 20' Generic BOURGOGNE AOP for villages in hills behind main C Do vyds. Attractive lighter reds, whites for early drinking, both putting on weight in warmer times. Sweet spots are villages of Arcenant, Meloisey, Nantoux and plateau above CÔTES DE NUITS. Look for Carré, Champy (Boris), CHEVROT, Devevey, DOM de la Douaix, Faure, Hoffmann-Jayer, Jacob, Naudin (Claire), Parigot, Vantey.

Haut-Médoc B'x ★★→★★★★ 14 15 16' 18 19' 20 Prime source of dry, digestible CAB/MERLOT reds. Usually gd value. Plenty of CRUS BOURGEOIS. Wines usually sturdier in n; finer in s. Five Classed Growths (BELGRAVE, CAMENSAC, *Cantemerle*, *La Lagune*, LA TOUR-CARNET). Eight CRUS BOURGEOIS Exceptionnels (Arnauld, BELLE-VUE, CAMBON LA PELOUSE, Charmail, D'AGASSAC, de Malleret, du Taillan, *Malescasse*). Other top CHX: CISSAC, CITRAN, COUFRAN, *de Lamarque*, Lamothe-Bergeron, LANESSAN, Larose Perganson, REYSSON, SÉNÉJAC, *Sociando-Mallet*.

Haut-Poitou Lo ★→★★ 20 21 Mainly early, easy drinking from isolated AC; CAB SAUV, CAB FR, CHARD, GAMAY, PINOT N, SAUV BL. Ampelidae (IGP) dominates. Also LaCheteau (Ohh! Poitou), Morgeau La Tour.

Heidsieck, Charles Champ ★★★★ Iconic house, small but beautiful, wines as exquisite as ever. Cyril Brun winemaker since 2012. *NV Brut* all purity, subtle ripe complexity. Peerless Blanc des Millénaires 06. Millénaires 07 tighter, fine tension. Great Vintage 12' lovely now, can keep to 2030; older Collection, esp **83 81**; CHAMP Charlie Prestige CUVÉE could be reintroduced from 2023. BLANC DE BLANCS NV nicely priced, delicious.

Hengst Al Marl-limestone-sandstone AL GC, powerful wines. Top PINOT N (ALBERT MANN); PINOT GR, AUXERROIS (JOSMEYER); GEWURZ (ZIND HUMBRECHT).

Henriot Champ Fine family house. Ace BLANC DE BLANCS de CHARD NV; BRUT **98' 02' 08**; Brut Rosé **09**. Exceptional prestige CUVÉE Hemera **05 06**. Old yrs of Les Enchanteleurs still going strong. New AVIZE **16** released 2022.

Hermitage N Rh ★★★→★★★★ **01' 05' 06' 07' 09' 10' 11' 12' 13' 15' 16' 17' 18' 19' 20'** (**10 15** brilliant.) Part granite hill on E Rhône bank with grandest, deepest, majestic SYRAH and complex, nutty/white-fruited, fascinating, v.-long-lived white (MARSANNE, some ROUSSANNE) best left for 6–7 yrs+. Best: Alexandrins, Belle (organic), *Chapoutier (bio, magic w)*, Colombier, DELAS, Faurie (pure, last vintage 20), GUIGAL, Habrard (w), *J-L Chave* (rich, elegant), M SORREL (mighty Le Gréal r, retired 2018, now G Sorrel, still gd), PAUL JABOULET AÎNÉ (refined), Philippe & Vincent Jaboulet (r/w), Tardieu-Laurent (oak). TAIN co-op gd (esp Gambert de Loche r, VIN DE PAILLE w).

Horizon, Dom de l' Rouss ★★★ Pure expression of rugged ROUSS terroir under IGP CÔTES CATALANES from old vines nr Calce. Mar y Muntanya gd-value SYRAH; gd gastronomic rosé.

Hortus, Dom de l' Ldoc ★★★→★★★★ Dynamic, family-run PIC ST-LOUP estate. Fine SYRAH-based reds: elegant Bergerie and oak-aged Grande CUVÉE (r). New Le Dit d'Hortus 100% SYRAH. Intriguing Bergerie IGP Val de Montferrand (w) with seven grapes. Also CLOS du Prieur (r) in cooler TERRASSES DU LARZAC.

Hospices de Beaune C d'O Spectacular medieval foundation with grand charity auction of CUVÉES from its 61 ha for Beaune's hospital, 3rd Sunday in Nov, now run by Sotheby's. Individuals can buy as well as trade. Winemaker Ludivine Griveau doing a great job. Quality high, prices too, charity is the point. Try BEAUNE CUVÉES, VOLNAYS or expensive GCS, (r) CORTON, ÉCHÉZEAUX, MAZIS-CHAMBERTIN, (w) BÂTARD-MONTRACHET.

Hudelot C d'O ★★★ VIGNERON family in CÔTE DE NUITS. H-NOËLLAT (VOUGEOT) is top class, esp GCS ROMANÉE-ST-VIVANT, RICHEBOURG, while H-Baillet (CHAMBOLLE) is challenging with punchy reds.

Huet Lo ★★★★ **16' 17 18' 19' 20'** (**21**) World, and VOUVRAY, reference for CHENIN BL. Anthony Hwang has Királyudvar in Tokaj, Hungary. Bio since 1990. Single vyds: CLOS du Bourg and Le Mont on Première Côte; Le Haut Lieu nearby, more clay. Wines v. age-worthy, esp sweet: from 1919 **47 59 03 05 07 08 10 11 15 18**.

Hugel & Fils Al ★★→★★★★ Famous Riquewihr address, 12 generations and counting, famed for late-harvest, esp RIES, GEWURZ VT, SGN (Johnny H "invented" SGN). Superb Ries Schoelhammer **10 13 17** from GC SCHOENENBOURG.

IGP (indication géographique protegée) Potentially most dynamic category in France (150+ regions), scope for experimentation. Replacing VdP, but new terminology still not accepted by every area. Zonal names most individual: eg. CÔTES DE GASCOGNE, Côtes de Thongue, Pays des Cévennes, Haute Vallée de l'Orb, among others. Enormous variety in taste, quality, never ceases to surprise.

Irancy Burg ★★ **15' 16 17 18'** 19 Structured red nr CHAB made from PINOT N and more rustic local César. Beware hot, dry vintages. Best vyds: Mazelots, Palotte. Best: Cantin, Ferrari, GOISOT, Maison Chapelle, Renaud and Richoux.

Irouléguy SW Fr ★→★★★ **15' 16 18'** 19 (**20**) From lusciously green hillsides in the Basque Country. Reds based on Axéria (CAB FR), TANNAT. Look out for rediscovered vines like Arrouya (Manseng N), Erremaxaoua. Best: ★★★ Arretxea, Bordaxuria, Brana, Ilarria. Fruity white based on Petit Courbu and both MANSENGS; ★★★ Brana, also for fruit-based spirits, gin. New CUVÉE, Liberum (pre-phylloxera Tannat planted 1850).

Jaboulet Aîné, Paul N Rh Grower-merchant at Tain. Organic vyds across HERMITAGE, CONDRIEU, CORNAS, CROZES-HERMITAGE, CÔTE-RÔTIE, ST-JOSEPH. Reds sleek, easy-fruited, would love more soul. Once-leading producer of HERMITAGE, esp ★★★★ La Chapelle (legendary 61 78 90), quality varied since 90s, some revival since 2010 on reds. Also CORNAS St-Pierre, Crozes Thalabert (can be stylish), Roure (sound). Merchant of other Rh, notably robust VACQUEYRAS, VENTOUX (r, quality/value). Whites lack true Rh body, drink most young, range incl new v. expensive La Chapelle (w, not every yr).

Jacquart Champ ★★★ Simplified range from co-op-turned-brand, concentrating on what it does best: PC Côte des Blancs CHARD from member growers. Fine range of Vintage BLANC DE BLANCS 13 17 19' 20'. Vintage Rosé 12 v.gd.

Jacquart, André Champ ★★★→★★★★ Marie Doyard has 24 ha incl 18 ha in GC LE MESNIL. Flagship is Mesnil Experience. Magic Vintage trio 18' 19' 20'. Ace BLANC DE BLANCS specialist.

Jacquesson Champ ★★★★ Ace Dizy house for precise, v. dry wines. Outstanding single-vyd Avize CHAMP Caïn 12'. Corne Bautray, all CHARD. Dizy 10 13' innovative **numbered NV cuvées** 730' 731 732 733 734 735 738 739 739 740 741 742 743 744. Focus on intrinsic character of each base-wine harvest rather than notional consistency yr on yr.

Jadot, Louis Burg ★★→★★★★ Dynamic BEAUNE merchant making powerful whites (Diam corks) and well-constructed age-worthy reds with significant vyd holdings in BEAUJ, C D'O, MÂCON inc POUILLY-FUISSÉ (*Dom Ferret*), MOULIN-À-VENT (CH des Jacques). On v.gd form at moment.

In N Rh. "Famille xxx" = bought-in grapes; "Dom xxx" = own grapes. Just be aware.

Jamet, Dom N Rh ★★★★ 05' 09' 10' 12 13' 14 15' 16' 17' 18' 19' 20' Must-buy. Illustrious CÔTE-RÔTIE, v.-long-lived, complex *vins de terroir* from many sites, mostly schist. Classic red intricate, dashing fruit, Côte Brune (r) is mighty, smoky, mysterious, 30 yrs+. Also high-grade CÔTES DU RH (r/w), COLLINES RHODANIENNES (r).

Jasnières Lo ★★→★★★ 19' 20' (21) Age-worthy AOP CHENIN BL, austerely dry to vibrantly sweet, s-facing slopes of Loir V. Try Ange Vin (also VDF), BELLIVIÈRE, Breton, Briseau, Gigou, Janvier, Maisons Rouges, Métais, Raderie, *Roche Bleue*, Ryke.

Jobard C d'O ★★★ VIGNERON family in MEURSAULT. Antoine J for esp long-lived Poruzots, Genevrières, CHARMES plus reds from former Dom Mussy (POMMARD) from 2019. Rémi J for immediately classy Meursaults, esp Poruzots. Valentin J impressing too.

Jones, Dom Rouss ★★→★★★ Englishwoman Katie J; consistently gd FITOU, Côtes du ROUSS. Terroir wines with modern twist. Lively social media, livestreamed vyd rambles.

Josmeyer Al ★★→★★★★ Run by sisters Isabelle (oenologist) and Celine (MD) Meyer; bio viticulture AL pioneer. Marvellous RIES, PINOT GR GC Hengst 13 14 16 17' 18' 19; Al's best AUXERROIS H, superb SYLVANER Peau Rouge (red-berried variant).

Juliénas Beauj ★★★ 15' 18' 19' 20' Source of dark-fruited, structured BEAUJ, esp for CLIMATS Beauvernay, Capitans etc. Try Audras (CLOS de Haute Combe), Aufranc, Besson, Burrier, CH BEAUREGARD, CH FUISSÉ, DOM Granit Doré, Perrachon.

Jurançon SW Fr ★→★★★ (sw) 15' 16' 18 19 20 (dr) 16 17' 18' 19 (20) Separate AOPs for sweet and dry whites. Balance of richness, acidity the key to quality. ★★★ DOMS *Cauhapé*, Guirardel, Lapeyre, Larrédya, Larrouyat. ★★ CH Jolys; Doms Bellegarde, Bordenave, Castéra, CLOS Benguères, Nigri, Uroulat. ★ Gan co-op gd value. *See also* CABIDOS.

Kientzler, Andre Al ★★→★★★★ Family DOM, 5th generation. Lush sensual GEWURZ GC Kirchberg 16 17' 18' 19 20; VT dessert wines. Exemplary care in vyds.

Kreydenweiss, Marc Al ★★→★★★★ Bio for decades. Rich soil diversity: GC Moenchberg on limestone for PINOT GR and majestic RIES Kastelberg on black schist, ages for up to 20 yrs; 10 17′ ★★★★ 18′ 19. Also in COSTIÈRES DE NÎMES.

Krug Champ Supremely prestigious deluxe house. ★★★★ Grande CUVÉE. Edition 168, based on superb 12; Edition 169, based on classic 13. Vintage 98 02 04; Rosé; CLOS D'AMBONNAY 95′ 98′ 00; CLOS DU MESNIL 02 98; Krug Collection 69 76′ 81 85. Rich, nutty wines, oak-fermented; highest quality, ditto price. Shame it never released 12, truly great yr.

Kuentz-Bas Al ★★→★★★ Organic/bio, serious yet accessible wines: drier RIES 13 17′ 20. GEWURZ VT CUVÉE Caroline 09 12 17.

Labet, Dom Jura ★★★ Key CÔTES DU JURA organic estate in Rotalier. Best-known for vibrant range of single-vyd CHARD whites, eg. En Chalasse, La Bardette, Les Varrons; gd PINOT N, Poulsard.

Ladoix C d'O ★★★ (r) 15′ 17 18′ 19′ 20′ (w) 14′ 15 17′ 18 19′ 20′ Exuberant whites, eg. PC Grechons, and juicy reds, esp PC Joyeuses. Try (w) Chevalier, FAIVELEY, Loichet; (r) Capitain-Gagnerot, CH DE MEURSAULT, Mallard, Naudin-Ferrand, Ravaut.

Lafarge, Michel C d'O ★★★★ Outstanding bio VOLNAY estate run by Frédéric L. Unbeatable PCS *Clos des Chênes*, CLOS du CH des Ducs. Also fine BEAUNE, esp Grèves (r) and Clos des Aigrots (w). Plus FLEURIE project, Lafarge-Vial.

Lafon, Dom des Comtes Burg ★★★★ Fabulous bio MEURSAULT DOM, esp PC Perrières, Genevrières and GC MONTRACHET, with long-lasting red VOLNAY *Santenots* equally outstanding. Value from excellent Mâconnais under Héritiers du Comte L label. Try also Dominique L's own label for BEAUNE, Volnay, Meursault.

Laguiche, Marquis de C d'O ★★★★ Largest owner of Le MONTRACHET and a fine PC CHASSAGNE, both excellently made by DROUHIN.

Lalande de Pomerol B'x ★★→★★★ 14 15 16 18 19 20 Satellite neighbour of POM; similar but without same depth, class; value. Largely MERLOT on clay, gravel and sandy soils. Top CHX: Ame de Musset, Belles-Graves, Chambrun, Enclos de Viaud, Garraud, Grand Ormeau, Haut-Chaigneau, Jean de Gué, La Chenade, LA FLEUR DE BOÜARD, La Sergue, Les Cruzelles, *Les Hauts Conseillants*, Pavillon Beauregard, Sabines, Samion, Siaurac, *Tournefeuille*.

Lallier Artisan CHAMP from GC AŸ PINOT N. Black Label R series, 12 and 13 base, ace expression of each yr. Lovely Oger CHARD too. Now owned by Campari group; Dominique Demarville chef de CAVE. Could affect hierarchy of *grandes maisons*.

Lamé Delisle Boucard Lo ★★ →★★★ Source of old vintages, and they age brilliantly. Long-est (1869) family BOURGUEIL DOM, now organic; v. consistent, gd value from fruity DOM des Chesnaies CUVÉE Prestige.

Lamy C d'O ★★★ DOM Hubert L is go-to address for ST-AUBIN. Breathtakingly fresh, concentrated whites, often from higher-density plantings. Reds now worthy of note. Also DOMS L-Caillat (intense w), more traditional L-Pillot in CHASSAGNE.

Landes SW Fr DYA IGP area in far sw, similar to CÔTES DE GASCOGNE. Better known for sand dunes. DOM de Lamballe is decent enough: try its Sables Fauves. Coteaux de Chalosse is separate IGP: lots of weird vines. Try TURSAN co-op.

Landron, Doms Lo ★★→★★★ 20′ 21 Fine, age-worthy MUSCADET SÈVRE ET MAINE from flamboyant winemaker. Try Fief du Breil and Atmosphères (sp).

Langlois-Chateau Lo ★★→★★★ Well made: fine CRÉMANT DE LO, SAUMUR (v.gd, age-worthy Saumur Bl VIEILLES VIGNES 16′ 17′ 18′), CHINON, POUILLY-FUMÉ, SANCERRE (Fontaine-Audon/Thauvenay), SAUMUR-CHAMPIGNY. BOLLINGER-owned.

Languedoc Region and sprawling AOP from Nîmes to Spanish border, inland to Carcassonne and LIMOUX. Theoretically bottom of the pyramid of L'doc AOPs with other levels more specific, based on region and terroir, eg. FAUGÈRES, MINERVOIS, ST-CHINIAN. Subregions incl Cabrières, Grès de Montpellier, PÉZENAS, Quatourze, St-Saturnin, St-Georges d'Orques. Clairette du L'doc tiny AOP for white CLAIRETTE. Top of hierarchy: crus incl CORBIÈRES-Boutenac, MINERVOIS la Livinière, PIC ST-

LOUP, LA CLAPE, TERRASSES DU LARZAC. Usual L'doc grapes: (r) CARIGNAN, CINSAULT, GRENACHE, MOURVÈDRE, SYRAH; (w) GRENACHE BL, ROUSSANNE, VERMENTINO, but many others. IGP d'Oc covers whole region, regional IGPs too. *See* box, below.

Lanson Champ ★★★ Major house, owned by BCC. Black Label NV; Rosé NV; Vintage BRUT on a roll, esp 02 08 12 15 18'. Ace prestige NV Noble CUVÉE BLANC DE BLANCS, rosé and Vintage. Single-vyd Brut Vintage CLOS Lanson 12 Extra Age multi-vintage, Blanc de Blancs esp gd. Experienced new winemaker Hervé Dantan (since 15) allows some malo for a rounder style.

Lapierre, Marcel Beauj ★★★ Mathieu and Camille L in vanguard of sulphur-free movement. Range of styles, CUVÉES of BEAUJOLAIS, MORGON.

Laplace, Dom SW Fr Oldest and at one time sole producer of MADIRAN. Still at top. Top wine ★★★ CH d'Aydie, needs time. Odie d'Aydie less so. Beautifully polished, less extracted than before; ★ Les Deux Vaches easy intro to TANNAT, lighter, rounder. Excellent ★★★ PACHERENCS (dr/sw). Sweet fortified Maydie (think BANYULS) gd with chocolate.

Laroche Chab ★★ Major player in CHAB with quality St Martin blend, Vieille Voye special CUVÉE, exceptional GC *Res de l'Obediencerie* named after historic HQ (worth a visit). Winemaker Grégory Viennois involved in NÉGOCIANT IRANCY project. Also Mas La Chevalière in L'DOC.

Latour, Louis Burg ★★→★★★ Traditional BEAUNE merchant making full-bodied whites from C D'O vyds (esp CORTON-CHARLEMAGNE), Mâconnais, Ardèche (all CHARD). Reds more controversial but top end is age-worthy: CORTON, ROMANÉE-ST-VIVANT. Also owns Henry Fessy in BEAUJ.

Latricières-Chambertin C d'O ★★★★ 99 05 09' 10' 12' 15' 16 17 18' 19' 20' GC next to CHAMBERTIN. Deep soil and cooler site gives rich earthy wines in warm dry yrs. Best: ARNOUX-LACHAUX, BIZE, Drouhin-Laroze, Duband, Duroché, FAIVELEY, LEROY, Remy, ROSSIGNOL-TRAPET, TRAPET.

Laudun S Rh ★→★★ 19' 20 CÔTES DU RH-VILLAGE. Clear-fruited, peppy, superior whites. Red-fruit, peppery reds (much SYRAH), lively rosés. Immediate flavours from Maison Sinnae (old name Laudun-CHUSCLAN) co-op. DOM Pelaquié best, esp fresh white. Also CHX Courac, de Bord, Juliette; Doms Duseigneur (bio), Carmélisa, Maravilhas (bio, character, r/w), Olibrius.

Laurens, J L'doc Specialist in LIMOUX bubbles. BLANQUETTE Le Moulin v.gd value. CLOS des Demoiselles Vintage CRÉMANT rivals many CHAMPS.

Laurent-Perrier Champ Important house; family presence resurrected. BRUT NV (CHARD led) perfect apéritif; v.gd skin-contact Rosé. Fine vintages 08 12'. Grand Siècle CUVÉE multi-vintage on form, peerless Grand Siècle Alexandra Rosé 12 ★★★★. Also owns DELAMOTTE, SALON. Much-respected new chef de CAVE for all three, Dominique Demarville, stayed 4 mths, went to Lallier. Reported as clash of personalities, issue of oak.

Lavantureux Chab ★★ Specialist source for PETIT CHAB, CHAB, esp single-vyd Vauprin. BOURGOGNE Epineuil reds from 19 look brilliant too.

Leccia, Yves Cors ★★★ Small bio DOM nr Bastia. Intense, precise fruit rather than oak. AOP Patrimonio El Croce, IGP Île de Beauté YL. Worth seeking out.

Languedoc grapes go back to the future

Growers in L'DOC are looking to new and old grapes to adapt to climate change. Terret (w) and Counoise, Piquepoul Noir, Rivairenc (r) well suited to heat but were abandoned in favour of trendy international CHARD, SYRAH etc. CLOS du Gravillas, DOM Mas de la Seranne reviving ancient varieties. Others look to Greek white ASSYRTIKO: Doms La Tasque, SARRAT DE GOUNDY, Pech Redon too, along with Calabrese (r) from Sicily. NERO D'AVOLA, PRIMITIVO also planted. Exciting stuff.

Leflaive, Dom Burg ★★★★ Reference PULIGNY-MONTRACHET DOM, back on top form since 2017 but with prices to match. Outstanding GC, incl MONTRACHET, CHEVALIER and *fabulous PCs*: Pucelles, Combettes, Folatières, etc. Also try S Burg range, eg. MÂCON Verzé.

Leflaive, Olivier C d'O ★★→★★★ White specialist NÉGOCIANT at PULIGNY-MONTRACHET. Outstanding BOURGOGNE Les Sétilles and all levels up to own GC vyds. Reds improving. Also La Maison d'Olivier, hotel, restaurant, tasting room.

Leroux, Benjamin C d'O ★★★ BEAUNE-based NÉGOCIANT equally at home in red or white; C D'O only, strengths (w) in MEURSAULT with increasing DOM and (r) BLAGNY, GEVREY, VOLNAY. Smart wines in every sense.

Leroy, Dom C d'O ★★★★ Legendary bio pioneer Lalou Bize L delivers extraordinary red burg from tiny yields in VOSNE-ROMANÉE and from DOM d'Auvenay (more whites). Both fiendishly expensive even ex-dom, as is amazing treasure trove of mature wines from family NÉGOCIANT Maison L.

Liger-Belair, Comte C d'O ★★★★ Comte Louis-Michel L-B makes brilliantly ethereal wines in VOSNE-ROMANÉE, ever-increasing stable headed by LA ROMANÉE. Try village La Colombière, PC Reignots, NUITS-ST-GEORGES crus. In Chile, Oregon.

Liger-Belair, Thibault C d'O ★★★→★★★★ New winery in NUITS-ST-GEORGES, for succulent bio burg from generics up to Les St-Georges and GC RICHEBOURG. Outstanding ALIGOTÉ. Also range of stellar old-vine single-vyd MOULIN-À-VENT.

Lignier C d'O ★★★ Family in MOREY-ST-DENIS. Whole range from Laurent L (DOM Hubert L) brilliant, esp CLOS DE LA ROCHE; v.gd PCS from Virgile L-Michelot, esp Faconnières, but Dom Georges L divides opinion.

Lilbert Champ Bijou DOM, highest standards. Young CHARD hard as diamond, age gracefully 30 yrs. Entry-level Perlé GC NV gd intro. Vintage 17 drinking well; 19 will be even better.

Limoux L'doc ★★→★★★ Bustling market town renowned for sparkling BLANQUETTE and CRÉMANT de Limoux. Don't miss stylish still white AOP Limoux from CHARD, CHENIN BL, Mauzac; must be barrel-aged. Red AC: MERLOT, plus SYRAH, GRENACHE, Cabs. PINOT N in Crémant and for IGP Haute Vallée de l'Aude. Growers: CH de Gaure; DOMS de Baronarques, Begude, de Fourn, de l'Aigle, Mouscaillo; RIVES-BLANQUES; Cathare, Jean-Louis Denois.

Liquière, Ch de la L'doc ★★ Unpretentious fruity AOP FAUGÈRES, family estate. Les Amandières (r/w/rosé) terrific value. Cistus old vines (r/w). Malpas top SYRAH.

Lirac S Rh ★★→★★★ 15' 16' 17' 18 19' 20 Four villages nr TAVEL, stony, quality soils. Full, spiced red (life 5 yrs+), gd impetus from CHÂTEAUNEUF owners via crisper fruit, more flair. Reds best, esp DOMS Carabiniers (bio), *de la Mordorée* (organic, best, r/w), Duseigneur (bio) Giraud, Joncier (bio, character), Lafond Roc-Epine (organic), La Lôyane, La Rocalière (organic, gd fruit), Maby (Fermade, gd w), Maravilhas (bio), Marcoux (stylish), Plateau des Chênes; CHX Boucarut (organic, revived), de Bouchassy (gd w), de Manissy (organic), de Montfaucon (v.gd w, incl CÔTES DU RH), Mont-Redon, St-Roch; Mas Isabelle (handmade), P Usseglio, Rocca Maura (esp w), R Sabon. Whites always gd, convey freshness, body, go 5 yrs. Decent table rosés.

Listrac-Médoc H-Méd ★★→★★★ 14 15 16' 18 19 20 Much-improved AOP for savoury red B'x; now more fruit, depth and MERLOT due to clay soils. Also gd whites under AOP B'x (eg. Le Cygne de Fonréaud). Best CHX: Cap Léon Veyrin, CLARKE, FONRÉAUD, Fourcas-Borie, FOURCAS DUPRÉ, FOURCAS-HOSTEN, l'Ermitage, LESTAGE, Liouner, MAYNE-LALANDE, Reverdi, SARANSOT-DUPRÉ.

Londe, La Prov ★★ Maritime vyds. Chanel has moved in here and bought DOM de l'Île; LVMH owns Galoupet. Others: CLOS Mireille, Dom Perzinsky, Léoube, St Marguerite. Coastal schist subzone of CÔTES DE PROV, incl island of Porqueyrolle (three vyds).

Long-Depaquit Chab ★★★ BICHOT-owned CHAB DOM with famous flagship brand, GC La Moutonne.

Lorentz, Gustave Al ★★→★★★ Grower-merchant at Bergheim. RIES is strength, best GC Altenberg de Bergheim, age-worthy and refined 12 13 14 16 18 19 but not among most concentrated. Entry-level wines (esp GEWURZ) passable.

Lot SW Fr ★→★★ DYA IGP of Lot département increasingly useful to CAHORS growers for rosés and whites not allowed in AOP (eg. CLOS de Gamot, CH DU CÈDRE). Look beyond AOP for ★★ DOMS Belmont, Sully, Tour de Belfort.

Loupiac B'x ★★ 15 16 17 18 19 Minor SÉM-dominant *liquoreux*. Lighter, fresher than SAUT across River Garonne. Top CHX: CLOS Jean, *Dauphiné-Rondillon*, *de Ricaud*, du Cros, Les Roques, *Loupiac-Gaudiet*, Noble.

Luberon S Rh ★→★★ 19' Hilly, tourist region e of S Rh; terroir is v. dry, can be no more than okay. Excess of technical wines. SYRAH plays lead role. Whites improving. Bright star: CH de la Canorgue (organic). Also gd: Chx Clapier, Edem, Fontvert (bio, gd w), Puy des Arts (w), Ravoire, St-Estève de Neri (improver), Tardieu-Laurent (rich, oak); DOMS de la Citadelle (organic), Fontenille (organic), La Cavale (swish), Le Novi (terroir), Marrenon, Maslauris (organic), Val-Joanis; Laura Aillaud; gd-value La Vieille Ferme (w/rosé, can be VDF).

Luneau-Papin, Dom Lo Eye-opening MUSCADET from top-class family bio DOM; impeccable, age-worthy wines. Reputation made by Pierre and Monique L; son Pierre-Marie and wife Marie continuing upwards trajectory.

Lussac-St-Émilion B'x ★★ 15 16 18 19 20 Most n of ST-ÉM satellites; lightest in style. Top CHX: Barbe Blanche, Bel-Air, Bellevue, Courlat, DE LUSSAC, La Grande Clotte, La Rose Perrière, Le Rival, LIONNAT, Mayne-Blanc.

Macle, Dom Jura ★★★ Legendary producer of CH-CHALON VIN JAUNE for long ageing. Best drunk 10 yrs+ after bottling. Also CÔTES DU JURA white.

Mâcon Burg ★ DYA Simple, juicy GAMAY reds and most basic rendition of Mâconnais whites from CHARD.

Mâcon-Villages Burg ★★ 14' 17' 18 19' 20' Chief appellation for Mâconnais whites. Individual villages may also use their own names eg. Mâcon-Lugny. Co-ops at Lugny, Terres Secrètes, Viré for quality/price ratio, plus *brilliant grower wines* from Guffens-Heynen, Guillot, Guillot-Broux, Maillet, Merlin. C D'o-based DOMS BOILLOT J-M, LAFON, LEFLAIVE. Also major NÉGOCIANTS: DROUHIN, LATOUR, etc.

Macvin Jura AOP, not Scottish. Grape juice fortified by oak-matured local marc to make off-dry apéritif of 16–20% alc. Usually white, can be red. Try over ice cream.

Madiran SW Fr ★★→★★★ 05' 10 12 15' 16 18 19 (20) Gascon AOP. France's home of TANNAT grape. Worthy reds in range of styles with many keepers. Look for ★★★ CHX BOUSCASSÉ, MONTUS (owner BRUMONT has 15% entire AOP), Laffitte-Teston, *Laplace*. Wide ranges from ★★★ Chx Arricaud-Bordès, de Gayon; DOMS Berthoumieu, Capmartin, Damiens, Dou Bernés, Labranche-Laffont, Laffont, Pichard; CLOS Basté. Doms ★★ Barréjat, ★★ Crampilh, Maouries not far behind.

Maillard, Nicolas Champ ★★★ Uses stainless steel, a little aerating oak. PC Platine 12 15; 19' a classic to age.

Mailly Grand Cru Champ Top co-op, all GC grapes. Prestige CUVÉE des *Echansons* 08 12' for long ageing. Sumptuous Echansons Rosé 12; refined, classy L'Intemporelle 15 18 19'. Sébastien Moncuit, cellarmaster since 14, a real talent.

Mann, Albert Al ★★→★★★★ Excellent Wettolsheim bio grower, run by Barthelmé family. Deft winemaking; superb GCS – HENGST, SCHLOSSBERG (esp 17'); RIES epicentre. One of two best Al PINOT N producers (Les Stes Claires 15 18' 19).

Maranges C d'O ★★ 15' 17 18' 19' 20' Name to watch. Robust well-priced reds from s end of CÔTE DE BEAUNE. Try PCS Boutière, Croix Moines, Fussière. Best: BACHELET-Monnot, Chevrot, Giroud, MATROT, Rouges Queues.

Marcillac SW Fr ★★ 15' 18 19 20 AVEYRON AOP based on Mansois (aka FER SERVADOU). Fruity, curranty/raspberry food wines; rustic, lowish alc. Best at 3 yrs. Try with strawberries as well as charcuterie or sausages, and of course *aligot* (local dish of mashed potato, garlic and an unhealthy amount of cheese). ★★ DOM du Cros largest independent grower (gd w IGPs too), also Doms des Boissières, Laurens. Excellent co-op. Recent heatwave vintages outstanding.

Margaux H-Méd ★★→★★★★ 09' 10' 14 15 16' 18 19' 20' Most s MÉD communal AC. Famous for elegance, fragrance; reality is more diverse. Top CHX: BRANE-CANTENAC, DAUZAC, DU TERTRE, FERRIÈRE, GISCOURS, ISSAN, KIRWAN, LASCOMBES, MALESCOT ST EXUPÉRY, MARGAUX, PALMER, RAUZAN-SÉGLA. Chx: ANGLUDET, Arsac, Deyrem Valentin, LABÉGORCE, LA TOUR DE MONS, Mongravey, Paveil de Luze, SIRAN all gd-value.

Marionnet, Henry Lo ★★→★★★ 19' 20' 21 TOURAINE DOM famous for ungrafted vyds (sandy soil), rare vines, esp Romorantin; now run by son Jean-Sébastien. Big range incl SAUV BL (ages well), GAMAY, La Pucelle de Romorantin, *Provignage* (Romorantin planted c.1820), Renaissance (ungrafted Gamay, no sulphur). Managing historic vyd at CH de Chambord.

Marsannay C d'O ★★→★★★ (r) 15' 17 18' 19' 20 Most n AOP of CÔTE DE NUITS, *still* hoping to get PCS (eg. Champ Salomon, CLOS du Roy, Longeroies). Further village-level vyds added in 2019, eg. Le Chapitre. Accessible, crunchy, fruit-laden reds, from energetic producers: Audoin, Bart, Bouvier, Charlopin, CH de M, CLAIR, Derey, Fournier, *Pataille*; v.gd unfashionable *rosé* needs 1–2 yrs; whites getting better.

Mas, Doms Paul L'doc ★★→★★★ Jean-Claude M directs huge empire from Grès de Montpellier to ROUSS. Mainly IGP. Working on organics, bio and low sulphur. Wine tourism, restaurant. Arrogant Frog range; also Côté Mas, La Forge, Les Tannes, Les VIGNES de Nicole and DOMS Ferrandière, Crès Ricards in TERRASSES DU LARZAC, Martinolles in LIMOUX; CHX Lauriga in ROUSS, Villegly in MINERVOIS.

Mas Amiel Rouss ★★★ Leading MAURY, Côtes du ROUSS, IGP. Look for Altaïr (w), Origine, Vers le Nord, Vol de Nuit from v. old CARIGNAN/GRENACHE, others. Plus excellent VDN from young, fruity *grenat*, to venerable RANCIO 20- to 40-yr-old Maury aged in 60-litre glass demijohns, intensely sweet and savoury.

Mas de Daumas Gassac L'doc ★★★→★★★★ Star since 80s; now run by 2nd-generation Samuel Guibert, who uses horses in vyd. CAB SAUV-based age-worthy reds from apparently unique soil. Perfumed white from CHENIN BL blend; super-CUVÉE Émile Peynaud (r); rosé Frizant; v.gd sweet Vin de Laurence (MUSCAT/SERCIAL).

Mas de l'Ecriture L'doc ★★→★★★ Father/daughter team; exquisite reds; small organic TERRASSES DU LARZAC estate.

Mas Gabriel L'doc ★★★ PÉZENAS. Small bio DOM, fine range; CLOS des Lièvres SYRAH/GRENACHE top-notch. Champ des Bleuets mostly VERMENTINO, v.gd.

Mas Jullien L'doc ★★★★ TERRASSES DU LARZAC star: typical Larzac freshness. MOURVÈDRE/CARIGNAN red: Autour de Jonquières, Carlan, États d'Âme, Lous Rougeos from L'DOC varieties. Carignan Bl and Gr, CHENIN BL for white.

Mas Llossanes Rouss ★★★ Highest vyds in ROUSS (700m/2297ft): freshness, elegance. Pure SYRAH, Pur CARIGNAN, super varietal expression. Dotrera, au Dolmen gd red blends. Pur Chasan (w) perhaps most thrilling of all. IGP CÔTES CATALANES.

Massif d'Uchaux S Rh ★★ 16' 18' 19' 20 A gd village, tree-fringed vyds, clear-fruited, fresh, spiced reds, not easy to sell, but best genuine, stylish. Note CH St-Estève (incl gd old-vine VIOGNIER), DOMS *Cros de la Mûre* (interesting, gd value), de la Guicharde, La Cabotte (bio, on great form), Renjarde (swish fruit).

Matrot Burg ★★→★★★ Since sisters Elsa and Adèle took over quality, consistency soared at MEURSAULT DOM. Equally gd in red (BLAGNY, VOLNAY Santenots) and white (MEURSAULT Perrières, Charmes, Blagny and PULIGNY PCS). Not fans of new wood or whole bunches.

Maury Rouss ★★→★★★ Sweet VDN from GRENACHES N/BL/Gr on island of schist. Ambré, tuilé and RANCIO styles. Don't confuse with AOP Maury SEC, dry red. MAS AMIEL leader for both styles. DOMS de Lavail, Lafage, de L'OU, Pouderoux, OF THE BEE; Maury Co-op.

Maxime Magnon L'doc ★★→★★★ Burgundian native farms steep, high vyds in rugged CORBIÈRES; bio. Rozeta (r): fruit-bomb with elegance, spice. Unfashionably dark, delicious Métisse rosé.

Mazel, Le S Rh ★★ 18′ 19′ 20 Mother lode for Rh natural wine, no sulphur, S ARDÈCHE. Gérald Oustric 1st vinified 1997, carbonic maceration, vats only (no wood), low-alc CARIGNAN, GRENACHE; whites CHARD, VIOGNIER notably. All VDF.

Mazis- (or Mazy-) Chambertin C d'O ★★★★ 90′ 93 96′ 99′ 05′ 09′ 10′ 12′ 15′ 16′ 17 18′ 19′ 20′ GC of GEVREY-CHAMBERTIN, top class in upper part; intense, *heavenly wines*. Best: Bernstein, DUGAT-PY, FAIVELEY, HOSPICES DE BEAUNE, LEROY, MORTET, Rebourseau (from 2019), ROUSSEAU, Tawse.

Mazoyères-Chambertin C d'O ★★★★ Usually sold as CHARMES-CHAMBERTIN, but style is different: less succulence, more stony structure. Try DUGAT-PY, MORTET, Perrot-Minot, Taupenot-Merme, Tawse.

Médoc B'x ★★ 15 16 18 19 20 AOP for reds in low-lying n part of Méd peninsula (aka Bas-Méd). Often more guts than grace. Can be gd value, but be selective. Top CHX: Castera, CLOS Manou, d'Escurac, Fleur La Mothe, *Goulée*, GREYSAC, La Cardonne, *La Tour-de-By*, LES ORMES-SORBET, LOUDENNE (Le Ch) Lousteauneuf, Potensac d'Aux, POTTEVIN, Potensac, PREUILLAC, Rollan-de-By (HAUT-CONDISSAS), TOUR HAUT-CAUSSAN, TOUR ST-BONNET, Vieux Robin.

Meffre, Gabriel S Rh ★★→★★★ Merchant with gd standards, owns GIGONDAS DOM Longue Toque (top Hommage GM). Fruit quality up, less oak. Makes CHÂTEAUNEUF (gd St-Théodoric, also small doms), VACQUEYRAS St-Barthélemy. Reliable-to-gd S/N Rh Laurus range, esp CONDRIEU, HERMITAGE (w), ST-JOSEPH.

Mellot, Alphonse Lo ★★→★★★★ 18′ 19′ 20′ (21) Impressive SANCERRE, bio, cellars under centre of Sancerre. Powerful reds. All handpicked: Cuvée Edmond, Génération XIX (r/w), La Moussière (r/w); gd single vyds Demoiselle, *En Grands Champs, Satellite*; Les Pénitents (Côtes de la Charité IGP) CHARD, PINOT N.

Menetou-Salon Lo ★★→★★★ 19′ 20′ 21 AOP nr SANCERRE (similar wines), SAUV BL; reds (PINOT N). Try Chatenoy, *Gilbert* (bio), *Henry Pellé* (organic), Prieuré de Saint-Céols, Jean-Max Roger, Joseph Mellot, Teiller, Tour St-Martin.

Méo-Camuzet C d'O ★★★★ Noted DOM in VOSNE-ROMANÉE: icons Brûlées, CROS PARANTOUX, RICHEBOURG. Value from M-C Frère et Soeur (NÉGOCIANT branch) and plenty of choice in between. Sturdy, oaky wines that age well. Interesting HAUTES-CÔTES white Clos St-Philibert.

Mercurey Burg ★★→★★★ 15′ 17 18′ 19′ 20′ Leading village of CÔTE CHALONNAISE, firmly muscled reds, aromatic whites. Try BICHOT, Champs de l'Abbaye, *Ch de Chamirey*, CH Philippe Le Hardi, DOM de Suremain, FAIVELEY, Génot-Boulanger, *Juillot-Theulot*, Lorenzon, M Juillot, Raquillet.

Merlin Burg ★★→★★★ Olivier M, wizard of the MÂCONNAIS, now joined by sons. Top wines MÂCON La Roche Vineuse Les Cras, expanding POUILLY-FUISSÉ range, MOULIN-À-VENT La Rochelle. Co-owners CH des Quarts with Dominique LAFON.

Mesnil-sur-Oger, Le Champ Top Côte des Blancs village, v. long-lived CHARD. Best: ANDRÉ JACQUART, JL Vergnon, PIERRE PÉTERS. Needs 10 yrs+ ageing. Top named vyd Les Chétillons. Moulin-à-Vent 19′ classic cooler vintage to watch.

Méthode champenoise Champ Traditional method of putting bubbles into CHAMP by re-fermenting wine in its bottle. Outside Champ region, makers must use terms "classic method" or *méthode traditionnelle*.

Meursault C d'O ★★★→★★★★ 09' 10' 12 14' 15 17' 18 19' 20' Potentially great full-bodied whites from PCS: Charmes, Genevrières, Perrières, more nervy from hillside vyds *Narvaux*, Tesson, *Tillets*. Producers: Ballot-Millot, BOILLOT, Boisson-Vadot, BOUZEREAU, *Ch de Meursault*, COCHE-DURY, *de Montille*, Ente, Fichet, Girardin, Javillier, JOBARD, *Lafon*, Latour-Giraud, LEROUX, *Matrot*, Michelot, Mikulski, *P Morey*, PRIEUR, Rougeot, *Roulot*. Try de Cherisey for Meursault-BLAGNY.

Meursault, Ch de C d'O ★★ Huge strides lately at this newly bio 61-ha estate of big-biz Halley family: improving reds from BEAUNE, POMMARD, VOLNAY. Some stunning whites, esp MEURSAULT, also v.gd BOURGOGNE Blanc, PULIGNY PC.

Minervois L'doc ★★→★★★ Lots to choose from in this undulating region ne of Carcassonne. Characterful, herbal reds, esp CHX COUPE-ROSES, d'Agel, Gourgazaud, de Homs, La Grave, La Tour Boisée, Oupia, *Paumarhel*, St-Jacques d'Albas, STE EULALIE, Senat, Villerambert-Julien; DOM CLOS Centeilles; Abbaye de Tholomiès, *Borie-de-Maurel*, PIERRE CROS, PIERRE FIL, Laville-Bertrou. Gros and Tollot (from Burg) raising bar. La Livinière on Black Mtn slopes cru for fine, long-lived red. Clos d'Ora (GERARD BERTRAND), COMBE BLANCHE, Gayda, Maris, de l'Ostal, Piccinini, STE EULALIE excel. St Jean de Minervois is delicious fresh MUSCAT VDN: BARROUBIO, Clos du Gravillas best.

Mis en bouteille au château / domaine Bottled at CH, property, or estate. Note *dans nos caves* (in our cellars) or *dans la région de production* (in the area of production) often used but mean little.

Moët & Chandon Champ By far largest CHAMP house, impressive quality for such a giant. Fresher, drier BRUT Imperial NV. Rare prestige CUVÉE MCIII "solera" concept, daunting complexity at sky-high prices. Certainly better value in Grand Vintages Collection, long lees-aged; 09 elegant, 08 a little severe. Ace 12 ★★★★, finely chiselled 13'. Outposts across New World. *See also* DOM PÉRIGNON.

Railways released wines of SW from iron grip of B'x trade: new routes to market.

Monbazillac SW Fr ★★→★★★ 15' 17 18 19 BERGERAC sub-AOP: ★★★★ *Tirecul-la-Gravière* up there with gd SAUTERNES. ★★★ CLOS des Verdots, L'Ancienne Cure, Les Hauts de Caillavel, co-op's *Ch de Monbazillac*. ★★ CHX de Belingard-Chayne, Grande Maison, Kalian, Le Faget, Monestier la Tour, Pech La Calevie, Pécoula. Neighbouring SAUSSIGNAC produces similar wines.

Monopole A vyd under single ownership.

Montagne-St-Émilion B'x ★★ 15 16 18 19 20 Largest ST-ÉM satellite. Solid reputation. Top CHX: Beauséjour, CLOS de Boüard, Corbin, Croix Beauséjour, Faizeau, La Couronne, Maison Blanche, Roc de Calon, Roudier, Simon Blanchard, Teyssier, Tour Bayard, Vieux Bonneau, Vieux Ch Palon, *Vieux Ch St-André*.

Montagny Burg ★★ 14' 17 18 19' 20' CÔTE CHALONNAISE village with crisp whites, mostly in hands of CAVES DE BUXY but gd NÉGOCIANTS too, incl LOUIS LATOUR, O LEFLAIVE. Top growers: *Aladame*, Berthenet, Cognard, *Feuillat-Juillot*, Lorenzon.

Monthélie C d'O ★★→★★★ 15' 16 17 18' 19' 20' Pretty reds, grown uphill from VOLNAY, but a touch more rustic. Les Duresses best PC. *Ch de Monthélie* (de Suremain), Changarnier, Dubuet, Dujardin, Garaudet. Whites interesting (hillside sites).

Montille, de C d'O ★★★ Dense, spicy, whole-bunch reds from BEAUNE, CÔTE DE NUITS (Malconsorts), POMMARD (Rugiens), VOLNAY (Taillepieds) and exceptional white from MEURSAULT, plus outstanding PULIGNY-MONTRACHET Caillerets. Since 2017 CH de Puligny wines are incl under de Montille. New projects in Sta Rita Hills (California) and Hokkaido (Japan).

Montlouis sur Loire Lo ★★→★★★★ 18 19' 20' 21 Top CHENIN BL, racier than Anjou Chenins (esp dr w); sp incl *Pétillant Originel*. Top: Berger, Chanson, CHIDAINE, Delecheneau, Jousset, Merias, Pierres Ecrites, Saumon, *Taille-aux-Loups*, Vallée Moray, *Weisskopf*. Small crop 2021.

Montpeyroux L'doc ★★★ Cru 40 km (25 miles) n of Montpellier dominated by Mt Baudile. Innovative growers. DOM D'AUPILHAC, also Chabanon, Divem, *Jasse-Castel*, Joncas, Mas d'Amile, Villa Dondona. Serious co-op.

Montrachet (or Le Montrachet) C d'O ★★★★ 02' 04 05 08 09' 10 12 14' 15 17 18 19' 20' The GC vyd that lent name to both PULIGNY and CHASSAGNE. Should be greatest white burg for intensity, richness of fruit and perfumed persistence. Top: BOUCHARD, COLIN, DRC, LAFON, LAGUICHE (DROUHIN), LEFLAIVE, Ramonet.

Montravel SW Fr ★★ (r) 15' 18' 19 20 (w/rosé) DYA Sub-AOP of BERGERAC. Modern-style reds must be oak-aged. CHX de Bloy, de Krevel. ★★ CHX Jonc Blanc, Masburel, Masmontet, Moulin Caresse. ★★ dry white from same and others.

Montus, Ch SW Fr ★★★★ 12' 14 15' 16 17 18 19 (20) Alain BRUMONT's flagship property, famous for long-extracted oak-aged wines. Long-lived all-TANNAT reds, much prized by lovers of old-fashioned MADIRAN. Classy sweet and dry white barrel-raised PACHERENC-DU-VIC-BILH. La Tyre, Prestige equal to Classed Growths.

Mordorée, Dom de la S Rh ★★★ 15' 16' 17' 18' 19' 20 Top estate at TAVEL, organic, rosés with flair, drive. Also LIRAC, La Reine des Bois (r/w); gd CHÂTEAUNEUF La Reine des Bois (incl 1929 GRENACHE), La Dame Voyageuse (r). Nifty VDF La Remise (r/w/rosé).

Moreau Chab ★★→★★★ Widespread family in CHAB, esp *Dom Christian M*, noted for PC Vaillons CUVÉE Guy M and GC Les CLOS des Hospices. Louis M, more commercial range; DOM M-Naudet, concentrated wines for longer keeping.

Moreau C d'O ★★→★★★★ At s end of C D'O. Outstanding CHASSAGNE PCS from DOM Bernard M; fine La Cardeuse (r). Tidy range of SANTENAY, MARANGES from David M. Neither related to CHAB dynasty.

Morey, Doms C d'O ★★★ VIGNERON family in CHASSAGNE. Look for Caroline M and husband Pierre-Yves COLIN-M, Marc (En Virondot), Sylvain, Thibault M-Coffinet (LA ROMANÉE), Thomas (v. fine pure whites), Vincent (plumper style). Also Pierre M in MEURSAULT for Perrières and BÂTARD.

Morey-St-Denis C d'O ★★★→★★★★ 99' 02' 05' 09' 10' 12' 15' 16' 17 18' 19' 20' Terrific source of top-grade red burg, to rival neighbours GEVREY-CHAMBERTIN, CHAMBOLLE-MUSIGNY. GCS CLOS DE LA ROCHE, CLOS DE LAMBRAYS, CLOS DE TART, CLOS ST-DENIS. Top producers: ARLAUD, CLOS DE TART, *Clos des Lambrays*, *Dujac*, H LIGNIER, Perrot-Minot, PONSOT, *Roumier*, Tremblay. Also try Amiot, Castagnier, Coquard-Loison-Fleurot, LIGNIER-Michelot, Magnien, Remy, Taupenot-Merme. Interesting whites too, esp PC Monts Luisants.

Morgon Beauj ★★★ 15' 17 18' 19' 20' Powerful BEAUJ cru; volcanic slate of Côte du Py makes meaty, age-worthy wine, clay of Les Charmes for earlier, smoother drinking. Grands Cras, Javernières of interest too. *Burgaud*, CH de Pizay, Ch des Lumières (JADOT), *Desvignes*, Foillard, Gaget, Godard, *Lapierre*, Piron, Sunier.

Mortet, Denis C d'O ★★★→★★★★ Arnaud M on song with powerful yet refined reds from BOURGOGNE Rouge to CHAMBERTIN. Key wines GEVREY-CHAMBERTIN Mes Cinq Terroirs, PCS Champeaux, Lavaut St-Jacques. From 2016 separate Arnaud M label, equally brilliant, incl CHARMES- and MAZOYÈRES-CHAMBERTIN.

Moueix, J-P et Cie B'x Respected Libourne-based NÉGOCIANT and proprietor named after founder Jean-Pierre. Still in family hands. CHX BÉLAIR-MONANGE, CLOS La Madeleine, HOSANNA, LA FLEUR-PÉTRUS, *La Grave à Pomerol*, LATOUR-À-POMEROL, *Trotanoy. See also* Dominus Estate, California.

Moulin-à-Vent Beauj ★★★ 09' 11' 15' 18' 19' 20' Grandest BEAUJ cru, transcending GAMAY grape. Weight, spiciness of Rh but matures towards rich, gamey PINOT flavours. Increasing interest in single-vyd bottlings from eg. *Ch des Jacques*, *Ch de Moulin-à-Vent*, DOMS Janin, Janodet, Labruyère, *Merlin* (La Rochelle), Rottiers. *See also* C D'O producers eg. BICHOT (Rochegrès), L BOILLOT (Brussellions), T LIGER-BELAIR (Rouchaux).

Moulin de la Gardette S Rh ★★★ 16′ 17′ 19′ 20 GIGONDAS: gd, organic, *garrigue* expression, benefit from time. Petite Gardette value, main red Tradition is GRENACHE with terroir, 20 yrs+ life. Oaked Ventabren substantial.

Moulis H-Méd ★★→★★★ 14 15 16 18 19 20 Tiny inland AOP w of MARGAUX. Honest, gd-value wines; best have ageing potential. Top CHX: Anthonic, Biston-Brillette, BRANAS GRAND POUJEAUX, BRILLETTE, Caroline, *Chasse-Spleen*, Dutruch Grand Poujeaux, *Gressier Grand Poujeaux*, MAUCAILLOU, *Mauvesin Barton*, *Poujeaux*.

Mourgues du Grès, Ch S Rh ★★→★★★ 19′ 20 21 Leading COSTIÈRES de NÎMES estate, organic, v.gd fun, early drinking range: racy rosé (Dorés, Galets Rouges, Rosés). Firmer Capitelles: Terre d'Argence (SYRAH), Terre de Feu (GRENACHE).

Moutard Champ Original champion of local Arbanne grape. Also eaux de vie. Greatly improved quality, esp CHARD Persin 14, CUVÉE des 6 CÉPAGES 11 15′ 18 19′.

Mugneret C d'O ★★★→★★★★ VIGNERON family in VOSNE-ROMANÉE. Sublime wines from Georges M-Gibourg (from BOURGOGNE to ÉCHÉZEAUX), now matched by cousins at DOM Gérard M. Also Dom Mongeard-M.

Mugnier, J-F C d'O ★★★★ Outstanding grower of CHAMBOLLE-MUSIGNY *Les Amoureuses*, *Musigny*. Do not miss PC Fuées. Finesse, not muscle. Equally at home with MONOPOLE NUITS-ST-GEORGES CLOS de la Maréchale. No longer sells young vintages of MUSIGNY to avoid infanticide.

Mumm, GH & Cie Champ Powerful house owned by Pernod Ricard. Chef de CAVE talented Laurent Fresnet (ex-HENRIOT): a renaissance? Mumm de Verzenay BLANC DE NOIRS 08, ★★★★ RSVR BLANC DE BLANCS 12′. Mumm de Cramant, renamed Blanc de Blancs, elegantly subtle. Cordon Rouge NV much improved: PINOT-led weight with élan, tension. Also impressive in California.

Muré / Clos Saint Landelin Al ★★→★★★★ Great AL name est 1650 in Westhalten, now in Rouffach, led by Véronique and Thomas. Exceptional, full-bodied PINOT GR, RIES 17′ 18 from iron-rich soils of Clos Saint Landelin, 12-ha monopole in GC Vorbourg. Best Al PINOT N (CUVÉE V 15′ 18′) and CRÉMANTS.

Muscadet Lo ★→★★★ 19′ 20′ 21 Popular, bone-dry w from nr Nantes. Made for fish, seafood. Choose zonal Muscadet AOPS: Coteaux de la Loire (Carroget, Guindon, Landron-Chartier), CÔTES de Grandlieu (Eric Chevalier, Haut-Bourg, Herbauges, Malidain), SÈVRE-ET-MAINE. MUSCADET CRUS COMMUNAUX v.gd.

Muscadet Crus Communaux Lo ★★→★★★ MUSCADET's remarkable top level. Long lees-ageing from specified soil sites. Complex, age-worthy, value. Seven crus: Clisson, Gorges, Goulaine, La Haye Fouassière, Le Pallet, Monnières-St Fiacre, Mouzillon-Tillières. Champtoceaux, Côtes de Grandlieu, Vallet in progress.

Muscadet Sèvre et Maine Lo ★→★★★ 18′ 19′ 20′ 21 Largest MUSCADET zone. Often excellent, v.gd value. Top incl *Bonnet-Huteau (organic)*, Brégeon, Briacé, Caillé, Chereau Carré, Cormerais, Delhommeau, Douillard, Gadais, Grenaudière, Grand Mouton, Gunther-Chereau, Haute Fevrie, Huchet, l'Ecu, Landron, *Lieubeau*, *Luneau-Papin (bio)*, Olivier, Pépière, Sauvion. 2021 severe frost.

Musigny C d'O ★★★★ 93 96′ 99′ 02′ 05′ 09′ 10′ 12′ 15′ 17 18 19′ 20′ Most beautiful red burg; GC lent its name to CHAMBOLLE-MUSIGNY. Hauntingly fragrant but with sinuous power beneath. Best: DE VOGÜE, DROUHIN, FAIVELEY, JADOT, LEROY, *Mugnier*, PRIEUR, ROUMIER, VOUGERAIE.

Nature Unsweetened, esp for CHAMP: no dosage. Fine if v. ripe grapes, raw otherwise. Vin Nature = natural wine; *see* A Little Learning.

Negly, Ch la L'doc ★★★→★★★★ Dynamic LA CLAPE estate, impressive range from salty Brise Marine (w), La Côte (r), and La Falaise (r) for everyday to icon La Porte du Ciel (SYRAH), MOURVÈDRE-based L'Ancely and extravagant CLOS des Truffiers from oldest Syrah vyd in L'DOC, nr PÉZENAS.

Négociant-éleveur Merchant who "brings up" (ie. matures) the wine.

Noblaie, Dom de la Lo ★★→★★★★ Jérôme Billard has made this family DOM a leader in CHINON. Excellent range from fruity Le Temps des Cerise to complex Pierre de Tuf via single vyds: Chiens Chiens, Les Blanc Manteaux. Fine CHENIN BL.

Noëllat C d'O ★★★ Noted VOSNE-ROMANÉE family. Maxime Cheurlin at DOM Georges N on top form: try NUITS Boudots, Vosne Petits-Monts and GC ÉCHÉZEAUX, also some gd-value lesser appellations. Cousins at Michel N starting to cause a stir. *See also* v. stylish HUDELOT-N in VOUGEOT.

Nuits-St-Georges C d'O ★★→★★★★ 99' 02' 05' 09' 10' 12' 15' 16 **17** 18' 19 20 Three parts to this major AC: Premeaux vyds for elegance (various CLOS: de la Maréchale, des Corvées, des Forêts, St-Marc), centre for dense dark plummy wines (Cailles, Les St-Georges, Vaucrains) and n side for the headiest (Boudots, Cras, Murgers, Richemone). Top names ARLOT, ARNOUX-LACHAUX, CATHIARD, CHEVILLON, *Faiveley*, GOUGES, GRIVOT, LEROY, *Liger-Belair*, *Mugnier*, but try also Ambroise, Chauvenet, Chicotot, Confuron, Gavignet, Lechéneaut, Ledy, Machard de Gramont, Michelot, Millot, Perdrix, *Rion*. Note also Hospices de Nuits sale, a more local affair than the BEAUNE version.

Ogier, Stéphane N Rh ★★★ Busy, expanding CÔTE-RÔTIE DOM; oak, modern, swish fruit style. Top: Belle Hélène, Côte Blonde, La Viaillière. IGP SYRAH, VIOGNIER, IGP Seyssuel (r) gd.

Oratoire St Martin, Dom de l' S Rh ★★★ 16' 17' 18' 19' 20 21 At CAIRANNE, choice vyds, now owned by CH Mont-Redon of CHÂTEAUNEUF. Top-grade bio reds, much purity, Haut Coustias (vines c.70 yrs), Les Douyes (1905 GRENACHE/ MOURVÈDRE); gd food-friendly white (esp CLAIRETTE, Rés Seigneurs).

Orchidées Lo ★→★★★ Group name for Ackerman. Entry-level wines esp gd. CH de SANCERRE, Celliers du Prieuré, Donatien-Bahuaud, Drouet Frères, Hardières (COTEAUX DU LAYON), Monmousseau, Perruche, Rémy-Pannier, Varière (ANJOU).

Osmin, Lionel SW Fr Well-respected NÉGOCIANT, full range of SW wines, all styles.

Ostertag, Dom Al ★★★ Ace bio grower André has handed reins to son Arthur. Terroir-driven RIES Muenchberg 10 **14** 18', trend-setting barrique-fermented, rich Muenchberg PINOT GR 15. Excellent SYLVANER VIEILLES VIGNES 15 18' 19'.

Ou, Ch de l' Rouss ★★★→★★★★ Go-ahead organic estate. Beautifully crafted, all colours and styles. Secret de Schistes elegance and power. Rhapsody GRENACHE icon wine.

Overnoy, Maison Jura AOP ARBOIS-Pupillin. Founded by natural-wine pioneer Pierre O, run by Emmanuel Houillon. Otherworldly CHARD, SAVAGNIN and ★★ Ploussard (Poulsard). Several other Jura Overnoy estates are cousins.

Pabiot, Jonathan Lo ★★★→★★★★ Top level of Pouilly-Fumé bio DOM with lovely, precise, vibrant wines. High above Les Loges hamlet Try Aubaine, Florilège, Prédilection, Prélude, Utopia.

Pacherenc du Vic-Bilh SW Fr ★★→★★★ White AOP contiguous with MADIRAN. Gros/ Petit MANSENG, sometimes Petit Courbu and local Arufiiac produce dry and sweet styles. Made by most Madiran growers but note too ★★ CH de Mascaaras. Dry DYA, but sweet, esp if oaked, can be matured.

Paillard, Bruno Champ ★★★→★★★★ Youngest *grande marque* (1981). Top-quality BRUT Première CUVÉE NV, Rosé Première Cuvée; refined style, esp in slow-aging Prestige NPU 95 04 02'. Maybe most accomplished yet is 08: elegant, precise, classic. BP assemblage 12' a sculpted great. Bruno P heads LANSON-BCC group of mainly family houses; daughter Alice taking over at Paillard.

Palette Prov ★★★ Characterful reds, tiny AOP nr Aix. GRENACHE, MOURVÈDRE; fragrant rosés, intriguing forest-scented whites; also oddities like FURMINT. Traditional, serious *Ch Simone*, Crémade, Henri Bonnaud.

Partagé, Dom Sav ★★ Tiny bio DOM of Gilles Berlioz in CHIGNIN, v.gd. Altesse, Jacquère, ★★ MONDEUSE and range of fun ★★★ ROUSSANNE incl Les Christine, Les Filles, Les Fripons.

> **Super-natural in Jura**
> The Jura, with its plethora of styles, is a darling of the natural-wine
> world. Traditional oxidative (w) SAVAGNIN, as well as reductive local (r)
> Poulsard (also Ploussard) and Trousseau lend themselves to natural
> methods using indigenous yeasts and low sulphite additions. For a
> natural experience try OVERNOY Ploussard, Pignier Léandre (r), Tournelle
> Fleur de Savagnin. *See also* ARBOIS, CÔTES DU JURA.

Pataille Burg ★★→★★★ Wild-haired guru Sylvain P deservedly earned cult following for red (single-vyd MARSANNAY), white (esp site-specific ALIGOTÉ) and rosé (Fleur de PINOT), low sulphur and long ageing. Cutting-edge and consistent quality.

Pauillac H-Méd ★★★→★★★★ 00′ 05′ 09′ 10′ 15 16′ 18′ 19′ 20 Communal AOP in N MÉD with 18 Classed Growths, incl LAFITE, LATOUR, MOUTON. Famous for long-lived wines, the acme of CAB SAUV. Other top CHX: CLERC MILON, DUHART-MILON, GRAND-PUY-LACOSTE, LYNCH-BAGES, PICHON BARON, PICHON LALANDE, PONTET-CANET. Relatively gd-value: BATAILLEY, PÉDESCLAUX, PIBRAN.

Pays d'Oc, IGP L'doc ★→★★★ Largest IGP, covering whole of L'DOC-ROUSS. Extremes of quality from simple, quaffing varietals to innovative, exciting; 58 different grapes allowed. CARIGNAN, esp old vines, increasingly popular. Big players incl Bruno Andreu, DOM PAUL MAS, GÉRARD BERTRAND, Jeanjean and co-op Foncalieu.

PC (Premier Cru) First Growth in B'X; 2nd rank of vyds (after GC) in Burg; 2nd rank in Lo: one so far, COTEAUX DU LAYON Chaume.

Pécharmant SW Fr ★★ 17 18 19 20 BERGERAC inner AOP on edge of town. Iron and manganese in soil generate biggest, longest-living wines of area. Veteran ★★★ CH deTiregand, DOM du Haut-Pécharmant, l'Ancienne Cure; ★★ Chx Beauportail, Corbiac, du Rooy, Terre Vieille; Dom des Bertranoux.

Peira, La L'doc ★★★★ Superlative TERRASSES DU LARZAC DOM making intense and complex reds. La Peira (SYRAH/GRENACHE), Las Flors de la Peira (Syrah/Grenache/MOURVÈDRE), Obriers de la Peira (CARIGNAN/CINSAULT).

Pélican, Dom du Jura ★★★ VOLNAY's MARQUIS D'ANGERVILLE venture nr ARBOIS; bio vyds incl those from retired legend Jacques Puffeney – giving pristine range.

Pernand-Vergelesses C d'O ★★★ (r) 09′ 10′ 12 15′ 17 18′ 19′ 20′ (w) 14′ 15′ 17′ 18′ 19′ 20′ Village hosting w-facing part of CORTON-CHARLEMAGNE. Now less austere. Different vyds thrive in each colour: (r) Île des Vergelesses; (w) Combottes, Sous Frétille. Local DOMS CHANDON DE BRIAILLES, Dubreuil-Fontaine, Rapet, Rollin lead way.

Perret, André N Rh ★★★ 15′ 16′ 17 18′ 19′ 20′ 21 CONDRIEU DOM, 1st class. Three wines, classic, stylish, clear CLOS Chanson, rich, musky Chéry. Also ST-JOSEPH (r/w), bright-fruit classic red, deep, flowing Les Grisières (r). Also gd COLLINES RHODANIENNES (r/w).

Perrier, Joseph Champ ★★★→★★★★ Fine family-run CHAMP house with v.gd PINOTS N/M vyds, esp in own Cumières DOM. Ace Prestige CUVÉE Joséphine 12′; BRUT Royale NV as generous as ever but more precise with less dosage. Distinctive, tangy BLANC DE BLANCS 08 13′ 17′ 19 20. Now drier, finer CUVÉE Royale Brut NV; older BLANC DE BLANCS vintages age well esp 95. Owner Jean-Claude Fourmon, one of Champ's great characters, easing reins to son.

Perrier-Jouët Champ The 1st (in C19) to make dry CHAMP for UK market; strong in GC CHARD, best for gd vintage and de luxe Belle Epoque 12′ ★★★★ in painted bottle. BRUT NV; Blason de France NV Brut and Rosé; Belle Epoque Rosé 12′; new BLANC DE BLANCS. A gd team of winemakers now, change for better.

Pessac-Léognan B'x ★★★→★★★★ 05′ 09′ 10′ 15 16 18 19′ 20 AOP created in 1987 for best part of n GRAV, incl all Crus Classés (1959): DOM DE CHEVALIER, HAUT-BAILLY, HAUT-BRION, LA MISSION HAUT-BRION, PAPE CLÉMENT, SMITH HAUT LAFITTE, etc.

Aspiring unclassified: LES CARMES HAUT-BRION. Firm, full-bodied, earthy reds; B'X's finest dry whites. Value from Brown, de la Solitude, LA LOUVIÈRE.

Péters, Pierre Champ ★★★★ Tiptop Côte des Blancs estate. Les Chétillons probably longest-lived CHARD in CHAMP: 05 07 13 14 17'. 19' a cooler classic.

Petit Chablis Chab ★ DYA Thirst-quenching mini-CHAB from outlying vyds mostly not on Kimmeridgian clay. Best wines from BILLAUD, BROCARD, DAUVISSAT, Defaix, LAVANTUREUX, Pommier, RAVENEAU and co-op LA CHABLISIENNE.

Pézenas L'doc Charming medieval town, birthplace Molière. AOP L'DOC, diverse soils, fun to explore. Big gun PAUL MAS and smaller estates MAS GABRIEL. DOMS Allegria, La Croix Gratiot, Magellan, Prieuré St Jean de Bebian, Turner-Pageot, Villa Tempora.

Philipponnat Champ ★★→★★★★ Small house, intense, esp in pure Mareuil-sur-AŸ CUVÉE under careful oak 18. Now owned by LANSON-BCC group. NV, Rosé NV, BRUT, Cuvée 1522 04. Famous for majestic single-vyd *Clos des Goisses* 04, CHARD-led 08; exceptional late-disgorged vintage 09.

Picpoul de Pinet L'doc ★→★★ DYA AOP for PICPOUL around Pinet, darling of sommeliers as different but oh-so-likeable; vyds overlooking oyster farms by Med, wines have salty tang, lemony freshness. Perfect with seafood. Best not oaked. Co-ops Pinet, Pomerols do gd job, also DOMS des Lauriers, FÉLINES-JOURDAN, Font-Mars, La Croix Gratiot, Petit Roubié, St Martin de la Garrigue.

Pic St-Loup L'doc ★★★ AOP n of Montpellier with dramatic scenery; some high vyds. Cooler, more rain gives elegance to wines; 50% min SYRAH plus GRENACHE, MOURVÈDRE. Reds for ageing; white potential considerable but still AOP L'DOC or IGP Val de Montferrand. Growers: Bergerie du Capucin, Cazeneuve, CH PUECH-HAUT, CLOS de la Matane, Clos Marie, de Lancyre, DOM DE L'HORTUS, Gourdou, Lascaux, Mas Bruguière, Mas Peyrolle, Pegaline, Valflaunès. Tiny co-op Hommes et Terres du Sud v.gd.

Pieretti, Dom Cors Small estate in windswept Cap CORSE, n of Bastia. Lina Pieretti Venturi makes top rosé, also delicious red and sweet MUSCAT.

Pierre-Bise, Ch Lo ★★→★★★★ 17 18' 19' 20' 21 Fine DOM in COTEAUX DU LAYON, QUARTS DE CHAUME (OPULENT), SAVENNIÈRES (incl ROCHE-AUX-MOINES, concentrated and precise, 18 esp gd). ANJOU BL Haut de la Garde (rich), ANJOU-VILLAGES, GAMAY (concentrated). Inspirational Claude Papin now retired, son René in charge. Excellent, varied single-vyd sweets.

Pierre Fil, Dom L'doc ★★★ MINERVOIS; USP MOURVÈDRE with carbonic maceration. Top Dolium and CUVÉE M superb, reward ageing. Drink *garrigue*-scented Orebus while waiting.

Pillot Burg ★★→★★★★ Smart family in CHASSAGNE-MONTRACHET, all branches on form: F&L P (sound all round), Jean-Marc P (esp CLOS St-Marc) and Paul P (Grandes Ruchottes, LA ROMANÉE, etc.). Look out also for red Chassagne PCS from all three.

Pinon, François Lo ★★★ 17 18' 19' 20' 21 Age-worthy VOUVRAY. Fame made by François (RIP 2021), son Julien in charge; v.gd sparkling.

Piper-Heidsieck Champ ★★★ On surging wave of quality. Dynamic Brut Essentiel with more age, less sugar, floral yet vigorous; great with sushi, sashimi. Prestige Rare, now made as separate brand in-house, is a jewel, precise, pure, refined texture 08 12' 13 to keep, 18' 19' 20'. Exceptional Rare Rosé 12'.

Plageoles, Dom SW Fr Defenders and rebels guarding true GAILLAC style. Rare local grapes rediscovered incl Ondenc (base of ace sweet ★★★ Vin d'Autan), ★★ Prunelard (r, deep fruity), Verdanel (dr w, oak-aged) and countless sub-varieties of Mauzac. More reds from Braucol (FER SERVADOU), Duras.

Plan de Dieu S Rh ★→★★★ 16' 18 19' 20 A gd village nr CAIRANNE with stony, vast, windswept *garrigue* plain. Full-throttle, peppery, mainly GRENACHE wines; drink with game, casseroles. Broad choice. Best: CH la Courançonne, CLOS St Antonin, Le Plaisir; DOMS APHILLANTHES (bio, character), Arnesque, Durieu (full), Espigouette, Favards (organic), La Bastide St Vincent, Longue Toque, Martin (traditional), Pasquiers (organic), St-Pierre.

Pol Roger Champ ★★★★ Family-owned Épernay house. BRUT Rés NV excels, dosage lowered since 2012; Brut 04 06 08', lovely 09 12' 13 15; Rosé 09; BLANC DE BLANCS 09. Fine *Pure* (no dosage). Sumptuous CUVÉE Sir Winston Churchill 02 09 12; a blue-chip choice for judicious ageing, best value of prestige cuvées.

Pomerol B'x ★★★→★★★★ 05' 09' 10' 15 16' 18' 19' 20' Tiny, pricey AC; MERLOT-led, plummy to voluptuous styles, but long life. Chemical weedkiller banned: other AOPs please note. Top CHX on clay, gravel plateau: CLINET, HOSANNA, L'ÉGLISE-CLINET, L'ÉVANGILE, LA CONSEILLANTE, LAFLEUR, LA FLEUR-PÉTRUS, LE PIN, PETRUS, TROTANOY, *Vieux Ch Certan*. Occasional value (BOURGNEUF, CLOS du Clocher, LA POINTE).

Pommard C d'O ★★★→★★★★ 90' 96' 99' 03 05' 09' 10' 12 15' 16' 18' 19' 20' Stand by for revolution as Pommard thrives in recent warmer conditions. Best PC vyds: Epenots for grace, Rugiens for power. Try Noizons at village level. Top producers: BICHOT (DOM du Pavillon), CH de Pommard, Clerget, Commaraine (from 2019), COMTE ARMAND, DE COURCEL, DE MONTILLE, HOSPICES DE BEAUNE, JM BOILLOT, Launay-Horiot, Lejeune, Parent, Rebourgeon-Mure, Violot-Guillemard.

Pommery Champ ★★ Historic house with spectacular cellars; brand owned by Vranken. BRUT NV steady bet, no fireworks; Rosé NV; Brut 04 08 09 12'. Once outstanding CUVÉE Louise 02 04 less striking recently. Now in England.

Perrier-Jouët 1874 sold for £42,875 in 2021, one bottle. Check your cellar.

Ponsot, Dom C d'O ★★→★★★★ Idiosyncratic MOREY-ST-DENIS DOM. Rose-Marie P in charge since departure of Laurent P (2016). No significant changes in style. Key wines: *Clos de la Roche*, PC Monts Luisants (ALIGOTÉ).

Ponsot, Laurent C d'O ★★→★★★★ The man who made DOM PONSOT wines for 30 yrs left family business to create own haute-couture label nearby (2016). Kept sharecropping contracts, incl amazing CLOS ST-DENIS, GRIOTTE-CHAMBERTIN. Buying vines and developing white wines, in new purpose-built winery.

Pouilly-Fuissé Burg ★★→★★★ 14' 15 17 18 19' 20' Top AOP of MÂCON; potent, rounded but intense whites from around Fuissé, more mineral style in Vergisson. Enjoy young or with age. PC classification finally in place for 2020 vintage; hurrah! Top: Barraud, Bret, Carette, *Ch de Beauregard*, CH DE FUISSÉ, CH des Quarts, Ch des Rontets, Cornin, *Ferret*, Forest, Lassarat, *Merlin*, Paquet, Renaud, Robert-Denogent, Rollet, Saumaize, Saumaize-Michelin, VERGET.

Pouilly-Fumé Lo ★★→★★★★ 19' 20' 21 SANCERRE's e-bank neighbour, wines a little softer, more variable quality, SAUV BL; 21 fresher than recent yrs. Best: Bain, Belair, Bouchié-Chatellier, Bourgeois, Cailbourdin, Ch de Favray, CH de Tracy, Chatelain, *Didier Dagueneau* (VDF from 2017), E&A Figeat, Jean Pabiot, *Jonathan Pabiot*, Jolivet, Joseph Mellot, Ladoucette, Landrat-Guyollot, Masson-Blondelet, Redde, Saget, Serge Dagueneau & Filles, Tabordet, Treuillet. AOP Pouilly-sur-Loire (CHASSELAS) now historical curiosity.

Pouilly-Loché Burg ★★ 14' 15 17 18 19' 20' Least known of Mâconnais's Pouilly family. Try CLOS des Rocs. Also Bret Bros, Tripoz, local CAVE des GCS Blancs.

Pouilly-Vinzelles Burg ★★ 14' 15 17 18 19' 20' Close to POUILLY-FUISSÉ geographically and in quality. Outstanding vyd: Les Quarts. CH de V, DROUHIN, Soufrandière (Bret), Valette best. Volume from CAVE des GCS Blancs.

Premières Côtes de Bordeaux B'x ★→★★ 16 18 19 20 Same zone as CADILLAC-CÔTES DE B'X but for sweet whites only; SÉM-dominated *moelleux*. Generally early drinking. Best CHX: Crabitan-Bellevue, du Juge, Faugas, Marsan.

Prieur, Dom Jacques C d'O ★★★ Major MEURSAULT estate with range of underplayed GCS from MONTRACHET to MUSIGNY. Style aims at weight from late picking and oak more than finesse. New project Labruyère-Prieur in Burg. Owners Famille Labruyère also have CHAMP and MOULIN-À-VENT projects plus CH ROUGET (B'X).

Primeur "Early" wine for refreshment and uplift; esp from BEAUJ; VDP too. Wine sold en primeur is still in barrel, for delivery when bottled.

Producteurs Plaimont SW Fr France's most dynamic co-op, bestriding SAINT MONT, MADIRAN and CÔTES DE GASCOGNE like the colossus it is. Has abandoned B'X varieties for grapes traditional to SW, incl some pre-phylloxera discoveries. All colours, styles, mostly ★★, all tastes, purses. *See* ANDRÉ DUBOSC.

Propriétaire-récoltant Champ Owner-operator, literally owner-harvester.

Puech-Haut, Ch L'doc ★★★ AOP L'DOC St Drézéry. Powerful Prestige (r/w), Tête de Belier (r/w/rosé) swish wine, packaging. IGP Argali rosé gd value. Owns ★★★ PIC ST-LOUP CH Lavabre.

Puisseguin St-Émilion B'x ★★ 15 16 18 19 20 Most e of four ST-ÉM satellites; MERLOT-led; meaty but firm. Top CHX: Beauséjour, Branda, Clarisse, DES LAURETS, Durand-Laplagne, Fongaban, Guibot la Fourvieille, Haut-Bernat, l'Anglais, La Mauriane, Le Bernat, Soleil.

Puligny-Montrachet C d'O ★★★→★★★★ 09' 10' 12 14' 15 17' 18 19' 20' Floral, fine-boned, tingling white burg. Decent at village level, Enseignières vyd exceptional, outstanding PCS, esp Caillerets, Champ Canet, Combettes, Folatières, Pucelles, plus amazing MONTRACHET GCS. Producers: *Carillon*, Chartron, CH de Puligny, Chavy, *Dom Leflaive*, *Drouhin*, Ente, *J-M Boillot*, *O Leflaive*, Pernot family, *Sauzet*, Thomas-Collardot.

Pyrénées-Atlantiques SW Fr Mostly DYA. IGP in far sw for wines outside local AOPS. ★★★ CABIDOS in middle of nowhere (superb dr/sw PETIT MANSENG w varietals that will age), ★★ DOM Moncaut (JURANÇON in all but name nr Pau).

Quartironi de Sars, Ch L'doc ★★→★★★ Scented, mineral ST-CHINIAN reflects vyds on schist among wild herbs. Haut-Priou top red. Skhistos (w) from Carignan Bl is a revelation.

Quarts de Chaume Lo ★★★→★★★★ 10' 11' 14' 15' 16 17 18' 20' (21) Remarkable site in Layon, CHENIN BL. Best richly textured, nearly immortal, but hard to sell so more and more dry Chenin made. Top: Baudouin, *Belargus*, Bellerive, Branchereau, FL, Guegniard, *Ogereau*, *Pierre-Bise*, Plaisance, Suronde.

Quenard Sav Six separate Q estates around CHIGNIN incl ★★ stalwart A&M Q; ★★ improving J-F Q and ★★ small-but-lovely organic P&A Q. All offer range of top ROUSSANNE (Chignin-Bergeron), Jacquère and MONDEUSE.

Quincy Lo ★★ 20' 21 Revived AOP, SAUV BL on low-lying sand/gravel banks by Cher. Citrus, usually less complex than SANCERRE. Try Ballandors, Eric Louis, l'Epine, Portier, Rouzé, *Siret-Courtaud*, Tremblay, Villalin.

Rancio Rouss Describes complex, evolved aromas from extended, oxidative ageing. Reminiscent of Tawny Port, or old Oloroso Sherry. Associated specifically with BANYULS, MAURY, RASTEAU, RIVESALTES. Can be a grand experience; don't miss.

Rangen Al Most s GC of AL at Thann; v. warm, steep (average 90%) volcanic soils. Top: ZIND HUMBRECHT PINOT GR CLOS St Urbain 08' 10' 17' (world's greatest dry Pinot Gr), SCHOFFIT RIES St-Théobald 15' 17'.

Rasteau S Rh ★★ 16' 17' 18 19' 20 Mostly GRENACHE, mostly full-on, some suave, early reds, mainly clay soils. Best in hot yrs, consistent quality. Note Beaurenard (bio, serious, age well), CAVE Ortas/Rhonéa, CH La Gardine, Ch du Trignon, Famille Perrin; DOMS Beau Mistral, Collière (stylish), Combe Julière (punchy), Coteaux des Travers (bio), Didier Charavin, Elodie Balme (soft), Escaravailles (style), Girasols, Gourt de Mautens (quirky, talented, IGP from 2010), Gramiller (organic), Grand Nicolet (depth, character), Grange Blanche, Rabasse-Charavin (full), M Boutin, Soumade (polished), **St Gayan**, Trapadis (bio). Grenache dessert VDN: quality on the up (Doms Banquettes, Combe Julière, Coteaux des Travers, Escaravailles, Trapadis). Rasteau doms also gd source CÔTES DU RH (r).

Raveneau Chab ★★★★ Topmost CHAB producers, using classic methods for *extraordinary long-lived wines*. A little more modern while still growing in stature of late. Excellent value (except in secondary market). Look for PC Butteaux, Chapelot, Vaillons and GC Blanchots, Les CLOS.

Rayas, Ch S Rh ★★★★ 05' 06' 07' 09' 10' 11' 15' 16' 17' 19' 20' Outstanding, lost-in-time, one-off CHÂTEAUNEUF estate, tiny yields, sandy soils, tree-sheltered plots. Sensuous reds (100% GRENACHE) age superbly. Save up! White Rayas (CLAIRETTE, GRENACHE BL) v.gd over 20 yrs+. Stylish second label Pignan. Supreme CH Fonsalette CÔTES DU RH, incl marvellous long-lived SYRAH. Decant all; each an occasion. No 18 (mildew). Ch des Tours VACQUEYRAS, gd VDP.

Regnié Beauj ★★ 18' 19 20' Most recent BEAUJ cru, lighter wines on sandy soil, meatier nr MORGON. Starting to get some gd growers now. Try Burgaud, Chemarin, de la Plaigne, Dupré, Rochette, Sunier (A), Sunier (J).

Reuilly Lo ★★→★★★ 20' 21 AOP w of QUINCY, similar SAUV BL. Delicate rosés, *Vin Gr* PINOT N and/or *Pinot Gr*, gd PINOT N reds can be comparable to SANCERRE. Try *Claude Lafond*, Cordaillat, *Jamain* (Bio), Pagerie, Renaudat, Rouzé, Sorbe.

Riceys, Les Champ DYA Key AOP in AUBE for notable PINOT N rosé. Producers: *A Bonnet*, BRICE, Jacques Defrance, Morize. Great 09, ace 15'.

Richebourg C d'O ★★★★ 90' 93' 96' 99' 02' 05' 09' 10' 12' 15' 16 17 18' 19' 20' VOSNE-ROMANÉE GC. Supreme burg with great depth of flavour; vastly expensive. Growers: DRC, GRIVOT, GROS, HUDELOT-NOËLLAT, LEROY, LIGER-BELAIR (T), MÉO-CAMUZET.

Rion C d'O ★★ →★★★ Related DOMS in NUITS-ST-GEORGES, VOSNE-ROMANÉE. Patrice R for excellent Nuits CLOS St-Marc, Clos des Argillières and CHAMBOLLE-MUSIGNY. Daniel R for Nuits and Vosne PCS, now being split between family members; A&B R Vosne-based. All fairly priced.

Rivesaltes Rouss ★★→★★★★ Underappreciated VDN with styles: Ambré, RANCIO/ Hors d'Age, Rosé, Tuilé. MUSCAT de Rivesaltes AOP fragrant, youthful. Look for Boucabeille, des Chênes, des Schistes, DOM CAZES, Puig-Parahy, Rancy, Roc des Anges, Sarda-Malet, Valmy, Vaquer. You won't be disappointed.

Rives-Blanques, Ch L'doc ★★★ Irish-Dutch couple with son Jean Ailbe make LIMOUX white (and rosé), incl unusual 100% MAUZAC, Occitania, blend Trilogie and age-worthy CHENIN BL Dédicace. BLANQUETTE and CRÉMANT too.

Roederer, Louis Champ ★★★★ Peerless family-owned house. Enviable vyds, largest bio DOM. BRUT Premier NV phased out in favour of MV Collection 242 in fight for freshness. Food-friendly BLANC DE BLANCS 13, superb Cristal Vinothèque Bl and Rosé 95; Brut NATURE Philippe Starck (all Cumières 09 12 18' 20'). Also owns CH PICHON-LALANDE, DEUTZ. See also California.

Rolland, Michel B'x Veteran French consultant winemaker and MERLOT specialist (B'x and worldwide). Owner of FONTENIL in FRON. See also Mariflor, Argentina.

Rolly Gassmann Al ★★★ Revered DOM, esp Moenchreben vyd. Almost endless range, mostly off-dry, rich, sinuous; maybe too sweet for some. Now into bio, more finesse. Mineral zesty RIES, rich SYLVANER 13 16 17' 18 19'.

Romain Portier L'doc ★★★ Ex-wine merchant making expressive, finely structured bio wines from myriad varieties in TERRASSES DE LARZAC.

Romanée, La C d'O ★★★★ 09' 10' 12' 15' 16' 17 18' 19' 20. Tiniest GC in VOSNE-ROMANÉE, MONOPOLE of COMTE LIGER-BELAIR. Exceptionally fine, perfumed, intense: now on peak form and understandably expensive.

Romanée-Conti, La C d'O ★★★★ 90' 93' 96' 99' 00 02' 05' 09' 10' 12' 14' 15' 16' 17 18' 19' 20' A GC in VOSNE-ROMANÉE, MONOPOLE of DRC. Most celebrated GC in Burg, a legend that delivers given patience (10–20 yrs). But beware geeks bringing fake gifts.

Romanée-St-Vivant C d'O ★★★★ 90' 99' 02' 05' 09' 10' 12' 15' 16' **17** 18' 19' 20' A GC in VOSNE-ROMANÉE, downslope from LA ROMANÉE-CONTI. Haunting perfume, delicate but intense. Ready a little earlier than famous neighbours. Growers: if you can't afford ARNOUX-LACHAUX, DRC or LEROY, or indeed CATHIARD or HUDELOT-NOËLLAT now, try ARLOT, Follin-Arbelet, JJ Confuron, LATOUR, Poisot. Nobody letting side down.

Romanin, Ch Prov ★★★ Leading bio DOM in BAUX-EN-PROVENCE, worth visiting for magnificent cathedral cellar. *Grand vin* (r/w), equally majestic. Also IGP Alpilles (r/w/rosé).

Roquemale, Dom de L'doc Grés de Montpellier. Cooling sea breezes give lift and elegance to ripe L'DOC fruit. Méli Mélo v.gd value IGP ALICANTE BOUSCHET. Mâle top red, brooding, peppery joy.

Rosacker Al A GC at Hunawihr. Limestone/clay makes some of longest-lived RIES in AL (CLOS STE-HUNE).

Rosé d'Anjou Lo ★→★★ DYA Lots of off-dry to sweet rosé, mainly Grolleau. Increasingly well made. Try Ackerman, Bougrier, CH de Brissac, Clau de Nell, Grandes VIGNES, Mark Angeli (VDF).

Rosé de Loire Lo ★→★★ Dry rosé: a big, baggy AOP. Grapes: Grolleau Gr/N, CAB FR, CAB SAUV, GAMAY; PINOT N. Try Bablut, Beaujardin, Bois Brinçon, Branchereau, CADY, Fontaines, Passavant, Robert et Marcel.

Grenache: naturally high alc, not naturally massive. Can be 14.5% and delicate

Rosette SW Fr ★★ Tiny AOP DYA. Birthplace of BERGERAC, now reviving traditional off-dry apéritif whites. Also gd with foie gras or mushrooms. Avoid oaked versions that deny the style. CLOS Romain; CHX Combrillac, de Peyrel, Monplaisir, Puypezat-Rosette, Spingulèbre; DOMS de Coutancie, de la Cardinolle, du Grand-Jaure.

Rossignol-Trapet C d'O ★★★ Equally bio cousins of DOM TRAPET, with healthy holdings of GC vyds, esp CHAMBERTIN; gd value across range from GEVREY VIEILLES VIGNES up. Also some BEAUNE vyds from Rossignol side. Wonderful 20 vintage.

Rostaing, Dom N Rh ★★★ 01' 05' 09' 10' 12' 13' 15' 16' 17' 18' 19' 20' High-level CÔTE-RÔTIE DOM: five tightly knit wines from top sites, all v. fine, pure, clear, low-key oak, wait 6 yrs+, decant. Complex, enticing, top-class Côte Blonde (5% VIOGNIER), Côte Brune (iron), also La Landonne (dark fruits, 20–25 yrs). Mineral, firm Condrieu, also IGP COLLINES RHODANIENNES (r/w), L'DOC Puech Noble (r/w).

Rouget, Dom C d'O ★★★★ Renamed as DOM R rather than Emmanuel R with new generation refreshing dom famed for Henri Jayer connection and CROS PARANTOUX vyd. Late picking means turbocharged wines in recent vintages.

Roulot, Dom C d'O ★★★→★★★★ Jean-Marc R leads outstanding MEURSAULT DOM, now cult status so beware secondary-market prices. Sign on chai, in gold: "I pick when I want". Great PCS, esp CLOS des Bouchères, Perrières; value from top village sites Luchets, Meix Chavaux, esp Clos du Haut Tesson.

Roumier, Georges C d'O ★★★★ Reference DOM for BONNES-MARES and other *brilliant Chambolle* wines (incl Amoureuses, Cras) from Christophe R. Long-lived wines but still attractive early. Cult status means now hard to find at sensible prices. Best value is MOREY CLOS de la Bussière.

Rouquette-sur-Mer, Ch L'doc ★★★★ Impressive LA CLAPE estate, vyds and *garrigue* right by Med. *Arpège* (w), crisp, herby, terrific value. CLOS de la Tour oaky MOURVÈDRE/SYRAH, also l'Absolu, both excellent.

Rousseau, Dom Armand C d'O ★★★★ Cyrielle R continuing family tradition with father Eric. Unmatchable GEVREY-CHAMBERTIN DOM: balanced, fragrant, refined, age-worthy wines from village to GC, plus legendary PC CLOS ST-JACQUES.

Roussette de Savoie Regional AOP, same area as AOP SAV; 100% ALTESSE. Age 5 yrs+. Try Curtet, ★ Dupasquier, Lupin, ★★ *St-Germain*, ★★★ Prieuré St-Christophe.

Roussillon Often linked with L'DOC, and incl in AOP L'doc but has distinct identity. Exciting region with different soils and topography, innovative producers. Lots of old vines. GRENACHE key variety. Largest AOP CÔTES DU ROUSS, gd-value spicy reds. Original, sometimes stunning, traditional VDN (eg. BANYULS, MAURY, RIVESALTES). Also serious age-worthy table wines (r/w). *See* COLLIOURE, CÔTES CATALANES, CÔTES DU ROUSS-VILLAGES, MAURY (Sec).

Ruchottes-Chambertin C d'O ★★★★ 99' 02' 05' 09' 10' 12' 15' 16 17' 18' 19' 20' Tiny GC neighbour of CHAMBERTIN. Less weighty but ethereal, intricate, lasting, great finesse. Top: MUGNERET-Gibourg, ROUMIER, ROUSSEAU. Also CH de MARSANNAY, F Esmonin, H Magnien, Marchand-Grillot, Pacalet. Plus Lambrays from 2021.

Ruinart Champ ★★★★ Oldest sparkling CHAMP house (1729). High standards going higher still. Rich, elegant. R de Ruinart BRUT NV; Ruinart Rosé NV; R de Ruinart Brut 08. Prestige cuvée *Dom Ruinart* one of two best Vintage BLANC DE BLANCS in Champ (90' esp in magnum, 02 04' 07'). DR Rosé also v. special 06'. NV Blanc de Blancs much improved. High hopes for 13', classic cool lateish vintage. Winemaker Fred Panaïotis top of his game.

Rully Burg ★★ (r) 15' 17 18' 19' 20' (w) 17' 18 19' 20' CÔTE CHALONNAISE village. *Light, fresh, tasty, gd-value whites*. Reds all about fruit, not structure. Best vyds: Grésigny, Pucelle, Rabourcé. Try Briday, Champs l'Abbaye, Devevey, *de Villaine*, DOM de la Folie, DROUHIN, *Dureuil-Janthial*, FAIVELEY, Jacqueson, Jaeger-Defaix, Jobard (C), Leflaive (O).

Sablet S Rh ★★ 19' 20 CÔTES DU RH-VILLAGE on mainly sandy soils nr GIGONDAS. Easy GRENACHE-led wines, neat red-berry reds, some deeper, esp CAVE CO-OP Gravillas, CH Cohola (organic), du Trignon; DOMS de Boissan (organic, full), Les Goubert (r/w), Pasquiers (organic, full), Piaugier (r/w). Full whites gd, for apéritifs, food: Boissan, ST GAYAN.

St-Amour Beauj ★★ 18' 19' 20' Most n BEAUJ cru: mixed soils, so variable character, but signs of revival. Try Cheveau, DOM de Fa, Patissier, *Pirolette*, Revillon.

St-Aubin C d'O ★★★ (w) 14' 15 17' 18 19' 20' Fine source for *lively, refreshing whites*. PC St A often both cheaper and better than village CHASSAGNE. Also pretty reds mostly for early drinking. Best vyds: Chatenière, *En Remilly*, Murgers Dents de Chien. Best growers: BACHELET (JC), COLIN (Joseph, Marc), COLIN-MOREY, *Lamy*, Larue. Value Prudhon.

St-Bris Burg ★ DYA. Unique AOP for SAUV BL in N Burg. Fresh, lively, but also worth keeping from GOISOT or de Moor. Try also Bersan, Davenne, Felix, Simonnet-Febvre, Verret.

St-Chinian L'doc ★→★★★ Large hilly area nr Béziers with schist in nw, clay and limestone in se. Sound reputation. Incl CRUS Berlou (mostly CARIGNAN), Roquebrun

---

Fortifying food matches

BANYULS, MAURY, RIVESALTES are magnificent and underappreciated. They're sweet, fortified VDNs in myriad styles, and sublime food matches. Try fresh, white MUSCAT (AOP Rivesaltes) with goats' cheese, apple tart. Other white GRENACHE BL/Gr, MACCABEO with pâté, apricot sorbet. (Not together.) Ambré is aged, oxidative. RANCIO extended ageing, complex, nutty; v.gd with hard cheese or blue. Reds based on GRENACHE – fruity, youthful Grenat or Banyuls Rimage, in contrast to aged Tuilé – all heaven with chocolate. Best from CH DE L'OU, MAS AMIEL, Parcé Frères; DOMS CAZES, Comelade, Gardiès, F Jaubert, Madeloc.

(mostly SYRAH) on schist. Warm, spicy reds, based on Syrah, GRENACHE, Carignan, MOURVÈDRE. Whites from ROUSSANNE, MARSANNE, VERMENTINO, GRENACHE BL. Roquebrun co-op gd; CH CASTIGNO; DOMS Borie la Vitarèle, des Jougla, La Dournie, La Madura, Les Eminades, Milhau-Lacuge, Navarre, LA LINQUIÈRE, Rimbert; CLOS Bagatelle, Mas Champart, Mas de Cynanques, QUARTIRONI DE SARS, Viranel.

St Cosme, Ch de S Rh ★★ 09′ 10′ 11′ 12′ 13′ 14′. 15′ 16′ 17′ 18′ 19′ 20′ 21 At GIGONDAS, high-quality bio estate; wine with flair, drive, oak. Plot-specific CÔTES DU RH Les Deux Albion (r), Gigondas Le Poste. Owner CH de Rouanne, VINSOBRES (2018). N Rh merchant range gd, esp CONDRIEU, CÔTE-RÔTIE.

Ste-Anne, Dom S Rh ★★ Estate at St-Gervais, old-vines, high-quality, robust reds incl Rouvières (mainly MOURVÈDRE), Mourillons (mainly SYRAH), live 15 yrs. VDF from 1977 VIOGNIER, gd, more subtle than most.

Ste-Croix-du-Mont B′x ★★ 15 16 18 19 20 AOP for sweet, white *liquoreux*. Soils consist of fossilised oysters. Best: rich, creamy, can age. Top CHX: Crabitan-Bellevue, des Arroucats, du Mont, La Caussade, La Rame, *Loubens*.

Ste Eulalie, Ch L'doc ★★★ Consistently fine MINERVOIS DOM in pretty valley above La Livinière. Not showy but with light touch. Top Cantilène v.gd. Printemps d'Eulalie gd value.

St-Émilion B′x ★★→★★★★ 05′ 09′ 10′ 15′ 16 18 19′ 20′ Big MERLOT-led district on B′x's Right Bank; CAB FR also strong. AOPS St-Ém and (lots of) St-Ém GC (geographically same area). Environmental certificate obligatory from 2023. Top designation St-Ém PREMIER GRAND CRU CLASSÉ. Warm, full, rounded style but much diversity due to terroir, winemaking and blend. Best firm, v. long-lived. Top CHX: ANGÉLUS, AUSONE, CANON, CHEVAL BLANC, FIGEAC, PAVIE.

St-Estèphe H-Méd ★★→★★★★ 00′ 05′ 09′ 10′ 15 16′ 18 19′ 20 Most n communal AOP in MÉD. Varied terroir (gravel, limestone); solid, structured wines for ageing. Five Classed Growths: CALON SÉGUR, COS D'ESTOURNEL, COS LABORY, LAFON-ROCHET, MONTROSE. Top unclassified estates: CAPBERN, DE PEZ, HAUT-MARBUZET, LE BOSCQ, LE CROCK, LILIAN LADOUYS, MEYNEY, ORMES-DE-PEZ, PHÉLAN SÉGUR.

Ste-Victoire Prov ★★★ Subzone of CÔTES DE PROV. Limestone slopes of Montagne Ste-Victoire: much-needed freshness in hotter vintages. DOMS de St Ser, Gassier, Mas de Cadenas, Mathilde CHAPOUTIER, St Pancrace. IGP Dom Richeaume.

St-Gall Champ Brand of Union-CHAMP, top co-op at AVIZE. BRUT NV; Extra Brut NV; Brut BLANC DE BLANCS NV; Brut Rosé NV; Brut Blanc de Blancs 08; CUVÉE Orpale Blanc de Blancs 09 13′ a keeper too. Fine-value PINOT-led *Pierre Vaudon NV*. Makes top base wines for some great Reims houses.

Saint Gayan, Dom S Rh ★★★ 01′ 05′ 06′ 07′ 10′ 15′ 16′ 17′ 18′ 19′ 20′ 21 Top GIGONDAS address, consistent, long-lived, little oak; 80% GRENACHE Origine great value. Full, v.gd RASTEAU Ilex (r), charming SABLET L'Oratory (w).

St-Georges-St-Émilion B′x ★★ 15 16 18 19 20 Smallest ST-ÉM satellite. Sturdy, structured. Best CHX: Calon, Cap St-Georges, CLOS Albertus, Macquin-St-Georges, St-André Corbin, St-Georges, Tour du Pas-St-Georges.

St-Joseph N Rh ★★→★★★ 09′ 10′ 12′ 15′ 16′ 17′ 18′ 19′ 20′ Mainly granite vyds, 64 km (40 miles) n–s, some high, along w bank of N Rh. SYRAH reds. Best, oldest vyds nr Tournon: stylish, red-fruited wines; further n nr Chavanay darker, peppery, younger oak. More complete, interesting than CROZES-HERMITAGE, esp CHAPOUTIER (Les Granits), *Gonon* (top), Gripa, GUIGAL (VIGNES de l'Hospice), **J-L Chave**. Plus Alexandrins, Amphores (bio), A PERRET (Grisières), Boissonnet, Chèze, Courbis (modern), Coursodon (racy, modern), DELAS, E Darnaud, Faury, FERRATON, F Villard, Gaillard, J Cécillon, J&E Durand, Monier-Perréol (bio), P-J Villa, P Marthouret, S Blachon, Vallet, Vins de Vienne, Y CUILLERON. Food-friendly *white (mainly Marsanne)*, esp A PERRET, Barge, *Chapoutier* (Les Granits), Curtat, DOM Faury, Gonon, Gouye, *Gripa*, GUIGAL, J Pilon, Vallet, Y Cuilleron.

St-Julien H-Méd ★★★→★★★★ 00 05' 09' 10' 15 16' 18 19' 20 Epitome of harmonious, fragrant, savoury red; v. consistent mid-MÉD communal AC; 11 classified (1855) estates own most of vyd area; incl BEYCHEVELLE, DUCRU-BEAUCAILLOU, GRUAUD-LAROSE, LANGOA BARTON, LAGRANGE, LÉOVILLES (x3), TALBOT.

Saint Mont SW Fr ★★ (r) 16 17 18 19 20 AOP from Gascon heartlands. Similar to Madiran but often softer, less intense. PRODUCTEURS PLAIMONT'S ANDRÉ DUBOSC largely responsible for creating this AOP, PP makes most of the wine. Try ★★★ CH de Sabazan. White is dry, PACHERENC-like, less intense and terrific value.

St Nicolas de Bourgueil Lo ★→★★★ 17' 18' 19' 20' 21 Contiguous with BOURGUEIL with similar soils. St N can be lighter than Bourgueil, but differences are slight. CAB FR. Try David, Delanoue, *Frédéric Mabileau* (now son Rémy), Jamet, Laurent Mabileau, Mabileau-Rezé, Ménard, Mortier, Taluau-Foltzenlogel, Vallée, *Yannick Amirault*, Xavier Amirault.

St-Péray N Rh ★★ 19' 20' 21' Elegant white (MARSANNE/ROUSSANNE) from hilly granite, some lime vyds opposite Valence, lots of rapid new planting. Once *famous for fizz;* classic-method bubbles well worth trying (A Voge, J-L Thiers, R Nodin, TAIN co-op). Still white should be fine-tuned, flinty. Best: CHAPOUTIER, Clape (pure), Colombo (stylish), CUILLERON, *du Tunnel* (v. elegant), Gripa (best, esp mainly ROUSSANNE Figuiers), J&E Durand, J-L Thiers, J Michel, L Fayolle, R Nodin, TAIN co-op, Vins de Vienne, Voge (oak).

St-Pourçain Lo, Mass C ★→★★ 19' 20' 21 AOP Refreshing, gd red and rosé (GAMAY, PINOT N); 100% Pinot N banned. White from local Tressalier and/or CHARD, SAUV BL Try *Bérioles* (complex, age-worthy), Bellevue, CLOS de Breuilly, Grosbot-Barbara, Nebout, Pétillat, Ray, Terres d'Ocre, Terres de Roa, VIGNERONS de St-Pourçain (La Ficelle brand).

St-Romain C d'O ★★ (w) 15 17' 18' 19' 20' *Crisp whites* from side valley of CÔTE DE BEAUNE. Excellent value by Burg standards. Best vyds: Combe Bazin, Sous la Roche, Sous le CH. Specialists Alain Gras, de Chassorney, outstanding H&G Buisson; most NÉGOCIANTS have a gd one. Some fresh reds too. Watch this space.

St-Véran Burg ★★ 17 18' 19 20' AOP in s, either side of POUILLY-FUISSÉ. Best sites in Davayé. Try CH de Beauregard, Chagnoleau, Corsin, Deux Roches, Litaud, Merlin; gd-value DUBOEUF, Poncetys, Terres Secretes co-op.

Salon Champ ★★★★ Original BLANC DE BLANCS, from LE MESNIL in Côte des Blancs. Tiny quantities. Awesome reputation for long-lived luxury-priced wines: in truth, inconsistent. On song recently, viz 83' 90 97', but 99 disappoints. *See also* DELAMOTTE. Both owned by LAURENT-PERRIER. Watch future developments after new chef de CAVE left.

Sancerre Lo ★→★★★★ 18 19' 20' 21 Reference for SAUV BL; PINOT N can be v.gd. Range: youthful, citrus easy-drinking to complex wines to age. Top producers meticulous, incl *A Mellot*, *Boulay*, *Bourgeois*, Cotat, C Riffault, Dezat, D Roger, Fleuriet, Fouassier, *François Crochet*, Jean-Max Roger, Jolivet, J Mellot, L Crochet, Natter, Neveu, P&N Reverdy, Paul Prieur, Pierre Martin, *Pinard*, Raimbault, Roblin, Thomas, Thomas Laballe, Vacheron, Vatan, Vattan, *V Delaporte*.

Sang des Cailloux, Dom Le S Rh ★★★ 10' 13' 15' 16' 17' 18' 19' 20' 21 Best VACQUEYRAS DOM, bio, *garrigue* thrust, heart-on-sleeve mainly GRENACHE reds. Classic red rotates name every 3 yrs, Azalaïs (18), Floureto (19), Doucinello (20). Top Lopy red ages well. Solid Un Sang Bl (w).

Santenay C d'O ★★→★★★ 09' 12 15' 16 17 18' 19' 20' The s end of COTE DE BEAUNE, fine reds. Best vyds: CLOS de Tavannes, Clos Rousseau, Gravières (r/*w*). Some gd whites too, eg. Charmes. Local producers: Bachey-Legros, CH Philippe le Hardi, CHEVROT, Girardin (Justin), Jessiaume, MOREAU, Muzard, *Vincent*. Try also Giroud, JADOT (incl DOM Prieur-Brunet), LAMY.

Sarrat de Goundy L'doc ★★→★★★ Innovative Olivier Calix makes v.gd-value

LA CLAPE du Planteur (r/w/rosé). CUVÉE Sans Titre (r) scented NIELLUCCIO; (w) 100% BOURBOULENC, floral, mineral.

Saumur Lo ★→★★★★ 17 18' 19' 20' 21 Whites, light to age-worthy; often easy reds except SAUMUR-CHAMPIGNY; Saumur Rosé (Cabs); Lo sparkling centre. Saumur-Le-Puy-Notre-Dame AOP for CAB FR (mainly). Also Coteaux du Saumur: late-harvest CHENIN BL. Try *Antoine Foucault*, Arnaud Lambert, BOUVET-LADUBAY, CH de Brézé, CLOS Mélaric, *Clos Rougeard*, Ditterie, Guiberteau, Nerleux, Paleine, Parnay, Robert et Marcel (co-op), Rocheville, Targé, *Villeneuve*, Yvonne.

Saumur-Champigny Lo ★★→★★★★ 17' 18' 19' 20' 21 Can be top CAB FR from nine-commune AC, usually more complex than SAUMUR, gd vintages 15–20 yrs+. Best: *Antoine Sanzay*, Arnaud Lambert, *Bonnelière (value)*, Bruno Dubois, Champs Fleuris, CLOS Cristal (now co-op-run), CLOS ROUGEARD, Cune, Ditterie, Filliatreau, Hureau, Nerleux, Petit St-Vincent, Robert et Marcel (co-op), Roches Neuves, Rocheville, St-Vincent, Seigneurie, Targé, Vadé, Val Brun, *Villeneuve*, Yvonne.

Saussignac SW Fr ★★ 16' 17 18' 19 (20) BERGERAC sub-AOP, adjoining MONBAZILLAC, producing similar sweet wines perhaps with shade more acidity. Best: ★★★ DOMS de Richard, La Maurigne, Les Miaudoux, Lestevénie; ★★ CHX Le Chabrier, Le Payral, Le Tap.

Sauternes B'x ★★→★★★★★ 09' 11' 13 14 15' 16' 18 19 20 AOP making France's best *liquoreux* from "noble rotted" grapes. Luscious, golden and age-worthy. Tiny volume 21, frost. Top classified (1855) CHX: D'YQUEM, GUIRAUD, *Lafaurie-Peyraguey*, LA TOUR BLANCHE, RIEUSSEC, SUDUIRAUT. Exceptional unclassified: *Fargues, Gilette, Raymond-Lafon*. Value from HAUT-BERGERON, LAVILLE.

Sav's red Persan grape has bounced back: 3 ha in 90s, now 20 ha+.

Sauzet, Etienne C d'O ★★★ Leading DOM in PULIGNY with superb range of PCS (Combettes, Champ Canet best) and GC BÂTARD-M. Concentrated, lively wines, certified bio, once again capable of ageing.

Savennières Lo ★★→★★★★ 16 18' 19' 20' (21) AOP for v. long-lived dry whites (CHENIN BL) with marked acidity; a few DEMI-SEC. Baudouin, BAUMARD, *Belargus*, *Bergerie*, Boudignon, Closel, DOM FL, Epiré, *Laureau*, Mahé, Mathieu-Tijou, Morgat, Ogereau, PIERRE-BISE, Soucherie. Seriously frost prone: wiped out 17, badly hit 21.

Savennières Roche-aux-Moines Lo ★★★→★★★★ 18' 19 20' (21) SAVENNIÈRES cru with stricter rules. Top CHENIN BL usually a step up from Savennières. Best: aux Moines, *Dom FL*, Forges, *Laureau*, PIERRE-BISE.

Savigny-lès-Beaune C d'O ★★★ 05' 09' 10' 15' 18' 19' 20' Important village next to BEAUNE; similar mid-weight wines, savoury touch (but can be rustic). Top vyds: Dominode, Guettes, Lavières, Vergelesses. Local growers incl A Guyon, *Bize*, Camus-Bruchon, *Chandon de Briailles*, Chenu, Girard, Guillemot (w), Pavelot, Rapet, *Tollot-Beaut*. Exceptional CUVÉES from CLAIR, DROUHIN, JP Guyon, LEROY.

Savoie Alpine wines, two-thirds white. AOP incl 20 crus, incl APRÉMONT, ARBIN, Ayze, Chautagne, CHIGNIN, Crépy, Jongieux. Regional AOPs: CRÉMANT de Sav, ROUSSETTE DE SAV (Altesse), SEYSSEL; 25 grapes (r) mainly GAMAY, MONDEUSE, Persan, PINOT N; (w) Altesse, CHARD, CHASSELAS, Gringet, Jacquère, ROUSSANNE. Try organic stars ★ Baraterie, ★ A Berlioz, Chevillard, Côtes Rousses, *St-Germain* or safe bets E Carrel, Perrier, ★ P Grisard, Viallet.

Schlossberg Al A GC at Kientzheim famed since C15. Glorious compelling RIES from WEINBACH 10 and new TRIMBACH; 15 great Ries yr here.

Schlumberger, Doms Al ★★★→★★★★ Vast, top-quality S AL DOM owning c.1% of all Al vyds; GCS Kitterlé, Saering RIES, superb 08 17' 18. But GEWURZ best by far, esp VT CUVÉE Christine (unforgettable 76') and wondrous SGN Cuvée Anne (masterpiece 89').

Schoenenbourg Al Riquewihr GC famous since forever for RIES (Voltaire owned vines here); v. fine VT, SGN DOPFF AU MOULIN. HUGEL Schoelhammer from here.

Schoffit, Dom Al ★★★★ Exceptional Colmar grower, rich, opulent (dr/sw). Contrast RIES RANGEN CLOS St-Théobald 10' 17' 18' 20' on volcanic soil and RIES GC Sonnenberg 13 15 16 17' on granite. Harth CHASSELAS VIEILLES VIGNES Al best. Superb GEWURZ, PINOT GR VT/SGN GC.

Sec Literally means dry, though CHAMP so called is medium-sweet (and can be welcome at breakfast, teatime, weddings).

Séguret S Rh ★★ 19' 20 Picturesque hillside village nr GIGONDAS in Rh-Villages top three; vyds on both warm plain and cool heights. Mainly GRENACHE, peppery, quite deep reds; well-fruited table whites. Try CH la Courançonne (gd w), DOMS Amandine, Crève Coeur (bio), de Cabasse (stylish), de l'Amauve (fine, gd w), Fontaine des Fées (organic), Garancière, J David (organic), Maison Plantevin (organic), Malmont (fine), Mourchon (depth), Pourra (intense), Soleil Romain.

Selosse, Anselme Champ ★★★★ Leading grower, an icon for many. Vinous, oxidative style, oak-fermented. Son Guillaume adding finesse: Version Originale still vibrant after 7 yrs on lees. Top probably MESNIL Les Carelles: saline, complex, akin to MEURSAULT Perrières with bubbles. 16' 19' a cool classic, 02 a baby, worth waiting for.

Sérafin, Dom C d'O ★★★ Now niece Frédérique in charge, continuing Christian S recipe: deep colour, intense flavours, new wood. Look for back vintages as need age. Try Cazetiers, CHARMES-CHAMBERTIN, GEVREY-CHAMBERTIN VIEILLES VIGNES.

Sérol, Dom Lo ★★★ Leading bio family DOM in CÔTE ROANNAISE (100% GAMAY). Series of characterful, v.gd single vyds with potential to age. Also well-made CHENIN BL, VIOGNIER. Long-standing partnership with Troisgros.

Seyssel Sav Small regional AOP. Light white, sparkling. Grapes: Altesse, Molette. Try ★ Lambert de Seyssel (esp Royal Seyssel), Mollex.

SGN (Sélection des Grains Nobles) Al Term coined by HUGEL for AL equivalent to German Beerenauslese. Today mostly made with "noble rot" affected grapes, not just air-dried v. sweet grapes that meet legal SGN requirement.

Sichel & Co B'x Notable B'x merchant est in 1883 (Sirius a top brand). Family-run, 7th generation at helm. Interests in CHX ANGLUDET, Argadens, Daviaud and PALMER.

Simone, Ch Prov ★★★ Historic estate outside Aix, where Churchill painted Mont STE-VICTOIRE. Rougier family for c.200 yrs. Virtually synonymous with AOP PALETTE; n-facing slopes on limestone with clay, gravel give freshness. Many vines over 100 yrs old. Seek age-worthy whites; characterful rosé, elegant reds from GRENACHE, MOURVÈDRE, with rare grape varieties Castet, Manosquin (r).

Sipp, Louis Al ★★→★★★ Trades in big volumes of young wines, but also two GCS: fine RIES GC Kirchberg 13 16; luscious GEWURZ GC Osterberg VT 09 15 16 18 19.

Sipp-Mack Al ★★→★★★ Large range of lively, easy-drinking, dry, mineral wines, though not last word in concentration. RIES GC ROSACKER, expansive PINOT GR.

Sorg, Bruno Al ★★★ Small grower at Eguisheim, GCS Florimont (RIES 13 14 16' great 17') and Pfersigberg (MUSCAT 18'). Immaculate eco-friendly vyds.

Sorrel, Dom N Rh ★★★ Rock-steady, quality, small HERMITAGE grower; old vines; also CROZES-HERMITAGE (r/w). Traditional style, depth, character. Top: HERMITAGE Le Gréal (r, v. deep, long life), Les Rocoules (w, fab at 10 yrs+).

Souffrène, La Prov ★★★ Diverse VDF range; IGP Var. BANDOL red best: Les Lauves.

Sur lie "On the lees". Most MUSCADET is bottled straight from the vat, for max zest, body, character.

Tâche, La C d'O ★★★★ 90' 93' 96' 99' 02' 03 05' 09' 10' 12' 15' 16' 17 18' 19' 20' A GC of VOSNE-ROMANÉE, MONOPOLE of DRC. Firm in its youth, but how glorious with age. More tannic than stablemates but becomes headily perfumed (rose petals, Alpine strawberries) with age.

FRANCE

Taille-aux-Loups, Dom de la Lo ★★★→★★★★ 17 18' (r) 19' 20' 21 Jacky Blot, top producer, wife Joëlle, son Jean-Philippe. Excellent MONTLOUIS, VDF (aka VOUVRAY) mainly bone-dry, finely balanced, esp single-vyds: CLOS Mosny, *Michet* (Montlouis), *Venise* (Vouvray); *Triple Zéro* Montlouis *pétillant* (w/rosé); v.gd BOURGUEIL DOM de la Butte. Age brilliantly (r/w); 19 white stunning.

Tain, Cave de N Rh ★★→★★★ Top N Rh co-op, many mature vyds, incl 25% of HERMITAGE. Steady-to-v.gd red Hermitage, esp Gambert de Loche (best), bountiful Hermitage Au Coeur des Siècles (w, value). ST-JOSEPH (r/w) gd, ST-PÉRAY, interesting Bio (organic) range (St-Joseph), others modern, mainstream. Distinguished, genuine MARSANNE VIN DE PAILLE.

Taittinger Champ ★★★★ Family-run Reims house, exquisite elegance. BRUT NV, Rosé NV, Brut Vintage Collection Brut 89 95'. Epitome of apéritif style, inimitable weightlessness. Ace luxury *Comtes de Champagne* 95', classic great 08. Opinions differ on Comtes maverick release in tricky 11; Rosé shines in 12'. New cellarmaster strives to fill Loïc Dupont's big shoes. New English bubbly project in Kent, DOM Evremond. *See also* DOM Carneros, California.

Tavel S Rh ★★ Mainly DYA. Historic GRENACHE rosé, aided by white grapes for texture, should be full red, deep, herbal, for vivid Med dishes. Best show well 3–4 yrs. Now some lighter, Prov-style, often for apéritif; a shame. Top: CHX Aquéria, de Manissy (organic), La Genestière (organic), Ségriès, *Trinquevedel* (fine, organic); DOM de l'Anglore (no sulphur); *Dom de la Mordorée* (top name, organic), Carabiniers (bio), Corne-Loup, GUIGAL, Lafond Roc-Epine (organic), Maby, Moulin-la-Viguerie (character), Prieuré de Montézargues (fine, organic), Rocalière (organic, fine), Tardieu-Laurent, VIDAL-FLEURY.

Southern growers trialling vines under solar panels: shade, cooler, lower frost risk.

Tempier, Dom Prov ★★★★ Iconic BANDOL: where Lucien and Lulu Peyraud revived AOP in 30s. Tops for elegance, concentration, longevity. Single-vyds Cabassaou, La Tourtine pure expressions MOURVÈDRE. Smart rosé.

Terrasses du Larzac L'doc ★★★ Great terroir. High AOP on limestone with cold nights makes fresh, stylish reds. Attracts innovative growers, small plots of vines; 50%+ organic/bio. Try Cal Demoura, Combarela, Jonquières, LA PEIRA, Mas Conscience, MAS DE L'ECRITURE, MAS JULLIEN, Malavielle, Montcalmès, Pas de l'Escalette. Neighbouring AOP L'DOC St-Saturnin: DOMS Archimbaud, Virgile Joly.

Terre des Dames L'doc ★★★→★★★★ Dutch mother and daughters make strikingly pure, focused wines nr ST-CHINIAN. La Dame (r/w/rosé) gd-value AOP L'DOC. La Diva (r/w) IGP d'Oc true class.

Thénard Dom Burg ★★→★★★★ Historic producer, large holding of MONTRACHET, mostly sold on to NÉGOCIANTS. Look out for v.gd reds from home base in GIVRY.

Thévenet, Jean Burg ★★★ Top MÂCONNAIS purveyor of rich, some semi-botrytized CHARD, eg. CUVÉE Levroutée at *Dom de la Bongran*. Also owns DOMS de Roally, Emilian Gillet.

Thézac-Perricard SW Fr ★★ 16 18' 19 IGP Lighter version of adjoining CAHORS (reds from MALBEC, MERLOT). Sandrine Annibal's ★★ DOM de Lancement the one independent. Lively co-op nearly as gd.

Thiénot, Alain Champ Young house, new generation Stan and Garance now in charge. Ever-improving quality; fairly priced ★★★ BRUT NV; Rosé NV Brut; Vintage Stanislas 02 04 06 08' 09 12' 13 15. Voluminous VIGNE aux Gamins (single-vyd AVIZE 02 04 06). CUVÉE Garance CHARD 07 sings, classic 08 for long haul. Also owns CANARD-DUCHÊNE, JOSEPH PERRIER and CH Ricaud in LOUPIAC.

Thivin, Ch Beauj ★★→★★★★ Eight generations of Geoffray family make great Côte de Brouilly. Single-vyd bottlings cover soil types. Sept VIGNES blend also a winner.

Tissot Jura Dominant family around ARBOIS. ★ Jacques T offers volume; ★★ Jean-

Louis T value. ★★★ Stéphane T (also as A&M Tissot), bio with top single-vyd CHARD, VIN JAUNE, brooding reds and classy CRÉMANT du Jura, Indigène.

Tollot-Beaut C d'O ★★★ Consistent CÔTE DE BEAUNE grower with 20 ha in BEAUNE (Grèves, CLOS du Roi), CORTON (Bressandes), SAVIGNY (esp MONOPOLE PC Champ Chevrey) and at CHOREY-LÈS-BEAUNE base (Note Pièce du Chapitre). Easy-to-love fruit-and-oak combo. CORTON-CHARLEMAGNE gd too.

Touraine Lo ★→★★★ 19' 20' 21 Many AOPS (eg. BOURGUEIL, CHINON, VOUVRAY), plus AOP with fruity to medium-bodied age-worthy reds (CAB FR, CÔT, GAMAY, PINOT N), whites (SAUV BL), rosés, fizz. Often v.gd value. Also Touraine Villages AOPs: AMBOISE, Azay-le-Rideaux, Chenonceaux, Mesland, Noble Joué (rosé), Oisly. Best: Aulée, Biet, Bois-Vaudons, Cellier de Beaujardin, Corbillières, Desroches-Manois, Echardières, Fontenay, Garrelière, *Gosseaume*, Grosbois, Joël Delaunay, Lacour, Mandard, Marionnet, Morantin, Presle, Prieuré, Ricard, Roche, *Roussely*, Rousseau, Sauvète, Tue-Boeuf, Villebois. Return to fresher style in 21.

Touraine-Amboise Lo ★→★★★ 19' 20' 21 Light François 1er blend (GAMAY/CÔT/CAB FR); Côt (MALBEC) and CHENIN BL top: best ripe, finely textured. Try Bessons, Closerie de Chanteloup, Gabillière, *Grange Tiphaine*, Mesliard, Montdomaine, Plou, T Frissant, X Frissant. Amboise cru (Chenin Bl, Côt) on-going.

Trapet C d'O ★★★ Long-est GEVREY-CHAMBERTIN DOM making sensual bio wines from eye-catching COTEAUX BOURGUIGNONS up to GC CHAMBERTIN plus AL whites by marriage. Next generation pushing boundaries. *See* cousins ROSSIGNOL-T.

Latest wine region: Brittany. Sea influence, less frost than Lo; 1st fizz 2023.

Trévallon, Dom de Prov ★★★ Famous DOM at LES BAUX, created by late Eloi Dürrbach; now daughter Ostiane. No GRENACHE, so must be IGP Alpilles. Huge reputation fully justified: meticulous viticulture, age-worthy wines. Intense CAB SAUV/SYRAH. Barrique-aged MARSANNE/ROUSSANNE, drop of CHARD and now GRENACHE BL. Terrific.

Trimbach, FE Al ★★★★ RIES CLOS STE-HUNE is without peer: 71' 75' 89' still great; 13 17' 18 classic; almost-as-gd (much cheaper) *Frédéric Emile* 10 12 13 14 16 17'. Dry, elegant wines for great cuisine. Underrated, fresh PINOT GR, knockout SYLVANER from Trottacker lieu-dit.

Trinités, Dom des L'doc ★★★ Brit Simon Coulshaw makes poised, complex FAUGÈRES, Le Portail. L'Etranger, 100% CINSAULT mere VDF, as is cheeky Coulsh Rôtie SYRAH/VIOGNIER.

Tursan SW Fr ★★ Mostly DYA. AOP in LANDES. Super-chef Michel Guérard makes ★★★ lovely wines in chapel-like cellar at CH de Bachen, but not traditional Tursan. Real thing from ★★ DOM de Perchade. Delicious dry white from ★★ Dom de Cazalet (two MANSENGS plus rare local Baroque). Worthy co-op rather outclassed.

Vacqueyras S Rh ★★ 09' 10' 15' 16' 17' 18 19' 20 Robust, spiced, GRENACHE-fuelled neighbour of GIGONDAS, hot, flat vyds; for game, big flavours. Lives 10 yrs+; v. bad 21 frost. Note CHX de Montmirail, *des Tours* (v. fine); CLOS de Caveau (organic, character), *Clos des Cazaux* (top value); DOMS Amouriers (organic), Archimbaud-Vache, Charbonnière, Couroulu (full, v.gd), Famille Perrin, Font de Papier (organic), Fourmone, Garrigue, Grapillon d'Or, Monardière (organic, v.gd), Montirius (bio), Montvac (organic), Roucas Toumba (organic, character), SANG DES CAILLOUX (organic, esp Lopy), Semelles de Vent, Verde; JABOULET. Whites substantial, table wines (Clos des Cazaux, Ch des Roques, SANG DES CAILLOUX).

Val de Loire Lo Mainly DYA. One of France's four regional IGPS, formerly Jardin de la France.

Valençay Lo ★→★★ 20' 21 Small AOP; enjoyable blends of SAUV BL/CHARD; CÔT,

GAMAY, PINOT N. Try Bardon, Delorme, Jourdain, *Lafond*, Preys, Sinson, Vaillant, VIGNERONS de Valençay.

Valréas S Rh ★★ 19' 20' (21) CÔTES DU RH-VILLAGE in windy N Vaucluse truffle area, late ripening, quality rising; large co-op. Spicy, can be heady, red-fruited mostly GRENACHE red, improving white. Esp CH la Décelle, CLOS Bellane incl Val des Rois (organic), Mas de Ste-Croix; DOMS des Grands Devers, du Séminaire (organic), *Gramenon* (bio), Prévosse (organic).

Vaugelas, Ch L'doc ★★→★★★ Created by Benedictine monks, owned by Bonfils family, classic CORBIÈRES with warmth, spice. Le Prieuré (r) gd value. CUVÉE V, ambitious, powerful.

Vazart-Coquart Champ Family CHARD DOM, est 1954. Special Club 14 ★★★★ captivating medley of yellow fruits, saline finish.

VdF (Vin de France) Replaces Vin de Table, but with mention of grape variety, vintage. Often blends of regions with brand name. Can be source of unexpected delights if talented winemaker uses this category to avoid bureaucractic hassle, eg. Mark Angeli, YVES CUILLERON VIOGNIER; S ARDÈCHE, ANJOU hotbeds of VdF.

VDN (vin doux naturel) Rouss Sweet wine fortified with wine alc, so sweetness natural, not strength. Speciality of ROUSS based on GRENACHES BL/Gr/N. Top, esp aged RANCIOS, can finish a meal on a sublime note. MUSCAT from BEAUMES DE VENISE, FRONTIGNAN, Lunel, RIVESALTES, ROUSS, St Jean de MINERVOIS gd.

VDP (Vin de Pays) *See* IGP.

Vendange Harvest. **VT (Vendange Tardive)** Late-harvest; AL equivalent to German Auslese but usually higher alc.

Venoge, de Champ ★★★ Venerable house, precise, more elegant under LANSON-BCC ownership. Gd niche blends: Cordon Bleu Extra-BRUT, Vintage BLANC DE BLANCS 00 04 06 08 12 13 14 16 17. Excellent Vintage Rosé 09 CUVÉE 20 Ans, Prestige Cuvée Louis XV 10-yr-old BLANC DE NOIRS.

Ventoux S Rh ★★ 19' 20' (21) Widespread AOP circles around much of Mont Ventoux between Rh and Prov. A few leading-edge DOMS: v.gd value reds, decent choice. Tangy red (GRENACHE/SYRAH), café-style to fuller, peppery, rising quality), rosé, gd white (more oak). Best: CH Unang (organic, gd w), Ch Valcombe, CLOS des Patris (organic), Gonnet, La Ferme St Pierre (w/rosé, organic), La Vieille Ferme (r, can be VDP), St-Marc, Terra Ventoux, VIGNERONS Mont Ventoux; Doms Allois (organic), Anges, Berane, Brusset, Cascavel, Champ-Long, Croix de Pins (gd w), du Tix, Fondrèche (organic), Grand Jacquet, Martinelle (organic, stylish), Murmurium, Olivier B (organic), PAUL JABOULET, Pesquié, Pigeade, St-Jean du Barroux (organic, character). Terres de Solence, Verrière, VIDAL-FLEURY, Vieux Lazaret, Vignobles Brunier and co-op Bédoin.

Vernay, Dom Georges N Rh ★★★★ 18' 19' 20' 21' Top CONDRIEU; three wines, balance, style, purity; Terrasses de l'Empire *apéritif de luxe*; Chaillées d'Enfer, richness; Coteau de Vernon, magic, intricate, supreme style, lives 20 yrs+. CÔTE-RÔTIE, ST-JOSEPH (r) clear fruit, restrained; v.gd IGP COLLINES RHODANIENNES (r/w).

Veuve Clicquot Champ ★★★★ Historic house, both traditional and creative. Improving Yellow Label NV through soupçon of 5% oak. Best DEMI-SEC NV. Sadly classy CUVÉE Extra BRUT Extra Age discontinued 2021: pandemic casualty. Vintage Rés 12 ★★★★ a magical wonder of perfect maturity, elegant acidity: best yr for PINOT N since 1952? Luxe La Grande Dame (GD) 12' is 92% Pinot N yet so graceful. Older vintages of GD stay course effortlessly in 04 (no 02, deemed too butch for GD) gloriously in 89, firmly in 71. GD Rosé 06 delicious, ready. Didier Mariotti new chef de CAVE since 2020.

Veuve Devaux Champ ★★★ Premium brand of powerful Union Auboise co-op. Excellent aged Grande Rés NV, and Œil de Perdrix Rosé, Prestige CUVÉE D 08, BRUT Vintage on top form 12 15 17 18' 19.

Vézelay Burg ★→★★ Age 1–2 yrs. Lovely location (with abbey) in NW Burg Promoted to full AOP for tasty whites from CHARD. Also try revived MELON (COTEAUX BOURGUIGNON) and light PINOT (generic BOURGOGNE). Best: DOM de la Cadette, des Faverelles, Elise Villiers, La Croix Montjoie.

Vidal-Fleury N Rh ★★→★★★ GUIGAL-owned merchant/grower of CÔTE-RÔTIE. Top-quality, accomplished *La Chatillonne* (from Blonde, 12% VIOGNIER, oak, wait min 7 yrs). Broad range of sound quality: gd CAIRANNE, CHÂTEAUNEUF (r), CÔTES DU RH (r/rosé), MUSCAT DE BEAUMES-DE-VENISE, ST-JOSEPH (r/w), TAVEL, VENTOUX.

Vieille Ferme, La S Rh ★→★★ Brand from Famille Perrin of CH DE BEAUCASTEL. v.gd value; much now labelled VDF, with VENTOUX (r), LUBÉRON (w/rosé) in some countries (France, Japan). Back on form recently (incl r/rosé).

Vieilles vignes Old vines, which can give particular depth and complexity. Eg. DE VOGÜÉ, MUSIGNY. But no rules about age and can be a tourist trap. Can a vine be old if it's younger than you are?

Vieux Télégraphe, Dom du S Rh ★★★ 01' 05' 07' 09' 10' 12' 14' 15' 16' 17' 18' 19' 20' 21 High-quality estate, in gd form; classic big-stone plateau soils, slow-burn red CHÂTEAUNEUF; top two wines La Crau (crunchy, packed), Pied Long et Pignan (sandy, top purity, elegant, no 18). Also rich, splendid *garrigue* white *La Crau* (v.gd 15 16 18 19 20), CLOS La Roquète (refined, great with food). Owns fine, slow-to-evolve, complex GIGONDAS DOM Les Pallières with US importer Kermit Lynch.

Vignelaure, Ch de Prov ★★★ Top estate for reds, AOP COTEAUX D'AIX EN PROVENCE. CAB SAUV, SYRAH at 300m (984ft) give depth, freshness. Will age 10 yrs. Gd rosé. Intriguing ROUSSANNE, Rolle, SÉM.

Red wines can taste of violets: Syrah in the Rh, Petit Verdot in B'x.

Vigne or vignoble Vineyard (vyd), vineyards (vyds).

Vigneron Vine-grower.

Vignoble du Rêveur Al ★★★ New project of Mathieu DEISS/Emmanuelle Milan using Mathieu's maternal family vines. Bio-farmed, co-planted vines in Bennwihr; dry, food-friendly, pure wines from uncommonly talented duo that won't make a horse faint, like many natural wines; superb Singulier, Pierres Sauvages 17' 18.

Vigouroux, Georges SW Fr Top name in CAHORS, instrumental in reviving the AOP. Estates incl Haute Serre and Mercues. Generations have changed; wines seem to be getting better after a slightly dull period.

Villeneuve, Ch de Lo ★★★→★★★★ 16' 17' 18' 19' 20' 21 Top DOM: balance, finesse is key (r/w age-worthy). SAUMUR Bl (CHENIN BL, esp Les Cormiers) and SAUMUR-CHAMPIGNY (esp Grand CLOS, VIEILLES VIGNES). Organic. Daughter Cécile now joined parents.

Vin de paille Wine from grapes dried several mths before pressing, so concentrated sweetness and acidity; esp in Jura. *See also* CHAVE.

Vin gris "Grey" wine is v. pale pink, made of red grapes pressed before fermentation begins – unlike rosé, which ferments briefly before pressing. Or from eg. PINOT GR, not-quite-white grapes. "Œil de Perdrix" means much the same; so does "blush".

Vin jaune Jura ★★→★★★★ Speciality of Jura; inimitable not-so-"yellow" wine. SAVAGNIN, 6 yrs+ in barrel without topping up, develops flor like Sherry but no added alc. Ages for decades. Top spot, CH-CHALON has own AOP. Sold in unique 62cl clavelin bottles. S TISSOT specializes in single-vyd bottlings, Bourdy in old vintages. *See also* ARBOIS, CÔTES DU JURA, L'ETOILE.

Vinsobres S Rh ★★ 16' 18' 19' 20 AOP starting to wake, notable for quality SYRAH, mix hillside, high-plateau vyds. Best reds have smooth depth, to drink with red meats, age 10 yrs. New DOMS emerging. Best: CAVE la Vinsobraise, CH Rouanne, CLOS Volabis (organic); DOMS Autrand, Chaume-Arnaud (bio, v.gd), Constant-

Duquesnoy, du Tave, Famille Perrin (*Hauts de Julien* top class, Cornuds value), Jaume (modern), Moulin (traditional, gd r/w), Péquélette (bio, character), Serre Besson, Vallot (bio).

Viré-Clessé Burg ★★ 14′ 17′ 18 19′ 20′ AOP based around two of best white villages of MÂCON. Known for exuberant rich style, esp from Quintaine area, sometimes late-harvest. Specialists Bonhomme, Chaland, DOM de la Verpaille, Gandines, Gondard-Perrin, *Guillemot-Michel*, J-P Michel, THÉVENET and all gd Mâconnais NÉGOCIANTS.

Visan S Rh ★★ 16′ 19′ 20′ (21) Progressive CÔTES DE RH-VILLAGE, now top three: peppery mainly GRENACHE reds, fine depth, clear fruit; some softer, plenty organic. Whites okay. Best: DOMS Bastide, Coste Chaude (organic), Dieulefit (bio, low sulphur), Florane (bio), Fourmente (bio esp Nature), Guintrandy (organic, full), Montmartel (organic), Philippe Plantevin, Roche-Audran (organic, style), VIGNOBLE Art Mas (organic).

Vogüé, Comte Georges de C d'O ★★★★ Aristo CHAMBOLLE estate with lion's share of MUSIGNY. Great from barrel, but takes many yrs in bottle to reveal glories. Unique white Musigny. Change of winemaker in 2021 so watch this space.

Volnay C d'O ★★★→★★★★ 99′ 05′ 09′ 10′ 15′ 16 17′ 18 19 20 Top CÔTE DE BEAUNE reds, except when it hails or gets too hot. Can be structured, should be silky, astonishing with age. Best vyds: Caillerets, Champans, CLOS des Chênes, Santenots (more clay), Taillepieds and MONOPOLES Clos de la Bousse d'Or, Clos de la Chapelle, Clos des Ducs, Clos du CH des Ducs. Reference growers: D'ANGERVILLE, *de Montille*, *Lafarge*, Pousse d'Or. Also v.gd Bitouzet-Prieur, BOULEY, Buffet, *Clerget*, Glantenay, H BOILLOT, HOSPICES DE BEAUNE, LAFON, Rossignol.

Vosne-Romanée C d'O ★★★→★★★★ 90′ 93′ 96′ 99′ 02′ 05′ 09′ 10′ 12 15′ 16′ 17 18′ 19′ 20′ Village with Burg's grandest crus (eg. ROMANÉE-CONTI, LA TÂCHE) and outstanding PCS Beaumonts, Brûlées, Malconsorts, etc. There are (or should be) no common wines in Vosne. Just question of price... Top names: ARNOUX-LACHAUX, Bizot, CATHIARD, Coquard-Loison-Fleurot, DRC, EUGÉNIE, GRIVOT, GROS, Lamarche, LEROY, LIGER-BELAIR, MÉO-CAMUZET, MUGNERET, NOËLLAT, ROUGET. Plus Clavelier, Forey, Guyon, Tardy.

Vougeot C d'O ★★★ 99′ 02′ 05′ 09′ 10′ 12′ 15′ 16 17 18′ 19′ 20 Mostly GC as CLOS DE VOUGEOT but also village and PC, Cras, Petits Vougeots, and outstanding white MONOPOLE *Clos Blanc de V.* Best: Clerget, Fourrier, HUDELOT-NOËLLAT, LEROUX, *Vougeraie*.

Vougeraie, Dom de la C d'O ★★★ ★★★★ Bio DOM uniting all BOISSET's vyd holdings. Fine-boned, perfumed wines, whole-bunch vinification, most noted for sensual GCS, esp BONNES-MARES, CHARMES-CHAMBERTIN, MUSIGNY. Fine whites too, with unique *Clos Blanc de Vougeot* and four GCs incl unique CHARLEMAGNE.

Vouvray Lo ★→★★★★ (dr) 16 17 18′ 19′ 20′ 21 (sw) 08 09′ 10 11 15′ 16 18′ 20′ AOP DEMI-SEC CHENIN BL is classic style, plus *moelleux* in best yrs, rich with balancing acidity, almost everlasting. Top sites on limestone bluffs. Fizz: *pétillant* a speciality. Best: *Aubuisières*, Autran, Bonneau, Brunet, *Carême*, Champalou, Clos Baudoin (VDF), Florent Cosme, Fontainerie, *Foreau* (a reference) Gaudrelle, *Huet* (a reference), Mathieu Cosme, Meslerie (Hahn), Perrault-Jadaud, Rouvre, Pinon, *Taille-aux-Loups* (VdF), Vigneau-Chevreau.

Weinbach, Dom Al ★★★→★★★★ Finest vyds in AL, exceptionally elegant and pure, dry gastronomic style: MUSCAT, SYLVANER, RIES GC SCHLOSSBERG, GEWURZ Furstentum and Mambourg, PINOT GR Altenbourg 10′ 13′ 17′ 18′ 19′. Pinot Gr SGN world's best.

Zind Humbrecht, Dom Al ★★★★ One of world's greatest. Stellar GCS Brand, HENGST, GOLDERT and volcanic RANGEN; 18′ outstanding, esp PINOT GR. MUSCAT GC Goldert best buy and one of two best Al MUSCATS: 19 more classic than sumptuous 18s.

Châteaux of Bordeaux

Abbreviations used in the text:

B'x	Bordeaux
Bar	Barsac
Cas	Castillon-Côtes de Bordeaux
E-2-M	Entre-Deux-Mers
Fron	Fronsac
Grav	Graves
H-Méd	Haut-Médoc
L de P	Lalande de Pomerol
List	Listrac
Marg	Margaux
Méd	Médoc
Mou	Moulis
Pau	Pauillac
Pe-Lé	Pessac-Léognan
Pom	Pomerol
Saut	Sauternes
St-Ém	St-Émilion
St-Est	St-Estèphe
St-Jul	St-Julien

AC	appellation contrôlée
ch(x)	château(x)
dom(s)	domaine(s)

The reputation of Bordeaux's red wines has long been based on their ability to age. Complexity and nuance come with time, and left to repose gracefully the best of Bordeaux's Left and Right Banks can evolve into something special. The reality these days, though, is that red wines are drunk ever earlier: lack of cellaring space, budgetary concerns or even tastes being the reasons for the new norm. Faced with this actuality, winemakers in Bordeaux are now obliged to produce wines that have both early appeal and the ability to age. Climate change has certainly helped with the former, the added ripeness emphasizing fruit and opulence and banishing anything green. Add in the skills and armoury

Châteaux of Bordeaux entries also cross-reference to France.

of the producer – greater precision in the vineyard, exactitude in picking dates, sorting and selection, temperature control and machinery that offers gentler handling – and the aggressive tannins of yore have now taken on a suppler nature. The key is the management of those tannins, necessary for structure and ageing potential but unacceptable if they are as hard and aggressive as they were in the past. Vintages from 2015 have shown a rising crescendo in the exactitude of texture and mouthfeel, though a more difficult vintage like 2021 may offer a different appraisal.

2021 concluded the run of "solar" years (20 19 18 17) with conditions that harked back to bygone times. Frost in April and rain throughout the growing season with accompanying mildew made for a complicated year. Yields are down and quality varied (thought not as bad as 13), the Merlot, in particular, suffering from the climatic difficulties. Chaptalization was permitted up to 1.5% abv, a rarity these days. So it's a year to be choosy and not necessarily one for the long haul. Luckily, there are plenty of vintages for drinking now. The luscious 09s are tempting at whatever level. The 08s have come into their own. The 10s are just opening (as are the "classic" 14s), although the Grands Crus need longer. For early drinking, try the often-charming 12s or underrated 11s, which have improved with bottle-age.

Of recent years, the softer 17s will be the first to broach. Mature vintages to look for are 96 (best Médoc), 98 (particularly Right Bank), 00 01 (Right Bank again), 04 and 06. Some of the splendid 05s are also opening, although patience is still a virtue here. Dry white Bordeaux remains consistent in quality and value, and despite conditions it fared well in 21. Remember that fine white Graves can age as well as white burgundy – sometimes better. And Sauternes continues to offer an array of remarkable years, 21 possibly another but in tiny quantity. The problem here is being spoilt for choice. Even moderate years like 06 08 and 14 offer approachability and a fresher touch, while the great years (09 11 15 16) have the concentration and hedonistic charm that make them indestructible.

A, Dom de L' Cas ★★ 14 15 16 18 19 20 Leading CAS property owned by STÉPHANE DEREANONCOURT and wife. Consistent quality.

Agassac, D' H-Méd ★★ 14 15 16' 18 19 20' CRU BOURGEOIS Exceptionnel in S H-MÉD. Modern, accessible style. Same stable as FOURCAS DUPRÉ since 2021.

Aiguilhe, D' Cas ★★ 12 14 15 16 18 19 20 Large von Neipperg-owned estate. Same stable as CANON-LA-GAFFELIÈRE, LA MONDOTTE. MERLOT-led *power and finesse*. Visitor-friendly.

Andron-Blanquet St-Est ★★ 14 15 16' 18 19 Sister to COS LABORY. Can be value.

Angélus St-Ém ★★★★ 05 09 10' 11 14 15' 16' 17 18' 19' 20 21 PREMIER GRAND CRU CLASSÉ (A) until 2022; declined classification. Pioneer of modern ST-ÉM; dark, rich, sumptuous. More finesse from 19. Lots of CAB FR (min 40%). Second label: Carillon d'Angélus. Also No 3 d'Angélus. New Hommage à Elisabeth Bouchet: old-vine CAB FR, top yrs (16).

Angludet Marg ★★ 09' 10' 14 15' 16' 18' 19 20 Property owned by B'X NÉGOCIANT SICHEL. CAB SAUV, MERLOT and 13% PETIT VERDOT. Fragrant and stylish. Often gd value.

Archambeau Grav ★★ (r) 15 16 18 19 20 (w) 19 20 Illats-based property owned by Dubourdieu family. Gd *fruity dry white* (SAUV BL/SÉM 60/40); fragrant MERLOT/CAB (50/50) reds. Also rosé.

Arche, D' Saut ★★ 09' 10' 11 14 15 16 17' 18 20 Second Growth steadily being overhauled. New management team (2020). Can be value.

Armailhac, D' Pau ★★★ 08 09' 10' 12 14 15' 16' 17 18 19 20 21 (MOUTON) ROTHSCHILD-owned Fifth Growth. Approachable earlier. On top form, fair value.

Aurelius St-Ém ★★ 16 17 18 Top CUVÉE from go-ahead ST-ÉM co-op; 16,000 bottles p.a. MERLOT-led, new oak, concentrated.

Ausone St-Ém ★★★★ 00' 01' 03' 04 05' 06' 08 09' 10' 12 14 15' 16' 17 18' 19' 20 21 Tiny, illustrious. PREMIER GRAND CRU CLASSÉ (A) until 2022; then it declined classification. Only C.1500 cases; vyds s- and se-facing, sheltered from winds; lots of CAB FR (50% in 2020). Long-lived wines with volume, texture, finesse. At a price. Second label: Chapelle d'Ausone (500 cases). *La Clotte*, FONBEL, MOULIN ST-GEORGES, Simard sister estates.

Balestard la Tonnelle St-Ém ★★ 10 14 15 16 18 19 20 Historic property on limestone plateau. New winery 2021. Modern, MERLOT-led.

Barde-Haut St-Ém ★★→★★★ 08 09 10 14 15' 16 18 19' 20 merlot-led GRAND CRU CLASSÉ. Rich, supple, modern. Environmental certification.

Bastor-Lamontagne Saut ★★ 10 11 14 15 16 17 18 19 21 Large SÉM-led Preignac estate. Earlier-drinking style. Organic certification. Second label: Les Remparts de Bastor. Also dry white B de B-L.

Batailley Pau ★★★ 05' 08 09' 10' 11 12 14 15 16 17 18 19 20 A gd-value, CAB SAUV-led Fifth Growth. Second label: Lions de Batailley.

Beaumont H-Méd ★★ 09' 10' 14 15 16 18 19 20 Large CRU BOURGEOIS Supérieur. Sister to BEYCHEVELLE; early maturing, easily enjoyable wines.

Beauregard Pom ★★★ 05' 09' 10' 12 14 15 16 18 19' 20' 21 Much-improved POM, revamped since 2014. Organic certification. Second label: Benjamin de Beauregard. Also Pavillon Beauregard in L DE P. B&B accommodation.

Beau-Séjour Bécot St-Ém ★★★ 05 08 09' 10' 12 14 15 16 17 18 19 20' 21 Distinguished PREMIER GRAND CRU CLASSÉ (B) on plateau. Old quarried cellars for bottle storage. Lighter touch these days but still gd ageing potential. Limestone terroir comes through.

Beauséjour Duffau St-Ém ★★★ 05' 08 09' 10' 12 14 15 16 17 18' 19' 20' Tiny PREMIER GRAND CRU CLASSÉ (B). Owned by Joséphine Duffau-Lagarrosse and Courtin family (of Clarins cosmetics) since 2021 (blended and aged 2020). Rich and cellar-worthy.

Beau-Site St-Est ★★ 09 10 15 16 18 19 20 Gd-value ST-EST cru owned by BORIE-MANOUX; C18 cellar. Supple, fresh, accessible.

Bélair-Monange St-Ém ★★★ 05' 08 09' 10' 11 12 14 15 16' 17 18 19 20 21 PREMIER GRAND CRU CLASSÉ (B) on limestone plateau and côtes. Huge investment in vyd and state-of-the-art winery in last 15 yrs. Refined style with more intensity and precision these days. Second label: Annonce de Bélair-Monange.

Belgrave H-Méd ★★ 05' 09' 10' 14 15 16 18 19 20 21 DOURTHE-owned n H-MÉD Fifth Growth. Environmental certification. Modern-classic in style. Can be value. Second label: Diane de Belgrave.

Bellefont-Belcier St-Ém ★★ 08 09' 10' 14 15' 16 17 18 19' 20 GRAND CRU CLASSÉ on s côtes, owned by Vignobles K. Quality on the up since 17.

Belle-Vue H-Méd ★★ 09 10 14 15' 16 17 18 19 CRU BOURGEOIS Exceptionnel in S H-MÉD. Dark, dense but firm; 15–25% PETIT VERDOT in blend.

Berliquet St-Ém ★★ 08 09 10 14 15' 16' 17 18 19' 20 21 Tiny GRAND CRU CLASSÉ on plateau and côtes. Same stable as CANON, RAUZAN-SÉGLA. On upward curve.

Bernadotte H-Méd ★★ 10' 14 15' 16' 17 18' CRU BOURGEOIS Supérieur in n H-MÉD. Owned by King Power of Hong Kong. Hubert de Boüard (ANGÉLUS) consults. Savoury style. Best age 15 yrs.

Beychevelle St-Jul ★★★ 05' 08 09' 10' 11 12 14 15' 16' 17 18 19' 20 21 Sizeable Fourth

Growth owned by Castel and Suntory. Wines of consistent *elegance* rather than power. CAB SAUV-led but 40% MERLOT. Second label: Amiral de Beychevelle.

Biston-Brillette Mou ★★ 14 15 16' 18 19 Family-owned CRU BOURGEOIS Supérieur. Gd-value, attractive, early drinking wines (50/50 MERLOT/CAB SAUV).

Bonalgue Pom ★★ 09 10 12 14 15 16 18 19 20 Accessible MERLOT-led (90%) POM from sand, gravel, clay soils. Gd value for AC. Owned by NÉGOCIANT JB Audy. Sisters CLOS du Clocher; CH Les Hauts-Conseillants in L DE P.

Bonnet B'x ★★ (r) 16 18 19 20 (w) DYA. Forged by André Lurton; now run by son Jacques. Big producer of some of best E-2-M and red (oak-aged Rés) B'x. LA LOUVIÈRE, *Couhins-Lurton*, ROCHEMORIN and Cruzeau in PE-LÉ same stable.

Bon Pasteur, Le Pom ★★★ 05' 09' 10' 11 12 14 15' 16 17 18 19 20 Tiny cru on ST-ÉM border made from 21 different plots. MICHEL ROLLAND makes the wine. Ripe, opulent, seductive. Second label: L'Etoile de Bon Pasteur.

Boscq, Le St-Est ★★ 09' 10 12 14 15' 16' 17 18' 19' DOURTHE-owned CRU BOURGEOIS Exceptionnel. Consistently great value.

Bourgneuf Pom ★★ 09 10 11 12 14 15' 16' 17 18 19' 20 21 Steady progress over the yrs. MERLOT-led (80%). Subtle, savoury wine. As gd value as it gets.

Bouscaut Pe-Lé ★★★ (r) 09 10' 12 14 15 16' 17 18 19 20 (w) 15 16 17 18 19 20 GRAV Classed Growth. Structured, MERLOT-based reds. Sappy, age-worthy SAUV BL/SÉM *whites*. Environmental certification.

Boyd-Cantenac Marg ★★★ 05' 09' 10' 12 14 15 16 18 Discreet Cantenac-based Third Growth. Owned by the Guillemet family since 1932. CAB SAUV-dominated. Needs time. POUGET same stable. Second label: Jacques Boyd.

Branaire-Ducru St-Jul ★★★ 05' 08 09' 10' 14 15 16' 17 18' 19 20 Consistent Fourth Growth; regularly gd value; ageing potential. Environmental certification. Second label: *Duluc*.

Branas Grand Poujeaux Mou ★★ 05' 09 10 12 14 15' 16' 17 18 19' 20' Neighbour of CHASSE-SPLEEN, POUJEAUX. Expanded vyd 2020. Investment. Rich, modern style. Hubert de Boüard consults. Second label: Les Eclats de Branas.

Brane-Cantenac Marg ★★★ →★★★★ 05' 08 09' 10' 11 12 14 15' 16' 17 18 19' 20 Second Growth owned by Henri Lurton. Classic, fragrant MARG with structure to age. Second label: *Baron de Brane*, value, consistency. Also B'x white from 2019.

Brillette Mou ★★ 10 14 15 16' 18 19 Reputable DOM on gravelly soils. Solid, understated. Owned by Flageul family since 1975. Brin de Brillette vegan.

Cabanne, La Pom ★★ 00' 06 09' 10' 14 15 16' 18 19 20 MERLOT-dominant (90%+) on deep clay soils. Firm when young; needs bottle-age. Owned by Estager family.

Caillou Saut ★★ 12 13 14 15' 16 18 Second Growth BAR for pure *liquoreux*. 100% SÉM. Also CUVÉE Reine, Cuvée Prestige.

Calon Ségur St-Est ★★★★ 00' 05' 08 09' 10' 11 12 14 15' 16' 17 18' 19' 20 21 Third Growth on flying form; more CAB SAUV these days (78% in 2020). Firm but fine, complex. Second label: Le Marquis de Calon.

Cambon la Pelouse H-Méd ★★ 10' 12 14 15 16' 17 18 19 20 Big, reliable CRU BOURGEOIS Exceptionnel. Owned by Aussie group TWE.

Camensac, De H-Méd ★★ 09 10' 11 14 15 16 18' 19 20 Fifth Growth in N H-MÉD.

All that glitters is gold
Land prices in B'x are a fair indication of what is hot and what isn't. In 2020 prices in PAU were up 22% to an average €2.8 million/ha. Likewise, land in ST-JUL increased by 23% to €1.6 million/ha and in PE-LÉ by 20% to €600,000/ha. At the other end of the scale, viticultural land in AC MÉD dropped by 20% to €40,000/ha and in the Côtes (BOURG, BLAYE, etc.) by 10% to €14,000–19,000/ha, with only gd-quality organic estates bucking the trend. Same region, but what disparity.

Owned by Merlaut family. Eric Boissenot consults. Steady improvement; recent vintages clearly better. Second label: La Closerie de Camensac.

Canon St-Ém ★★★★ 05' 06 08' 09' 10' 12 13 14 15' 16' 17 18' 19' 20 21 Esteemed PREMIER GRAND CRU CLASSÉ (B) with vyd on limestone plateau. Wertheimer-owned, alongside BERLIQUET and RAUZAN-SÉGLA. Now flying; elegant, complex wines for long ageing. Second label: Croix Canon (separate winery formerly a C12 chapel).

Canon-la-Gaffelière St-Ém ★★★ 05' 08 09' 10' 12 13 14 15' 16 17 18 19' 20 PREMIER GRAND CRU CLASSÉ (B) on s foot slope. Same stable as CLOS DE L'ORATOIRE, D'AIGUILHE, LA MONDOTTE. Lots of CABS FR (40%) and SAUV (10%). Stylish, impressive. Organic certification.

Cantemerle H-Méd ★★★ 05' 08 09' 10' 14 15 16 18 19' 20 Large Fifth Growth in s H-MÉD. Laure Canu new (2021) general manager. Renovated and replanted over last 40 yrs. CAB SAUV-led. On gd form and gd value too.

Cantenac-Brown Marg ★★★ 05' 08 09' 10' 11 14 15 16' 17 18 19' 20 Third Growth owned by Le Lous family; vyd expanded 2020. New eco winery built entirely of raw earth due 2023. Also, in 2022, NFT of drone footage of ephemeral art project based on vyd soil. Have your cryptocurrency ready. More voluptuous, refined these days. Second label: BriO de Cantenac-Brown. Also SAUV BL-led dry white Alto.

Capbern St-Est ★★ 08' 09' 10' 11 12 14 15' 16' 18 19 20 21 Capbern Gasqueton until 2013. Same team as CALON SÉGUR. CAB SAUV-led; gd form and value.

Cap de Mourlin St-Ém ★★ 09 10 11 14 15 16 18 19 20 GRAND CRU CLASSÉ on n slopes. Limestone-clay and sandy soils. MERLOT-led (65%) with 25% CAB FR and 10% CAB SAUV. Firm, tannic wines.

Carbonnieux Pe-Lé ★★★ 05' 08 09' 10 12 14 15' 16' 18 19' 20 GRAV Classed Growth owned by Perrin family. Sterling red/white; large volumes of both. Fresh *whites*, 65% SAUV BL, eg. 18 19 20. Red can age. Visitor-friendly. Second label: La Croix de Carbonnieux. Also CH Tour Léognan.

Carles, De Fron ★★ 11 12 14 15 16 18 19 Haut-Carles is prestige CUVÉE; 90% MERLOT. Opulent, modern style.

Carmes Haut-Brion, Les Pe-Lé ★★★ 05' 08 09' 10' 12' 14 15 16' 17 18' 19' 20' 21 Tiny, high-flying property in heart of Bordeaux city. CABS FR (40%+) and SAUV-led wines: structured but suave. Ageing in barrels, *foudres* and amphorae. Second label: Le C des Carmes Haut-Brion from vines in Léognan.

Caronne Ste Gemme H-Méd ★★ 09' 10' 14 15 16 18 19 20 Historic estate. CAB SAUV-led (60%) wines; fresh, structured. Can be gd value.

Carruades de Lafite Pau ★★★ Second label of CH LAFITE. 20,000 cases/yr. Second Growth prices. Refined, smooth, savoury. Accessible but age-worthy.

Carteau Côtes-Daugay St-Ém ★★ 09 10' 14 15 16' 18 19 Gd-value ST-ÉM GRAND CRU; full-flavoured, supple wines from 70% MERLOT.

Certan de May Pom ★★★ 05' 08 09' 10' 11 12' 14 15' 16' 17 18 19' 20 21 Tiny neighbour of VIEUX CH CERTAN. Elegant, complex, long-ageing.

Chantegrive, De Grav ★★→★★★ 10' 12 14 15 16 18 19 20' Leading, large estate; v.gd quality, value. Hubert de Boüard (ANGÉLUS) consults. CUVÉE Caroline is top, *fragrant white* 18 19 20. Also a CÉRONS.

Chasse-Spleen Mou ★★★ 05' 08 09' 10' 14 15 16' 17 18 19' 20 Big (100 ha), well known. Often impressive, long-maturing; classical structure, fragrance. Second label: L'Oratoire de Chasse-Spleen. Environmental certification.

Chauvin St-Ém ★★ 09 10' 11 12 14 15 16' 18' 19' 20' 21 GRAND CRU CLASSÉ on POM border. MERLOT-led; constant progression. Second label: Folie de Chauvin.

Cheval Blanc St-Ém ★★★★ 01' 05' 06 08 09' 10' 11 12 13 14 15' 16' 17 18' 19' 20' 21 PREMIER GRAND CRU CLASSÉ (A) until 2022; opted out of future classification.

Superstar, easier to love than buy. High CAB FR (60%). Firm, fragrant wines verging on POM. Delicious young; lasts a generation. Second label: Le Petit Cheval. Also 100% SAUV BL Le Petit Cheval Blanc.

Chevalier, Dom de Pe-Lé ★★★★ 02 04 05' 06 08 09' 10' 12 13 14 15' 16' 17 18' 19' 20 Reliable Classed Growth. Pure, dense, finely textured red. Impressive, complex, long-ageing white (drink early on the fruit or later for more complexity) 15' 16' 17' 18 19' 20. DOM de la Solitude same stable. Second label (r/w): l'Esprit de Chevalier.

Cissac H-Méd ★★ 09 11 12 14 15 16' 17 18 19 CRU BOURGEOIS Supérieur in n H-MÉD. Classic CAB SAUV-led wines; used to be austere, now purer fruit. Second label: Reflets du CH Cissac.

Citran H-Méd ★★ 10' 14 15 16 18 19 20 Sizeable S H-MÉD estate owned by Merlaut family. Medium-weight, accessible early.

Clarence de Haut-Brion, Le Pe-Lé ★★★ 05 06 08 09' 10' 11 12 14 15 16' 17 18 19' 20 21 Second label of CH HAUT-BRION, previously known as Bahans Haut-Brion. Blend varies each vintage (usually MERLOT-led), same suave texture, elegance as *grand vin*. More approachable but can age.

Clarke List ★★→★★★ 09' 10' 12 14 15 16' 17 18 19 20 RIP Baron Benjamin de Rothschild. MERLOT-based (70%) red is v.gd. Eric Boissenot consults from 2016; style change; more length and precision. Also dry white: Le Merle Blanc du CH Clarke.

Clauzet St-Est ★★ 09 10 12 14 15 16' 17 CAB SAUV-led, consistent quality, was gd value. Property sold 2018; vines acquired by CH LILIAN LADOUYS, brand and buildings by CH La Haye.

Climens has contributed casks for the ageing of Salcombe Gin's Phantom gin.

Clerc Milon Pau ★★★ 05' 06 08 09 10' 11 12 14 15 16' 17 18' 19' 20 21 Fifth Growth owned by (MOUTON) ROTHSCHILD since 1970. Powerful but harmonious: consistent quality but prices up. Terroir incl sandy-gravel and limestone-clay.

Climens Bar ★★★★ 01' 02 05 06 07 09' 10' 11' 12' 13' 14 15 16' 19 Classed Growth. Concentrated wines with vibrant acidity; ageing potential guaranteed. Certified bio. No 17 18 20 21 (frost). Second label: Les Cyprès (gd value). Atypical dry white Asphodèle in 18 19 20 (100% SÉM).

Clinet Pom ★★★ 05' 06 08 09' 10' 11 12 14 15' 16' 17 18 19 20 21 Family property run by Ronan Laborde. Well located on plateau. 80% MERLOT, 20% CAB SAUV. Sumptuous, modern, to age.

Clos des Jacobins St-Ém ★★→★★★ 05' 09 10' 12 14 15' 16' 18 19 20 Côtes GRAND CRU CLASSÉ owned by Decoster family. Renovated, modernized, great consistency; powerful, modern style. Hubert de Boüard (ANGÉLUS) consults.

Clos du Marquis St-Jul ★★→★★★ 05' 08 09' 10' 11 12 14 15' 16' 17 18' 19' 20 21 Same stable as LÉOVILLE LAS CASES, NÉNIN, POTENSAC; gd ST-JUL character. Second label: La Petite Marquise.

Clos de l'Oratoire St-Ém ★★ 05' 09 10' 14 15 16' 17 18 19 Supple GRAND CRU CLASSÉ in von Neipperg stable (CANON-LA-GAFFELIÈRE). 80% MERLOT.

Clos Floridène Grav ★★ (r) 10' 12 14 15' 16 18' 19 20 (w) 17' 18 19 20 Dubourdieu family-owned/run DOM. SAUV BL/SÉM from limestone provides *fine modern white* GRAV; vibrant, CAB SAUV-led red. Own winery completed 2018. CH REYNON same stable. Second label (r/w) Drapeaux de Floridène.

Clos Fourtet St-Ém ★★★ 05' 08 09' 10' 11 12 14 15' 16' 17 18 19 20 21 PREMIER GRAND CRU CLASSÉ (B) on limestone plateau owned by Cuvelier family. Classic, stylish ST-ÉM. Consistently gd form. 13 ha of underground cellars. CH POUJEAUX in MOU same stable. Second label: La Closerie de Fourtet.

Clos Haut-Peyraguey Saut ★★★ 05' 07 08 10' 11' 12 13 14 15 16 17 18 19 20 21 First

---

A sign of climate change

CRU BOURGEOIS Exceptionnel CH de Malleret is planting hundreds of olive trees with the intention of producing B'x's 1st commercial olive oil. It will be vintaged like the wine. Located at the s end of AOP HAUT-MÉD in Le Pian Médoc, the organically run property has planted older olive trees sourced from the s of France and hopes to produce 500 litres of olive oil, with the 1st crop in 2021. What's next? Lavender?

Growth in Bommes. Owned by magnate Bernard Magrez (FOMBRAUGE, PAPE CLÉMENT same stable). SÉM-led (95%), harmonious wines; can age. Second label: Symphonie.

Clos l'Église Pom ★★★ 05' 06 08 09' 10 12 14 15' 16 18 19 20 Top-flight, consistent, on edge of plateau. 80% MERLOT, 20% CAB FR on clay-gravel soils. Seductive wine, will age. Environmental certification.

Clos Puy Arnaud Cas ★★ 10 11 12 14 15 16' 17 18 19' 20' Leading CAS estate run with passion by Thierry Valette. Certified bio. Vibrant wines with plenty of energy. Les Acacias is 100% CAB FR.

Clos René Pom ★★ 08 09 10 11 12 14 15' 16 18 19 20 Family-owned for generations. MERLOT-led with a little spicy MALBEC; sand and gravel soils. Classical rather than modern, but gd value for AC.

Clotte, La St-Ém ★★→★★★ 09' 10' 11 12 14 15' 16' 17 18 19' 20 21 Steady improvement under AUSONE ownership. Powerful but balanced.

Conseillante, La Pom ★★★★ 01 05' 06' 07 08 09' 10' 11 12 13 14 15' 16' 17 18 19' 20' Owned by Nicolas family since 1871; vyd surface area same since this time. Some of noblest, most fragrant POM with structure to age. Environmental certification.

Corbin St-Ém ★★ 08 09 10' 12 14 15' 16' 18 19 20 Consistent, gd-value GRAND CRU CLASSÉ close to POM border. Gourmand, approachable.

Cos d'Estournel St-Est ★★★★ 00 01 05' 06 08 09' 10' 11 12 13 14 15' 16' 17 18 19' 20' 21 Big Second Growth. Refined, suave, high-scoring. Specially engraved bottle in 2020; 20 yrs of Michel Reybier ownership. Pricey SAUV BL-dominated white; now more refined. Second label (r/w): Les Pagodes de Cos.

Cos Labory St-Est ★★→★★★ 00 05 06 08 09' 10' 11 12 14 15 16' 18 19 Small Fifth Growth neighbour of COS D'ESTOURNEL. Savoury and firm. Usually gd value. Second label: Charme de Cos Labory.

Coufran H-Méd ★★ 09' 10 11 12 14 15' 16' 18 19 20 Atypical n HAUT-MÉD estate with 85% MERLOT. Vyd in one block. Supple wine with some ageing potential. Second label: No2 de Coufran.

Couhins-Lurton Pe-Lé ★★→★★★ (r) 09 10' 14 15 16 17 18 19 20 (w) 05 06 08' 09 15' 16 17 18 19 20 Fine, tense, long-lived Classed Growth white from SAUV BL (100%). Polished, MERLOT-led (up to 100%) red.

Couspaude, La St-Ém ★★★ 05 06 08 09' 10' 14 15 16 18 19 20 GRAND CRU CLASSÉ on limestone plateau. Rich, creamy, MERLOT-led, lashings of spicy oak.

Coutet Saut ★★★ 01' 05 07 09' 10' 11' 13 14' 16 17' 18 19' 20 21 Baly family-owned (Aline is 3rd generation). Consistently v. fine. CUVÉE Madame: v. rich, old-vine selection 01 03 09. Second label: La Chartreuse de Coutet. Dry white, Opalie v.gd.

Couvent des Jacobins St-Ém ★★ 05 09 10 12 14 15 16' 18 19' 20 GRAND CRU CLASSÉ vinified within walls of town. Organic certification (2020). Ample but fresh. Also micro-CUVÉE Calicem.

Crabitey Grav ★★ (r) 14 15 16 17 18 19 20 (w) 17 18 19 20 Located on the gravelly soils of Portets. Owner Arnaud de Butler makes harmonious CAB SAUV-led reds; small volume of lively SAUV BL (70%), SÉM.

Crock, Le St-Est ★★ 09' 10 11 14 15 16' 18 19' 20 CRU BOURGEOIS Exceptionnel. Same stable as LÉOVILLE POYFERRÉ. Firm but polished, can age. Usually gd value.

Croix, La Pom ★★ 09 10 11 12 14 15 16 18 19 (20) Owned by NÉGOCIANT Janoueix since 1960. Rich, MERLOT-led, plenty of oak. Also HAUT-SARPE.

Croix du Casse, La Pom ★★ 12 14 15' 16 18 19 20 Supple and early drinking. MERLOT-based (90%+), sandy/gravel soils. Gd value. Part of BORIE-MANOUX stable since 2005.

Croix de Gay, La Pom ★★★ 05 08 09' 10' 12 14 15 16 17 18 19 20 Tiny MERLOT-dominant (95%) vyd. Rich, round wines. La Fleur de Gay from separate parcels.

Croizet-Bages Pau ★★ 05 08 09 10' 12 14 15 16' 18 19 20 Striving, CAB SAUV-led Fifth Growth. More consistency since 2015 but can improve further. Second label: Alias de Croizet-Bages.

Cru Bourgeois Méd Three-tier classification in 2020 (CB, CB Supérieur, CB Exceptionnel); runs for vintages 2018–22. 249 CHX all told.

Cruzelles, Les L de P ★★ 09 10 12 14 15' 16' 17 18 19' 20 21 Consistent, expressive, gd-value, MERLOT-led (90%) wine. Ageing potential in top yrs. Same stable as L'ÉGLISE-CLINET, Montlandrie.

Dalem Fron ★★ 08 09' 10' 12 14 15 16 17 18' 19 20 MERLOT-dominated (90%) property. Smooth, ripe, fresh. De la Huste same stable.

Dassault St-Ém ★★ 09' 10 12 14 15 16 18' 19 20 Rich, modern GRAND CRU CLASSÉ. 70% MERLOT, 30% CABS FR/SAUV; 70% new oak. CHX La Fleur, Trimoulet, Faurie dc Souchard same stable. Second label: D de Dassault.

Dauphine, De la Fron ★★→★★★ 09' 10' 12 14 15 16 17 18' 19 20 Leading FRON estate. Expansion and renovation over last 20 yrs. Vrai Canon Bouché added 2021. Organic. Now more substance, finesse. Second label: Delphis. Also Le Blanc de La Dauphine (80% SAUV BL, 20% SÉM).

Dauzac Marg ★★→★★★ 09' 10' 11 12 14 15 16' 17 18' 19 20 Busy Fifth Growth at Labarde. Owned by Roulleau family. Eric Boissenot consults. Sustainable approach. Dense, rich, dark wines. Second label: La Bastide Dauzac. Also fruity Aurore de Dauzac. D de Dauzac is vegan.

Desmirail Marg ★★→★★★ 05 09' 10' 12 14 15 16' 18 19 20 Discreet Third Growth in Lurton hands. Fine, delicate style. Second label: Initial de Desmirail.

Destieux St-Ém ★★ 05' 08 09' 10 12 14 15 16 18 19 20 GRAND CRU CLASSÉ. MERLOT-based but 30%+ CABS FR, SAUV. Firm, powerful, modern. Needs time.

Doisy-Daëne Bar ★★★ 05' 07 08 09 10' 11' 12 13' 14 15' 17' 18' 19 20 21 A gem in the Dubourdieu family stable (REYNON). *Fine and sweet*. L'Extravagant 16 17' 18' 19' intensely rich, expensive, 100% SAUV BL CUVÉE. Also Doisy-Daëne SEC dry white.

Doisy-Védrines Bar ★★★ 04 05' 09 10' 11' 13 14 15' 16' 17 18' 19 20 Owned by Castéja family. *Long-term fave*; delicious, gd value. Second label: Petit Védrines.

Dôme, Le St-Ém ★★★ 09 10' 11 12 14 15 16 17 18 19 20 Micro-wine; rich, modern, powerful. Two-thirds old-vine CAB FR, 80% new oak. State-of-the-art winery close to ANGÉLUS. CH Teyssier (value) same stable.

Dominique, La St-Ém ★★★ 05' 08 09' 10' 11 12 14 15 16' 17 18' 19 20 21 GRAND CRU CLASSÉ owned by Clément Fayat. Rich, juicy. MERLOT-led (81%). Ex-HAUT-BAILLY winemaker. Rooftop restaurant (La Terrasse Rouge), shop. Second label: Relais de la Dominique.

Ducru-Beaucaillou St-Jul ★★★★ 04 05' 06 07 08 09' 10' 11 12 13 14 15' 16 17 18' 19' 20' 21 Outstanding Second Growth owned by Bruno Borie. Majors in CAB SAUV (85%+). Excellent form; classic cedar-scented claret for long ageing. Special label for 2020 vintage to celebrate 300 yrs of the property.

Duhart-Milon Rothschild Pau ★★★ 00' 05' 06 08 09' 10' 11 12 14 15 16' 17 18' 19' 20 21 Fourth Growth owned by LAFITE ROTHSCHILD. Later-ripening terroir. CAB SAUV-dominated (70%); v. fine quality. On the up and up.

Durfort-Vivens Marg ★★★ 05 06 08 09' 10' 11 13 14 15' 16' 17 18 19' 20' 21 Much-improved Second Growth; recent vintages tiptop. CAB SAUV-dominated (90%). Organic, bio certification. Amphorae for ageing. Second labels: Vivens, Relais de Durfort-Vivens.

Eglise, Dom de l' Pom ★★ 05′ 06 08 09 10′ 11 12 14 15 16 18 19 20 Neighbour of L'ÉGLISE-CLINET. Owned by BORIE-MANOUX. Clay/gravel soils of plateau. Consistent, fleshy of late. CROIX DU CASSE same stable.

Église-Clinet, L' Pom ★★★★ 00′ 01′ 04 05′ 06 08 09′ 10′ 11′ 12 13 14 15′ 16′ 17 18 19′ 20′ 21 Tiny, top-flight POM. Great consistency; full, concentrated and fleshy, but expensive. Noémie Durantou now the winemaker. Second label: La Petite Église.

Evangile, L' Pom ★★★★ 05 06 08 09′ 10′ 11 12 13 14 15′ 16′ 17 18′ 19′ 20′ 21 Rothschild (LAFITE)-owned since 1990. Juliette Couderc now winemaker. MERLOT-dominated (80%) with CAB FR. Consistently rich, opulent. Second label: Blason de L'Evangile.

Fargues, De Saut ★★★ 03′ 04 05′ 06 07 08 09′ 10′ 11′ 13 14 15′ 16′ 17′ 18 19 20 21 Unclassified but top-quality (and price). Lur-Saluces family owners since 1472. Rich, unctuous but refined. Age-worthy.

Faugères St-Ém ★★→★★★ 08 09′ 10′ 11 12 14 15 16′ 18 19′ 20 Sizeable GRAND CRU CLASSÉ owned by Silvio Denz. Rich, bold, modern wines. Sister CHX Péby Faugères (also classified), Cap de Faugères (CAS).

Ferrand, De St-Ém ★★→★★★ 05 06 08 09 10′ 12 14 15 16 18 19 20 Big St-Hippolyte GRAND CRU CLASSÉ. MERLOT-led (75%): fresh, firm, expressive. On an upward curve. Environmental certification.

Ferrande Grav ★★ 10 14 15 16′ 17 18 19 20 Substantial property owned by NÉGOCIANT Castel. Much improved; enjoyable red; creamy SÉM/SAUV BL and Gr white.

Ferrière Marg ★★★ 05 06 08 09 10′ 12 14 15 16′ 17 18 19 20 Confidential Third Growth in MARG village. Organic, bio certification. Dark, firm, perfumed wines.

Feytit-Clinet Pom ★★→★★★ 05′ 06 08 09′ 10′ 11 12 14 15 16′ 17 18 19′ 20 Tiny 6-ha property owned by the Chasseuil family. 90% MERLOT on clay-gravel soils. Top, consistent form; rich, seductive. Relatively gd value.

Fieuzal Pe-Lé ★★★ (r) 06 08 09′ 10′ 11 12 14 15 16′ 18′ 19′ 20 (w) 15 16 18 19 20 Classified estate owned by Irish Quinn family. Rich, ageable, SAUV BL-led white; rich, firm red. Second label (r/w): L'Abeille de Fieuzal.

Figeac St-Ém ★★★★ 00′ 01′ 04 05′ 06 07 08 09′ 10′ 11 14 15 16′ 17′ 18′ 19′ 20′ 21 Large PREMIER GRAND CRU CLASSÉ (B) currently on roll (magnificent 16 to 20). Gravelly vyd with unusual 70% CABS FR/SAUV. Now richer but always elegant wines; need long ageing. New winery complex (2021); visitor-friendly. Second label: Petit-Figeac.

Filhot Saut ★★ 01 05 07 09′ 10′ 11′ 13 14 15 16 17 18 19 20 Second Growth owned by de Vaucelles family; 60% SÉM, 36% SAUV BL, 4% MUSCADELLE. Richer, purer in recent yrs.

Fleur Cardinale St-Ém ★★ 09′ 10′ 11 12 14 15 16′ 18 19′ 20′ 21 GRAND CRU CLASSÉ

Eclectic Place

La Place de Bordeaux, the virtual stock exchange used by B'x's merchants and producers, built its reputation on its ability to distribute large volumes of B'x around the world. Now foreign producers and those from elsewhere in France are harnessing this potential. Among the wines sold on the Place today you will find BEAUCASTEL's Hommage à Jacques Perrin and PHILIPPONNAT's Clos des Goisses, as well as Inglenook Rubicon (California), Klein Constantia's Vin de Constance (S Africa), Masseto (Tuscany), Michele Chiarlo (Barolo), Yjar from Rioja and a whole lot more. Commerce, it's in the blood.

on flying form. Ripe, unctuous and modern style. Second label: Intuition. Conversion to organic (2021).

Fleur de Boüard, La L de P ★★→★★★ 09 10 11 12 13 14 15 16' 17 18 19 20 Leading estate in L DE P. Owned by de Boüard family (ANGÉLUS). Dark, dense and modern. Special CUVÉE, Le Plus: 100% MERLOT; more extreme. Environmental certification. Second label: Le Lion.

Fleur-Pétrus, La Pom ★★★★ 05' 08 09' 10' 11 12 14 15 16 17' 18 19' 20' 21 Top-of-range J-P MOUEIX property on plateau. Sizeable: 18.7 ha. 91% MERLOT, 6% CAB FR, 3% PETIT VERDOT. Refined and long ageing.

Fombrauge St-Ém ★★→★★★ 05 06' 08 09 10 14 15 16' 18 19 20 Substantial GRAND CRU CLASSÉ owned by Bernard Magrez (PAPE CLÉMENT). Rich, dark, creamy, opulent. Magrez Fombrauge is special red CUVÉE; also name for dry white B'X.

Fonbadet Pau ★★ 05' 08 09' 10' 14 15 16' 18 19' 20' Small non-classified estate; gd value, CAB SAUV-led (60%). Less long-lived but reliable. Excellent 19 20.

Fonbel, De St-Ém ★★ 10 12 14 15 16 18 19 20 Consistent, juicy, fresh, gd value. MERLOT-led; incl CARMENÈRE. Vauthier (AUSONE)-owned.

Fonplégade St-Ém ★★ 08 09' 10 12 14 15 16' 18' 19 20' 21 American-owned GRAND CRU CLASSÉ on the côtes. Previously concentrated, modern; now more fruit, balance. Bio certification.

Fonréaud List ★★ 05 06 08 09' 10' 12 14 15 16' 17 18 19 20 CRU BOURGEOIS Supérieur. One of bigger, better LIST for satisfying, savoury wines. Small-volume v.gd dry white: Le Cygne. Same stable as LESTAGE.

Fonroque St-Ém ★★★ 05 06 08 09' 10' 12 14 15 16 18 19 20' Côtes GRAND CRU CLASSÉ nw of ST-ÉM town. Bio certification. Medium-bodied, minerally, fresh.

Fontenil Fron ★★ 09' 10' 11 12 14 15' 16 18' 19 20 Leading FRON, owned by MICHEL ROLLAND. 100% MERLOT. Ripe, chocolatey, opulent.

Forts de Latour, Les Pau ★★★★ 05' 06 08 09 10 11 12 14 15 16' 17 18 19' 20 Second label of CH LATOUR; authentic PAU flavour in slightly lighter format; high price. No more en PRIMEUR sales; only released when deemed ready to drink (15 in 2021) but another 10 yrs often pays off.

Fourcas Dupré List ★★ 09 10' 12 15' 16' 17 18 19 20 Well-run property, fairly consistent; medium-bodied, dry, fresh. D'AGASSAC same stable. Also a dry white.

Fourcas-Hosten List ★★→★★★ 09 10' 12 14 15 16 17 18 19 20 Large estate owned by Hermès connections. Lots of investment. Organic certification. Finesse and precision these days. Also SAUV BL-led dry white.

France, De Pe-Lé ★★ (r) 09 10 12 14 15 16' 18 19 20 (w) 15 16 17 18 19 20 Unclassified neighbour of FIEUZAL owned by Thomassin family. Ripe, modern reds. White fresh, balanced. Value.

Franc Mayne St-Ém ★★ 05 08 09 10' 12 14 15 16 17 19' Tiny GRAND CRU CLASSÉ on côtes. Fresh, structured. Second label (from 2018): Ilex.

Gaby Fron ★★ 08 09 10 12 14 15 16 18 19' (20) Well-sited CANON-FRON estate. MERLOT-dominated. Can age. Organic certification. Also special CUVÉE Gaby.

Gaffelière, La St-Ém ★★★ 05' 08 09 10' 12 14 15 16' 18 19' 20 21 First Growth owned by the de Malet-Roquefort family. Investment, improvement; part of vyd replanted. Elegant, long-ageing wines. Lots of CAB FR (40%). Second label: CLOS la Gaffelière.

Garde, La Pe-Lé ★★ (r) 09' 10' 12 14 15 16' 18 19 20 (w) 17 18 19 20 Unclassified PE-LÉ in Martillac; supple, CAB SAUV/MERLOT reds. Tiny production of SAUV BL (90%), SÉM white. 30 yrs of DOURTHE ownership in 2020; new winery.

Gay, Le Pom ★★★ 05 08 09' 10' 11 12 14 15 16 17 18' 19' 20 Neighbour of LAFLEUR. MERLOT dominant (90%). Rich, suave with ageing potential. CH Montviel and La Violette same stable. Second label: Manoir de Gay.

Gazin Pom ★★★ 05 08 09 10' 11 12 14 15' 16' 18' 19' 20 Owned by 5th generation of de Bailliencourt family. On v.gd form; generous (90% MERLOT), long ageing. Environmental certification.

Gilette Saut ★★★ 75 86 88 89 90 96 97 99 Extraordinary small Preignac CH. Owned by Médeville family since C18. Vintages back to 1953. Stores its sumptuous wines in concrete vats for 16–20 yrs. Ch Les Justices (SAUT) same stable.

Giscours Marg ★★★ 05 08 09 10′ 11 12 14 15 16′ 17 18′ 19′ 20′ Substantial Third Growth. CAB SAUV-led (60%). Full-bodied, long-ageing MARG; recent vintages on song. Thomas Duclos consultant oenologist since 2019. Second label: La Sirène de Giscours. Little B′x rosé.

19,952 ha in B′x are organically certified or in conversion (2020).

Glana, Du St-Jul ★★ 08 09 10 15 16′ 17 18 19′ 20 Big, unclassified estate. CAB SAUV-led (65%). Undemanding; robust; value. Owned by Meffre family. Second label: Pavillon du Glana.

Gloria St-Jul ★★ →★★★ 05 06 08 09′ 10′ 12 14 15 16′ 17 18 19′ 20 21 Owned by Triaud family. CAB SAUV-dominant (65%). Unclassified but sells at Fourth Growth prices. Same stable as ST-PIERRE. Superb form recently.

Grand Corbin-Despagne St-Ém ★★ →★★★ 05 08 09′ 10′ 11 12 14 15 16′ 18′ 19′ 20′ 21 Gd-value GRAND CRU CLASSÉ discerningly run by François Despagne. Aromatic wines now with riper, fuller edge. Organic certification; bio trials. Ampélia (CAS) sister estate. Second label: Petit Corbin-Despagne.

Grand Cru Classé St-Ém 2012: 64 classified; reviewed every 10 yrs (next in 2022).

Grand Mayne St-Ém ★★★ 08 09′ 10′ 12 14 15 16′ 17 18 19′ 20 21 Impressive GRAND CRU CLASSÉ on the côtes and pied de côtes. Consistent, full-bodied, structured wines. Second label: Filia de Grand Mayne.

Grand-Puy Ducasse Pau ★★★ 05′ 06 08 09′ 10′ 12 14 15′ 16′ 18 19′ 20 21 Fifth Growth, steady rise in quality (19 best buy yet). 60% CAB SAUV, 40% MERLOT. Sister to MEYNEY. Second label: Prélude à Grand-Puy Ducasse.

Grand-Puy-Lacoste Pau ★★★ 05′ 06 08 09′ 10′ 12 14 15′ 16′ 17′ 18′ 19′ 20 Fifth Growth famous for CAB SAUV-driven (75%+) PAU to lay down. Owned by François-Xavier Borie; vyd in one block around CH. Second label: Lacoste Borie.

Grave à Pomerol, La Pom ★★★ 05 08 09′ 10 12 13 14 15 16′ 17 18 19′ 20 21 Small J-P MOUEIX property acquired in 1971. Mainly gravel soils; gd value, MERLOT-dominant (85%). Refined; can age.

Greysac Méd ★★ 10′ 12 14 15 16′ 17 18 19 (20) CRU BOURGEOIS Supérieur. C18 CH. MERLOT-led (65%), fine, fresh, consistent quality.

Gruaud-Larose St-Jul ★★★★ 05′ 06 08 09′ 10′ 12 14 15′ 16′ 17 18′ 19′ 20′ One of biggest, best-loved Second Growths. Vigorous claret to age. Cracking 18 19 20. Organic certification (2022). Second label: *Sarget de Gruaud-Larose.*

Guadet St-Ém ★★ 09 10 12 14 15 16′ 18 19 20 Tiny GRAND CRU CLASSÉ. Better form recently; 7th generation at helm. DERENONCOURT consults. Bio certification.

Guiraud Saut ★★★ 05′ 06 07 08 09 10′ 11′ 13 14 15′ 16′ 17′ 19 20 Classed Growth SAUT; organic certification. Unusual 35% SAUV BL. Luc Planty successor to father, Xavier, at helm. Dry white G de Guiraud. Second label: Petit Guiraud.

Gurgue, La Marg ★★ 09′ 10 14 15 16′ 18 19 20 Same ownership as FERRIÈRE. Organic, bio certification. Accessible earlier. Gd value.

Hanteillan, D H-Méd ★★ 09′ 10 14 15 16 18 19 (20) Large CRU BOURGEOIS at Cissac. CAB SAUV-led (50%). Reliable; early drinking. Second label: CH Laborde.

Haut-Bages-Libéral Pau ★★★ 05′ 06 08 09′ 10′ 12 14 15′ 16 18 19′ 20′ Improving Fifth Growth (next to LATOUR). 19 20 best yet. CAB SAUV-led (70%). Organic certification. Reasonable value. Second label: Le Pauillac de Haut-Bages-Libéral.

Haut-Bailly Pe-Lé ★★★★ 05′ 06 08′ 09′ 10′ 11 12 14 15′ 16′ 17 18′ 19′ 20′ 21 Top-quality Classed Growth. Refined, elegant, CAB SAUV-led (parcel of 100-yr+

mixed vines). New circular winery (2021). Second label: Haut-Bailly II (previously La Parde de H-B). CH Le Pape (PE-LÉ) same stable.

Haut-Batailley Pau ★★★ 05 06 08 09' 10' 11 12 14 15 16' 17 18' 19' 20' Fifth Growth owned by Cazes family (LYNCH-BAGES). Upward curve since 2017; elegant edge. Second label: Verso.

Haut-Beauséjour St-Est ★★ 09 10 14 15 16 17 19 Property created, improved by CHAMP ROEDERER; sold in 2017 to Laffitte Carcasset. MERLOT-led (60%+).

Haut-Bergeron Saut ★★ 05 07 09 10 11 13 14 15' 16' 17' 18 19 20 Consistent, unclassified; 9th-generation family ownership. Vines in SAUT, BAR. Mainly SÉM (90%). Rich, opulent, gd value.

Haut-Bergey Pe-Lé ★★→★★★ (r) 09 10 12 15 16' 18' 19 20' (w) 18 19 20 Non-classified property in Léognan. Rich, bold red (from all five B'X varieties). Fresh, SAUV BL-led dry white. Organic, bio certification.

Haut-Brion Pe-Lé ★★★★ 00' 01 04 05 06 07 08 09' 10' 11' 12 13 14 15' 16' 17' 18' 19' 20' 21 Only non-MÉD First Growth in list of 1855, owned by American Dillon family since 1935. Prince Robert de Luxembourg the titular head today. Documented evidence of wine from 1521. Deeply harmonious, wonderful texture, for many No 1 or 2 choice of all clarets. Constant renovation: next project the *cuvier*. A little *sumptuous dry white* (SAUV BL/SÉM) for tycoons: 15' 16' 17 18 19 20 21. Also new La Clarté (w) from both H-B and LA MISSION HAUT-BRION. Second label: LE CLARENCE DE HAUT-BRION (previously Bahans H-B).

Haut Condissas Méd ★★★ 09' 10' 12 14 15 16 17 18 19' (20) Top wine from Jean Guyon stable (GREYSAC, Rollan-de-By). 5000 cases p.a. Rich, exuberant, modern. MERLOT-LED plus 20% PETIT VERDOT.

Haut-Marbuzet St-Est ★★→★★★ 05' 08' 09 10' 12 15 16 18 19' 20' Started in 1952 with 7 ha; now 70; owned by Duboscq family. Easy to love, but unclassified; 60%+ sold directly by CH. Scented, unctuous wines matured in 100% new oak barrels. CAB SAUV-led with MERLOT, CAB FR, PETIT VERDOT. Second label: Mac Carthy.

Haut-Sarpe St-Ém ★★ 08 09 10 12 14 15' 16' 18 19 20 GRAND CRU CLASSÉ on limestone plateau. MERLOT-led (70%). Ripe, modern style.

Hosanna Pom ★★★★ 05' 06 08 09 10' 11 12 14 15 16' 17 18 19' 20' 21 Tiny vyd in heart of plateau. PETRUS, VIEUX CH CERTAN neighbours. MERLOT (70%), old-vine CAB FR (30%). Created by J-P MOUEIX (1999). Power, purity, balance; needs time.

Issan, D' Marg ★★★ 05' 06 08 09' 10' 11 12 14 15' 16' 17 18' 19' 20 21 Third Growth with moated CH. Fragrant, CAB SAUV-led (66%) wines that age; on top form; vyd extended 2020; CAB FR, Malbec and old-vine PETIT VERDOT added. Second label: Blason.

Jean Faure St-Ém ★★ 08 09 10' 11 12 14 15 16 18' 19' 20' GRAND CRU CLASSÉ on clay, sand, gravel soils nr CHEVAL BLANC. Organic certification, bio methods. CAB FR-led with a touch of MALBEC; gives fresh, elegant style.

Kirwan Marg ★★★ 05' 06 08 09 10' 12 14 15' 16' 18 19' 20' Third Growth. CAB SAUV-dominant (60%). Dense, fleshy in 90s; now less forced, more finesse. Visitor-friendly. Second label: Charmes de Kirwan.

More than 65% of B'x's vyd area has a certified environmental approach.

Labégorce Marg ★★→★★★ 05' 08 09 10' 12 14 15 16' 17 18' 19 20 Substantial unclassified CH owned by Perrodo family. Considerable investment, progression. Ripe, modern but fresh. Marquis d'Alesme same stable.

Lafaurie-Peyraguey Saut ▲▲▲ 01' 05' 06 07 09' 10' 11 13 14 15' 16' 17' 18 19 20 21 Leading Classed Growth owned by Lalique crystal-owner Silvio Denz (*see* FAUGÈRES). Rich, harmonious, sweet 90% SÉM. Only 7hl/ha in 2020. Luxury hotel and restaurant. Also Sém-led dry white B'x.

Lafite Rothschild Pau ★★★★ 02 05' 06 07 08' 09' 10' 11' 12 13 14' 15' 16' 17 18' 19' 20' 21 Big (112 ha) First Growth of famously elusive perfume and style, never great weight, although more dense, sleek these days. Great vintages need keeping for decades. Saskia de Rothschild now at helm; sustainable action; transition to organics. Plans for new winery extension. Second label: CARRUADES DE LAFITE. Also owns CHX DUHART-MILON, L'ÉVANGILE, RIEUSSEC.

Budget for Lafite's new winery extension is €18 million.

Lafleur Pom ★★★★ 04' 05' 06 07 08 09' 10' 11 12 13 14 15' 16' 17' 18' 19' 20 21 Superb but tiny family-owned property. Elegant, intense wine for maturing. Expensive. New cellar extension. Second label: *Pensées de Lafleur*. Also v.gd Les Champs Libres 100% SAUV BL.

Lafleur-Gazin Pom ★★ 09 10 12 14 15 16' 17 18' 19 20' Small, gd-value J-P MOUEIX estate. 85% MERLOT, 15% CAB FR. Elegant and fresh.

Lafon-Rochet St-Est ★★★ 00' 01 04 05' 06 07 08 09' 10' 12 13 14 15' 16' 17 18' 19 20' Fourth Growth neighbour of COS LABORY. Same stable as LILIAN LADOUYS, PÉDESCLAUX since 2021. On gd form; firm but zesty. Eye-catching canary-yellow buildings and label. Second label: Les Pélerins de Lafon-Rochet.

Lagrange St-Jul ★★★ 05' 06' 08' 09' 10' 12 13 14 15' 16' 17 18 19' 20 21 Substantial (118 ha) Third Growth owned by Suntory. Much investment in vyd, cellars. Consistent, fresh. Dry white Les Arums de Lagrange. Second label: Les Fiefs de Lagrange (gd value). Pairings with Japanese dishes proposed.

Lagrange Pom ★★ 09 10 14 15 16' 18 19 20 Tiny vyd on n edge of plateau. Owned by J-P MOUEIX (1953). 100% MERLOT. Supple, round, accessible.

Lagune, La H-Méd ★★★ 05' 08 09' 10' 11 14 15' 16' 17 19 20 21 Third Growth in v. s of MÉD. Dipped in 90s; now on form. Fine-edged with more structure and depth. No 18 (hail). Bio certification. Second label: Moulin de La Lagune.

Lamarque, De H-Méd ★★ 08 09' 10' 14 15 16 18 19' 20 Medium-sized estate with medieval fortress. Competent, savoury, mid-term wines; value. Lots of PETIT VERDOT. Second label: D de Lamarque.

Lanessan H-Méd ★★ 09 10' 12 14 15' 16' 17 18 19' 20 Gd-value classic claret; accessible. Property located just s of ST-JUL. Eric Boissenot consults.

Langoa Barton St-Jul ★★★ 05' 06 08' 09' 10' 11 12 14 15' 16' 17 18' 19' 20 21 Small Third Growth sister CH to LÉOVILLE BARTON; CAB SAUV-led (55%+); charm, elegance; 9th-generation Barton at helm.

Larcis Ducasse St-Ém ★★★ 05' 06 08 09' 10 12 14 15' 16' 17 18 19' 20 21 PREMIER GRAND CRU CLASSÉ on top form. Wine to mature; s-facing terraced vyd; 80% MERLOT. Second label: Murmure de Larcis Ducasse.

Larmande St-Ém ★★ 05 08 09' 10 12 14 15 16 18' 19' 20 GRAND CRU CLASSÉ in same stable as SOUTARD. Sound but lighter weight and frame; accessible. Environmental certification.

Laroque St-Ém ★★→★★★ 08 09' 10' 11 12 14 15 16 17 18' 19' 20' 21 Large GRAND CRU CLASSÉ. Terroir-driven wines. More finesse since 2017.

Larose-Trintaudon H-Méd ★★ 10 12 14 15 16 18 19' (20) CRU BOURGEOIS Supérieur. Largest vyd in MÉD (165 ha). 75,000 cases p.a. Environmental certification. Generally for early drinking. Second label: Les Hauts de Trintaudon.

Laroze St-Ém ★★ 09' 10' 12 14 15 16' 18' 19' 20 GRAND CRU CLASSÉ owned by Guy Meslin. MERLOT-led (70%). Environmental certification. Lighter-framed wines but fruit, balance. Second label: La Fleur Laroze.

Larrivet Haut-Brion Pe-Lé ★★★ (r) 05' 06 08 09 10' 12 14 16' 18' 19 20 Unclassified property owned by Gervoson family since 1987; vyd expanded from 17 to 75 ha. Visitor-friendly (decorated concrete-egg tanks). Opulent, seductive red. Rich, creamy *white* 18 19 20. Second label (r/w): Les Demoiselles.

Lascombes Marg ★★★ 05' 06 08 09 10' 11 12 14 15' 16' 17 18 19' 20 Large (120-ha) Second Growth with chequered history. Wines rich, dark, opulent, modern with touch of MARG perfume. Lots of MERLOT (50%+). Second label: Chevalier de Lascombes. Also a HAUT-MÉD.

Latour Pau ★★★★ 01 02 04 05' 06 07 08 09' 10' 11 12 13 14 15' 16' 17 18' 19' 20' First Growth considered grandest statement of B'X. Profound, intense, almost immortal wines in great yrs; even weaker vintages have unique taste and run for many yrs. Sustainable and organic practices. Ceased en PRIMEUR sales in 2012; wines now only released when considered ready to drink (13 in 2021: just about ready but decant). New cellars for more storage. Owned by Pinault family; vines also in Burgundy, Rhône, Napa. Second label: LES FORTS DE LATOUR; *third label: Pauillac;* even this can age 20 yrs.

Latour à Pomerol Pom ★★★ 05' 06 08 09' 10' 12 14 15' 16' 17 18 19' 20 21 J-P MOUEIX property. Mix of clay-gravel and loamy soils. Extremely consistent, well-structured wines that age.

Latour-Martillac Pe-Lé ★★-→★★★ (r) 05' 08 09' 10' 12 14 15' 16 17 18' 19 20 GRAV Cru Classé owned by Kressmann family. Environmental certification. Fresh, fragrant red. Appetizing *white* 18 19 20. Usually gd value.

Laurets, Des St-Ém ★★ 14 15 16 18 19 (20) RIP Benjamin de Rothschild. Large estate: MERLOT-led. Sélection Parcellaire from old-vine plots.

Laville Saut ★★ 11' 13 14 15 16 18 Non-classified Preignac estate run by Jean-Christophe Barbe; also lectures at Bordeaux University. SÉM-dominated (85%) with a little SAUV BL, MUSCADELLE. Lush, gd-value, botrytized wine.

Léoville Barton St-Jul ★★★★ 04 05' 06 07 08 09' 10' 11 12 13 14 15' 16' 17 18' 19' 20' 21 Second Growth owned by Anglo-Irish Bartons since 1826, Lilian Barton now at helm. RIP Anthony Barton 2022. Harmonious, classic claret; CAB SAUV-dominant (74%). Ongoing renovation/extension of cellars incl new gravity-flow *cuverie*. Second label: La Rés de Léoville Barton.

Léoville Las Cases St-Jul ★★★★ 00' 04' 05' 06 08 09' 10' 11' 12 13 14 15' 16' 17' 18' 19' 20' 21 Largest Léoville and original "Super Second" owned by Jean-Hubert Delon. CAB SAUV dominant but also CAB FR. Elegant, complex wines built for long ageing. On-going cellar extension. Second label: Le Petit Lion.

Léoville Poyferré St-Jul ★★★★ 04 05' 06 07 08 09' 10' 11' 12 13 14 15' 16' 18' 19' 20' 21 In Cuvelier family hands since 1920; Sara Lecompte Cuvelier now at helm. "Super Second" level; dark, rich, spicy, long-ageing. *Ch Moulin Riche* a separate 21-ha parcel. LE CROCK same stable.

Lestage List ★★ 09 10 12 14 15 16' 18 19' 20 CRU BOURGEOIS Exceptionnel; 3rd-generation Chanfreau family taking over. Firm, slightly austere.

Lilian Ladouys St-Est ★★ 09' 10' 12 14 15 16' 17 18' 19' 20 Sizeable CRU BOURGEOIS Exceptionnel. Owned by Lorenzetti family (*see* LAFON-ROCHET, PÉDESCLAUX). More finesse in recent vintages. Second label: La Devise de Lilian.

Liversan H-Méd ★★ 10 14 15 16' 18 19 20 CRU BOURGEOIS in n H-MÉD; vyd in single block; owned by Advini group. Round, savoury, early drinking.

Estimated cost of square metre of vyd on St-Ém plateau is €800; €30 on the plain.

Loudenne Méd ★★ 14 15 16 18 19' 20 Large estate, now Chinese-owned. Landmark C17 pink-washed *chartreuse* by river. Visitor-friendly. Supple 50/50 MERLOT/CAB SAUV reds for early drinking. SAUV BL-led white.

Louvière, La Pe-Lé ★★★ (r) 08 09' 10' 12 14 15 16' 17 18 19 20 (w) 16 17 18 19 20 Vignobles André Lurton property; Jacques Lurton now in charge. Excellent white (100% SAUV BL), savoury red that can age (generally CAB SAUV-led).

Lussac, De St-Ém ★★ 14 15 16 18 19 (20) Top estate in LUSSAC-ST-ÉM. Supple red, rosé. B&B.

Lynch-Bages Pau ★★★★ 02 04′ 05′ 06 07 08 09′ 10′ 11 12 13 14 15 16′ 17 18 19′ 20′ Always popular, now a star, far higher than its Fifth Growth rank. Rich, dense CAB SAUV-led wine for ageing. Owned by Cazes family (Jean-Charles runs it). Second label: Echo de Lynch-Bages. Gd white, ***Blanc de Lynch-Bages***. Impressive new gravity-fed winery.

Lynch-Moussas Pau ★★ 08 09 10′ 12 14 15 16 18 19′ 20 Fifth Growth owned by Castéja family. Medium-bodied; less gravitas than top PAU (75% CAB SAUV), but much improved.

Lyonnat St-Ém ★★ 15 16 18 19 (20) Leading LUSSAC-ST-ÉM. MERLOT-led; more precision lately. Special CUVÉE Emotion, cuvée Sans Sulfites (without sulphur).

Malartic Lagravière Pe-Lé ★★★ (r) 09 10 11 14 15′ 16′ 17 18 19′ 20′ (w) 17 19 20 Classed Growth. Loads of investment. Visitor-friendly. Rich, modern, CAB SAUV-led red; more racy from 2019; appetizing *white* (majority SAUV BL). CH Gazin Rocquencourt (PE-LÉ) same stable.

Malescasse H-Méd ★★ 09 10 12 14 15′ 16′ 17 18 19 20 CRU BOURGEOIS Exceptionnel. Investment, upgrade. Ripe, fleshy, polished; early drinking.

Malescot St Exupéry Marg ★★★ 05′ 08 09′ 10′ 12 14 15′ 16′ 17 18′ 19 20 Third Growth owned by Jean-Luc Zuger. CAB SAUV, MERLOT, CAB FR, PETIT VERDOT in blend. Ripe, fragrant, finely structured.

Malle, De Saut ★★★ (w/sw) 01′ 05 06 09 10′ 11′ 13 14 15′ 16′ 17′ 18 19 Second Growth owned by de Bournazel family. Fine, medium-bodied SAUT. Second label: Les Fleurs de Malle.

Margaux, Ch Marg ★★★★ 04′ 05′ 06′ 07 08 09′ 10′ 11 12 13 14 15 16′ 17 18′ 19′ 20′ 21 First Growth; most seductive, fabulously perfumed, consistent. Owned by Mentzelopoulos family (1977). CAB SAUV-dominated (as much as 90%). The *grand vin* is c.37% of total production. Second label: Pavillon Rouge 09 10′ 11 16′ 17 18′ 20 21. Third label: Margaux du CH Margaux. *Pavillon Blanc* (100% SAUV BL): best white of MÉD, recent vintages fresher 17 18′ 19′ 20 21.

Marojallia Marg ★★★ 08 09′ 10 12 15 16 17 18′ 19′ 20 Micro-CH looking for big prices for big, rich, un-MARG-like wines. CAB SAUV-led (70%). Second label: CLOS Margalaine.

Marquis de Terme Marg ★★→★★★ 06 08 09′ 10′ 12 14 15 16′ 17 18′ 19′ 20 Fourth Growth with vyd dispersed around MARG. Investment, progression in recent yrs but solid side of Marg. Second label: La Couronne. Also a rosé.

Maucaillou Mou ★★ 09 10 11 12 14 15′ 16′ 18 19′ 20 Large, consistent estate. CAB SAUV-led but 7% PETIT VERDOT. Clean, value. Second label: No2 de Maucaillou.

Mayne Lalande List ★★ 10 12 14 15 16 18 19 (20) RIP Bernard Lartigue, creator (1982) and owner. Full, finely textured. B&B.

Mazeyres Pom ★★ 09 10 14 15 16′ 18 19′ 20 Lightish but consistent. Earlier-drinking, MERLOT-led. Organic, bio certification. Alain Moueix at helm.

Meyney St-Est ★★→★★★ 04 05′ 06 08 09′ 10′ 11 12 14 15′ 16′ 17 18′ 19′ 20′ Big river-slope vyd, superb site next to MONTROSE. Structured, age-worthy; CAB SAUV-led (60%) but lots of PETIT VERDOT as well (13% in 2020); gd value at this level. Second label: Prieur de Meyney.

Mission Haut-Brion, La Pe-Lé ★★★★ 02 04 05′ 06 07 08 09′ 10′ 11 12 13 14 15′ 16′

St-Émilion armed for hail

A new collective hail-control system is being used by producers in ST-ÉM and satellites PUISSEGUIN and LUSSAC. Forewarned by a radar that detects storm fronts, vignerons can activate the 37 launchers spread across the territory when the storm is 7 km (4.3 miles) away. These release helium-filled balloons complete with hygroscopic salts to seed the clouds, so causing precipitation. Hail to the new system if it vanquishes hail.

17' 18' 19' 20' 21 As with neighbouring HAUT-BRION, owned by Dillon family. Considerable investment over last 35 yrs. Consistently grand-scale, full-blooded, long-maturing wine. Second label: La Chapelle de la Mission. Magnificent SÉM-dominated white (more SAUV BL in 2020): previously Laville Haut-Brion; renamed La Mission Haut-Brion Blanc 17' 18 19' 20' 21. Second label: La Clarté.

Monbousquet St-Ém ★★★ 05' 08 09' 10' 12 14 15 16' 18 19 20 GRAND CRU CLASSÉ on sand and gravel plain. Transformed by Gérard Perse (*see* PAVIE). Concentrated, oaky and voluptuous. Rare *v.gd Sauv Bl/Gr* (AC B'X). Second label: Angélique de Monbousquet.

Cab Fr accounts for just 9% of red plantings in B'x.

Monbrison Marg ★★→★★★ 08 09' 10' 12 14 15' 16 17 18 19 20 Tiny property owned by Vonderheyden family. 75%+ CAB SAUV in blend. Delicate, fragrant.

Mondotte, La St-Ém ★★★→★★★★ 05' 06 08 09' 10' 11 12 14 15' 16 17 18 19' 20' Tiny (4.5 ha) PREMIER GRAND CRU CLASSÉ on limestone-clay plateau; 75% MERLOT, 25% CAB FR; intense, powerful wines. Organic certification. Same stable as D'AIGUILHE, CANON-LA-GAFFELIÈRE.

Montrose St-Est ★★★★ 04 05' 06 07 08 09' 10' 12 13 14 15' 16' 17 18' 19' 20' 21 Second Growth with riverside vyd. Famed for forceful, long-ageing claret. Vintages 1979–85 were lighter. Bouyges brothers owners. Massive environmental programme: biodiversity, carbon neutral by 2050, compost, electric tractors. Second label: *La Dame de Montrose*.

Moulin du Cadet St-Ém ★★ 09 10' 12 14 15' 16 18' 19' 20' Tiny (2.85 ha) GRAND CRU CLASSÉ on plateau; 100% MERLOT. Modern with an undertow of terroir. Same owner as SANSONNET.

Moulinet Pom ★★ 09 10 12 15 18 19 20 Large CH for POM: 18 ha. MERLOT-led (85%). Lighter style. DERENONCOURT consults.

Moulin Haut-Laroque Fron ★★ 09' 10' 11 12 14 15' 16 17 18' 19' 20' RIP Jean-Noël Hervé: put CH on map. Consistent, structured, can age.

Moulin Pey-Labrie Fron ★★ 09' 10' 12 15 16 18 19 20' MERLOT-led CANON-FRON. Organic certification. Sturdy, well-structured wines that can age.

Moulin St-Georges St-Ém ★★★ 09' 10' 12 14 15' 16' 17 18' 19' 20 Lively, fresh, harmonious. Gd value at this level. Same stable as AUSONE, FONBEL.

Mouton Rothschild Pau ★★★★ 02 04' 05' 06 07 08' 09' 10' 11 12 13 14 15' 16' 17' 18' 19' 20' 21 Rothschild-owned (1853). Most exotic, voluptuous of PAU First Growths; at top of game. 2018 label created by Chinese artist Xu Ding. New winemaker in 2020, Jean-Emmanuel Danjoy (ex-CLERC MILON). White Aile d'Argent (SAUV BL/SÉM) now more graceful. Second label: *Le Petit Mouton*. *See also* D'ARMAILHAC.

Nairac Bar ★★ 05' 06 07 09 10 11 13 15 16 17 Second Growth run by brother and sister, Nicolas and Eloïse Tari-Heeter. SÉM-led (90%). Rich but fresh. Second label: Esquisse de Nairac.

Nénin Pom ★★★ 05 06 08 09' 10' 12 14 15' 16' 17 18' 19' 20' 21 Large estate; same stable as LÉOVILLE LAS CASES. Plenty of investment; wines now restrained but generous, precise, built to age (20 lusher). Increase in CAB FR (40%). Gd-value second label: Fugue de Nénin.

Olivier Pe-Lé ★★★ (r) 08 09' 10' 12 14 15 16' 18 19' 20 (w) 17 18 19' 20 Vast DOM owned de Bethmann family. Structured red (55% CAB SAUV), juicy SAUV BL-led (75%) white; gd value at this level.

Ormes de Pez St-Est ★★ 05 06 08' 09' 10' 12 14 15' 16' 17 18' 19 20' Same stable as LYNCH-BAGES (Cazes family). Cool, classic, age worthy.

Ormes Sorbet, Les Méd ★★ 10' 12 14 15 16' 18 19 (20) Reliably consistent MÉD cru; 9th-generation ownership. Environmental certification. CAB SAUV-LED (65%). Elegant, gently oaked wines.

Palmer Marg ★★★★ 04 05' 06' 07 08 09' 10' 11 12 13 14 15' 16' 17 18' 19' 20' 21 Third Growth on par with "Super Seconds" (occasionally Firsts). Voluptuous wine of power, complexity and much MERLOT (40%+). Dutch (Mähler-Besse) and British (SICHEL) owners. Philosophy and certification bio. Sells 50% en PRIMEUR and releases vintage at 10 yrs (11 in 2021). Second label: *Alter Ego de Palmer*.

Pape Clément Pe-Lé ★★★★ (r) 04 05 06 08 09' 10' 12 13 14 15' 16' 17 18' 19' 20 (w) 18 19 20 Historic estate in Bordeaux suburbs. Owned by magnate Bernard Magrez (*see* FOMBRAUGE, TOUR-CARNET). Dense, long-ageing reds. Tiny production of rich, oaky white. Second label (r/w): Clémentin.

Patache d'Aux Méd ★★ 12 14 15 16' 18 19 20 Sizeable CRU BOURGEOIS. Classic, savoury. Fairly consistent; value. DERENONCOURT consults.

Pavie St-Ém ★★★★ 05' 06 07 08 09' 10' 11 12 13 14 15' 16' 17 18 19' 20' 21 PREMIER GRAND CRU CLASSÉ (A) splendidly sited on plateau and s côtes. Imposing winery. Intense, powerful wines for long ageing; recent vintages less extreme. MERLOT-led but now more CABS FR/SAUV (50% in 19 20). Second label: Arômes de Pavie.

Pavie Decesse St-Ém ★★★ 05 06 08 09' 10' 12 14 15' 16 17 18 19' 20' 21 Tiny GRAND CRU CLASSÉ. 90% MERLOT. Tight, tannic, needs time.

Pavie Macquin St-Ém ★★★ 05' 06 08 09' 10 12 15 16' 17 18' 19' 20' 21 PREMIER GRAND CRU CLASSÉ (B); vyd on limestone plateau; 80% MERLOT, 20% CABS FR/SAUV. Managed by Nicolas Thienpont. New cellar in 2022. Sturdy, full-bodied wines need time. Second label: Les Chênes de Macquin.

Pédesclaux Pau ★★ 09 10' 12 14 15' 16' 17 18 19' 20 Underachieving Fifth Growth revolutionized by owner Jacky Lorenzetti. Extensive investment since 2014; gravity-fed cellar, more vyds. Lighter than top PAU but well defined.

Petit Village Pom ★★★ 05 06 08 09' 10' 12 14 15' 16' 17 18' 19' 20' 21 Much-improved estate on plateau. Atypical 40% CABS FR and SAUV. Organic conversion from 2020. Cellars renovated. Suave, dense, increasingly finer tannins.

Petrus Pom ★★★★ 01 02 04 05' 06 07 08 09' 10' 11' 12 13 14 15' 16' 17 18' 19' 20' 21 (Unofficial) First Growth of POM: MERLOT solo *in excelsis*. 11.5-ha vyd on blue clay gives 2500 cases of massively rich, concentrated wine for long ageing; at a price. Owner Jean-François Moueix. Winemaker Olivier Berrouet. No second label.

Pey La Tour B'x ★★ 15 16 18 19 20 Large (176 ha) DOURTHE property. Quality-driven B'X SUPÉRIEUR. Three red CUVÉES: Rés du CH (MERLOT-led) top; rosé, dry white.

Peyrabon H-Méd ★★ 09' 10 12 15 16 17 18 19 20 Savoury CRU BOURGEOIS Supérieur in n HT-MÉD; NÉGOCIANT Millésima owner.

Pez, De St-Est ★★★ 08 09' 10' 11 12 14 15' 16' 17 18' 19' (20) Same owner as HAUT-BRION (Pontac) in C16. Dense, reliable. Now in PICHON L COMTESSE stable.

Phélan Ségur St-Est ★★★ 05' 06 08 09 10' 11 12 14 15' 16' 17 18' 19' 20 Reliable, top-notch, unclassified CH with Irish origins; long, supple style. Belgian owner, Philippe Van de Vyvere. Second label: Frank Phélan.

Pibran Pau ★★ 08 09' 10' 11 12 14 15 16' 17 18' 19' 20 21 Earlier-drinking PAU allied to PICHON BARON. Almost 50/50 MERLOT/CAB SAUV.

Beam me down, Scotty

Twelve bottles of PETRUS 2000 spent 14 mths in orbit on the International Space Station, returning to earth in Jan 2021. Tastings were then carried out to see if there was any difference between terrestrial and space-aged wines. The conclusion was: some variation in colour but deviation in taste was more dependent on the individual taster's sensitivity. Biggest change is the price the space-wine hopes to achieve at auction. Estimated value is $1 million. However, it's all in a gd cause as the proceeds go towards further experiments in space linked to agriculture.

Pichon Baron Pau ★★★★ 04 05′ 06 07 08 09′ 10′ 11 12 13 14 15′ 16′ 17 18′ 19′ 20 21 Owned by AXA. Second Growth on flying form. Powerful, long-ageing PAU at a price. New *cuvier* and visitor centre 2023. Second labels: Les Tourelles de Longueville (approachable: more MERLOT); Les Griffons de Pichon Baron (generally CAB SAUV-dominant).

Pichon Longueville Comtesse de Lalande (Pichon Lalande) Pau ★★★★ 02 04 05′ 06 07 08 09′ 10′ 11 12 13 14 15′ 16′ 17 18′ 19′ 20′ 21 ROEDERER-owned Second Growth (investment), neighbour of LATOUR. Always among top performers; long-lived wine of famous breed. MERLOT-marked in 80s, 90s; more CAB SAUV in recent yrs (77% in 20 with 6% CAB FR). Second label: *Rés de la Comtesse.*

There are 138 wine-growers for 782 ha in Pom.

Pin, Le Pom ★★★★ 04 05′ 06′ 07 08′ 09′ 10′ 11 12 14 15 16′ 17 18′ 19′ 20′ 21 The original B'x cult wine owned by Jacques Thienpont. Only 2.8 ha. Neighbour of TROTANOY. 100% MERLOT; almost as rich as its drinkers, prices out of sight. Ageing potential. L'If (ST-ÉM), L'Hêtre (CAS) are stablemates. New MD from 2022 vintage: Diana Berrouet Garcia, ex-PETIT VILLAGE.

Plince Pom ★★ 08 09 10 12 14 15 16′ 18 19 (20) Lighter style of POM on sandy soils. 84% MERLOT. Oaky when young.

Pointe, La Pom ★★ 08 09′ 10 11 12 14 15′ 16′ 18 19 20 Large, well-run estate; gd value, consistent but less intensity than top POM.

Poitevin Méd ★★ 12 14 15 16 17 18 19 (20) CRU BOURGEOIS Supérieur. Consistent quality, value. Also dry white B'x.

Pontet-Canet Pau ★★★★ 02 04 05′ 06′ 07 08 09′ 10′ 12 14 15 16′ 17 18′ 19′ 20 21 Large, bio-certified, Tesseron family-owned Fifth Growth. Radical improvement has seen prices soar. CAB SAUV-dominant (60%+). Amphorae for ageing from 2012 (with 50% new oak barrels). Classic PAU but generous, refined.

Potensac Méd ★★→★★★ 09′ 10′ 11 12 14 15 16′ 17 18′ 19′ 20′ 21 Same stable as LÉOVILLE LAS CASES. Firm, long-ageing wines; gd value. MERLOT-led (45%) but lots of old-vine CAB FR (up to 20%). Second label: Chapelle de Potensac.

Pouget Marg ★★ 08 09′ 10′ 12 14 15′ 16 18 19 (20) Tiny (10 ha) Fourth Growth sister of BOYD-CANTENAC. Blend varies. Sturdy; needs time.

Poujeaux Mou ★★ 08 09 10 12 14 15′ 16′ 17 18′ 19 20 Same stable as CLOS FOURTET. DERENONCOURT consults. Full, robust wines with ageing potential. Second label: La Salle de Poujeaux.

Premier Grand Cru Classé St-Ém 2012: 18 classified: new classification in 2022. See box, p.118

Pressac, De St-Ém ★★ 09 10 12 14 15′ 16′ 17 18′ 19′ 20′ GRAND CRU CLASSÉ e of St-Ém town. Five B'x grape varieties. Ripe, full, engaging. Value.

Preuillac Méd ★★ 10 15 16 18 19′ 20 Savoury, structured CRU BOURGEOIS Supérieur. MERLOT-led (54%). DERENONCOURT consults.

Prieuré-Lichine Marg ★★★ 09′ 10′ 11 12 14 15′ 16′ 17 18′ 19′ 20′ Fourth Growth put on map in 60s by Alexis Lichine. CAB SAUV-led (60%). Parcels in all five MARG communes. Environmental certification. Fragrant, currently on gd form. Visitor-friendly. Gd white SAUV BL/SÉM too.

Puygueraud B'x ★★ 10 14 15′ 16′ 17 18′ 19 20 Leading CH of tiny FRANCS-CÔTES DE B'X AC. Owned by Thienpont family. MERLOT-led wines of surprising class. Small crop 20 (hail). Also MALBEC/CAB FR-based cuvée George. A little dry white (SAUV BL/Gr).

Quinault L'Enclos St-Ém ★★→★★★ 10 12 14 15 16′ 17 18′ 19′ 20 GRAND CRU in Libourne. Same team/owners as CHEVAL BLANC. Plenty of investment. MERLOT-led but 26% CABS SAUV/FR; more freshness, finesse.

Quintus St-Ém ★★★ 11 12 14 15 16′ 17 18′ 19′ 20′ Created by Dillons of HAUT-BRION

St-Émilion classification – likely change in 2022
The 2012 classification incl a total of 82 CHX: 18 PREMIERS GRANDS
CRUS CLASSÉS and 64 GRANDS CRUS CLASSÉS. Chx ANGÉLUS and PAVIE were
upgraded to Premier Grand Cru Classé (A), while added to the rank of
Premier Grand Cru Classé (B) were CANON-LA-GAFFELIÈRE, LA MONDOTTE,
LARCIS DUCASSE and VALANDRAUD. New to status of Grand Cru Classé
were Chx BARDE-HAUT, CLOS de Sarpe, Clos La Madeleine, Côte de
Baleau, DE FERRAND, DE PRESSAC, FAUGÈRES, FOMBRAUGE, JEAN FAURE, La
Commanderie, La Fleur Morange, Le Chatelet, Péby Faugères, QUINAULT
L'ENCLOS, Rochebelle and SANSONNET. The classification is reviewed every
10 yrs, with latest edition in 2022. One thing is sure: it will not incl CHX
ANGÉLUS, AUSONE and CHEVAL BLANC (or siblings LA CLOTTE and QUINAULT
L'ENCLOS) as they have now officially withdrawn from the classification.
Another change touted is the promotion of FIGEAC to Premier Grand
Cru Classé (A). All will be revealed in due course.

from former Tertre Daugay and L'Arrosée vyds. Grand Pontet vyd added in 2021.
Gaining in stature but expensive. Second label: Le Dragon de Quintus.

Rabaud-Promis Saut ★★→★★★ 09' 10 11 13 14 15' 16 17' 18 19 First Growth (1855).
Quality, gd value. Special bottle in 18.

Rahoul Grav ★★ (r) 14 15 16' 17 18 19 20 DOURTHE property; reliable, MERLOT-led red.
SÉM-dominated white 18 19' 20. Environmental certification; gd value.

Ramage la Batisse H-Méd ★★ 10 15 16 18 19 (20) Widely distributed CRU BOURGEOIS
Supérieur. CAB SAUV, MERLOT and lots of PETIT VERDOT (16% in 18).

Rauzan-Gassies Marg ★★★ 08 09' 10 14 15 16' 18 19 20' Second Growth. Same
stable as CROIZET-BAGES. Improvement but still lags behind top MARGS (20 best
yet). Second label: Gassies.

Rauzan-Ségla Marg ★★★★ 00' 05' 06 08 09' 10' 11 12 14 15' 16' 17 18' 19' 20' Leading
Second Growth; fragrant, structured; owned by Wertheimers of Chanel. CAB
SAUV-led (60%). Second label: Ségla (value). BERLIQUET, CANON same stable.

Raymond-Lafon Saut ★★★ 05' 06 07' 09' 10 11' 13 14 15' 16' 17' 18 19 20 Unclassified
SAUT but First Growth quality. 80% SÉM. Rich and complex wines that age;
gd value.

Rayne Vigneau, De Saut ★★★ 05' 07 09' 10' 11' 13 14 15' 16' 17 18' 19' 20 21
Substantial First Growth. Suave, age-worthy (74% SÉM). Special CUVÉE Gold.
Visitor-friendly. Environmental certification. Second label: Madame de Rayne.

Respide Médeville Grav ★★★ (r) 14 15 16 18 19 20 (w) 18 19 20 Top GRAV property. Same
ownership as GILETTE. Elegant red, complex *white*.

Reynon B'x ★★ Leading CADILLAC-CÔTES DE B'x estate owned by Denis Dubourdieu
Doms. MERLOT-led red 16' 18 19' 20'; B'x white SAUV BL (DYA).

Reysson H-Méd ★★ 10' 14 15 16 18 19 20 CRU BOURGEOIS Supérieur. Mainly MERLOT
(90%), CAB FR, PETIT VERDOT. Consistent; rich but fragrant.

Rieussec Saut ★★★★ 05' 06 07 09' 10' 11' 13 14 15' 16' 17 18' 19 First Growth with
substantial vyd in Fargues, owned by (LAFITE) Rothschild. Regularly powerful,
opulent. New bottle marked with yellow crown from 2019. Second label:
Carmes de Rieussec.

Rivière, De la Fron ★★ 14 15 16' 18 19 20 Largest (65 ha), most impressive FRON
property with C16 CH. MERLOT-led, CABS SAUV and FR, MALBEC. Formerly big,
tannic; now more refined. Visitor-friendly; B&B.

Roc de Cambes B'x ★★★ 05 06 08 09 10' 11 12 14 15' 16' 17 18 19 20 21 Undisputed
leader in CÔTES DE BOURG; savoury, opulent but pricey. Same Mitjavile stable and
winemaking as TERTRE-ROTEBOEUF. Also DOM de Cambes.

Rochemorin, De Pe-Lé ★★ (r) 14 15 16 18' 19 20 (w) 18 19 20 Large property at

Martillac owned by VIGNOBLES André Lurton. Fleshy, dark fruit red (50/50 MERLOT/CAB SAUV); aromatic white (100% SAUV BL). Fairly consistent quality.

Rol Valentin St-Ém ★★ 09' 10' 12 14 15' 16' 18 19 20' With two separate vyd sites. Ripe, concentrated, structured. MERLOT-led (85%) with 5% MALBEC.

Rouget Pom ★★ 08 09' 10' 12 14 **15** 16' 18 19 20 Sizeable (for POM) go-ahead estate on n edge of plateau. Labruyère family owned. Rich, powerful, unctuous wines. Organic conversion. Second label: Le Carillon de Rouget.

St-Georges St-Ém ★★ 10 12 14 **15 16** 18 19 (20) Vyd represents 25% of ST-GEORGES AC. MERLOT-led (80%) with 10% each CABS FR and SAUV; gd wine sold direct to public. Second label: Puy St-Georges. Also Trilogie, old-vine MERLOT.

St-Pierre St-Jul ★★★ 05' 06 08 09' 10' 11 12 14 15' 16' 18' 19' 20 21 Tiny Fourth Growth to follow. CAB SAUV-led (75%). Stylish, consistent and classic. Same ownership as GLORIA.

Sales, De Pom ★★ 09 10' 12 **15 16 17** 18 19' 20' Biggest vyd of POM (47.6 ha). New impetus since 2017; more depth and polish. Honest, drinkable; can age. Second label: CH Chantalouette.

Sansonnet St-Ém ★★→★★★ 08 09' 10' 12 14 **15** 16' **17** 18 19' 20 21 Plateau-based GRAND CRU CLASSÉ. Modern but refreshing. Same owner as MOULIN DU CADET and Villemaurine.

Saransot-Dupré List ★★ 09' 10' 12 **15 16** 18 19 (20) CRU BOURGEOIS Supérieur. MERLOT-led with PETIT VERDOT (20% in 18) and CABS FR and SAUV. Firm, generous.

Sénéjac H-Méd ★★ 09 10' 12 14 **15 16' 18** 19' 20' Cru in S H-MÉD (Pian). Consistent, well-balanced wines. Drink young or age. TALBOT same stable.

Serre, La St-Ém ★★ 09' 10 12 14 **15 16' 17** 18' 19' 20' 21 Small GRAND CRU CLASSÉ on plateau. Fresh, stylish wines with fruit. Consistent.

Sigalas Rabaud Saut ★★★ 05 07' 09' 10' 11' 13 14 15' 16 17' 18 19' 21 Tiny First Growth; *v. fragrant and lovely*. Second label: Le Lieutenant de Sigalas (only SAUT made in 20). La Sémillante dry white.

Siran Marg ★★→★★★ 05 08 09' 10' 12 14 15' 16' **17** 18' 19' 20 Unclassified MARG estate in Labarde. Run by Edouard Miailhe. Wines have substance, fragrance. Hubert de Boüard consults. Visitor-friendly. Second label: S de Siran.

Smith Haut Lafitte Pe-Lé ★★★★ (r) 05' 06 08 09' 10' 11 12 14 15' 16' **17** 18 19' (w) 18 19' 20 Celebrated Classed Growth with spa hotel (Caudalie), regularly one of PE-LÉ stars. White is full, ripe, sappy; red precise, generous. Cathiard owners 30th vintage in 2020. Second label: Les Hauts de Smith. Also CAB SAUV-based Le Petit Haut Lafitte. Organic certification.

Sociando-Mallet H-Méd ★★★ 05 06' 08 09' 10' 11 12 14 15' 16' 18 19' 20 Large H-MÉD estate in St-Seurin-de-Cadourne founded by Jean Gautreau in 1969. 54% MERLOT, 46% CABS SAUV/FR. Classed Growth quality. Wines for ageing. Second label: La Demoiselle de Sociando-Mallet.

Sours, De B'x ★★ Valid reputation for popular B'x rosé (DYA); gd white; acceptable B'x red. Big transformation under ownership of Jack Ma of Alibaba fame.

Embrace a tree

Agroforestry, or the practice of planting trees within or around a vyd, is back in vogue in B'x. In the distant past it was common to find peach or pear trees planted within vine rows, but with increased mechanization the practice went out of fashion. Now with more and more CHX adopting an environmental approach, trees are again seen as a positive factor, providing windbreaks and shade and ensuring biodiversity as well as soil stability and fertility via microbial activity. Among the converts are CHEVAL BLANC with 2000 trees planted and Claire Villars-Lurton of HAUT-BAGES-LIBÉRAL. It's all v. green.

Soutard St-Ém ★★★ 09 10 12 14 15 16' 17 18' 19' 20' 21 *Potentially excellent* GRAND CRU CLASSÉ on limestone plateau. Massive investment, making strides from 2018. MERLOT-led (63%). Visitor-friendly. Second label: Petit Soutard. LARMANDE same stable.

Suduiraut Saut ★★★★ 05' 06 07' 09 10' 11' 13 14 15' 16' 17' 18 19' 20' 21 One of v. best SAUT. SÉM-dominant; Pierre Montégut long-time winemaker. Luscious quality, v. consistent. Second labels: Castelnau de Suduiraut; Lions de Suduiraut (fruitier). Dry wines "S" and entry-level Le Blanc Sec.

Sweet whites represent 1% of production in B'x, same as sparkling Crémant de B'x.

Taillefer Pom ★★ 08 09' 10 12 14 15 16 18' 19 20' Claire Moueix at the helm. MERLOT-led (75%). Sandier soils. Lighter weight but polished, refined.

Talbot St-Jul ★★★ 04 05' 08 09' 10' 11 12 14 15 16 17 18 19 21 Substantial (110 ha) Fourth Growth in heart of AC ST-JUL. Wine rich, *consummately charming*, *reliable*. DERENONCOURT consults. Second label: Connétable de Talbot. Approachable SAUV BL-based: Caillou Blanc.

Tertre, Du Marg ★★★ 08 09' 10' 12 14 15' 16' 18' 19' 20 Fifth Growth isolated s of MARG. Grands Chais de France new owner (2021). Fragrant (20% CAB FR), fresh (5% PETIT VERDOT), fruity but structured wines. Visitor-friendly. Second label: Les Hauts du Tertre. Also Tertre Blanc VDF dry white.

Tertre-Roteboeuf St-Ém ★★★★ 04 05' 06' 07 08 09' 10' 11 12 14' 15' 16' 17 18' 19 20 21 Tiny, unclassified, ST-ÉM côtes star; concentrated, exotic; can age. Work of François Mitjavile; frightening prices; v.gd ROC DE CAMBES.

Thieuley B'x ★★ E-2-M supplier of consistent quality AC B'x (r/w); oak-aged CUVÉE Francis Courselle (r/w). Run by sisters Marie and Sylvie Courselle.

Tour Blanche, La Saut ★★★ 05' 06 07' 08 09' 10' 11' 13 14 15 16' 17 18' 19' 20 Excellent First Growth SAUT; rich, bold, powerful wines on sweeter end of scale. SÉM-dominant (83%); only 1.8 hl/ha in 2020. Second label: Les Charmilles de La Tour Blanche.

Tour Carnet, La H-Méd ★★★ 04 05' 06 08 09' 10' 12 14 15 16' 18' 19' 20 21 Large H-MÉD Fourth Growth owned by Bernard Magrez (*see* FOMBRAUGE, PAPE CLÉMENT). Concentrated, opulent wines. Private collection of 75 different grape varieties. Visitor-friendly. Second label: Les Pensées de La Tour Carnet.

Tour de By, La Méd ★★ 09 10 12 14 15' 16' 18 19' 20' Substantial family-run estate in N MÉD. Popular, sturdy, reliable, CAB SAUV-led (60%) wines; can age. Environmental certification. Also a rosé.

Tour de Mons, La Marg ★★ 08 09' 10 12 14 15' 16' 18' 19 20 CRU BOURGEOIS Supérieur. Same owner as LABÉGORCE (2020). Upward curve; gd value.

Tour du Pas St-Georges St-Ém ★★ 10 12 14 15 16 18 19 (20) ST-GEORGES-ST-ÉM estate run by Delbeck family. Classic style. Special CUVÉE: Âme.

Tour du Haut Moulin H-Méd ★★ 05' 06 08 09 10 14 15' 16' 18 19 (20) CRU BOURGEOIS in n H-MÉD. Classic wines; intense, structured, to age.

Tour Figeac, La St-Ém ★★ 09' 10' 12 14 15' 16' 18' 19' 20 21 GRAND CRU CLASSÉ in "graves" sector of ST-ÉM. Fine, floral. Bio practices.

Tour Haut-Caussan Méd ★★ 09 10' 12 14 15 16' 18 19 (20) Consistent, family-run property. 50/50 CAB SAUV/MERLOT on clay-limestone. Value.

Tournefeuille L de P ★★ 09 10' 12 14 15' 16 18 19 20 Reliable L DE P on clay and gravel soils. MERLOT-led. Round, firm, fleshy wine.

Tour St Bonnet Méd ★★ 10' 15 16 18 19 (20) CRU BOURGEOIS. MERLOT, CABS SAUV/FR, PETIT VERDOT, MALBEC. Best in top yrs. Usually value.

Trois Croix, Les Fron ★★ 09 10 12 14 15 16' 17 18' 19' 20 Léon family-owned since 1995; Bertrand winemaker. Fine, balanced, gd-value wines from consistent producer. Clay-limestone soils. MERLOT-led (80%).

Tronquoy-Lalande St-Est ★★★ 09' 10' 11 12 14 15' 16' 17 18' 19' 20 21 Same stable as MONTROSE; MERLOT-led (49%). Consistent, dark, satisfying. Organic leaning. Second label: Tronquoy de Ste-Anne. A little B'X white.

Troplong Mondot St-Ém ★★★ 08 09 10 12 14 15' 16 17 18' 19' 20 21 PREMIER GRAND CRU CLASSÉ (B) on limestone plateau. Lots of investment since 2017; new winery, restaurant. *Wines of power, depth*; major style change in 2017; more elegance, freshness (earlier picking, less new oak). Second label: Mondot.

Trotanoy Pom ★★★★ 05' 06 08 09' 10' 11 12 14 15' 16' 17 18' 19' 20 21 One of jewels in J-P MOUEIX crown. 90% MERLOT, 10% CAB FR (but 100% Merlot in 19 20). Dense, powerful, long ageing.

Trottevieille St-Ém ★★★ 08' 09' 10' 12 14 15 16' 17 18' 19' 20 PREMIER GRAND CRU CLASSÉ (B) owned by BORIE-MANOUX. Greater consistency; wines long, fresh, structured. Lots of CAB FR (50%+). Cellar recently renovated. Second label: La Vieille Dame de Trottevieille.

Valandraud St-Ém ★★★★ 05' 06 08 09' 10 11 12 13 14 15' 16' 17 18 19' 20 21 PREMIER GRAND CRU CLASSÉ (B). Garage wonder turned First Growth. Formerly super-concentrated; now rich, dense, balanced. New sustainable cellar in 2021. Also Virginie de Valandraud and Valandraud Blanc.

Vieille Cure, La Fron ★★ 10' 14 15 16' 17 18' 19' 20 Leading FRON estate; appetizing wines. Value. J-L Thunevin (VALANDRAUD) consults.

Vieux Château Certan Pom ★★★★ 05' 06 07 08 09' 10' 11' 12 13 14 15' 16' 17 18' 19' 20' 21 One of the great POMS. Different in style to neighbour PETRUS; *elegance, harmony and fragrance*. Plenty of old-vine CABS FR/SAUV (30%) is one of the reasons. Great consistency; long-ageing. Alexandre Thienpont and son Guillaume at helm.

Vieux Ch St-André St-Ém ★★ 10 14 15' 16' 17 18 19 (20) Small MERLOT-based vyd in MONTAGNE-ST-ÉM; *gd value*. Property of ex-PETRUS winemaker.

Villegeorge, De H-Méd ★★ 09' 10 14 15 16 18 19 20 Tiny S H-MÉD owned by Marie-Laure Lurton. CAB SAUV-led (63%). Light but elegant wines.

Vray Croix de Gay Pom ★★★ 08 09' 10 12 14 15 16' 17 18' 19' 20' Tiny vyd in best part of POM (seven parcels altogether). More finesse of late. Organic certification; bio practices. Owner Suravenir (CALON SÉGUR).

Yquem Saut ★★★★ 01' 02 03' 04 05' 06' 07 08 09' 10' 11' 14 15' 16' 17' 18' (19) King of sweet, *liquoreux*, wines. Strong, intense, luscious; kept 2 yrs in barrel. Most vintages improve for 15 yrs+, some live 100 yrs+ in transcendent splendour. 100 ha in production (75% SÉM/25% SAUV BL). No Yquem made 51 52 64 72 74 92 2012. Makes small amount of off-dry (5g/l sugar) Sauv Bl (75%), Sém (25%) "Y" (pronounced "ygrec"). New technical director 2021. Cellarmaster Sandrine Garbay has reduced ageing, freshened wines. Yquem now promoted in top restaurants by the glass, young.

In the money – now

Top CHX now seem to have unlimited money to spend; but it wasn't always so. Until the early 80s many were so unprofitable that owners often had to have a day job to pay the bills: Jean-Michel Cazes of LYNCH-BAGES worked as insurance agent. When the Cathiards bought SMITH HAUT LAFITTE in 1990 there were holes in the roof. When Anthony Barton arrived at LANGOA BARTON in 1951, the ch (where his uncle Ronald lived) was so cold that a vase of mimosas in the dining room froze solid and shattered. When May Eliane de Lencquesaing took over PICHON LONGUEVILLE COMTESSE DE LALANDE in 1978, the roof leaked and the furniture was all broken. This might not make today's prices seem more palatable, but it does put them in context.

Italy

More heavily shaded areas are
the wine-growing regions.

Abbreviations used in
the text:

Ab	Abruzzo	Mol	Molise
Bas	Basilicata	Pie	Piedmont
Cal	Calabria	Pu	Puglia
Cam	Campania	Sar	Sardinia
E-R	Emilia-Romagna	Si	Sicily
FVG	Friuli Venezia Giulia	T-AA	Trentino-Alto Adige
Lat	Latium	Tus	Tuscany
Lig	Liguria	Umb	Umbria
Lom	Lombardy	VdA	Valle d'Aosta
Mar	Marches	Ven	Veneto

VALLE D'AOSTA
L Con
L Maggiore
Milan
Turin
PIEDMONT
LOMBAR
Genoa
Po
LIGURIA
Ligurian Sea

It's all about the terroir. Why is it that Barolo from the Vignarionda vineyard can be so distinct from the same producer's Barolo Brunate, even if the wines are made the same way? Why do wines from the Fiano grape from Australia taste so different from those made in Campania? All grape varieties deliver a signature of place, but some (Riesling, Pinot Noir, Nebbiolo) seem to do so more than others. Italy's native grapes are no slouches in this department, though because they are less well known, people don't always realize just how terroir-specific many – most – of them are. The ancient Romans aptly spoke of the *genius loci*, or "genius of place", and it applies to foodstuffs: they speak of where they grow and those who make them. This is the reason why two 100% Sangiovese Chianti Classicos from Gaiole and Radda (two specific subzones of the denomination) are usually sleeker, with higher acidity, than those of Castelnuovo Berardenga (another subzone), which is warmer and has richer iron-containing soils with less limestone.

But Italy has a large array of different habitats, in which exposures, temperatures, altitudes, rainfall and geological origins of the soils differ greatly. Vineyard locations range from the mountainous areas of the Alps with high diurnal temperature variations, to hillsides with a continental climate, to Mediterranean climate zones (where the majority of Italy's vineyards are found). The innumerable wine terroirs that typify Italy are a direct consequence of its topography and geological origin. Blanc de Morgex et de la Salle (in Valle d'Aosta) and the wines of Valtellina and Etna (white and red) are mountain wines. High hillside wines would be those of Piedmont's Langhe or Matelica (Marches); while wines of lower hillside areas would be those of the Collio in Friuli Venezia Giulia and parts of Franciacorta. The wines of the flatlands include some Lambrusco wines of Emilia-Romagna, the incredibly intense Nero d'Avola wines from Pachino (Sicily) and the great Vernaccia di Oristano from Sardinia. Clearly, wines made from one specific terroir are not "better" than those made from another, just different.

Recent vintages

Amarone, Veneto & Friuli

2021 Ven: dry, v.gd quality, less quantity. Friuli: vgd quality, uneven volume.
2020 Ven: balance, gd quality/quantity. Friuli: wet June, uneven, better whites.
2019 Low quantity. Ven: gd for Amarone, Soave. Friuli: fair quality.
2018 Optimal weather conditions. Gd for quantity, quality, esp fresh whites.
2017 V. difficult (non-stop rain); weedy Amarones a risk, green reds in general.
2016 V. hot summer; round but low-acid, big reds, chunky whites.
2015 Quantity gd, quality better; v.gd Friuli reds. Fresher than initially thought.
2014 Not memorable for Amarone, better Soave, Friuli whites, Valpolicellas v.gd.

Campania & Basilicata

2021 Cam: balanced, gd quality, less quantity (best for r). Bas: late harvest, classic, v.gd vintage.
2020 Similar to 19. Classic, balanced but lower quantity. Gd for reds, whites.
2019 Classic, balanced vintage. Best for Aglianico, gd for whites.
2018 Rainy, but whites fresh, lively; sleek reds (Aglianico best).
2017 Low-volume yr of reds plagued by gritty tannins. Whites flat; Greco best.
2016 Cold spring delayed flowering, hot summer allowed catch-up. Fiano best.
2015 Hot, dry early summer: ripe (at times tough) reds, broad whites; drink up.
2014 Rain; spotty quality. Avoid green Aglianicos, Piedirossos. Whites pretty gd.

Marches & Abruzzo

2021 Hot, dry late vintage; gd quality but low yields; gd for late-ripening grapes.
2020 Uneven, quality peaks; wet June, fresh weather; gd for late-ripening grapes.
2019 Difficult yr. Rainy, cold spring. Low quantity, medium quality.
2018 Patchy spring. More balanced than 17; v. high volume, gd quality (r/w).
2017 Best to forget: hot, droughty. Reds gritty, whites overripe. Low volume.
2016 Rain, cold, lack of sun = difficult yr. Lemony, figgy Pecorino probably best.
2015 Hot summer, whites fresher than expected (esp Trebbiano), reds ripe.
2014 Mar: sleek reds, classic whites. Ab: best for Pecorino.

Piedmont

2021 Gd vintage, despite warm dry summer. Gd quality, less quantity.
2020 Classic. Slightly less balance than 19, elegant; better Barbera than Nebbiolo.
2019 Classic vintage, more balanced than 18. Lower quantity but higher quality.
2018 Despite difficult spring, potentially classic Barolo, Barbaresco.
2017 Among earliest harvests in living memory. Can lack depth.
2016 Potentially top vintage; classic, perfumed, age-worthy Barolo/Barbaresco.
2015 Outstanding Barolo/Barbaresco. Should be long-lived. Barbera, Dolcetto gd; Grignolino less so.
2014 Rain: later-picked grapes thrived, Barbaresco (not Barolo) best.
Earlier fine vintages: 10 08 06 04 01 00 99 98 97 96 95 90. Vintages to keep: 10 06 01 99 96. Vintages to drink up: 11 09 03 00 97 90 88.

Tuscany

2021 V.gd vintage (even if warm, dry) mainly for Sangiovese (best: fresher sites).
2020 V.gd vintage, but complicated. Less power, alc than 19, elegant.
2019 Maybe one of best since 2000. Classic, balanced. Gd quality, quantity.
2018 Reds gd: of steely personality, age-worthiness.
2017 Hot, v. difficult vintage. Better in Chianti Classico than coastal Maremma.
2016 Hot summer, fresh autumn; success across the region. Small crop.
2015 Rich flavourful reds of gd ripeness; some soft.
2014 Quantity up, quality patchy, buyer beware.
Earlier fine vintages: 11 10 08 07 06 04 01 99 97 95 90. Vintages to keep: 10 01 99. Vintages to drink up: 07 06 04 03 00 97 95 90.

Abrigo, Orlando Pie ★★★ Some of most mineral, steely, refined BARBARESCOS. Top: Meruzzano, Montersino, new Barbaresco III (100% NEBBIOLO rosé).

Accornero Pie ★★★ Some of Italy's best medium-bodied reds. GRIGNOLINO sings in Bricco del Bosco (steel-aged) and Bricco del Bosco Vigne Vecchie (oak-aged). Also v.gd BARBERA Cima.

Adriano, Marco e Vittorio Pie ★★★ High quality, low prices, BARBARESCO full of early appeal. Top: Basarin, Sanadaive.

Aglianico del Taburno Cam ★→★★★ DOCG Around Benevento. Generally cooler microclimate than TAURASI. Spicier notes (leather, tobacco), herbs, higher acidity than other AGLIANICOS. CANTINA del Taburno, Fontanavecchia, La Rivolta gd.

Aglianico del Vulture Bas ★→★★★ DOC(G) **12 13** 15 16 18 (19) (20) DOC after 1 yr, SUPERIORE after 3 yrs, RISERVA after 5. From slopes of extinct volcano Monte Vulture. Volcanic soils with high content of clay-rich volcanic tuff. More floral (violet), dark fruits (plum), smoke, spice than other AGLIANICOS. Top: ELENA FUCCI, GRIFALCO. Also gd: CANTINA di Venosa, Cantine del Notaio, D'Angelo, Mastrodomenico, Paternoster, Re Manfredi.

Alba Pie Truffles, hazelnuts and PIE's, if not Italy's, most prestigious wines: BARBARESCO, BARBERA D'ALBA, BAROLO, DOGLIANI (DOLCETTO), LANGHE, NEBBIOLO D'ALBA, ROERO.

Albana di Romagna E-R ★→★★★ DYA Italy's 1st white DOCG, justified only by sweet PASSITO; dry and sparkling often unremarkable. Best: *Fattori Zerbina Passito*, Giovanna Madonia, PODERE Morini (Cuore Matto PASSITO), Tre Monti.

Alessandria, Fratelli Pie ★★★ Since 1870 top BAROLO producer in VERDUNO. Best crus: Monvigliero, San Lorenzo; v.gd Verduno Speziale; new Barolo di Verduno.

Allegrini Ven ★★ Popular VALPOLICELLA producer. Elegant AMARONE best. Owner of POGGIO al Tesoro and Poggio San Polo in TUS.

Almondo, Giovanni Pie ★★★→★★★★ Top ROERO estate. Best: Roero ARNEIS Bricco delle Ciliegie, Rive del Bricco. Outstanding Freisa; v.gd Roero Bric Valdiana (r).

Alta Langa Pie ★★→★★★★ DOCG The 1st METODO CLASSICO made in Italy, since mid-C19 in "underground cathedrals". Vintage only, and simply PINOT N, CHARD. Best: BANFI (Cuvée Aurora and Aurora 100 Mesi), COCCHI-BAVA, Contratto, Enrico Serafino (RISERVA Zero 140 and Riserva Zero), ETTORE GERMANO, FONTANAFREDDA, GANCIA, RIZZI.

Altare, Elio Pie ★★★ Now run by Silvia, Elio's daughter. Try BAROLOS: Arborina, Cannubi, Cerretta VIGNA Bricco, Unoperuno (selection from Arborina).

Alto Adige (Sudtirol) T-AA DOC Mountainous region with Bolzano its chief city (Austrian until 1919); arguably best Italian whites today but also underrated reds. Germanic vines dominate. GEWURZ, KERNER, SYLVANER, but PINOT GRIGIO too. GRÜNER V, RIES hopelessly overrated; probably world's best PINOT BIANCO. PINOT N often overoaked; *Lagrein* in gd yrs.

Alto Piemonte Pie Cradle of PIE quality in C19 (40,000 ha). Acidic soil, exposure, climate and altitude diversity, ideal for many different NEBBIOLO expressions (here called Spanna). Main DOC(G): BOCA, BRAMATERRA, Colline Novaresi, Coste della Sesia, Fara, GATTINARA, GHEMME, LESSONA, Sizzano, Valli Ossolane. Many outstanding wines. Actually, rarely 100% Nebbiolo: small additions of Croatina, Uva Rara, Vespolina common.

Ama, Castello di Tus ★★★★ Among 1st to produce single-vyd CHIANTI CLASSICO and push for quality in 80s. Gran Selezione VIGNETO Bellavista best; La Casuccia close 2nd; v.gd San Lorenzo. MERLOT L'Apparita one of Italy's three best.

Amarone della Valpolicella Ven ★★→★★★★ DOCG **10** 11' 13 15 16 (17) (18) CLASSICO area (from historic zone), Val d'Illasi and Valpantena (from extended zone) can make unique world-class reds from raisined grapes. Alas, many less than what they should be, despite hype. Choose carefully. (*See* VALPOLICELLA, and box p.151.)

Ambra, D' Cam ★★★ On ISCHIA; fosters rare local native grapes. Best: single-vyd Frassitelli (w, 100% Biancolella); v.gd Dedicato a Mario d'Ambra (r), Forastera (w), La VIGNA dei Mille Anni and new Le Ninfe (w blend).

Antinori, Marchesi L&P Tus ★★→★★★★ This family has been protagonist of C20 Italian wine renaissance. Top CHIANTI CLASSICO (TENUTE and *Badia a Passignano*), Cervaro (Umb *Castello della Sala*), the two excellent SUPER TUSCANS (TIGNANELLO, SOLAIA) and Prunotto (PIE BAROLOS). Also FVG (Jermann), MONTALCINO (Pian delle Vigne), MONTEPULCIANO (La Braccesca), PUG (Tormaresca), TUS MAREMMA (Fattoria Aldobrandesca).

Antoniolo Pie ★★★ Age-worthy benchmark GATTINARA. Outstanding RISERVAS Le Castelle, Osso San, San Francesco.

Argiano, Castello di Tus ★★★ Distinctive, refined yet flavourful BRUNELLOS from Sesti family estate. CLASSICO Brunello and RISERVA Phenomena equally gd, if different.

Argiolas Sar ★★→★★★ Top producer, native island grapes. Outstanding crus: Iselis Monica, Iselis Nasco, *Turriga* (★★★), *Vermentino* di Sardegna (Merì) and top Angialis (sw, mainly local Nasco grape). Bovale Korem, CANNONAU RISERVA Senes, CARIGNANO DEL SULCIS (Is Solinas and Cardenera) and new (sp) Argiolas METODO CLASSICO (from Nuragus grapes) v.gd.

Asti Pie ★→★★ NV sparkler from MOSCATO Bianco grapes, inferior to MOSCATO D'ASTI, not really worth its DOCG. Try Bera, Cascina Fonda, Caudrina, Vignaioli di Santo Stefano. Now dry version, Asti Secco.

Avignonesi Tus ★★★ Large bio estate. *Italy's best Vin Santo.* Top is VINO NOBILE single-vyd Poggetto di Sopra and Grandi Annate, but MERLOT Desiderio, 50&50 (Merlot/SANGIOVESE), CHARD Il Marzocco creditable internationals.

Azelia Pie ★★★ Distinctive, elegant BAROLOS from Luigi Scavino and son Lorenzo. Some of best crus. Top: Bricco Fiasco, Cerretta, San Rocco and RISERVA Bricco Voghera. Also v.gd LANGHE NEBBIOLO.

Azienda agricola / agraria Estate (large or small) making wine from its own grapes.

Badia a Coltibuono Tus ★★★ One of top CHIANTI CLASSICO producers; every wine worth buying: great terroirs spell non-stop success. SANGIOVETO v.gd. Organic.

Banfi (Castello or Villa) Tus ★→★★★ Giant of MONTALCINO, but top, limited-production POGGIO all'Oro is great BRUNELLO; v.gd Moscadello and ALTA LANGA.

Barbaresco Pie ★★→★★★★ DOCG 11 12 13 14 15 16 18 19 (20) Often better than BAROLO, Barbaresco's lesser reputation is undeserved. When spot-on, the gracefulness, age-worthiness and perfumed intensity is like that of no other wine in Italy – the world, really. Min 26 mths ageing, 9 mths in wood; at 4 yrs becomes RISERVA. (For top crus/producers, *see* box, below.) Two main types of soil: Serravalian (alternate sandy levels and grey silty marls), on average, elegant and v. perfumed wines, not too fleshy and with perhaps a lower propensity

---

Top Barbaresco by MGA

BARBARESCO has four main communes, four distinct styles: **Barbaresco** most complete, balanced. Growers incl Asili (BRUNO Giacosa, CA' DEL BAIO, CERETTO, PRODUTTORI DEL BARBARESCO), Martinenga (MARCHESI DI GRESY), Montefico (Produttori del B, ROAGNA), Montestefano (Giordano Luigi, Produttori del B, Rivella Serafino), Ovello (CANTINA del Pino, ROCCA ALBINO), Pajè (Roagna); Pora (Ca' del Baio, MUSSO, Produttori del B), Rabaja (Bruno Giacosa, BRUNO ROCCA, CASTELLO DI VERDUNO, GIUSEPPE CORTESE, Produttori del B), Rio Sordo (Cascina Bruciata, Cascina delle Rose, Musso, Produttori del B), Roncaglie (PODERI COLLA). **Neive** most powerful, fleshiest. Albesani (Cantina del Pino, Castello di Neive), Basarin (ADRIANO MARCO E VITTORIO, Giacosa Fratelli, Negro Angelo, PAITIN, SOTTIMANO), Bordini (La Spinetta), Currà (Bruno Rocca, Sottimano), Gallina (Castello di Neive, CERETTO, La Spinetta, Lequio Ugo, ODDERO), Serraboella (Cigliuti, Paitin), Starderi (La Spinetta). **San Rocco Seno d'Elvio** readiest to drink, soft. Rocche Massalupo (Lano, TENUTA Barac), Sanadaive (Adriano Marco e Vittorio). **Treiso** freshest, most refined. Bernardot (Ceretto), Bricco di Treiso (PIO CESARE), Marcarini (Ca' del Baio), Montersino (ORLANDO ABRIGO, Rocca Albino), Nervo (RIZZI), Pajorè (Rizzi, Sottimano), Rombone (Fiorenzo Nada, Luigi Oddero).

to age (mainly in San Rocco Seno d'Elvio, Treiso); Tortonian (blue-grey marls with more or less sand), v. structured wines that age (mainly in Barbaresco, Neive).

Barbera d'Alba Pie DOC Unique, luscious, sultry BARBERA, quite different from D'ASTI's more nervy, higher-acid version. AZELIA (Punta), BREZZA, CAVALLOTTO (Cuculo), CLERICO DOMENICO (Tre Vigne), CONTERNO FANTINO (Vignota), GERMANO ETTORE (della Madre), GIACOMO CONTERNO, PODERI COLLA (Costa Bruna), SOTTIMANO, VIETTI (Scarrone) make benchmarks. COGNO's Pre-Phylloxera is *hors classe*.

Barbera d'Asti Pie Huge DOCG, encompassing v. different soils (from sandy to marly) and climatic characteristics. Wines vary, but characterized by high acidity and fruity notes. SUPERIORE: higher quality (but often overoaked). Nizza is best: 100% BARBERA only from best sites and needs time. Try BAVA, BRAIDA, Cascina Castlet, Dacapo, Marchesi di Gresi (Monte Colombo), SCARPA (La Bogliona), Spertino, TENUTA Olim Bauda (Le Rocchette), *Vietti* (La Crena).

Barbera del Monferrato Superiore Pie DOCG From soils rich in limestone, full-bodied BARBERA with sharpish tannins, gd acidity. Top: Accornero (Bricco Battista and Cima), Castello di Uviglie, Iuli (Barabba, Rossore).

Barberani Umb ★★→★★★ Organic estate on slopes of Lago di Corbara, gd to excellent ORVIETO. Cru Luigi e Giovanna is star, also Polvento, Calcaia (noble rot).

Bardolino Ven DOC(G) DYA Light-bodied fresh red made with VALPOLICELLA's typical grapes; hit with tourists on Lake Garda. Ill-advised plantings of CAB SAUV in order to make bigger wines a tragic idea. Best today are rosés, called Chiaretto in this neck of the Italian woods. Look for Cavalchina, *Guerrieri Rizzardi*, Zeni.

Barolo Pie DOCG 09 10' **11 12** 13 15 16' 17 (18) (19) "King of wines and wine of kings" 100% NEBBIOLO. 2000-ha zone. Must age 38 mths before release (5 yrs for RISERVA), of which 18 mths in wood. Best are age-worthy, able to join power and elegance, with alluring floral scent and sour red-cherry flavour. Traditionally blend of vyds from different communes, but now most is single-vyd. The concept of "cru" is replaced (in Barolo and BARBARESCO at least) by subzones known officially as MGA (menzioni geografiche aggiuntive). Currently Barolo has 11 village mentions and 170(!) additional geographical mentions. Often underrated: Village MGA ("Barolo del Comune di...") best way to understand different Barolo terroirs. Best: "di Barolo" (Virna), "di Grinzane" (Canonica), "di La Morra" (Ciabot Berton), "di Serralunga" (Ettore Germano, RIVETTO). Three main types of soil: Messinian (chalk-sulphur formation), less interesting wines, w slope of La Morra; Serravalian (greyish yellow/red looser calcareous marl soil and sands), Castiglione Falletto (bold), Monforte (structure), Serralunga (power); Tortonian (blue-grey compact marl soil and calcareous sands, younger), Barolo (grace), La Morra (fragrance). (For top crus/producers, *see* box, p.128.)

Bartoli, Marco De Si ★★★★ One of best estates in all Italy. Marco spent life promoting "real" MARSALA and his must be tried. Top: VECCHIO SAMPERI, 20-yr-old Ventennale. Delicious table wines (eg. GRILLO Vignaverde and Grappoli del Grillo, ZIBIBBO Pietranera and Pignatello), outstanding sweet Zibibbo di PANTELLERIA *Bukkuram*.

Bastianich FVG ★★→★★★ FRIULANO plus one of Italy's 30 best whites; Vespa Bianco not far behind. Reds (Calabrone, REFOSCO del Peduncolo Rosso RISERVA Solo, Vespa Rosso) can be weedy and tough in poor yrs (air-drying unripe grapes is never a gd idea), but in gd vintages can be memorable.

Belisario Mar ★★→★★★ Quality co-op, gd quality/price. Largest producer of VERDICCHIO DI MATELICA. Top: RISERVA Cambrugiano

Benanti Si ★★★ Benanti family turned world on to ETNA. Bianco SUPERIORE *Pietramarina* one of Italy's best whites and new Etna Bianco Superiore Contrada Rinazzo. Top: (r) Etna RISERVA (Rovittello, Serra della Contessa); Selezione

Top Barolos by MGA

Here are a few top crus and their best producers: **Bricco Boschis** (Castiglione Falletto) CAVALLOTTO (RISERVA VIGNA San Giuseppe); **Bricco delle Viole** (BAROLO) GD VAJRA; **Bricco Rocche** (Castiglione Falletto) CERETTO; **Briccolina** (Serralunga) RIVETTO; **Brunate** (La Morra, Barolo) Ceretto, GIUSEPPE RINALDI, ODDERO, VIETTI; **Bussia** (Monforte) ALDO CONTERNO (Gran Bussia e Romirasco), GIACOMO FENOCCHIO (also Riserva 90 Dì), ODDERO (Bussia Vigna Mondoca), PODERI COLLA (Dardi Le Rose); **Cannubi** (Barolo) BREZZA, Giacomo Fenocchio, LUCIANO SANDRONE, PIRA E FIGLI – CHIARA BOSCHIS, EINAUDI, Virna; **Cerequio** (La Morra, Barolo) BOROLI, Roberto Voerzio; **Falletto** (Serralunga) BRUNO GIACOSA (Riserva Vigna Le Rocche); **Francia** (Serralunga) GIACOMO CONTERNO (Barolo Cascina Francia and Monfortino); **Ginestra** (Monforte) CONTERNO FANTINO (Sorì Ginestra and Vigna del Gris), DOMENICO CLERICO (Ciabot Mentin); **Lazzarito** (Serralunga) ETTORE GERMANO (Riserva), Vietti; **Monprivato** (Castiglione Falletto) GIUSEPPE MASCARELLO (Mauro); **Monvigliero** (VERDUNO) CASTELLO DI VERDUNO, FRATELLI ALESSANDRIA, GB BURLOTTO, PAOLO SCAVINO; **Mosconi** (Monforte) Conterno Fantino, Domenico Clerico, Pira e Figli – Chiara Boschis, PIO CESARE; **Ornato** (Serralunga) Pio Cesare; **Ravera** (Novello) ELVIO COGNO (Bricco Pernice), GD Vajra, Vietti; **Rocche dell'Annunziata** (La Morra) PAOLO SCAVINO (Riserva), RATTI, Roberto Voerzio, Rocche Costamagna, TREDIBERRI; **Rocche di Castiglione** (Castiglione Falletto) BROVIA, Oddero, ROAGNA, Vietti; **Vigna Rionda** (Serralunga) Ettore Germano, Figli Luigi Oddero, MASSOLINO VR, Oddero; **Villero** (Castiglione Falletto) Boroli, Brovia, Giacomo Fenocchio, Giuseppe Mascarello. And the Barolo of Bartolo Mascarello blends together Cannubi San Lorenzo, Ruè and Rocche dell'Annunziata. A sommelier's delight.

Contrade (Cavaliere, Dafara Galluzzo, Monte Serra, Rinazzo). Also v.gd Lamorèmio (NERELLO MASCALESE), Noblesse 48 Mesi (sp, CARRICANTE).

Berlucchi, Guido Lom ★★ Makes millions of bottles of METODO CLASSICO fizz. FRANCIACORTA Brut Cuvée Imperiale is flagship; v.gd Nature Dosaggio Zero '61.

Bersano Pie ★★→★★★ Large volume but gd quality. BARBERA D'ASTI, GRIGNOLINO, Ruchè, all inexpensive, delightful.

Bertani Doms Tus ★★→★★★ Long-est producer of VALPOLICELLA, SOAVE. Also owns Fazi Battaglia (Mar).

Bertinga Tus ★★★ Estate in Gaiole, CHIANTI. Best Bertinga (SANGIOVESE/MERLOT), Punta di Adine (Sangiovese).

Biondi-Santi Tus ★★★★ Invented BRUNELLO. High quality, high in acid, tannin requiring decades to develop fully.

Bisol Ven ★★★ Owned by FERRARI's Lunelli family; quality leader in PROSECCO. Top: CARTIZZE and Relio Rive di Guia, but Crede and Molera two of Italy's best buys.

Boca Pie DOC *See* ALTO PIEMONTE. Potentially among greatest reds. NEBBIOLO (70–90%), incl up to 30% Uva Rara and/or Vespolina. Volcanic quartz porphyry soils. Highest, freshest denomination of Alto Piemonte. Needs long ageing. *Le Piane* best. Carlone Davide, Castello Conti gd.

Bolgheri Tus DOC Mid-Maremma, on w coast, cradle of many expensive SUPER TUSCANS, mainly French variety-based. Big name and excellent quality: ANTINORI (Guado al Tasso), GAJA (CA' MARCANDA), Grattamacco. Le Macchiole, MICHELE SATTA (Marianova, Piastraia), SAN GUIDO (SASSICAIA, original Super Tuscan) and ORNELLAIA (FRESCOBALDI) are best (incl iconic 100% MERLOT Masseto.)

Borgo del Tiglio FVG ★★★→★★★★ Nicola Manferrari (one of Italy's top white winemakers) and his son Mattia run one of best COLLIO estates. Top is Black Label collection: Collio FRIULANO RONCO della Chiesa, MALVASIA Selezione, Rosso della Centa, Studio di Bianco esp impressive.

Borgogno, Virna Pie ★★★ Family estate run by Virna B and sister Ivana. Great-value BAROLO from famous crus: Cannubi and Sarmassa. Barolo del Comune di Barolo, Barolo Noi, RISERVA and new Sto Fuori (Timorasso grape) v.gd.

Boroli Pie ★★★ Achille now runs this winery, totally devoted to BAROLOS. Best: Brunella (monopole), Cerequio, Villero. New v.gd Langhe NEBBIOLO 1661.

Boscarelli, Poderi Tus ★★★ Small estate with reliably high-standard VINO NOBILE DI MONTEPULCIANO. Best: Costa Grande (100% SANGIOVESE single vyd), Nocio dei Boscarelli, RISERVA Sotto Casa.

Bosco, Tenute Si ★★★ Small ETNA estate owned by Sofia B, high quality. Best: Etna Rosso VIGNA Vico Prephylloxera. Piano dei Daini (r/w/rosé) v.gd.

Botte Big barrel, anything from 6 to 250 hl, usually 20–50, traditionally of Slavonian but increasingly of French oak. To traditionalists, ideal vessel for ageing wines without adding too much oak smell/taste.

Braida Pie ★★★ If BARBERA D'ASTI is known today, it's thanks to the Bologna family, world ambassadors for this wine. Top: Bricco dell'Uccellone, Bricco della Bigotta and Ai Suma. Monella, Montebruna v.gd. GRIGNOLINO d'Asti Limonte one of Italy's five best.

Bramaterra Pie DOC Volcanic porphyry and marine deposits. Most variegated in terms of soils. Wines tend to be lighter and less massive than the other ALTO PIEMONTE denominations. Top: Antoniotti Odilio, PROPRIETÀ SPERINO.

Brezza Pie ★★★ Organic. Certainty for those who love traditional BAROLOS. Great value, from famous crus incl Cannubi, Castellero, Sarmassa (esp RISERVA VIGNA Bricco); v.gd BARBERA D'ALBA SUPERIORE, LANGHE NEBBIOLO, Freisa (outstanding).

Brigaldara Ven ★★★ Elegant but powerful benchmark AMARONE from estate of Stefano Cesari. Top: Case Vecie.

Brolio, Castello di Tus ★★→★★★ Since 1141 run by RICASOLI family; historic estate – largest, oldest of CHIANTI CLASSICO. Outstanding Gran Selezione Castello di Brolio, Colledilà, Ceniprimo and Roncicone. Also v.gd Casalferro (MERLOT).

Brovia Pie ★★★→★★★★ Since 1863, classic BAROLOS in Castiglione Falletto. Organic certified. Top: Ca' Mia, Garblèt Sue, Rocche, Villero. After 10 yrs again iconic DOLCETTO D'ALBA Solatio. Waiting for new METODO CLASSICO (sp).

Brunelli, Gianni Tus ★★★ Lovely refined user-friendly BRUNELLOS (top RISERVA) and Rossos from two sites: Le Chiuse di Sotto n of MONTALCINO, Podernovone to s.

Brunello di Montalcino Tus 07 09 10' 12 13 15' 16 (18) (19) DOCG World-famous, but quality all over the shop. When gd, memorable and ageless, archetypal SANGIOVESE. Problems derive mostly from greedily enlarged production area (a ridiculous 2000 ha+), much less than ideal for fickle Sangiovese and world-class wines. Recent push to turn wine into a blend has been successfully stopped (thus far, at least). Generally speaking, MONTALCINO's soils vary from the limestone-rich Galestro of the n sector (wines are sleeker, more floral and mineral, generally higher acidity) to the loamier, siltier sandy clays of the s (wines broader, more powerful). (For top producers, *see* box, p.130.)

Bucci Mar ★★★★ Villa Bucci RISERVA one of Italy's ten best whites. All wines quasi-Burgundian, esp complex VERDICCHIOS, slow to mature but all age splendidly; v.gd red Pongelli. Look for Vintage Collection.

Burlotto, Commendatore GB Pie ★★★★ Fabio Alessandria maintains his ancestor Commander GB Burlotto's (among 1st to make/bottle BAROLO in 1880) high quality and focus. BAROLO Monvigliero and superlative Freisa best. Barolos Acclivi and Cannubi, VERDUNO Pelaverga also outstanding.

Bussola, Tommaso Ven ★★★★ Self-taught maker of some of the great AMARONES,

> **Best of Brunello**
> Any of the below provide satisfying BRUNELLO DI MONTALCINO; the top get
> a ★: Altesino ★, Baricci ★, BIONDI-SANTI ★, Campogiovanni, Canalicchio
> di Sopra ★, Canalicchio di Sotto, Caparzo, Casanova di Neri ★,
> CASE BASSE ★, Castelgiocondo, CASTELLO DI ARGIANO ★, Castiglion del
> Bosco, Ciacci Piccolomini, COL D'ORCIA ★, Collemattoni ★, Colombini,
> Costanti ★, Cupano ★, Donatella Cinelli, Eredi, Fossacolle, Franco
> Pacenti ★, FULIGNI ★, GIANNI BRUNELLI ★, Giodo, Il Colle, Il Marroneto ★,
> Il Paradiso di Manfredi, La Gerla, La Magia, La Poderina,
> Le Potazzine ★, Le Ragnaie ★, Le Ripi, LISINI ★, Mastrojanni ★,
> PIAN DELL'ORINO ★, Piancornello, Pieri Agostina, Pieve di Santa Restituta,
> POGGIO ANTICO, POGGIO DI SOTTO ★, San Filippo, Salvioni ★, Sesta di
> Sopra, Silvio Nardi, Siro Pacenti, Stella di Campalto ★, Talenti,
> TENUTA IL POGGIONE ★, TENUTA di Sesta, Uccelliera, Val di Suga.

RECIOTOS, RIPASSOS of our time. The great Bepi QUINTARELLI steered him; he steers his two sons. Top TB selection.

Ca' dei Frati Lom ★★★ Foremost quality estate of revitalized DOC LUGANA, I Frati a fine example at entry level; *Brolettino* a superior cru.

Ca' del Baio Pie ★★★ Family estate; best-value producer in BARBARESCO. Outstanding Asili (RISERVA too) and Pora; v.gd Autinbej, LANGHE RIES, Vallegrande.

Ca' del Bosco Lom ★★★★ Arguably Italy's best METODO CLASSICO sparkling, famous FRANCIACORTA estate owned by Zanella family and Santa Margherita Group. Outstanding and unforgettable Dosage Zéro Annamaria Clementi RISERVA (rosé too). Great Dosage Zéro and Dosage Zéro Noir Vintage Collection practically as gd; also excellent B'x-style Maurizio Zanella (r), PINOT N, CHARD.

Caiarossa Tus ★★★ Dutch-owned (Ch Giscours; *see* B'x) estate, n of BOLGHERI. Excellent Caiarossa Rosso plus reds Aria, Pergolaia.

Calcagno Si ★★★→★★★★ Lilliputian size and Brobdingnagian quality from ETNA family estate. Outstanding mineral NERELLO MASCALESE reds. Top (r) Arcurìa, Feudo di Mezzo (w), v.gd Ginestra (w), Nireddu (r), Romice delle Sciare (rosé).

Calì, Paolo Si ★★★ Passionate Paolo C makes numerous wines highly typical of Vittoria's terroir. Top: CERASUOLO DI VITTORIA Forfice and Frappato. GRILLO Blues (w), Manene (r) and Mood (rosé/sp) all v.gd.

Caluso / Erbaluce di Caluso Pie ★→★★★ DOCG Morainic soils giving v. interesting and mineral wines. Can be still, sparkling and sweet (Caluso PASSITO). Top: Ferrando (Cariola), Giacometto, TAPPERO MERLO (Kin, sp) Cieck (Misobolo), Favaro (Le Chiusure), Orsolani all v.gd.

Ca' Marcanda Tus ★★★★ BOLGHERI estate of GAJA. In order of price (high, higher, highest): Promis, Magari, Ca' Marcanda. Grapes mainly international.

Campania Some of Italy's greatest and most age-worthy whites, terroir-driven, full of character. Reds unfortunately less consistent due to combination of overripe grapes and too much oak, with some remarkable exceptions. Few international varieties cloud native grapes panorama. FIANO DI AVELLINO may well be Italy's best white zone, TAURASI makes better and better reds. Try BENITO FERRARA, Caggiano, CANTINE Lonardo, COLLI DI LAPIO, D'AMBRA, De Angelis, Fattoria La Rivolta, *Feudi di San Gregorio*, Galardi, Guastaferro, La Sibilla, Luigi Maffini, Luigi Tecce, MARISA CUOMO, *Mastroberardino*, Molettieri, Nicola Mazzella, Perillo, Pierlingeri, Pietracupa, QUINTODECIMO, Reale, Rocca del Principe, Sarno 1860, Terredora, Vadiaperti. MONTEVETRANO: world-class international mostly CAB wine.

Canalicchio di Sopra Tus ★★★→★★★★ Top-ten BRUNELLO estate. Owner Francesco Ripaccioli keen observer of terroir and MONTALCINO typicity; RISERVA usually spectacular; v.gd Brunello, single-vyd La Casaccia and ROSSO DI MONTALCINO.

ITALY

Cantina A cellar, winery or even a wine bar.

Capezzana, Tenuta di Tus ★★★ Certified organic production from noble Bonacossi family that made Carmignano's reputation. Now run by founder's children. Excellent Carmignano (Villa di Capezzana and Trefiano), exceptional VIN SANTO, one of Italy's five best.

Capichera Sar ★★★★ No better VERMENTINO anywhere than that of Ragnedda family, talented enough also to make top reds (Mantènghja CARIGNANO 100%). Top Capichera, Isola dei Nuraghi Bianco Santigaìni, VT. Assajè (Carignano, SYRAH) and Vign'angena (Vermentino) v.g.d.

Caprai Umb ★★★→★★★★ MONTEFALCO leader thanks to Marco C. Many outstanding wines (eg. 25 Anni); v.gd Collepiano (less oak), GRECHETTO Grecante, Rosso di Montefalco (smooth, elegant).

Carema Pie ★★→★★★DOC 10 11 13 15 16′ 18 (19) Only 22 ha, n of Turin. Steep terraces, morainic agglomerate soils for light, mineral, intense, outstanding NEBBIOLO. Top: Ferrando (esp Black Label), Monte Maletto, Murajè, Sorpasso. Also v.gd Chiussuma, Cella Grande, Milanesio, Produttori Nebbiolo di Carema.

Carignano del Sulcis Sar ★★→★★★ DOC 14 15 16 19 (20) From SW SAR, world-class CARIGNANO that ages gracefully but always boasts early appeal, accessibility. Best: Rocca Rubia and *Terre Brune* from ARGIOLAS, Mesa, SANTADI.

Carpineti, Marco Lat ★★★ Phenomenal bio whites from little-known Bellone and GRECO Moro, Greco Giallo varieties. Benchmark Moro and Ludum, one of Italy's best stickies. New Collesanti 2 (100% Bellone single-vyd), Kius Pas Dosé (100% Nero Buono).

Cartizze Ven ★★★→★★★★ DOCG At 107 ha, this PROSECCO super-cru is reportedly 2nd most expensive vyd land in Italy, after BAROLO; v. steep hills in heart of Valdobbiadene showcase just how great Prosecco DOCG can be. Usually on sweet side due to fully ripe grapes. Best: BISOL, Bortolomiol, Col Vetoraz, Le Colture, Merotto, NINO FRANCO, Ruggeri.

Case Basse Tus ★★★★ Gianfranco Soldera's children keep up similar lofty level of iconic BRUNELLO. Long-oak-aged. Rare, precious.

Castel del Monte Pug ★→★★(r) 15 16 18 19 (w/rosé) DOC DYA Dry, fresh, increasingly serious, from mid-PUG DOC, esp *Bocca di Lupo* from Tormaresca (ANTINORI). Il Falcone RISERVA (Rivera) is iconic; v.gd Torrevento (Riserva).

Castel Juval, Unterortl T-AA ★★★ Owned by mountaineer Reinhold Messner. Distinctive, crystalline. WEISSBURGUNDER best; RIES (Windbichel), PINOT N v.gd.

Castell' in Villa Tus ★★★ Traditional CHIANTI CLASSICO estate in extreme sw of zone. Wines of class, excellence, v. age-worthy. Top RISERVA.

Cataldi Madonna Ab ★★★ Organic. Top PECORINO Frontone (Ab's oldest Pecorino vines); v.gd CERASUOLO D'ABRUZZO Piè delle vigne, MONTEPULCIANO D'ABRUZZO.

Caudrina Pie ★★★→★★★★ Romano Dogliotti one of best for MOSCATO D'ASTI. Top: ASTI La Selvatica, La Galeisa. La Caudrina v.gd.

Cavallotto Pie ★★★→★★★★ Organic certified. Solid reference for traditional BAROLO, in Castiglione Falletto. Outstanding RISERVA VIGNA San Giuseppe and Riserva Vignolo, v.gd BARBERA D'ALBA SUPERIORE Vigna Cuculo, LANGHE NEBBIOLO. Surprisingly gd GRIGNOLINO, Freisa too.

Cave Mont Blanc VdA ★★★ Quality co-op at foot of Mont Blanc, with ungrafted indigenous 60–100-yr-old Prié Bl vines. Organic certified. Outstanding sparkling. Top Blanc de Morgex et de la Salle Rayon, sparkling (Brut Nature Cuvée des Guides, Cuvée du Prince, X.T.).

Cerasuolo d'Abruzzo Ab ★ DOC DYA ROSATO version of MONTEPULCIANO D'ABRUZZO, don't confuse with red CERASUOLO DI VITTORIA from SI. Can be brilliant; best (by far): CATALDI MADONNA (Pié delle vigne), EMIDIO PEPE, Praesidium, TIBERIO, VALENTINI.

> **Who makes really good Chianti Classico?**
> CHIANTI CLASSICO is a large zone with hundreds of producers, so picking
> the best is tricky. Top get a ★: AMA ★, ANTINORI, BADIA A COLTIBUONO ★,
> BROLIO, Capraia, Casaloste, Casa Sola, Castellare, CASTELL' IN VILLA,
> CASTELLO DI VOLPAIA ★, FELSINA ★, FONTERUTOLI, FONTODI ★, Gagliole,
> I Fabbri ★, Il Molino di Grace, ISOLE E OLENA ★, Le Boncie,
> Le Cinciole ★, Le Corti, Le Filigare, Lilliano, Mannucci Droandi, Meleto,
> Monsanto ★, Monte Bernardi, Monteraponi ★, NITTARDI, NOZZOLE,
> Palazzino, Paneretta, Poggerino, Poggiopiano, QUERCIABELLA ★, Rampolla,
> RIECINE, Rocca di Castagnoli, Rocca di Montegrossi ★, RUFFINO, San
> Fabiano Calcinaia, SAN FELICE, SAN GIUSTO A RENTENNANO ★, Paolina
> Savignola, Selvole, Tenuta Perano – FRESCOBALDI, Vecchie Terre di
> Montefili, Verrazzano, Vicchiomaggio, Vignamaggio, Villa Calcinaia ★,
> Villa La Rosa ★, Viticcio.

Cerasuolo di Vittoria Si ★★ 16 17 18 19 Blend of Frappato/NERO D'AVOLA. Only SI
DOCG, in se, around city of Vittoria, best terroir for Frappato. Try COS, GULFI,
OCCHIPINTI ARIANNA, PAOLO CALÌ, PLANETA, Valle dell'Acate (Iri da Iri).

Ceretto Pie ★★★→★★★★ Leading producer of BARBARESCO (Asili, Bernadot, Gallina),
BAROLO (Bricco Rocche, Brunate, Bussia, Cannubi San Lorenzo, Prapò), plus
LANGHE Bianco Blange (ARNEIS). Organic/bio (2015). Wines recently more classic.

Cerruti, Ezio Pie ★★★ Small estate in gd area for MOSCATO: best sweet Sol, naturally
dried Moscato. Also v.gd Fol (dr), new Mac Fol (on skins), Rosso (Freisa).

Cesanese (Comune and di Affile) Lat ★→★★★ Two grapes: Comune (more common
in the Olevano Romano area, s of Lat), d'Affile (in Affile and Piglio). Three wines:
Cesanese del Piglio, top Casale della Ioria (Torre del Piano); Cesanese di Affile,
top Colline di Affile (Le Cese); Olevano Romano Cesanese, top DAMIANO CIOLLI.

Chianti Tus ★→★★★ DOCG When gd, delightfully delicious, easy-going, food-
friendly fresh red. Modern-day production zone covers most of TUS; v. big
differences in the topography, climate, and soils among various Chianti
denominations. RÙFINA is the only terroir of quality comparable to historic
Chianti production zone (now called CHIANTI CLASSICO).

Chianti Classico Tus ★★→★★★ DOCG 15 16' 17 18' 19' (20) No wine in Italy has
improved more over last 20 yrs, now often 100% SANGIOVESE. More than 500
producers means inconsistent quality, but best are among Italy's greatest. Made
in historic (high, rocky) CHIANTI production zone between Florence and Siena
in nine townships. Climate varies greatly from n to s sectors, three main soil
types: Alberese (whitish marls), Galestro (clay schist) and macigno (mix of sands
and compacted sands) so potentially wines are v. different from each other. Gran
Selezione is new top level, above RISERVA. (*See also* box above.)

Ciabot Berton Pie ★★★ Marco and Paola Oberto, following their father, have turned
this La Morra estate into one of best-value producers; v.gd blended BAROLO
("1961" and Barolo di La Morra), crus Roggeri, Rocchettevino have distinctive
single-vyd characters, LANGHE NEBBIOLO.

Cinque Terre Lig ★★ DOC Dry VERMENTINO-based whites from vertiginous LIG
coast. Sweet version: Sciacchetrà. Try Arrigoni, Bisson, Buranco, De Battè.

Ciolli, Damiano Lat ★★★ One of most interesting wineries in central Italy. Best is
Cirsium, 100% CESANESE D'AFFILE, 80-yr-old vines; v.gd Silene. New Botte 22,
Trebbiano Verde (aka VERDICCHIO)/Ottonese.

Cirò Cal ★→★★★ DOC Brisk strong red from Cal's main grape, Gaglioppo, or
light, fruity white from GRECO (DYA). Best: 'A Vita, Caparra & Siciliani (RISERVA
Volvito), IPPOLITO 1845, *Librandi* (Duca San Felice ★★★), San Francesco (Donna
Madda, RONCO dei Quattroventi), Santa Venere.

Classico Term for wines from a restricted, usually historic and superior-quality area within limits of a commercially expanded DOC. *See* CHIANTI CLASSICO, SOAVE, VALPOLICELLA, VERDICCHIO, numerous others.

Clerico, Domenico Pie ★★★ Influential BAROLO innovator, modernist producer of Monforte d'ALBA, esp crus Ginestra (Ciabot Mentin, Pajana), Mosconi (Percristina only in best vintages, don't miss **10**). Barolo Aeroplanservaj (from Baudana cru) v.gd. Also v.gd BARBERA D'ALBA (Trevigne), LANGHE NEBBIOLO (Capisme-e).

Clivi, I FVG ★★★ Wealth of old vines (in FRIULI COLLI ORIENTALI and COLLIO), even up to 90 yrs of age. Some of FVG's purest, most age-worthy and mineral wines, ranking with Italy's best white wines. Top: MALVASIA and FRIULANO cru wines and also an outstanding dry VERDUZZO.

Cocchi-Bava Pie ★★★ Since 1891 producer of vermouth. Top ALTA LANGA Pas Dosé, Toto Corde, Vermouth RISERVA La Venaria and Storico di Torino; v.gd BARBERA D'ASTI Stradivario, NIZZA Piano Alto, Ruchè and BAROLO Scarrone.

Cogno, Elvio Pie ★★★★ Top estate; super-classy, austere, elegant BAROLOS from Ravera cru. Best: Bricco Pernice, RISERVA VIGNA Elena (NEBBIOLO Rosé clone); v.gd Anas-Cëtta (100% Nascetta), BARBERA D'ALBA Pre-Phylloxera (100-yr-old vines).

Col d'Orcia Tus ★★★ Top-quality MONTALCINO estate (3rd-largest) owned by Francesco Marone Cinzano. Best: BRUNELLO RISERVA POGGIO al Vento, new Brunello Nastagio. Col d'Orcia is valley between Montalcino and Monte Amiata.

Colla, Poderi Pie ★★★★ Family-run winery based on experience of Beppe Colla. Classic, traditional, age-worthy. Top: BARBARESCO Roncaglie, BAROLO Bussia Dardi Le Rose, LANGHE Bricco del Drago; v.gd NEBBIOLO D'ALBA, PINOT NERO (Campo Romano), sparkling Pietro Colla, RIES.

Colli = hills; singular: colle. **Colline** (singular collina) = smaller hills. *See also* COLLIO, POGGIO.

Colli di Catone Lat ★→★★★ Top producer of FRASCATI and IGT. Outstanding aged whites from MALVASIA del Lazio (aka Malvasia Puntinata) and GRECHETTO.

Colli di Lapio Cam ★★★ Clelia Romano's estate is Italy's **best Fiano** producer. Also v.gd GRECO DI TUFO Alèxandros, TAURASI Andrea and new FIANO Clelia.

Colli di Luni Lig, Tus ★★→★★★ DOC Nr Spezia. VERMENTINO, Albarola whites; SANGIOVESE-based reds easy to drink, charming. Top: Bisson (VIGNA Erta), Giacomelli (Boboli), La Baia del Sole (Oro d'Isée, Sarticola), LUNAE, Ottaviano Lambruschi (Costa Marina, Il Maggiore).

Collio FVG ★★→★★★★ DOC Famous white denomination, unfortunately moved steadily to nonsensical Collio Bianco blend rather than highlight terroir differences of its communes, incl coolish San Floriano and Dolegna, warmer Capriva. Happily, Collio boasts glut of talented producers: BORGO DEL TIGLIO, Castello di Spessa, GRAVNER, La Castellada, Livon, MARCO FELLUGA, Podversic, Primosic, Princic, *Radikon*, Renato Keber, RONCO dei Tassi, RUSSIZ SUPERIORE, *Schiopetto*, Venica & Venica, VILLA RUSSIZ.

Colterenzio CS / Schreckbichl T-AA ★★★ Cornaiano-based main player among ALTO ADIGE co-ops. Top whites (SAUV BL Lafoa, PINOT BIANCO Berg and new LR) and Lafoa reds (CAB SAUV, PINOT N).

Conegliano Valdobbiadene Ven ★→★★ DOCG DYA Name for top PROSECCO, may be used separately or together. Extremely steep hills; quality should be better.

Conero Mar ★★→★★★★ DOCG 15 17 aka ROSSO CONERO. Small zone making powerful, at times too oaky, MONTEPULCIANO. Try GAROFOLI (Grosso Agontano), Le Terrazze (Praeludium), Marchetti, Moncaro, Monteschiavo, Moroder (Dorico), UMANI RONCHI (Campo San Giorgio).

Conterno, Aldo Pie ★★★★ Top estate of Monforte d'ALBA, 25 ha for only 80,000 bottles of highest quality. Top BAROLOS Cicala, Colonello, Granbussia and esp Romirasco; v.gd CHARD Bussiador, LANGHE NEBBIOLO Il Favot.

Conterno, Giacomo Pie ★★★★ For many, Monfortino is best wine of Italy. Roberto recently acquired NERVI estate in GATTINARA. Outstanding BARBERAS. Top BAROLO Cascina Francia, Cerretta and now Arione.

Conterno Fantino Pie ★★★ Organic certified. One of top producers of excellent modern-style BAROLO crus at Monforte: Ginestra (Sorì Ginestra and VIGNA del Gris), Mosconi (Vigna Ped) and Castelletto (Vigna Pressenda). Also gd Ginestrino, NEBBIOLO/BARBERA blend Monprà, CHARD Bastia (one of PIE's best).

11 MGAs in Chianti Classico from 2022. Just what you wanted: more to remember.

Contini Sar ★★★ Benchmark VERNACCIA DI ORISTANO, oxidative-styled whites not unlike v.gd Amontillado or Oloroso. Antico Gregori one of Italy's best whites. Amazing Flor 22 and RISERVA; gd I Giganti (r).

Coroncino, Fattoria Mar ★★★ Organic traditional estate run by Canestrari family. Top: Gaiospino (also Fumè), Stragaio, VERDICCHIO DEI CASTELLI DI JESI CLASSICO SUPERIORE; v.gd Coroncino.

Correggia, Matteo Pie ★★★ Organic certified. Leading producer of ROERO (RISERVA Rochè d'Ampsej, Val dei Preti), Roero ARNEIS, plus BARBERA D'ALBA (Marun) and Roero Arneis Val dei Preti (aged 6 yrs). New Apapà (100% NEBBIOLO dedicated to Matteo).

Cretes, Les VdA ★★★ Costantino Charrère is father of modern VALLE D'AOSTA viticulture and saved many forgotten varieties. *Outstanding Petite Arvine*, two of Italy's best CHARDS; v.gd Fumin, Torrette, Neblu Brut Nature Rosé (blend), PINOT N Revei.

Cristo di Campobello, Baglio del Si ★★→★★★ Bonetta family estate just e of Agrigento. Top: Grillo La Luci, NERO D'AVOLA Lu Patri.

Crotta di Vegneron, La VdA ★★→★★★ Quality co-op in Chambave. Top La Griffe des Lions line (Fumin, Nus MALVOISIE); v.gd Chambave Muscat Attente, Chambave Superieur and sparkling line Quatremillemètres Vins d'Altitude.

CS (cantina sociale) Cooperative winery.

Cuomo, Marisa Cam ★★★ Fiorduva is *one of Italy's greatest whites*; v.gd red RISERVAS (Furore, Ravello) and Costa d'Amalfi.

Custodi delle Vigne dell'Etna, I Si ★★★→★★★★ Family estate run by Mario Paoluzi. Member of the consortium I VIGNERI. Outstanding red ETNA ROSSO RISERVA Saeculare and Aetneus; v.gd Etna Biancos (Ante, Contrada Muganazzi, SUPERIORE Contrada Caselle).

Dal Forno Romano Ven ★★★★ VALPOLICELLA, AMARONE, RECIOTO (latter not identified as such any more) v.-high-quality; vyds outside CLASSICO zone but wines great.

D'Attimis Maniago FVG ★★★ One of the best, traditional FVG estates; standout MALVASIA, RIBOLLA, Pignolo and Schioppettino; Tazzelenghe knockout.

Dei Pie, Tus ★★★ Estate in MONTEPULCIANO, making v.gd VINO NOBILES. Best: Bossona and cru Madonna della Querce.

Derthona Pie ★→★★★ Wine from Timorasso grapes grown only in COLLI Tortonesi. One of Italy's most *interesting whites*, like v. dry RIES from Rheinhessen. Best: La Colombera (Il Montino), Mariotto, Mutti, Poggio Paolo, Ricci, ROAGNA (Montemarzino), VIGNETI MASSA, VIETTI.

Di Barrò VdA ★★★ Small family estate, top-quality wines from typical VDA grapes. Best: (r) Mayolet and Torrette Sup (Ostro), (w) Petite ARVINE.

DOC / DOCG Quality wine designation: *see* box, p.153.

Dogliani Pie ★→★★★ DOCG Varietal DOLCETTO. Some to drink young, some for moderate ageing. Chionetti, Clavesana, EINAUDI, Pecchenino, TREDIBERRI all gd.

Donnafugata Si ★★→★★★ Classy range: (r) ETNA Rosso (Contrada Fragore, Marchesa and Sul Vulcano), Mille e Una Notte, NERO D'AVOLA Sherazade, Tancredi; (w) Chiaranda, Kebir. Also v. fine MOSCATO PASSITO di PANTELLERIA Ben Ryé.

Duca di Salaparuta Si ★★ Top Duca Enrico and Bianca di Valguarnera. White Kados from GRILLO grapes also gd.

Due Terre, Le FVG ★★★ Small family-run FRIULI COLLI ORIENTALI estate: Top Sacrisassi Rosso (SCHIOPPETTINO/REFOSCO); v.gd MERLOT, Sacrisassi Bianco (w).

Einaudi, Luigi Pie ★★★ Founded late C19 by ex-president of Italy, 52-ha estate in DOGLIANI. Solid BAROLOS from Cannubi, Terlo and outstanding Bussia. Top Dogliani (DOLCETTO) from VIGNA Tecc. New Barolo Monvigliero and BARBARESCO.

Elba Tus ★→★★ DYA Island's white, TREBBIANO/ANSONICA, can be v. drinkable with fish. Dry reds based on SANGIOVESE; gd sweet white (MOSCATO) and red (*Aleatico Passito DOCG*). Try Acquabona, La Mola, Ripalte, Sapereta.

Enoteca Wine library; also shop or restaurant with ambitious wine list. There is a national enoteca at the *fortezza* in Siena.

Etna Si ★★→★★★ DOC (r) 12 13 14 15 16 17 (18) One of hottest areas, remarkable development in last decade; 900 ha on n slopes, high-altitude, volcanic soils. Etna Rosso typically 90/10 blend of NERELLOS MASCALESE/Cappuccio, while Etna Bianco can be pure CARRICANTE or incl CATARATTOS Comune or Lucido. (*See also* box, below.)

Falerno del Massico Cam ★★→★★★ DOC ★★ (r) 15 16 18 (19) Soils can be volcanic tuffs, clay, loam. Antiquity's most famous wine, today average only. Best: elegant AGLIANICO (r), fruity dry FALANGHINA (w). Try Masseria Felicia, Villa Matilde.

Faro Si ★★★ DOC 13 14 15 16 17 Intense, harmonious red from NERELLO MASCALESE, NERO D'AVOLA and Nocera in hills behind Messina. Top: Bonavita. Palari most famous, but Le Casematte just as gd if not better.

Fay Lom ★★★ Run by Fay family in VALTELLINA since 1971. Top: SFORZATO (RONCO del Picchio), Valgella (Camerei, Carteria RISERVA).

Felline Pug ★★★→★★★★ Gregory Perucci was pioneer in rediscovery of PRIMITIVO and Sussumaniello vines. Top PRIMITIVO DI MANDURIA (Cuvèe Anniversario, Dunico, Giravolta, ZIN); v.gd Sum (Sussumaniello) and METODO CLASSICO Edmont Dantes (sp VERMENTINO).

Felluga, Livio FVG ★★★ Consistently fine FRIULI COLLI ORIENTALI, esp blends Terre Alte and Abbazia di Rosazzo; Bianco Illivio, *Pinot Gr*, PICOLIT (Italy's best?), MERLOT/REFOSCO blend Sossó.

Felluga, Marco FVG *See* RUSSIZ SUPERIORE.

Felsina Tus ★★★ CHIANTI CLASSICO estate of distinction in se of zone: classic RISERVA Rancia and IGT Fontalloro, both 100% SANGIOVESE; v.gd Gran Selezione Colonia.

Fenocchio, Giacomo Pie ★★★→★★★★ Small but outstanding Monforte d'ALBA-based

---

Summit of Etna

Some of Italy's most exciting wines come from this famous volcano. Vines grow up to 1000m (3281ft) on si's e coast. *Contrada* is Si's way to express cru: differences in soil, altitude and age of lava flows. Top growers incl Alberelli di Giodo, BENANTI ★ (r Rovittello, Serra della Contessa, w Pietra Marina), Calabretta, Calcagno ★ (r Arcuria, Feudo di Mezzo, w Ginestra), Cottanera (w Calderara, Zottorinoto), GIROLAMO RUSSO (r Calderara Sottana, Feudo di Mezzo, San Lorenzo; w San Lorenzo), Graci (Arcuria, Feudo di Mezzo), GULFI (Reseca), I CUSTODI DELL'ETNA ★ (r Aetneus, Saeculare, w Ante), I VIGNERI ★ (r Vinupetra, w Aurora), Le Vigne di Eli; PIETRADOLCE ★ (r/w Archineri, Barbagalli, Rampante, Santo Spirito), TASCA D'ALMERITA (Rampante, Sciaranuova, w Buonora), TENUTE BOSCO (Vigna Vico), TENUTA DI FESSINA ★ (r/w Il Musmeci, w A' Puddara), TENUTA DELLE TERRE NERE ★ (r/w Caldera Sottana, Guardiola, San Lorenzo, Santo Spirito), Tornatore (Pietrarizzo, Trimarchisa), Vini Franchetti ★ (Guardiola, Rampante, Sciaranuova).

BAROLO cellar. Traditional style. Crus: Bussia (also RISERVA 90 Dì), Cannubi, Villero. Outstanding Freisa, one of Italy's 2–3 best.

Ferrara, Benito Cam ★★★ Maybe Italy's best GRECO DI TUFO producer (Terra d'Uva, VIGNA Cicogna). Talent shows in excellent TAURASI (Vigna Quattro Confini). Wait for a Vigna Cicogna RISERVA.

Ferrari – Tenute Lunelli T-AA ★★★→★★★★ TRENTO maker of one of two best Italian METODO CLASSICOS. Outstanding Giulio Ferrari (RISERVA del Fondatore, Rosé and Selezione); v.gd CHARD-based Brut Riserva Lunelli, Perlè Bianco (gd value), Perlè Zero, PINOT N-based Perlè Nero. TENUTE Lunelli: Castelbuono Umb (Carapace MONTEFALCO), Margon T-AA (Chard Villa Margon, Pinot N Maso Montalto), Podernuovo Tus (Solenida, Teuto).

Fessina, Tenuta di Si ★★★→★★★★ Silvia Maestrelli has one of youngest and best estates in Rovittello area of ETNA. Elegant wines. Best: EB A' Puddara (w), Il Musmeci (r/w) and new Il Musmeci RISERVA Speciale RS 2017 (r). Erse Moscamento 1911 (r) v.gd.

Feudi di San Gregorio Cam ★★★ Much-hyped CAM producer, with DOCGS FIANO DI AVELLINO Pietracalda, GRECO DI TUFO Cutizzi and Goleto, TAURASI Piano di Montevergine and new RISERVA Gulielmus. Also gd: red Serpico (AGLIANICO); whites *Campanaro* (Fiano/Greco), FALANGHINA. Look for new *FeudiStudi* line: most expressive vyds, selected every yr among 700 sites depending on vintage.

Feudo di San Maurizio VdA ★★★★ Outstanding wines from rare native grapes CORNALIN, Mayolet and Vuillermin; last two rank among Italy's greatest reds. Try (r) Saro Djablo, Torrette (and SUPERIORE); (w) GEWURZ, Petite ARVINE; (sp) Trei (CHARD, PINOT N).

With VdA's valpellinentze soup, drink Torrette Superieur (r), Petite Arvine (w).

Feudo Montoni Si ★★★★ Exceptional estate in upland e SI. Best: NERO D'AVOLA Lagnusa, Vrucara; v.gd GRILLO della Timpa (w), CATARRATTO del Masso (w), Perricone del Core (r), PASSITO Bianco (sw).

Fiano di Avellino Cam ★★→★★★★ DOCG 12 15 16 18 19 Can be either steely (most typical) or lush. Best: Cantina del Barone, Ciro Picariello, COLLI DI LAPIO, FEUDI SAN GREGORIO, I Favati (Pietramara), Joaquin, MASTROBERARDINO, Pietracupa, Rocca del Principe, QUINTODECIMO, TENUTA Sarno, Traerte (Aipierti).

Fino, Gianfranco Pug ★★★★ Greatest PRIMITIVO, from old, low-yielding bush vines. Outstanding Es among Italy's top 20 reds and Es Selezione Red Label; v.gd Jo (NEGROAMARO) and Se (Primitivo).

Florio Si ★★→★★★ Historic quality maker of MARSALA. Specialist in Marsala Vergine Secco. Best: RISERVA Donna Franca, Targa; v.gd Baglio Florio.

Fongaro Ven ★★★★ Classic-method fizz Lessini Durello (Durello = grape). High quality, even higher acidity, age-worthy. Top RISERVAS (Pas Dosé and Brut); v.gd Pas Dosé.

Fontanafredda Pie ★★ Since 1858, former royal estates. Large producer of PIE. Try BAROLO La Rosa, Alta Langa Brut Nature VIGNA Gatinera and Contessa Rosa Brut Rosé.

Fonterutoli Tus ★★★ Historic CHIANTI CLASSICO estate of Mazzei family at Castellina. Notable: Chianti Classico Gran Selezione (Badiola, Castello Fonterutoli, Vicoregio 36), IGT Siepi (SANGIOVESE/MERLOT). Also owns Tenuta di Belguardo in MAREMMA and Zisola in SI.

Fontodi Tus ★★★★ One of v. best CHIANTI CLASSICOS. Top: Flaccianello (100% SANGIOVESE), Gran Selezione VIGNA del Sorbo. IGT SYRAH Case Via among best of that variety in TUS.

Foradori T-AA ★★★ Much-loved Elisabetta F is "lady of TRENTINO wine" making outstanding **Teroldego** but lovely macerated, amphora-aged Incrocio Manzoni and Nosiola. Look for TEROLDEGO Morei and Sgarzon (also amphora-aged Cilindrica), white Nosiola Fontanasanta. Top remains Teroldego-based Granato.

Franchetti, Vini Si ★★★→★★★★ ETNA estate (former Passopisciaro) run by Franchetti (*see* TENUTA DI TRINORO), contributor to fame of Etna. Outstanding Rosso Franchetti (PETIT VERDOT/Cesanese d'Affile) and NERELLO MASCALESE single-*contrada*. Best: Contrada C, Contrada G, Contrada S; v.gd Contrada R. New Contrada PC (CHARD).

Franciacorta Lom ★★→★★★★ DOCG Italy's zone for top-quality METODO CLASSICO fizz. Soils extremely complex (50+ different types). Two large sectors: eastern, generally more elegant, freshest wines; western, generally broadest, richest. Top: CA' DEL BOSCO, Cavalleri (RISERVA Giovanni Cavalleri, Blanc de Blancs, Collezione Grandi Cru), MOSNEL, UBERTI, VILLA, VILLA CRESPIA. Also v.gd: Barone Pizzini, Ca' del Vent, Cola-Battista (Millesimato Dosaggio Zero, Extra Brut), Majolini, Monte Rossa, Ricci Curbastro, TERRA MORETTI (Bellavista, Contadi Castaldi).

Frascati Lat ★→★★ DOC DYA Once proud name (Rome's favourite) sadly debased. Buy these small producers only: Borgo del Cedro (SUPERIORE), Castel de Paolis (Superiore), De Sanctis (Abelos Bio), Merumalia (Primo) and Villa Simone (RISERVA Filonardi).

Frascole Tus ★★→★★★★ Most n winery of most n CHIANTI RÙFINA zone, small estate run organically by Enrico Lippi, with an eye for typicity. Chianti Rùfina main driver, but VIN SANTO to die for.

Frescobaldi Tus ★★→★★★★ Ancient noble family, leading CHIANTI RÙFINA pioneer at NIPOZZANO estate (look for ★★★ *Montesodi*), also BRUNELLO from Castelgiocondo estate, MONTALCINO. Sole owner of LUCE estate (Montalcino), ORNELLAIA (BOLGHERI), TENUTA Perano (top: CHIANTI CLASSICO Gran Selezione Rialzi). Vyds also in COLLIO (Attems), MAREMMA (Ammiraglia), Montespertoli (Castiglioni), Gorgona Island (state prison).

Friuli Colli Orientali FVG ★★→★★★★ DOC 15 16 18 (was COLLI Orientali del Friuli) Hilly area of FVG next to COLLIO. Unlike latter, not just whites, but outstanding reds, v.gd stickies from likes of red Pignolo, SCHIOPPETTINO, Tazzelenghe and white PICOLIT, VERDUZZO Friulano. Top: Aquila del Torre, d'Attimis, Ermacora, Gigante, La Sclusa, LA VIARTE, LE DUE TERRE, LIVIO FELLUGA, MIANI, Meroi, RONCHI DI CIALLA, VIGNA PETRUSSA. Ramandolo DOCG is best sweet Verduzzo (look for Anna Berra) PICOLIT can be Italy's best sweet: Aquila del Torre, LIVIO FELLUGA, Marco Sara, Ronchi di Cialla, VIGNA Petrussa often amazing.

Friuli Grave FVG ★→★★ DOC (was Grave del Friuli) Largest DOC of FVG, mostly on plains, often v. rainy, making quality red production tricky. Soils are quite gravelly, as is suggested by the DOC's name. Mostly big volumes, whites best. Look for Borgo Magredo, Di Lenardo, Le Monde, RONCO Cliona, Villa Chiopris.

Friuli Isonzo FVG ★★★ DOC (previously just Isonzo). One of world's best flatland wine denominations. High-alc, powerful, rich, unfailingly complex whites, from gravel-rich and clay-loam soils and a downright hot mesoclimate. Best: LIS NERIS, RONCO DEL GELSO, VIE DI ROMANS. Also gd: Borgo Conventi, Pierpaolo Pecorari.

Friuli Venezia Giulia A ne region hugging Slovenian border, home to Italy's best whites (alongside ALTO ADIGE). Hills to ne best, but alluvial seaside regions (DOC from Annia, Aquilea, Latisana) improving markedly. DOCs Carso, COLLI ORIENTALI, COLLIO, ISONZO best. All have Collio preceded on label by "Friuli".

Frizzante Semi-sparkling, up to 2.5 atmospheres, eg. MOSCATO D'ASTI, much PROSECCO, LAMBRUSCO and the like.

Fucci, Elena Bas ★★★★ AGLIANICO DEL VULTURE Titolo from 55–70-yr-old vines in Mt Vulture's Grand Cru; one of Italy's 20 best, now also RISERVA and SUPERIORE.

Organic. Outstanding 13 15 17 Anniversary. Also v.gd Titolo by Amphora (18 mths in terracotta amphora) and Titolo Pink Edition.

Fuligni Tus ★★★★ Outstanding: BRUNELLO (Top RISERVA), ROSSO DI MONTALCINO.

Gaja Pie ★★★★ Old family firm at BARBARESCO led by eloquent Angelo Gaja; daughter Gaia G following. High quality, higher prices. Top: Barbaresco (Costa Russi, Sorì San Lorenzo, Sorì Tildìn), BAROLO (Conteisa, Sperss). Splendid CHARD (Gaia e Rey). Also owns CA' MARCANDA in BOLGHERI, Pieve di Santa Restituta in MONTALCINO. New acquisition on ETNA (with Graci).

Gancia Pie Famous old brand of MUSCAT fizz. Best new Alta Langa (★★★ Cuvée 120, Cuvée 60).

Garofoli Mar ★★→★★★ Quality leader in the Mar, specialist in CONERO (Grosso Agontano), VERDICCHIO (Podium, Serra Fiorese, sp Brut RISERVA).

Gattinara Pie 13 15 16 17 (18) Best known of a cluster of ALTO PIE DOC(G)s based on NEBBIOLO. Steep hills; wines suitable for long ageing. Best: ANTONIOLO, CANTINA del Signore, Iarretti Paride, NERVI, Torraccia del Piantavigna, TRAVAGLINI. *See also* ALTO PIEMONTE.

Gavi / Cortese di Gavi Pie ★→★★★ DOCG DYA Overhyped, but at best subtle dry white of Cortese grapes. Best: *Bruno Broglia*/La Meirana, Castellari Bergaglio, La Giustiniana, La Scolca, Martinetti, Villa Sparina.

Germano, Ettore Pie ★★★ Small family Serralunga estate run by Sergio G and wife Elena. Top BAROLOS: RISERVA Lazzarito, Cerretta and new VIGNA Rionda; v.gd Alta Langa (also outstanding RISERVA BdN – 65 mths), BARBERA D'ALBA SUPERIORE della Madre, BAROLO Prapò, Del Comune di Serralunga, Langhe RIES Hérzu.

Barolo's largest MGA: Bricco San Pietro in Monforte d'Alba, a massive 380.9 ha.

Ghemme Pie ★★→★★★ DOCG NEBBIOLO (at least 85%), incl up to 15% Uva Rara and/or Vespolina. Mainly morainic agglomerate soils of friable pebbles and rich in minerals, poor and not v. fertile (actually, youngest among those of Alto Piemonte). Top: *Antichi Vigneti di Cantalupo* (Collis Braclemae, Collis Carellae), Ioppa (Balsina), Rovellotti (RISERVA); v.gd Torraccia del Piantavigna (VIGNA Pelizzane). *See also* ALTO PIEMONTE.

Ghizzano, Tenuta di Tus ★★★ Historical bio estate on Pisa's hills. Best: (r) Nambrot (B'x blend), Il Ghizzano, new Mimesi Project (r/w).

Giacosa, Bruno Pie ★★★★ Now run by daughter Bruna; wines still top. Splendid traditional-style BARBARESCOS (Asili, Rabajà), BAROLOS (Falletto, Falletto VIGNA Rocche). Top wines (ie. RISERVAS) get famous red label. Amazing METODO CLASSICO Brut, ROERO ARNEIS (w), Valmaggiore (r).

Girlan, Cantina T-AA ★★→★★★ Quality co-op. Top Le Selezioni line: PINOT N RISERVA (VIGNA Ganger, Trattmann, new Curlan), PINOT BL Flora, VERNATSCH Alte Reben Gschleier; v.gd Pinot Bl Platt & Riegl.

Giuseppe Cortese Pie ★★★ Traditional producer of outstanding BARBARESCO Rabajà (also RISERVA). Also gd LANGHE NEBBIOLO, Scapulin (CHARD).

GIV (Gruppo Italiano Vini) Complex of co-ops and wineries, biggest vyd holders in Italy: Bigi, BOLLA, Melini, Negri. Also in s: Bas, SI.

Grappa Pungent spirit made from grape pomace (skins, etc., after pressing), can be anything from disgusting to inspirational. What the French call "marc".

Gravner, Josko FVG ★★★ Controversial but talented COLLIO producer, vinifies on skins (r/w) in buried amphorae without temperature control. Best Breg (r/w), RIBOLLA GIALLA.

Greco di Tufo Cam DOCG DYA Tannic, oily whites made with local GRECO variety (different to Cal's also outstanding Greco Bianco). Best: Bambinuto (Picoli), BENITO FERRARA (VIGNA Cicogna), Caggiano (Devon), COLLI di Lapio (Alexandros), Donnachiara, FEUDI DI SAN GREGORIO (Cutizzi, Goleto e FeudiStudi), I Favati

(Terrantica), Macchialupa, **Mastroberadino** (Nova Serra, Stilema), Pietracupa, QUINTODECIMO, Terredora (Loggia della Serra), Vadiaperti (Tornante).

Grifalco Bas ★★★ Small estate in high-quality Ginestra and Maschito subzone. Best: AGLIANICO DEL VULTURE Daginestra, Damaschito (SUPERIORE DOCG 15).

Grignolino Pie DYA Two DOCS: GRIGNOLINO d'ASTI, Grignolino del Monferrato Casalese. At best, light, perfumed, crisp, high in acidity, tannin. D'Asti: BRAIDA, Cascina Tavijin, Crivelli, Incisa della Rocchetta, Spertino, TENUTA Garetto. Monferrato C: Accornero (Bricco del Bosco and Bricco del Bosco Vigne Vecchie – vinified like BAROLO), Bricco Mondalino, Castello di Uviglie, PIO CESARE.

Grosjean VdA ★★★ Top quality; best CORNALIN, Premetta. Vigne Rovettaz one of VDA's oldest, largest.

Guerrieri Rizzardi Ven ★★→★★★ Noble family making top AMARONE, BARDOLINO: Amarones Villa Rizzardi and Calcarole (cru); v.gd SOAVE CLASSICO Costeggiola.

Gulfi Si ★★★★ Best producer of NERO D'AVOLA in SI; 1st to bottle single-*contrada* (cru) wines. Organic certified. Outstanding: Nerobufaleffj, Nerosanlorè, iconic Nerojbleo, new RISERVAS. Also v.gd CERASUOLO DI VITTORIA CLASSICO, Nerobaronj, Neromaccarj and Carjcanti (w). Interesting Pinò (PINOT N), ETNA red Reseca.

Haas, Franz T-AA ★★→★★★ Top: PINOT N, LAGREIN (Schweizer), MOSCATO Rosa and IGT blend Manna (w).

Hofstätter T-AA ★★★ Top quality; gd PINOT N. Look for Barthenau VIGNA Sant'Urbano, Vigna Roccolo. Also whites, mainly GEWURZ (esp *Kolbenhof*).

IGT (indicazione geografica tipica) Increasingly known as IGP (indicazione geografica protetta). *See also* box, p.153.

Inama Ven ★★★ One of SOAVE's and COLLI Berici's most important producers, making some of denomination's best. Top: Soave CLASSICO (Carbonare, Du Lot, Foscarino) and CARMENÈRE RISERVA; v.gd Carmenère (Carminum), SAUV BL Vulcaia Fumè.

Ischia Cam ★→★★★ DOC DYA Island off Naples, green volcanic tuff soils, own grape varieties (w: Biancolella, Forastera; r: Piedirosso, also found in CAM). Frassitelli vyd best for Biancolella. Best: Antonio Mazzella (VIGNA del Lume), Cenatiempo (Kalimera), D'AMBRA (Biancolella Frassitelli, Forastera).

Isole e Olena Tus ★★★★ Top CHIANTI CLASSICO estate run by Paolo de Marchi: superb red IGT Cepparello. Outstanding VIN SANTO, Chianti Classico; v.gd CHARD, CAB SAUV, SYRAH. Also owns fantastic PROPRIETÀ SPERINO in LESSONA.

Kaltern, Cantina T-AA ★★→★★★ Quality co-op close to Caldaro lake. Top Quintessenz line. Look for limited-edition Kunst Stück.

Köfererhof T-AA ★★★→★★★★ Great whites: KERNER, SYLVANER; MÜLLER-T excellent.

Lageder, Alois T-AA ★★★ Famous ALTO ADIGE producer. Most exciting are single-vyd varietals: CAB SAUV Cor Römigberg, CHARD Löwengang, GEWURZ Am Sand, LAGREIN Lindenberg, PINOT N Krafuss, MCM (MERLOT), *Sauv Bl Lehenhof.*

Lagrein Alto Adige T-AA ★★→★★★ DOC 13 15 16' 18 Alpine red with deep colour, rich palate (plus a bitter hit at back); refreshing pink *Kretzerrosé* made with LAGREIN. Top ALTO ADIGE: CANTINA Bolzano (Taber), Cantina Santa Maddalena, CS Andriano, CS Tramin, Elena Walch, IGNAZ NIEDRIST, LAGEDER, MURI GRIES (Abtei, VIGNA Klosteranger), Putzenhof, TIEFENBRUNNER. From TRENTINO try Francesco Moser's Dearnater.

Lambrusco E-R ★→★★★ DYA The 17 different Lambrusco grapes (five mostly planted) make for highly distinct wines, ie. Lambrusco wine does not exist. Plethora of different denominations usually linked to one of five main grapes, so each has its defining characteristcs. When gd, delightful fizzy fresh, lively red that pairs divinely with rich, fatty fare. DOCS: L Grasparossa di Castelvetro, L Salamino di Santa Croce, L di Sorbara. Best: Grasparossa – Cleto Chiarli (Enrico Cialdini), Moretto (Monovitigno and VIGNA Canova), Pederzana (Canto

Libero Semi Secco), *Vittorio Graziano* (Fontana dei Boschi); Maestri – Ceci (Nero di Lambrusco Otello), Dall'Asta (Mefistofele); Marani – Ermete Medici (Quercioli); Salamino – Cavicchioli (Tre Medaglie Semi Secco), Luciano Saetti (Vigneto Saetti), Medici Ermete (Concerto Granconcerto); Sorbara – Cavicchioli (Cristo Secco and Cristo Rose), Cleto Chiarli (Antica Modena Premium), Medici Ermete (Phermento, ancestral method), Paltrinieri.

Langhe Pie The hills of central PIE, home of BAROLO, BARBARESCO, etc. DOC name for several Pie varietals plus Bianco and Rosso blends. Those wishing to blend other grapes with NEBBIOLO can at up to 15% as LANGHE NEBBIOLO – a label to follow.

Langhe Nebbiolo Pie ★★→★★★ Like NEBBIOLO D'ALBA (Nebbiolo 85%+) but from a wider area: LANGHE hills. Unlike N d'Alba, may be used as a downgrade from BAROLO or BARBARESCO. Try ALDO CONTERNO, Boroli, BREZZA, BURLOTTO, CIABOT BERTON, CLERICO, ETTORE GERMANO, FRATELLI ALESSANDRIA, GIACOMO FENOCCHIO, GIUSEPPE RINALDI, MASSOLINO, PIO CESARE, TREDIBERRI, VAJRA.

Lessona Pie DOCG *See also* ALTO PIEMONTE. NEBBIOLO (at least 85%). Mostly clay-free, marine sandy soils, rich in minerals. Elegant, age-worthy, fine bouquet, long savoury taste. Best: PROPRIETÀ SPERINO; gd Cassina, Colombera & Garella, La Prevostura, TENUTE Sella.

Librandi Cal ★★★ Top producer pioneering research into Cal varieties; v.gd red CIRÒ (*Riserva Duca San Felice* is ★★★), IGT Gravello and Terre Lontane (CAB SAUV/ Gaglioppo blend), Magno Megonio (r) from Magliocco grape, IGT Efeso (w) from Mantonico.

Liguria ★→★★ Narrow ribbon of extreme mtn viticulture produces memorable (w) PIGATO, VERMENTINO, (r) Rossese di Dolceacqua varieties. Look for Alessandri, Bio Vio, Bruna, TENUTA di Selvadolce (PIGATO), Terre Rosse; Giacomelli, La Baia del Sole, Ottaviano Lambruschi (Vermentino). CINQUE TERRE is beautiful and Sciacchetrà (sw) one of Italy's best stickies; Ormeasco di Pornassio (r) made with Ligurian biotype of DOLCETTO.

Lisini Tus ★★★→★★★★ Some of finest, longest-lasting BRUNELLO (RISERVA Ugolaia).

Lis Neris FVG ★★★ Top ISONZO estate for whites. Best: PINOT GR (Gris), SAUV BL (Picol), FRIULANO (Fiore di Campo), Confini and Lis. Also v.gd Lis Neris Rosso (MERLOT/CAB SAUV), sweet Tal Luc (VERDUZZO/RIES).

Lo Triolet VdA ★★★ Top PINOT GR producer, v.gd Fumin, Coteau Barrage and new Heritage (both SYRAH/Fumin), GEWURZ, MUSCAT.

Luce Tus ★★★ FRESCOBALDI's estate. Luce (SANGIOVESE/MERLOT blend for oligarchs). Lovely Luce BRUNELLO DI MONTALCINO.

Lugana ★★→★★★ DOC DYA Much-improved white of s Lake Garda, rivals gd SOAVE next door. Main grape Turbiana (was TREBBIANO di Lugana). Best: CA' DEI FRATI (I Frati, esp *Brolettino*), Domini Veneti, Le Morette, Monte del Frà, Ottella, Roveglia, Tommasi, Zenato (oaked), Zeni (Vigne Alte).

Lunae, Cantine Lig ★★★ Owned by Bosoni family in COLLI DI LUNI. Best: VERMENTINO (Cavagino, Etichetta Nera, Numero Chiuso). Plus v.gd rare Vermentino Nero.

Lungarotti Umb ★★→★★★ Leading producer of TORGIANO. Star wines DOC Rubesco, DOCG RISERVA *Monticchio*. Also Giubilante, MONTEFALCO SAGRANTINO, Sangiorgio (SANGIOVESE/CAB SAUV), VIGNA Il Pino (VERMENTINO/GRECHETTO/TREBBIANO).

Macchiole, Le Tus ★★★★ Organic. One of few native-owned wineries of BOLGHERI; one of 1st to emerge after SASSICAIA, makes *Italy's best Cab Fr* (Paleo Rosso), one of best MERLOTS (Messorio), SYRAHS (Scrio).

Maculan Ven ★★★ Quality pioneer of Ven. Excellent CAB SAUV (Fratta, Palazzotto). Best known for sweet Torcolato (esp RISERVA Acininobili).

Majo Norante, Di Mol ★★→★★★ Best known of Mol with decent Biferno Rosso Ramitello, Don Luigi Molise Rosso RISERVA, Mol AGLIANICO Contado.

Malvasia delle Lipari Si ★★★ DOC Luscious sweet, from one of many MALVASIA varieties. Best: TENUTA Capofaro (Dydime, VIGNA di Paola), Caravaglio, Fenech, Lantieri, Marchetta; gd Hauner.

Malvirà Pie ★★★→★★★★ Top ROERO producer. Organic certified. Best Roero single-vyd: (r/w) Renesio, Trinità; (w) Saglietto. New ARNEIS RISERVA Saglietto.

Manduria (Primitivo di) Pug ★★→★★★ DOC Cradle of PRIMITIVO, alias ZIN, so expect gutsy, alc, sometimes Porty wines. Best: FELLINE, GIANFRANCO FINO, MORELLA. Try producers, located in Manduria or not: Cantele, Pietraventosa, Polvanera, TENUTE Chiaromonte, Vetrere.

Marchesi di Gresy Pie ★★★ Historical BARBARESCO producer since 1797, from Martinenga cru (monopole). Best: RISERVA Camp Gros and Gajun; v.gd BARBERA D'ASTI Monte Colombo.

Maremma Tus Coastal region of TUS boomed in C20. Drier, hotter climate and shorter growing season produce broader wines, delicious (often too overripe aromas) SANGIOVESE-based reds from DOC(G)s: Monteregio, MORELLINO DI SCANSANO, Parrina, Pitigliano, Sovana.

Marrone, Agricola Pie ★★★ Small estate, gd-value BAROLOS. Top: Bussia, Pichemej; gd ARNEIS, BARBERA D'ALBA SUPERIORE, Favorita.

Marsala Si ★→★★★★ DOC SI's once-famous fortified, created by Woodhouse Bros of Liverpool in 1773. Can be dry to v. sweet; best is bone-dry Marsala Vergine. *See also* MARCO DE BARTOLI.

19 recognized Malvasia grapes in Italy: red, white, pink, unrelated to each other.

Mascarello Pie ★★★★ Two top producers of BAROLO: the late Bartolo M, of Barolo, whose daughter Maria Teresa continues her father's highly traditional path (v.gd Freisa and LANGHE NEBBIOLO); and Giuseppe M, of Monchiero, whose son Mauro makes v. fine, traditional-style Barolo from the great **Monprivato** vyd in Castiglione Falletto. Both deservedly iconic.

Masi Ven ★★→★★★ Archetypal yet innovative producer of Verona, led by inspirational Sandro Boscaini; v.gd Rosso Veronese **Campo Fiorin**, Osar (Oseleta). Top AMARONES Campolongo di Torbe, Costasera. Masi Wine Estates: Canevel (CARTIZZE, Valdobbiadene Campofalco), Conti Bossi Fedrigotti (Fojaneghe B'x blend, Trento Conte Federico), Serego Alighieri (Amarone Vaio Armaron).

Massa, Vigneti Pie ★★★ Walter M brought Timorasso (w) grape back from nr extinction. Top: Coste del Vento, Montecitorio, Sterpi; Anarchia Costituzionale (MOSCATO Bianco), Avvelenata (Freisa), BARBERAS Monleale and Bigolla v.gd.

Massolino Vigna Rionda Pie ★★★ One of finest BAROLO estates, in Serralunga. Excellent Parafada, Margheria (firm structure, fruity drinkability); top long-ageing VIGNA Rionda; v.gd LANGHE NEBBIOLO, Parussi. Waiting for new BARBARESCOS.

Mastroberardino Cam ★★★ Historic top-quality producer of mtn Avellino province. Top Stilema Line (Taurasi, Fiano and Greco), **Taurasi** (Historia Naturalis, Radici RISERVA). FIANO DI AVELLINO (More Maiorum, Radici), GRECO DI TUFO Nova Serra.

Meroi FVG ★★★ Dynamic estate. Top FRIULANO, MALVASIA Zittelle Durì, RIBOLLA GIALLA, SAUV BL Zittelle Barchetta.

Metodo classico or tradizionale Italian for "Champagne method".

Miani FVG ★★★★ Enzo Pontoni is Italy's best white winemaker. Top: FRIULANO (Buri and Filip), RIBOLLA GIALLA Pettarin, SAUV BL Zittelle. Also v.gd Sauv Bl Saurint, CHARD Zitelle, MERLOT, REFOSCO Buri.

Mirizzi and Montecappone Mar ★★★ Two estates owned by Gianluca Mirizzi focus on VERDICCHIO DEI CASTELLI DI JESI. Former on marly sandstone soils, deep slopes, organic and traditional winemaking (top: Ergo, Ergo Sum RISERVA), latter on calcareous clay, reductive winemaking (top: Federico II, Riserva Utopia).

Molettieri, Salvatore Cam ★★★ Outstanding RISERVA VIGNA Cinque Querce and TAURASI; gd FIANO DI AVELLINO Apianum.

Monaci Pug ★★→★★★ Part of GIV. Characterful NEGROAMARO Kreos (rosé), PRIMITIVO Artas and SALICE SALENTINO Aiace (r).

Monica di Sardegna Sar ★→★★★ DOC DYA Delightfully perfumed, medium weight. Best: ARGIOLAS (Iselis), Cantina di Mogoro, CONTINI, Dettori (Chimbanta) Ferruccio Deiana (Karel), Josto Puddu (Torremora), SANTADI (Antigua).

Montalcino Tus Hilltop town in province of Siena, fashionable and famous for concentrated, expensive BRUNELLO and more approachable, better-value ROSSO DI MONTALCINO, both still 100% SANGIOVESE.

Monte Carrubo Si ★★★ Pioneer Peter Vinding-Diers planted SYRAH on a former volcano s of Etna. Exciting, complex results.

Monte del Frà Ven ★★→★★★ Owned by Bonomo family; v.gd value. Top: AMARONE Lena di Mezzo and RISERVA, Custoza (Bonomo Sexaginta, Ca' del Magro).

Montefalco Sagrantino Umb ★★★→★★★★ DOCG Once sweet PASSITO only (still best wine of area), drier version is Italy's most powerfully tannic red that requires optimal growing seasons to show best. Top: Adanti, Antonelli (regular, Chiusa di Pannone), CAPRAI (25 Anni, Collepiano), Colleallodole, LUNGAROTTI, Pardi (Sacrantino), TENUTA Castelbuono. Also v.gd Bocale, Perticaia, Ruggeri Sportoletti, Tabarrini, Villa Mongalli.

Montepulciano d'Abruzzo Ab ★★→★★★ DOC (r) 14 15 18 Thanks to new generation of winemakers, Ab's wines (MONTEPULCIANO and TREBBIANO D'ABRUZZO too) never been better. Reds can be either light, easy-going or structured, rich. Best: Cataldi Madonna (Piè delle Vigne, Tonì), EMIDIO PEPE, Filomusi Guelfi, Praesidium, TIBERIO (regular, Colle Vota), Torre dei Beati (Cocciapazza Mazzamurello), Valle Reale and of course *Valentini* (best, age-worthy).

Montevertine Tus ★★★★ Organic certified estate in Radda. Outstanding IGT Le Pergole Torte, world-class, pure, long-ageing SANGIOVESE; v.gd Montevertine.

Montevetrano Cam ★★★ Iconic CAM AZIENDA. Superb IGT Montevetrano (AGLIANICO/CAB SAUV/MERLOT); v.gd Core Rosso (Aglianico) and Core Bianco (FIANO, GRECO).

Morella Pug ★★★→★★★★ Gaetano M and wife Lisa Gilbee make outstanding PRIMITIVO (La Signora, Mondo Nuovo, Old Vines) from c.90-yr-old vines. Also v.gd Mezzarosa rosé (Primitivo/NEGROAMARO), Mezzogiorno (FIANO).

Morellino di Scansano Tus ★→★★★ DOCG 11 13 15 16' 17 18 MAREMMA's famous SANGIOVESE-based red: best cheerful, light than overoaked, gritty. *Le Pupille* (regular, RISERVA), Moris Farms, PODERE 414, POGGIO ARGENTIERA (Bellamarsilia) Roccapesta (Calestaia), TENUTA Belguardo, Terenzi (Purosangue) best.

Moris Farms Tus ★★★ One of 1st new-age producers of TUS's MAREMMA. Top: iconic IGT Avvoltore (SANGIOVESE/CAB SAUV/SYRAH) and MORELLINO DI SCANSANO (basic, RISERVA). But try VERMENTINO and Rosato Rosamundi.

Moscato d'Asti Pie ★★→★★★ DYA Similar to DOCG ASTI, but usually better grapes; lower alc, lower pressure, sweeter, fruitier, often from small producers. Best DOCG MOSCATO: Ca' d'Gal, *Caudrina* (La Galeisa), Forteto della Luja, Mongioia, *Saracco*, Vajra, Vignaioli di Santo Stefano. Plus v.gd: Braida, Cascina Fonda, Il Falchetto, L'Armangia, Perrone, RIZZI, Vico.

Mosnel Lom ★★★ Barboglio family-run organic winery in FRANCIACORTA, since 1836. Top: EBB, Pas Dosé RISERVA, Nature and the EBB 2010 Riedizione 2020.

Muri Gries T-AA ★★→★★★ Monastery in Bolzano suburb of Gries; traditional and still top producer of LAGREIN ALTO ADIGE DOC. Esp cru Abtei-Muri, Klosteranger.

Musso Pie ★★★ Musso family-run BARBARESCO estate, since 1929. Top: Barbarescos Pora (also RISERVA), Rio Sordo.

Nals Margreid T-AA ★★★ Small quality co-op making mtn-fresh whites (esp PINOT BIANCO Sirmian), CHARD RISERVA, new cuvée (w) Nama.

Nebbiolo d'Alba Pie ★★→★★★ DOC 13 14 15 16' 18 (19) (100% NEBBIOLO) Sometimes a worthy replacement for BAROLO/BARBARESCO, though it comes from a distinct area between the two. Best: BREZZA, BRUNO GIACOSA, CERETTO, Hilberg-Pasquero, LUCIANO SANDRONE, ORLANDO ABRIGO, PAITIN, PODERI COLLA.

Nervi Pie ★★★ Historical winery in GATTINARA now owned by Roberto CONTERNO. Best: Molsino, Valferana; v.gd Rosato.

Niedrist, Ignaz T-AA ★★★ LAGREIN Berger Gei RISERVA is reference. So are RIES, WEISSBURGUNDER (Limes), BLAUBURGUNDER Riserva; v.gd SAUV BL Limes and Trias (w blend), CHARD vom Kalk.

Nino Franco Ven ★★★→★★★★ Owner Primo Franco makes large volumes of top-notch PROSECCO that age surprisingly well. Among finest: Grave di Stecca Brut, Primo Franco Dry, Riva di San Floriano Brut, Rustico. Excellent CARTIZZE.

Nipozzano, Castello di Tus ★★★→★★★★ FRESCOBALDI estate in RÙFINA, e of Florence, making excellent CHIANTI Rùfina. Top Nipozzano RISERVA (esp Vecchie Viti) and IGT *Montesodi*; v.gd Mormoreto (B'x blend).

Nittardi Tus ★★→★★★ Reliable source of quality modern CHIANTI CLASSICO (esp Casanuova di Nittardi, RISERVA). German-owned; oenologist Carlo Ferrini.

Nossing, Manni T-AA ★★★★ Outstanding KERNER, MÜLLER-T Sass Rigais, SYLVANER. Benchmark wines.

Nozzole Tus ★★→★★★ Famous estate in heart of CHIANTI CLASSICO, n of Greve, owned by Folonari; v.gd Chianti Classico RISERVA, excellent CAB SAUV Pareto.

Nuragus di Cagliari Sar ★★ DOC DYA Lively, uncomplicated, from Nuragus grape, finally gaining visibility. Best: ARGIOLAS (S'Elegas), Mogoro (Ajò), Pala (I Fiori).

Occhio di Pernice Tus "Partridge's Eye". A type of VIN SANTO made predominantly from black grapes, mainly SANGIOVESE. *Avignonesi's is definitive*. Also an obscure black variety found in RÙFINA and elsewhere.

Montalcino, from 2021, has an "urban vineyard" inside the ancient fortress.

Occhipinti, Arianna Si ★★★ Cult producer, deservedly so. Organic certified. Top: Il Frappato and CERASUOLO DI VITTORIA CLASSICO Grotte Alte.

Oddero Pie ★★★→★★★★ Traditionalist La Morra estate for excellent BAROLO (Brunate, Bussia RISERVA, VIGNA Rionda RISERVA, Villero), BARBARESCO (Gallina) crus, plus other serious PIE wines. Also v.gd value Barolo, RIES. Monvigliero from 2022.

Oltrepò Pavese Lom ★→★★★ Multi-DOC, numerous varietals, blends from Pavia province: SPUMANTE best. Anteo, Barbacarlo, Castello di Cigognola, Conte Vistarino (METODO CLASSICO 1865, PINOT N Bertone), Giorgi, Mazzolino, Travaglino.

Ornellaia Tus ★★★★ 12 13 15 16 (18) Fashionable, indeed cult, estate nr BOLGHERI now owned by FRESCOBALDI Top wines of B'x grapes/method: Bolgheri DOC Ornellaia, IGT Masseto (MERLOT), Ornellaia Bianco (SAUV BL/VIOGNIER); gd Bolgheri DOC Le Serre Nuove and POGGIO alle Gazze (w).

Orvieto Umb ★→★★★ DOC DYA One of few areas of Italy where noble rot occurs spontaneously and often. Characterized by four soil types (sandy clay, volcanic, alluvial sands, and a mix of yellow compacted sands and conglomerates). Sweet late-harvest can be memorable though cheap-and-cheerful dry white v. popular too. Off-dry Amabile less in favour today but delicious. Top: BARBERANI (Luigi e Giovanna) but sweet Calcaia just as gd. Other gd: Bigi, Cardeto, *Castello della Sala*, Decugnano dei Barbi, Palazzone, Sergio Mottura (Lat).

Pacenti, Siro Tus ★★★ Modern-style BRUNELLO, ROSSO DI MONTALCINO.

Paitin Pie ★★★ Pasquero-Elia family has been bottling BARBARESCO since C19. Today back on track making "real" Barbaresco from cru Serraboella in large barrels. Sorì Paitin Vecchie Vigne is star; v.gd new Barbaresco Basarin.

Paltrinieri E-R ★★→★★★ One of top three LAMBRUSCO producers. Among 1st to produce 100% Lambrusco di Sorbara. Best: Leclisse, Secco Radice; v.gd La RISERVA and METODO CLASSICO Grosso.

Pantelleria Si ★★★ Windswept, black (volcanic) earth si island off Tunisian coast, famous for superb MOSCATO d'Alessandria stickies. PASSITO versions dense/intense. Try MARCO DE BARTOLI (Bukkuram), DONNAFUGATA (Ben Ryé), Ferrandes.

Passito Tus, Ven One of Italy's most ancient and characteristic styles, from grapes dried briefly under harvest sun (in s) or over a period of weeks or mths in airy attics – a process called *appassimento*. Best-known versions: AMARONE/RECIOTO, VALPOLICELLA/SOAVE (Ven); VIN SANTO (TUS). Try Loazzolo, MONTEFALCO, ORVIETO, Torcolato, VALLONE. Never cheap.

Pavese Ermes VdA ★★★ One estate, one grape: the Prié Bl, with ungrafted vines up to 1219m (4000ft). Top: Blanc de Morgex et de la Salle (Nathan and Le 7 Scalinate), Ninive (sw), Pavese XXXIV (sp).

Pepe, Emidio Ab ★★★ Artisanal winery, 15 ha, bio/organic certified. Top MONTEPULCIANO D'ABRUZZO; gd PECORINO, TREBBIANO D'ABRUZZO (Old Vines).

Petrussa, Vigna FVG ★★★ Small family estate: high-quality wines. Best: SCHIOPPETTINO di Prepotto (also RISERVA), PICOLIT and Richenza (cuvée of w indigenous grapes, old vines).

Pian dell'Orino Tus ★★★★ Small MONTALCINO estate, committed to bio. BRUNELLO seductive, technically perfect, Rosso nearly as gd. Many epic wines.

Piane, Le Pie ★★★→★★★★ BOCA DOC resurfaced thanks to Christoph Kunzli; v.gd (r) Maggiorina, Mimmo (NEBBIOLO/Croatina), Nebbiolo, Piane (Croatina); (w) Bianco (Erbaluce).

Picolit FVG ★★→★★★ DOCG 12 13 15 16 (18) Potentially Italy's best sweet (most from air-dried grapes; rare late-harvests even better), but plagued by poor versions that don't speak of the grape. Texture ranges from light/sweet (rare) to super-thick (PASSITO). Best: Aquila del Torre, d'Attimis, I Comelli, LIVIO FELLUGA, Marco Sara, Perusini, RONCHI DI CIALLA, Valentino Butussi, VIGNA PETRUSSA. Also v.gd: Ermacora, Girolamo Dorigo, Paolo Rodaro.

Barolo's smallest MGA: Bricco Rocche in Castiglione Falletto, just 1.46 ha.

Piedmont / Piemonte In ne, bordering France to the w. Turin is capital. Monferrato, LANGHE, ROERO and ALTO PIEMONTE main areas. With TUS, Italy's most important region for quality (10% of all DOC[G] wines). No IGTS allowed. Grapes incl BARBERA, Brachetto, Cortese, DOLCETTO, Freisa, GRIGNOLINO, MALVASIA di Casorzo, Malvasia di Schierano, MOSCATO, NEBBIOLO, Ruchè, Timorasso. *See also* BARBARESCO, BAROLO.

Pieropan Ven ★★★★ Andrea and Dario, Leonildo's sons, now run winery. Organic certified. Cru *La Rocca* still ultimate oaked SOAVE; Calvarino best of all. Plus v.gd AMARONE, PASSITO della Rocca.

Pietradolce Si ★★★→★★★★ Faro bros own vyds in key ETNA crus, often pre-phylloxera vines. Top: Etna Rosso Barbagalli, Rampante and white Sant'Andrea (100% Carricante); v.gd Archineri (r/w).

Pio Cesare Pie ★★★ Veteran ALBA producer; BAROLO, BARBARESCO in modern (barrique) and traditional (large cask-aged) versions. Particularly gd NEBBIOLO D'ALBA, *a little Barolo at half the price*. Best: Single-vyd and Classic collections (Pio) dedicated to Pio Boffa.

Pira e Figli – Chiara Boschis Pie ★★★→★★★★ Organic certified. Must-visit estate. Top: Cannubi, Mosconi, Via Nuova; v.gd BARBERA D'ALBA.

Planeta Si ★★★ Leading si estate with vyds all over island, incl Menfi (GRILLO Terebinto), Noto (NERO D'AVOLA Santa Cecilia), Vittoria (CERASUOLO Dorilli), most recently on Etna (Carricante, NERELLO MASCALESE Eruzione 1614). Also gd Nocera (r), La Segreta (r/w), Cometa (FIANO).

Podere Tus Small TUS farm, once part of a big estate.

Poggio Tus Means "hill" in TUS dialect. **Poggione** means "big hill".

Poggio Antico Tus ★★★ Paola Gloder looks after 32 ha estate, one of highest in MONTALCINO at c.500m (1640ft). Restrained, consistent, at times too herbal style.

Poggio Argentiera Tus ★★★ MAREMMA estate owned by TUA RITA. Best: Capatosta (95% SANGIOVESE); v.gd MORELLINO DI SCANSANO and Poggioraso (CAB FR).

Poggio di Sotto ★★★★ Small MONTALCINO estate with a big reputation recently. Has purchased adjacent vyds. Top BRUNELLO, RISERVA and Rosso of traditional character with idiosyncratic twist.

Poggione, Tenuta Il Tus ★★★ MONTALCINO estate, in s; consistently excellent BRUNELLO, ROSSO. Top Brunello RISERVA VIGNA Paganelli; v.gd VIN SANTO.

Poggiopiano Tus ★★ Opulent CHIANTI CLASSICO from Bartoli family. Chiantis are pure SANGIOVESE, but SUPER TUSCAN Rosso di Sera incl up to 15% Colorino; v.gd Colorino Taffe Ta'.

Poggio Scalette Tus ★★ Vittorio Fiore and son Jury run CHIANTI organic estate at Greve. Top: Il Carbonaione (100% SANGIOVESE); needs several yrs bottle-age. Above-average CHIANTI CLASSICO and B'x-blend Capogatto.

Poliziano Tus ★★★ MONTEPULCIANO organic estate of Federico Carletti. Top: VINO NOBILE (esp cru Asinone, new Le Caggiole). Also gd IGT Le Stanze (CAB SAUV/MERLOT).

Pomino Tus ★★★ DOC (r) 12 13 15 16 17 (18) Appendage of RÙFINA, with fine red and white blends (esp Il Benefizio). Virtually a FRESCOBALDI exclusivity.

Potazzine, Le Tus ★★★★ Organic estate of Gorelli family just s of MONTALCINO; vyd is quite high. Outstanding BRUNELLOS (also RISERVA) and Rossos: serious and v. drinkable. Try them at family's restaurant in town.

Prà Ven ★★★★ Leading SOAVE CLASSICO producer, esp crus Colle Sant'Antonio, Monte Grande, Staforte. Excellent AMARONE (15 top), VALPOLICELLA La Morandina.

Produttori del Barbaresco Pie ★★★ One of Italy's earliest co-ops, perhaps best in the world, makes excellent traditional straight BARBARESCO plus crus Asili, Montefico, Montestefano, Ovello, Pora, Rio Sordo. Super values.

Proprietà Sperino Pie ★★★→★★★★ Top estate of LESSONA. One of best of ALTO PIEMONTE run by Luca De Marchi (see ISOLE E OLENA). Outstanding: Lessona; v.gd L Franc (one of best Italian CAB FR), Rosa del Rosa (NEBBIOLO/Vespolina rosé) and Uvaggio (r). Waiting for BRAMATERRA, RISERVA.

Prosecco Ven ★→★★ DOC(G) DYA Prosecco is the wine, GLERA the grape variety. Quality is higher in the Valdobbiadene. Look for: Adami, Biancavigna, BISOL, Bortolin, Canevel, Carpenè-Malvolti, Case Bianche, Col Salice, Col Vetoraz, Gregoletto, La Riva dei Frati, Le Colture, Mionetto, NINO FRANCO, Ruggeri, Silvano Follador, Zardetto.

Puglia The "heel" of Italy. Many gd-value reds from likes of Bombino Nero, NEGROAMARO, PRIMITIVO, Susumaniello and Uva di Troia grapes. Bombino Bianco, aromatic Minutolo and Verdeca most interesting whites. Castel del Monte, Gioia del Colle Primitivo, PRIMITIVO DI MANDURIA, SALICE SALENTINO best denominations.

Quartomoro Sar ★★→★★★ Piero Cella works with old vines, rare local varieties. Best: Memorie di Vite line (Bovale, Monica, Semidano).

Querciabella Tus ★★★ Top CHIANTI CLASSICO estate, bio since 2000. Top IGT Camartina (CAB SAUV/SANGIOVESE), Batàr (CHARD/PINOT BL) and new single-commune wines (Gaiole, Greve in CHIANTI, Radda in Chianti); v.gd Chianti Classico (and RISERVA). From 2022 new Gran Selezione.

Quintarelli, Giuseppe Ven ★★★★ Arch-traditionalist artisan producer of sublime VALPOLICELLA, RECIOTO, AMARONE; plus a fine Bianco Secco, a blend of various grapes. Daughter Fiorenza and sons now in charge, altering nothing, incl the ban on spitting when tasting.

Quintodecimo Cam ★★★→★★★★ Oenology professor and winemaker Luigi Moio's beautiful estate. Outstanding: TAURASI VIGNA Grande Cerzito and Vigna Quintodecimo; great AGLIANICO (Terra d'Eclano), GRECO DI TUFO (Giallo d'Arles).

Ratti, Renato Pie ★★→★★★ Iconic BAROLO estate. Modern wines; short maceration but plenty of substance, esp Barolos Rocche dell'Annunziata and Conca.

Recioto della Valpolicella Ven ★★★→★★★★ DOCG Sweet-wine marketing problems mean this traditional Italian beauty is being made less and less. Shame, esp as always much better than many disappointing overly sweet/tannic AMARONES.

Recioto di Soave Ven ★★★→★★★★★ DOCG SOAVE from half-dried grapes: sweet, fruity, slightly almondy; sweetness is cut by high acidity. *Drink with cheese*. Best: Anselmi, Coffele, Gini, PIEROPAN, Tamellini; often v.gd from Ca' Rugate, Pasqua, PRÀ, Suavia, Trabuchi.

Refosco (dal Peduncolo Rosso) FVG ★★ **12 13** 15 16 (18) Most-planted native red grape of region. Best from FRIULI COLLI ORIENTALI DOC. Top: MIANI, VIGNA PETRUSSA, Volpe Pasini. Also gd Ca' Bolani, D'Attimis, La Viarte, LIVIO FELLUGA, MEROI, Valchiarò, Vignai da Duline, Zorzettig (Myò).

Ricasoli Tus Historic Tuscan family; 1st Italian prime minister Bettino R devised the classic CHIANTI blend. Main branch occupies medieval CASTELLO DI BROLIO.

Riecine Tus ★★★→★★★★ SANGIOVESE specialist at Gaiole since 70s. Riecine di Riecine, La Gioia (100% Sangiovese) potentially outstanding; gd Tresette (MERLOT).

Rinaldi, Francesco Pie ★★★ Paola and Piera R make classic, elegant BAROLOS. Top Cannubi (also RISERVA) and Brunate; v.gd Classic Barolo and LANGHE NEBBIOLO.

Rinaldi, Giuseppe Pie ★★★ Beppe R's daughters Marta and Carlotta continue father's highly traditional path. Top: BARBERA D'ALBA, BAROLOS and Freisa.

Italy's most planted red grape? Sangiovese: 54,000 ha.

Riserva Wine aged for a statutory period, usually in casks or barrels.

Rivetti, Giorgio (La Spinetta) Pie ★★★ Fine MOSCATO D'ASTI, excellent BARBERA, series of super-concentrated, oaky BARBARESCOS. Also owns vyds in BAROLO, CHIANTI COLLI Pisane DOCGS, traditional SPUMANTE house Contratto (Cuvée Novecento, For England ALTA LANGA).

Rivetto Pie ★★★ Bio estate; Enrico R one of most talented young winemakers. Top: BAROLO Briccolina; v.gd BARBERA D'ALBA, Barolo Serralunga, LANGHE NEBBIOLO.

Rizzi Pie ★★★→★★★★ Sub-area of Treiso, BARBARESCO commune, where Dellapiana family looks after 35 ha. Organic. Top crus: Barbaresco Pajorè and Rizzi RISERVA Boito. Also v.gd: ALTA LANGA, Barbaresco (Nervo, Rizzi), MOSCATO D'ASTI.

Roagna Pie ★★★★ Old vines, massal selection, organic, wild yeast, long maceration and long ageing in large oak casks. Outstanding BARBARESCO Crichet Pajet, BAROLO and Barbaresco Vecchie Viti ("old vines") line; v.gd Barbaresco Pajè, Barolo Pira, Timorasso Montemarzino and new Barolo Rocche di Castiglione.

Rocca, Albino Pie ★★★→★★★★ A foremost producer of elegant, sophisticated BARBARESCO: top crus Cottà, Ovello VIGNA Loreto, Ronchi, RISERVA.

Rocca, Bruno Pie ★★★→★★★★ Family estate run by Francesco and Luisa, Bruno's children. BARBARESCOS with a more traditional style. More elegance than power. Top: Maria Adelaide, RISERVA. Also v.gd: BARBERA D'ASTI, Currá, Rabajà.

Rocca delle Macie Tus ★★ Large estate in Castellina in CHIANTI, run by Sergio Zingarelli. Best: Fizzano, Gran Selezione Sergio Zingarelli.

Roero Pie ★★→★★★★ DOCG **13** 15 16 18 (19') Wilder, cooler compared to LANGHE. Wines have typically moderate alc and tannin, gd elegance, freshness and aromatic profile. NEBBIOLO (r) and ARNEIS (w). Best: BRUNO GIACOSA ★, Ca' Rossa, Cascina Chicco, Cornarea, GIOVANNI ALMONDO, MALVIRÀ ★, MATTEO CORREGGIA ★, Morra, Negro, Rosso, Taliano, Val del Prete, Valfaccenda.

Romagna Sangiovese Mar ★★→★★★ DOC At times too herbal and oaky, but often well-made, even classy SANGIOVESE red. Try Ca' di Sopra, Cesari, Condello,

Drei Donà, FATTORIA ZERBINA, Nicolucci, Papiano, Paradiso, Tre Monti, Trere, Villa Venti (Primo Segno). Also IGT RONCO delle Ginestre and Ronco dei Ciliegi (Castelluccio).

Ronchi di Cialla FVG ★★★→★★★★ Leading FVG estate, in Cialla subzone of FRIULI COLLI ORIENTALI, run by Rapuzzi family, devoted to local old native grapes. Best: Ciallabianco (blend of RIBOLLA GIALLA/VERDUZZO/PICOLIT), Picolit di Cialla, SCHIOPPETTINO di Cialla; v.gd REFOSCO dal Peduncolo Rosso and Sol (Picolit dry).

Ronco Term for a hillside vyd in ne Italy, esp FVG.

Ronco del Gelso FVG ★★★→★★★★ Tight, pure ISONZO: FRIULANO Toc Bas, MALVASIA VIGNA della Permuta, PINOT GR Sot lis Rivis are regional benchmarks; v.gd Latimis (w blend).

Rosato General Italian name for rosé. Other rosé names incl Chiaretto from Lake Garda; CERASUOLO from Ab; Kretzer from ALTO ADIGE.

Rossese di Dolceacqua or Dolceacqua Lig ★★→★★★ DOC Interesting reds. Intense, salty, spicy; greater depth of fruit than most. Best: Maccario-Dringenberg (Curli, Luvaira, Posaù, Sette Cammini), Terre Bianche (Bricco Arcagna). Also v.gd Ka' Mancinè, Poggi dell'Elmo, TENUTA Anfosso.

Rosso di Montalcino Tus ★★→★★★ 13 15 16 18 19 (20) DOC for earlier-maturing BRUNELLO (grape) wines, usually from younger or lesser vyd sites; bargains exist.

Rosso Piceno / Piceno Mar ★ DOC 15 18 (19) Blend of MONTEPULCIANO (35%+) and SANGIOVESE (15%+). SUPERIORE means it comes only from far s of region. Best: BUCCI, GAROFOLI, MONTECAPPONE (Utopia); v.gd Boccadigabbia, Moncaro, Monte Schiavo, Saladini Pilastri, Santa Barbara, TENUTA di Tavignano, Velenosi.

Ruffino Tus ★→★★★ Venerable CHIANTI firm, at Pontassieve nr Florence, produces reliable wines such as CHIANTI CLASSICO RISERVA Ducale and Ducale Oro.

Rùfina Tus ★★→★★★ Most n subzone of CHIANTI, e of Florence, is by far best and most interesting non-CHIANTI CLASSICO denomination. Highest of all Chiantis, meaning refined, age-worthy wines. Soils are varied and incl limestone, sand, Galestro, Alberese, marly clays and others. Top: Colognole, Fattoria SELVAPIANA Bucerchiale, Frascole, Il Pozzo, Lavacchio, NIPOZZANO (Riserva, Vecchie Viti *see* FRESCOBALDI); vgd Castello del Trebbio, Grignano, Travignoli, Vetrice. Don't confuse with RUFFINO, which has HQ in Pontassieve, Rùfina's main town.

Russiz Superiore FVG ★★→★★★ LIVIO FELLUGA's brother, Marco, est vyds in various parts of FVG. Now run by Marco's son Roberto. Wide range; best is PINOT GRIGIO, COLLIO Bianco blend Col Disôre; v.gd PINOT BIANCO RISERVA.

Russo, Girolamo Si ★★★→★★★★ Giuseppe R is one of three or four best ETNA producers. Among 1st on Etna to bottle crus separately. Top: (r) Etna Rosso (Feudo, Feudo di Mezzo, San Lorenzo); (w) Etna Bianco (Nerina, San Lorenzo).

Salento Pug Home to Italy's best rosé from NEGROAMARO (alongside Ab's CERASUOLO from MONTEPULCIANO). Plus v.gd red Negroamaro, with a bit of help from MALVASIA Nera, and now red from local grape Sussumaniello. *See also* PUG, SALICE SALENTINO.

Salice Salentino Pug ★★→★★★ DOC 15 16 17 19 Best known of Salento's too many NEGROAMARO-based DOCS. RISERVA after 2 yrs. Try Cantele, Conti Zecca (Cantalupi), Cosimo Taurino, Leone de Castris (Riserva), Mocavero, Palamà, Vallone (Vereto Riserva).

Salvioni Tus ★★★★ aka La Cerbaiola. Iconic small, highest-quality MONTALCINO estate run by father-and-daughter team. BRUNELLO, ROSSO DI MONTALCINO among v. best available and worthy of their high prices.

Sandrone, Luciano Pie ★★★→★★★★ Modern-style ALBA. Deep BAROLOS: Alèste (was Cannubi Boschis), Le Vigne, Vite Talin. Also gd NEBBIOLO D'ALBA Valmaggiore.

San Felice Tus ★★★ Important historic TUS grower, owned by Gruppo Allianz, run by Leonardo Bellaccini. Fine CHIANTI CLASSICO and RISERVA POGGIO Rosso from estate in Castelnuovo Berardenga. Also gd BRUNELLO DI MONTALCINO Campogiovanni, IGT *Vigorello* (1st SUPER TUSCAN, from 1968).

San Giusto a Rentennano Tus ★★★★ Top CHIANTI CLASSICO estate. Organic certified. Outstanding MERLOT (La Ricolma), SANGIOVESE IGT Percarlo, Vin San Giusto (PASSITO); v.gd Chianti Classico, RISERVA Le Baroncole.

San Guido, Tenuta Tus *See* SASSICAIA.

Vermouth di Torino: historic drink of the late C18. Bitter, herbal, wonderful.

San Leonardo T-AA ★★★★ Top TRENTINO estate of Marchesi Guerrieri Gonzaga. Main wine is B'x blend, *San Leonardo*, Italy's most claret-like wine; v.gd CARMENÈRE, Villa Gresti (MERLOT/Carmenère).

San Lorenzo, Fattoria Mar ★★★ Bio/organic; age-worthy VERDICCHIOS from Montecarotto. Top: Campo delle Oche (also Integrale) and Il San Lorenzo Bianco; v.gd Il San Lorenzo Rosso (100% MONTEPULCIANO), Le Oche.

San Michele Appiano T-AA ★★★ Historic co-op. *Mtn-fresh whites* a speciality brimming with varietal typicity, drinkability. Best: Appius (selected by Hans Terzer), The Wine Collection; v.gd PINOT BL Schulthauser, Sanct Valentin line.

Santadi Sar ★★★ Best Sar co-op (and one of Italy's), esp for CARIGNANO-based reds Rocca Rubia RISERVA, *Terre Brune* (all DOC CARIGNANO DEL SULCIS). Also v.gd MONICA DI SARDEGNA Antigua, Shardana, PASSITO Latinia (w).

Santa Maddalena / St-Magdalener T-AA ★→★★★ DOC DYA Teutonic-style red from SCHIAVA grapes from v. steep slopes behind ALTO ADIGE capital Bolzano. Notable: CS St-Magdalena (Huck am Bach), CS Tramin, Gojer, Rottensteiner, Waldgries.

Saracco, Paolo Pie ★★★★ Top MOSCATO D'ASTI; v.gd CHARD, LANGHE RIES, PINOT N.

Sardinia / Sardegna Italy's 2nd-largest island is home to world-class whites and reds. Look for VERMENTINO DI GALLURA DOCG, VERMENTINO DI SARDEGNA (fruitier, less mineral), Sherry-like VERNACCIA DI ORISTANO, NURAGUS among whites, late-harvest sweet Nasco; forgotten Semidano deserves much better. CANNONAU (GARNACHA), CARIGNANO most famous reds, but Bovale Sardo, Pascale just as gd.

Sassicaia Tus ★★★★ 06 07' 08 09 10 13 15' 16 (18) (19) Italy's sole single-vyd DOC (BOLGHERI), a CAB (SAUV/FR) made on First Growth lines by Marchese Incisa della Rocchetta at TENUTA SAN GUIDO. More elegant than lush, made for age – and often bought for investment, but hugely influential in giving Italy a top-quality image; 16 extremely elegant, one of best recent yrs.

Satta, Michele Tus ★★★ Virtually only BOLGHERI grower to succeed with 100% SANGIOVESE (Cavaliere). Bolgheri DOC red blends Piastraia, SUPERIORE I Castagni.

Scarpa Pie ★★★ Historic traditional winery in Nizza Monferrato. Top BARBERA D'ASTI La Bogliona, Rouchet (Ruchè); v.gd BARBARESCO Tettineive, Freisa.

Scavino, Paolo Pie ★★★ Modernist BAROLO producer of Castiglione Falletto, esp crus Rocche dell'Annunziata, Bric del Fiasc, Monvigliero, Ravera and Prapò. Waiting for new Bussia Fantini.

Schiava Alto Adige T-AA ★ DOC DYA Schiava (VERNATSCH in German): blend of the three main Schiava varieties (Gentile, Grigia, Grossa). Used to make DOC wines called Lago di Caldaro, St-Magdalener (Santa Maddalena). Light- to medium-bodied, v. fresh, easy-glugging red. Best: Cantina Bolzano, Cantina Girlan, Produttori Merano, Produttori Nalles-Magré, San Michele Appiano, Terlan.

Schiopetto, Mario FVG ★★★ Legendary late COLLIO pioneering estate now owned by Rotolo family; v.gd DOC FRIULANO, SAUV BL, *Pinot Bl*, RIBOLLA GIALLA and IGT blend Blanc des Rosis, etc.

Sella & Mosca Sar ★★ Major SAR grower/merchant. *See* TERRA MORETTI.

Selvapiana Tus ★★★★ RÙFINA organic estate among Italian greats. Best: RISERVA Bucerchiale, IGT Fornace; but even *basic Chianti Rùfina is a treat*. Also fine red Petrognano, POMINO, Riserva VIGNETO Erchi.

Sforzato / Sfursat Lom ★★★ DOCG Sforzato di VALTELLINA is made AMARONE-like, from air-dried NEBBIOLO grapes. Ages beautifully. Best: FAY (RONCO del Picchio); v.gd Dirupi (Vino Sbagliato), Mamete Prevostini (Albareda), Nino Negri (Cinque Stelle). See VALTELLINA.

Sicily The Med's largest island, modern source of exciting original wines and value. Native grapes (r Frappato, NERELLO MASCALESE, NERO D'AVOLA; W CATARRATTO, Grecanico, GRILLO, INZOLIA), plus internationals. The vyds are on flatlands in w, hills in centre, volcanic altitudes on Mt Etna.

Siddùra Sar ★★★ In Gallura area, NW SAR. Best: Cagnulari Bacco, CANNONAU RISERVA Fòla, VERMENTINO di Gallura SUPERIORES Beru, Maia.

Soave Ven ★→★★★ DOC Famous, hitherto underrated, Veronese white. Soils are a mix of mainly volcanic or calcareous elements. Wines from volcanic soils of CLASSICO zone can be intense, saline, v. fine, quite long-lived. See also RECIOTO. Best: Gini, INAMA, PIEROPAN, PRÀ; v.gd Coffele, Nardello, ROCCOLO GRASSI, Suavia.

Solaia Tus ★★★★ 11 12 13 15 16 (19) CAB SAUV/SANGIOVESE by ANTINORI; needs age.

Sottimano Pie ★★★→★★★★ Family estate. One of most inspired in BARBARESCO (crus: Basarin, Cottá, Currá, Fausoni, Pajorè); v.gd BARBERA D'ALBA, DOLCETTO D'ALBA, LANGHE NEBBIOLO.

Speri Ven ★★★ VALPOLICELLA family estate. Organic certified. Traditional style. Top: AMARONE Sant'Urbano.

Spumante Sparkling.

Südtirol T-AA German name for ALTO ADIGE.

Superiore Wine with more ageing than normal DOC and 0.5–1% more alc. May indicate a restricted production zone, eg. ROSSO PICENO Superiore.

Super Tuscan Tus Wines of high quality and price developed in 70s/80s to get around silly laws then prevailing. Now, esp with Gran Selezione on up, scarcely relevant. Wines still generally considered in Super Tuscan category, strictly unofficially: BERTINGA, CA' MARCANDA, Flaccianello, Guado al Tasso, Messorio, ORNELLAIA, Redigaffi, SASSICAIA, SOLAIA, TIGNANELLO.

Sylla Sebaste Pie ★★★ Illustrates merits of rare NEBBIOLO rosé variety: lighter, v. perfumed BAROLO. A beauty.

Tappero Merlo Domenico Pie ★★★ Roots in the past, eyes to the future and Erbaluce grape as life-partner. Top: Acini Perduti (blend of 80% rare MALVASIA Moscata/20% Erbaluce), ERBALUCE DI CALUSO (Cuvée des Paladins, Kin). Plus v.gd Bohemien (PASSITO).

Tasca d'Almerita Si ★★★ New generation runs historic, still prestigious ETNA estate. High-altitude vyds; balanced IGTS under Regaleali label. Top: NERO D'AVOLA-based **Rosso del Conte** and Etna wines Rampante, Tascante Sciaranuova. Also v.gd GRILLO Mozia TENUTA Whitaker, MALVASIA delle Lipari Capofaro.

Taurasi Cam ★★★ DOCG 10 11 12 13 15 16 (18) (19) The 1st DOCG in S Italy. Best AGLIANICO of CAM: none so potentially *complex, demanding, ultimately rewarding*. With 17 communes, four subzones (nw, w, Taurasi, s). Top: Contrade di Taurasi (Vigne d'Alto, Coste), I Favati (Terzotratto), Guastaferro (Primum), MASTROBERARDINO (Radici, Stilema), SALVATORE MOLETTIERI (VIGNA Cinque Querce), QUINTODECIMO (Quintodecimo, Vigna Gran Cerzito). Also v.gd: BENITO FERRARA (Vigna Quattro Confini), FEUDI DI SAN GREGORIO (Piano di Montevergine, Rosamilia, Candriano), Perillo.

Italy's most planted white grape? Glera (for Prosecco): 27,000 ha.

Tedeschi Ven ★★★ Bevy of v. fine VALPOLICELLA, AMARONE RECIOTO Capitel Monte Fontana, Amarone Capitel Monte Olmi best.

Tenuta An agricultural holding (see under name – eg. SAN GUIDO, TENUTA).

Terlano, Cantina di T-AA ★★★→★★★★ High-quality co-op, benchmark PINOT BL.

Outstanding Rarity special editions of mature white (aged min 10 yrs) and Primo Terlaner I Grande Cuvée (PINOT BL/SAUV BL/CHARD); v.gd LAGREIN RISERVA Porphyr, Pinot Bl Vorberg, Sauv Bl Quarz.

Teroldego Rotaliano T-AA ★★→★★★ DOC TRENTINO's best local grape makes seriously tasty wine on flat Campo Rotaliano. *Foradori* top. Also gd Dorigati, Endrizzi, MEZZACORONA'S riserva Nos, Zeni

Terra Moretti Lom, Sar, Tus ★★★ Three regions, six estates. In FRANCIACORTA Bellavista (Alma Non Dosato, Teatro alla Scala Brut), Contadi Castaldi (PINOT NERO, Zéro); in SAR Sella & Mosca (w Torbato Terre Bianche, VERMENTINO Cala Reale, Monteoro); in TUS Acquagiusta La Baiola, Petra (Quercegobbe), Teruzzi.

Terre Nere, Tenuta delle Si ★★★★ Marc de Grazia shows great wine can be made from NERELLO and CARRICANTE grapes, on coveted n side of Mt Etna. Top: Cuvée delle Vigne Niche wines, Guardiola and pre-phylloxera La Vigna di Don Peppino; v.gd Le Vigne di Eli. Look for new ER Bocca d'Orzo (monopole) and EB Montalto.

Terriccio, Castello del Tus ★★★ Large estate s of Livorno: excellent, v. expensive B'x-style IGT Lupicaia, v.gd IGTs Tassinaia, Terriccio (blend of mainly Rhône grapes).

Tiberio Ab ★★★★ Outstanding TREBBIANO D'ABRUZZO Fonte Canale (60-yr-old vines) one of Italy's best whites and MONTEPULCIANO D'ABRUZZO Colle Vota and Archivio (both single-vyd); CERASUOLO D'ABRUZZO (one of the best ROSATOS in Italy), PECORINO also exceptional.

Tiefenbrunner T-AA ★★★→★★★★ Grower-merchant in Teutonic castle (*Turmhof*) in S ALTO ADIGE. Wide range of mtn-fresh white and well-defined red varietals, esp 1000m (3281ft)-high MÜLLER-T *Feldmarschall*, one of Italy's best whites. Also v.gd CAB SAUV VIGNA Toren, SAUV BL RISERVA Rachtl.

Tignanello Tus ★★★★ 11 12 13 15 16 17 18 (19) SANGIOVESE/CAB SAUV blend, barrique-aged, from Antinori family.

Tommasi Ven ★★★ Now 4th generation in charge. Top: VALPOLICELLA Rafael, AMARONE (RISERVA Ca' Florian). Other estates in Bas (Paternoster), OLTREPÒ PAVESE (TENUTA Caseo), PUG (Masseria Surani), Ven (Filodora).

Torgiano Umb ★★ DOC and **Torgiano, Rosso Riserva** ★★→★★★ DOCG 11 12 13 15 16 18 (19) Red from Umb, gd to excellent. Top: LUNGAROTTI *Vigna Monticchio* Rubesco RISERVA. Keeps many yrs.

Torrette VdA ★→★★★ DOC Blend based on Petit Rouge and other local varieties. Best: Torrette Superieur; gd: Anselmet, D&D, DI BARRÒ, Didier Gerbelle, Elio Ottin, FEUDO DI SAN MAURIZIO, GROSJEAN, LES CRETES.

Tramin, Cantina T-AA ★★★ Quality co-op with benchmark GEWURZ. Outstanding Epokale, Nussbaumer and Terminum; v.gd PINOT GR Unterebner and CHARD Le Selezioni line, Troy.

Travaglini Pie ★★★ Solid producer of N PIE NEBBIOLO: v.gd GATTINARA RISERVA, Gattinara Tre VIGNE; gd Coste della Sesia, METODO CLASSICO Nebolé (Nebbiolo).

Trebbiano d'Abruzzo Ab ★→★★★★ DOC DYA Generally crisp, simple, but TIBERIO's (Fonte Canale) and VALENTINI's are *two of Italy's greatest* whites; also v.gd EMIDIO PEPE and Valle Reale.

Trediberri Pie ★★★ Dynamic estate, top BAROLO Rocche dell'Annunziata (best value); v.gd BARBERA D'ALBA, DOGLIANI (DOLCETTO), LANGHE NEBBIOLO.

Trentino T-AA ★→★★★ Varietally named DOC wines; best are perfumed, flavourful, inexpensive. Less successful ones dilute, boring, neutral. Best: GEWURZ, Marzemino, MÜLLER-T, Nosiola, SCHIAVA, TEROLDEGO. TRENTO DOC is name of potentially high-quality METODO CLASSICO wines.

Trento T-AA ★★→★★★★ DOC High-quality METODO CLASSICO fizz. Top: Abate Nero (by Roberta Lunelli, artisanal production, esp RISERVA Cuvée dell'Abate, Domini and Domini Nero), FERRARI, Maso Martis (by Antonio Stelzer, esp Blanc de Blancs Brut, Madame Martis Riserva); v.gd Letrari.

Trinoro, Tenuta di Tus ★★★★ Individualist TUS red estate, pioneer in DOC Val d'Orcia between MONTEPULCIANO and MONTALCINO. Heavy accent on B'x grapes in flagship TENUTA di Trinoro, also in Camagi, Palazzi, Magnacosta, Tenagli. *See also* VINI FRANCHETTI (ETNA).

Tua Rita Tus ★★★★ As new BOLGHERI in 90s, 1st producer of possibly Italy's greatest MERLOT in Redigaffi (19, 25th anniversary), also outstanding B'x blend *Giusto di Notri*, SYRAH Per Sempre and Keir (amphora). Also owns POGGIO ARGENTIERA in MAREMMA (best: Capatosta, MORELLINO DI SCANSANO).

Tuscany / Toscana Home of world's top SANGIOVESE. CHIANTI CLASSICO, BRUNELLO DI MONTALCINO, RÙFINA best, but BOLGHERI just as gd and world-class for international grapes (esp CAB FR, MERLOT), but SASSICAIA (CAB SAUV) most famous.

Uberti Lom ★★★→★★★★ Historical estate, excellent interpreter of FRANCIACORTA's terroir. Outstanding Comarì del Salem, Dequinque (blend of 10 yrs), Quinque (blend of 5 yrs); v.gd Dosaggio Zero Sublimis, Francesco I.

Umani Ronchi Mar ★★→★★★ Leading Mar producer, esp for VERDICCHIO (Casal di Serra, Plenio), CONERO Cumaro, IGTS Le Busche (w), Pelago (r).

Vajra, GD Pie ★★★→★★★★ Leading BAROLO producer in Vergne. Outstanding Bricco delle Viole and LANGHE Freisa Kyè; gd Langhe RIES Petracine, Barolos (Coste di Rose, Ravera), Serralunga's Luigi Baudana Barolos (Cerretta), DOLCETTO Coste & Fossati and BARBERA D'ALBA SUPERIORE.

Val di Cornia Tus ★★→★★★ DOC(G) 13 15 16 18 (19) Quality zone s of BOLGHERI. MERLOT, MONTEPULCIANO, SYRAH. Try Bulichella, Casadei, Gualdo del Re, Terricciola, TUA RITA.

Valentini, Edoardo Ab ★★★★ Collectors seek CERASUOLO D'ABRUZZO, MONTEPULCIANO D'ABRUZZO, TREBBIANO D'ABRUZZO: among Italy's v. best. Traditional, age-worthy.

Valle d'Aosta ★★→★★★ DOC Italy's smallest region makes some of its best reds/ whites, but hard to find. Soil of morainic origin (heterogeneous glacial deposits of gravel sands). DOCS mostly varietal; famous names incl NEBBIOLO-based Arnad Montjovet, Donnas; Torrette (r mostly Petit Rouge), Blanc de Morgex, made with Prié (w/sp); Chambave (made with local biotype of w MUSCAT); MALVOISIE (made with PINOT GR), Cornalin, Mayolet, Premetta a lovely light red.

Valle Isarco T-AA ★★ DYA ALTO ADIGE DOC for seven Germanic varietal whites made along Isarco (Eisack) River ne of Bolzano. Top: Abbazia di Novacella, Eisacktaler, KÖFERERHOF, Kuenhof, MANNI NÖSSING; gd GEWURZ, MÜLLER-T, RIES, SILVANER.

Valpolicella Ven ★→★★★★ DOC(G) Light, easy-going red (Valpolicella), medium-bodied to powerfully alc, rich, tannic (AMARONE) and supersweet (RECIOTO). Popular RIPASSO between Valpolicella and Amarone, but few noteworthy. (*See box, below.*)

Valpolicella Ripasso Ven ★★→★★★ DOC 11 12 13 15 16 18 (19) In huge demand, so changes from 2016. Used to be only from VALPOLICELLA SUPERIORE re-fermented (only once) on RECIOTO or AMARONE grapeskins to make a more age-worthy wine.

Valpolicella: the best

Time to take gd VALPOLICELLA more seriously. AMARONE DELLA VALPOLICELLA and RECIOTO DELLA VALPOLICELLA are now DOCG, while RIPASSO has new rules. Following producers make gd to great wine: ALLEGRINI ★, Begali, Bertani, BOLLA, Boscaini, Brigaldara ★, BRUNELLI, BUSSOLA ★, Ca' la Bianca, Ca' Rugate, Campagnola, CANTINA Valpolicella, Castellani, Corteforte, Corte Sant'Alda, CS Valpantena, DAL FORNO ★, GUERRIERI-RIZZARDI ★, Le Ragose, Le Salette, MASI ★, Mazzi ★, MONTE DEL FRÀ, Nicolis, PRÀ, QUINTARELLI ★, Roccolo Grassi ★, Serego Alighieri ★, SPERI ★, Stefano Accordini ★, TEDESCHI ★, TOMMASI ★, Valentina Cubi, Venturini, Viviani, ZENATO, Zeni.

Now can blend 10% Amarone with standard Valpolicella and call it Ripasso. Best: BUSSOLA, Castellani, DAL FORNO, QUINTARELLI, ZENATO.

Valtellina Lom ★→★★★ DOC/DOCG Rare e-w valley, just s of Swiss border. Soils are sandy-loamy. Home of CHIAVENNASCA. Best labelled Valtellina SUPERIORE (five subzones: Grumello, Inferno, Maroggia, Sassella, Valgella), *see* SFORZATO. Top: ArPePe (Grumello RISERVA Sant'Antonio, Inferno Fiamme Antiche, Rocce Rosse, Sassella Riserva Nuova Regina), Dirupi (Grumello Riserva), FAY, Mamete Prevostini (Sassella Sommarovina).

Vecchio Samperi Si *See* MARCO DE BARTOLI.

Verdicchio dei Castelli di Jesi Mar ★★→★★★ DOC DYA Versatile white from nr Ancona on Adriatic; light and quaffable or sparkling or structured, complex, long-lived (esp RISERVA DOCG, min 2 yrs old), usually lighter and more floral than those of Matelica. Montecarotto more structured. Cupramontana more vibrant. Also CLASSICO. Best: *Bucci* (Riserva), Coroncino (Gaiospino, Stracacio e Stragaio), Fattoria San Lorenzo, GAROFOLI (Podium), Marotti Campi (Salmariano), MONTECAPPONE (MIRIZZI Ergo Sum, Riserva Utopia), Sartarelli (Balciana, rare late-harvest, Tralivio). Also v.gd: Casalefarneto, Colonnara (Cuprese), La Staffa, Monte Schiavo, Santa Barbara, TENUTA di Tavignano (Misco), UMANI RONCHI.

Verdicchio di Matelica Mar ★★→★★★ DOC DYA Higher acidity level, but also more body and alc, intense minerality, than wines of Jesi; so longer-lasting though less easy-drinking young. Lower-lying vyds grow on alluvial soils, while higher hillsides have complex soils of calcarenites, marl, limestone, gravel and conglomerates. RISERVA is likewise DOCG. Try Belisario, Bisci, Borgo Paglianetto, Collestefano, La Monacesca (Mirum).

Verduno Pie ★★★ DOC DYA Berry, herbal flavours. Top CASTELLO DI VERDUNO (Basadone), Fratelli Alessandria (Speziale), GB BURLOTTO; Bel Colle, Reverdito gd.

Verduno, Castello di Pie ★★★ Husband/wife team, v.gd BARBARESCO Rabajà and Rabajà Bas, BAROLO Massara and Monvigliero (also RISERVA), VERDUNO Basadone.

Verduzzo FVG ★★→★★★ DOC (FRIULI COLLI ORIENTALI) Full-bodied white from local variety. Ramandolo (DOCG) is well-regarded subzone for sweet wine. Top: I Clivi, Marco Sara, Scubla.

Vermentino di Gallura Sar ★★★ DOCG More flinty, saline than VERMENTINO DI SARDEGNA. Different soils characterized by pink granite (rarity in Italy) with high acidity and minerality. Maybe best in Italy. Top: Capichera, CS del Vermentino (Funtanaliras), CANTINA di Gallura (Canayli), Masone Mannu (Petrizza), Mura (Sienda), Pala (Stellato), Paolo Depperu (Ruinas), SIDDURA, Surrau.

Vermentino di Sardegna Lig ★★ DOC DYA From anywhere on SAR; generally fruitier (less structured) and offer earlier, uncomplicated appeal compared to VERMENTINO DI GALLURA. Best: ARGIOLAS, Deiana, Mora e Memo, QUARTOMORO, *Santadi, Sella & Mosca*.

Vernaccia di Oristano Sar ★→★★★★ DOC Flor-affected wine, similar to light Sherry, a touch bitter, full-bodied. Delicious with *bottarga*. Must try. Top: CONTINI (Antico Gregori ★, RISERVA and Flor 22); gd Orro, Serra, Silvio Carta.

Vernaccia di San Gimignano Tus ★★→★★★ Better at entry-level than RISERVA. Top: Fontaleoni (VIGNA Casanuova), Giovanni Panizzi, Guicciardini Strozzi/Fattoria di Cusona (1933), La Lastra, Sono Montenidoli (Fiore).

Viarte, La FVG ★★→★★★ Organic estate of COLLI ORIENTALI; v.gd FRIULANO, SCHIOPPETTINO di Prepotto, Tazzelenghe.

Vie di Romans FVG ★★★→★★★★ Gianfranco Gallo has built up his father's ISONZO estate to top status. Outstanding Flors di Uis blend, Isonzo PINOT GR Dessimis, MALVASIA, SAUV BL Piere and Vieris (oaked). Also v.gd Pinot Gr Dessimis.

Vietti Pie ★★★★ Organic estate at Castiglione Falletto owned by Krause Group but still run by Luca Currado and Mario Cordero. Characterful. Textbook BAROLOS:

Brunate, Lazzarito, Ravera, Rocche di Castiglione, Villero RISERVA and new Cerequio and Monvigliero. Plus v.gd BARBERA D'ALBA Scarrone, BARBERA D'ASTI la Crena, BARBARESCO Masseria, new DERTHONA. Await Barbaresco Rabajà in 2023.

Vigna (or vigneto) A single vyd, generally indicating superior quality.

Vigneri, I Si ★★★→★★★★ Consortium of growers, also ETNA estate within that consortium, run by Salvo Foti, greatest expert on NERELLO MASCALESE and all Etna varieties. Consortium focus on bush-trained vines, native grape varieties, respect for the land. Outstanding estate wines: Vinupetra (r), Etna Bianco SUPERIORE, VIGNA di Milo; v.gd Aurora (w), I Vigneri (r), Vinudilice (ROSATO).

Villa Lom ★★★ Owned by Bianchi family, in small medieval hamlet (C15) in FRANCIACORTA. Best sparkling: Diamant Pas Dosè, Emozione Brut Millesimato and RISERVA, Selezione Riserva.

Villa Crespia Lom ★★★ Family estate in FRANCIACORTA; organic. Top: Brut Millè RISERVA, Dosaggio Zero Numero Zero and Riserva del Gelso; v.gd Brut Millè.

Villa Russiz FVG ★★★ Historic estate for DOC COLLIO. SAUV BL and MERLOT (esp de la Tour selections), CHARD, FRIULANO, PINOT BL, PINOT GR all v.gd.

Vino Nobile di Montepulciano Tus ★★→★★★ 13 15 16 17 18 (19) The 1st Italian DOCG (1980). Prugnolo Gentile (SANGIOVESE)-based, from TUS MONTEPULCIANO (distinct from grape). Recent focus on single-vyd wines. Complex, long-lasting Sangiovese expression, often tough with drying tannins. Top: AVIGNONESI, BOSCARELLI, DEI, La Braccesca, POLIZIANO, Salcheto. Also gd Bindella, Fattoria del Cerro, Fattoria della Talosa, Montemercurio, Valdipiatta. RISERVA after 3 yrs.

Vin Santo / Vinsanto / Vin(o) Santo T-AA, Tus ★★→★★★★ DOC Sweet PASSITO, usually TREBBIANO, MALVASIA and/or SANGIOVESE in TUS (Vin Santo), Nosiola in TRENTINO (Vino Santo). Tus versions extremely variable, anything from off-dry and Sherry-like to sweet and v. rich. May spend 3–10 unracked yrs in small barrels called *caratelli*. *Avignonesi's is legendary*; plus CAPEZZANA, FELSINA, FRASCOLE, ISOLE E OLENA, Rocca di Montegrossi, SAN GIUSTO A RENTENNANO, SELVAPIANA, Villa Sant'Anna, Villa di Vetrice. *See also* OCCHIO DI PERNICE.

Volpaia, Castello di Tus ★★→★★★ CHIANTI CLASSICO estate at Radda; v.gd, organic certified. Top Balifico (SANGIOVESE/CAB SAUV), Chianti Classico RISERVA, Gran Selezione Coltassala (Sangiovese/Mammolo).

Zenato Ven ★★★ Garda wines: v. reliable, sometimes inspired. Also AMARONE, LUGANA, SOAVE, VALPOLICELLA. Look for labels RISERVA Sergio Zenato.

Zerbina, Fattoria E-R ★★★ Leader in Romagna. Best: sweet ALBANA DOCG (AR, Scacco Matto); v.gd SANGIOVESE (Pietramora); barrique-aged IGT Marzieno.

Zibibbo Si ★★★ Alluring sweet MUSCAT d'Alessandria, most associated with PANTELLERIA and extreme W SI. Dry version exemplified by MARCO DE BARTOLI.

Zonin Ven ★→★★ Huge estate owner, based at Gambellara in Ven, but also big in FVG, TUS, PUG, SI and elsewhere in world (eg. Virginia, US).

What do the initials mean?

DOC (denominazione di origine controllata) controlled denomination of origin, cf. AOC in France.

DOCG (denominazione di origine controllata e garantita) "G" = "guaranteed". Italy's highest quality designation. Guarantee? It's still caveat emptor.

IGT (indicazione geografica tipica) geographic indication of type. Broader and more vague than DOC, cf. Vin de Pays in France.

DOP / IGP (denominazione di origine protetta / indicazione geografica protetta) "P" = "protected". The EU's DOP/IGP trump Italy's DOC/IGT.

MGA (menzione geografica aggiuntiva) or UGA (additional geographical units), eg. subzones, cf. crus in France.

Germany

Abbreviations used in the text:

Bad	Baden
Frank	Franken
Hess	Hessische Bergstrasse
M-M	Mittelmosel
M Rh	Mittelrhein
Mos	Mosel
Na	Nahe
Pfz	Pfalz
Rhg	Rheingau
Rhh	Rheinhessen
Sa-Un	Saale-Unstrut
Sachs	Sachsen
Würt	Württemberg

More heavily shaded areas are the wine-growing regions.

In 2020, for the first time in the history of the Federal Republic, all non-city federal states had commercial vineyards. Just a few years ago, it would have been considered an April Fool if someone had said there was a vineyard in, say, Ostfriesland on the North Sea coast near the border with the Netherlands, or on the islands of Sylt or Rügen, or in eastern Brandenburg, on the border with Poland. Now vines grow in all these. Sure, only a few hectares. But the growers there are just as ambitious and hard-working as their colleagues in the Rheingau and on the Mosel. They are not hobby growers; they press wine for a living. Certainly, these new plantings will not turn German wine upside down. The familiar flavours of Riesling and Pinot Noir, Silvaner and others, all taut acidity and ripe fruit, continue, from the familiar places. But these new chilly vineyards are the most visible expression of upheavals that are also quietly taking place in the established regions: there, too, cool side valleys are being (re-)planted, vineyards are being extended upwards, or varieties originating from southern Europe are being tried out. In warm 2018 in Pfalz, for example, even Cabernet Sauvignon and Merlot became overripe. German Bordeaux blends with cooked flavours and 15% alcohol? Hmm. Let's see what German growers come up with next. They will come up with something.

Recent vintages

Mosel

Don't drink them too young. Certainly, Mosels are delicious v. soon after bottling, but to enjoy their max complexity, allow Grosses Gewächs (GG) a min of 3 yrs, Kabinett 3–5 yrs, Spätlese 5–7 yrs, Auslese 7–10 yrs.

2021 Mildew, rain, harvest into Nov, lighter wines with v.gd balance.
2020 3rd yr running of drought, gd quality, quantity.
2019 Low quantity (-25%, Ruwer -40%: frost), but textbook raciness.
2018 Powerful, low acidity (but better than 2003). Crystal clear TBA.
2017 Low yield (frost) = high extract. Brilliant Kabinett, Spätlese.
2016 Balanced wines.
2015 Warm yr, rich Trocken, Spätlesen, Auslesen to keep.
2012 Classic, discreet wines, might turn out to be long-lived.
2011 Brilliant vintage, particularly Saar, Ruwer, sensational TBAs.
Earlier fine vintages: 09 08 07 05 04 03 01 99 97 95 94 93 90 89 88 76 71 69 64 59 53 49 45 37 34 21.

Rheinhessen, Nahe, Pfalz, Rheingau, Ahr

Rhg wines tend to be longest-lived of all German regions, improving for 15 yrs or more, but best wines from Rhh, Na and Pfz can last as long – and this applies not only to Spätlese and Auslese; GGs undoubtedly have potential to age for 10 yrs+ too. The same holds for Ahr V reds and their peers from Bad and other regions of the s.

2021 Classical yr like those of 80s, gd acidity, moderate alc, intense flavours.
2020 Early and fast harvest, mid-weight wines.
2019 Drought/heatwaves in July; rain in August/Sept. Excellent Ries, gd acidity.
2018 Record summer, powerful wines. Growers allowed to acidify.
2017 Roter Hang and Mittelhaardt outstanding: freshness combined with extract.
2016 Quality and quantity mixed.
2015 Hot, dry summer. Rhg excellent, both dry and nobly sweet.
2012 Quantities below average, but v.gd, classical at every level.
Earlier fine vintages: 11 09 08 05 03 02 01 99 98 97 96 93 90 83 76 71 69 67 64 59 53 49 45 37 34 21.

Adams, Weingut Rhh ★★★ Simone A is PhD oenologist putting emphasis on vibrant freshness that INGELHEIM's calcareous soils can give. Best: single-vyds Auf dem Haun and Pares.

Adelmann, Weingut Graf Würt ★★→★★★ Kept young by count Felix A, est in 1297. Classical LEMBERGER, RIES, but also delightful extravagances like Muskattrollinger EISWEIN.

Ahr ★★→★★★★ Small river valley s of Bonn, elegant, fruit-driven PINOT N from slate. In July 2021 devastated by flood (water 10m/33ft high; *see* box, p.156). Best: Adeneuer, BERTRAM, Brogsitter, Burggarten, Deutzerhof, Kreuzberg, MEYER-NÄKEL, Nelles, Riske, SCHUMACHER, Sermann, STODDEN. Two gd co-ops too (Dagernova, Mayschoss-Altenahr).

> **Disaster in the Ahr**
> On 14 July 2021, after days of heavy rainfall, meteorologists predicted a
> flood in the AHR V, a peak 2m (6ft) above normal. While residents placed
> sandbags in front of their homes, they recognized it wouldn't help
> much. The flood came faster than expected, and 8–10m (26–33ft) high.
> Many people spent the night on the rooftops of their houses, smelling
> the gasoline from smashed cars, hearing walls and houses collapse;
> 141 people died in the water. Of the 65 wineries in the valley, 62 were
> completely destroyed. Barrels of 2020 PINOT N floated away or leaked;
> the bottled 19s were broken or buried in mud. "Our whole life has
> swum away," commented Meike Näkel of MEYER-NÄKEL. It's a miracle
> that, only 10 wks after the flood, most estates were able to press a 2021
> vintage. Solidarity among growers played a decisive role: hundreds
> came from all over Germany and neighbouring countries to help with
> the clean-up and to lend equipment; many wineries even sent their own
> vyd team to the Ahr to tend the vines, and others collected donations.
> Luckiest of all in all this bad luck: 90% of vyds are steep slopes and
> were untouched by the waters.

Aldinger, Gerhard Würt ★★★→★★★★ Both white and red have density, tension. Outstanding Brut Nature SEKT. Newest stroke of genius is TROLLINGER ALTE REBEN Blanc de Noirs reminiscent of Côte de Beaune (*see* France) CHARD.

Alte Reben Old vines. But no min age.

Alter Satz Frank Wines from old co-planted (different varieties all mixed up) vyds, esp in FRANK, often more than 100 yrs old and ungrafted. Try (w) Otmar Zang, Scheuring, Scholtens – or Stritzinger (r).

Amtliche Prüfungsnummer (APNr) Official test number for quality wine. Useful for discerning different lots of AUSLESE a producer has made from the same vyd.

Assmannshausen Rhg ★★→★★★★ RHG hotspot for *Spätburgunder*. Steep Höllenberg (45 ha on slate) is emblematic, neighbouring vyds (Frankenthal, Hinterkirch) profit from climate change. Growers: Allendorf, BISCHÖFLICHES WEINGUT RÜDESHEIM, CHAT SAUVAGE, HESSISCHE STAATSWEINGÜTER, K Berg, KESSELER, König, KRONE, KÜNSTLER, SOLVEIGS.

Auslese Wines from selective picking of super-ripe bunches affected by noble rot (*Edelfäule*). Unctuous, but – traditionally – elegant rather than super-concentrated. 99% are sweet, but specialists (JB BECKER, Koehler-Ruprecht) show Auslese TROCKEN can be elegant too.

Aust, Karl Friedrich Sachs ★★★ Fine family estate at Radebeul nr Dresden with important property in steep, terraced Goldener Wagen vyd. Karl Friedrich also a stonemason: can repair vyd walls. Outstanding TRAMINER, PINOT N.

Ayl Mos ★→★★★ All vyds known since 1971 by name of historically best site: Kupp. Growers: BISCHÖFLICHE WEINGÜTER TRIER, *Lauer*, Vols, ZILLIKEN.

BA (Beerenauslese) Luscious sweet wine from exceptionally ripe, individually selected berries concentrated by noble rot.

Bacharach M Rh ★→★★★ Small, idyllic Rhine-side town; centre of M RH RIES. GROSSE LAGE: Hahn, Posten, Wolfshöhle. Growers: Bastian, JOST, Kauer, RATZENBERGER.

Baden Huge sw region and former Grand Duchy, 15,000 ha stretch over 300 km (186 miles), best known for PINOT N, GRAU- and WEISSBURGUNDER and pockets of RIES, usually dry. Two-thirds of crop goes to co-ops.

Bassermann-Jordan Pfz ★★★ Famous historical estate producing powerful RIES from FORST and DEIDESHEIM. Spontaneously fermented, oak-aged and off-dry Ries Ancestrale (from Pechstein vyd) and amphora CAB SAUV.

Battenfeld-Spanier Rhh ★★★ Passionate HO Spanier (*see also* KÜHLING-GILLOT) is leading grower in calcareous sites at Hohen-Sülzen, Mölsheim and ZELLERTAL (auction RIES about to get cultish), bio. Brilliant Brut Nature.

Becker, Friedrich Pfz ★★→★★★★ Outstanding SPÄTBURGUNDER (Heydenreich, Kammerberg, Sankt Paul) from most s part of PFZ; some vyds actually lie across border in Alsace. Wines need 5–10 yrs. Whites gd (esp CHARD, WEISSBURGUNDER).

Becker, JB Rhg ★★→★★★ Delightfully old-fashioned, cask-aged (and long-lived) dry RIES, SPÄTBURGUNDER at WALLUF and Martinsthal. Mature vintages (back to 90s) great value.

Bercher Bad ★★★ KAISERSTUHL family estate, v. reliable from ORTSWEIN up to GG, long experience in barrique ageing, best usually Haslen GRAUBURGUNDER and Kesselberg SPÄTBURGUNDER,

Bergdolt Pfz ★★★ Organic estate at Duttweiler, known for food-friendly WEISSBURGUNDER GG Mandelberg, mineral RIES, and taut SPÄTBURGUNDER. Stunning SEKT (Brut Nature Fluxus).

Bernkastel M-M ★→★★★★ Centre of M-M: timbered houses and flowery, balsamic RIES. Most prestige has DOCTOR, but Badstube, Graben, Johannisbrünnchen, Lay vyds outstanding too. Top: JJ PRÜM, Kerpen, Lauerburg, LOOSEN, MOLITOR, SCHLOSS LIESER, Studert-Prüm, THANISCH (both estates), WEGELER. Kurfürstlay GROSSLAGE name is a deception: avoid.

Bernkasteler Ring Mos One of two MOS growers' associations organizing an auction every yr mid–end Sept. Other is GROSSER RING.

Bertram-Baltes, Weingut Ahr ★★★ Shooting stars in AHR V. Early picking, moderate use of new oak bring fresh, dense, racy SPÄTBURGUNDER from prime vyds. Badly affected by Ahr flood (see box, left).

Bischöfliches Weingut Rüdesheim Rhg ★★★ 8 ha of best sites in ASSMANNSHAUSEN, RÜDESHEIM; vault cellar in Hildegard von Bingen's historic monastery. Peter Perabo (ex-KRONE) is *Pinot N specialist*, RIES also v.gd.

Bischöfliche Weingüter Trier Mos ★★ 130 ha of potentially 1st-class historical donations. Not v. reliable; do not buy without prior tasting.

Bocksbeutel Frank Belly-shaped bottle dating back to C18, today only permitted in FRANK and village of Neuweier, BAD. New: modernized, stackable "Bocksbeutel PS" – at last.

Bodensee Bad Idyllic district of S BAD and on Bavarian shore of Lake Constance, at altitude: 400–580m (1312–1903ft). Dry MÜLLER-T with elegance, light but delicate SPÄTBURGUNDER.

Boppard M Rh ★→★★★ Wine town of M RH with GROSSE LAGE Hamm, an amphitheatre of vines. Growers: Heilig Grab, Lorenz, M Müller, Perll, WEINGART. Unbeatable *value*.

Brauneberg M-M ★★★→★★★★ Top village on M-M; excellent full-flavoured RIES of great raciness. GROSSE LAGE vyds Juffer, Juffer-SONNENUHR. Growers: M Conrad, *F Haag*, KESSELSTATT, MF RICHTER, Paulinshof, Sankt Nikolaus Hospital, SCHLOSS LIESER, THANISCH, *W Haag*.

Bremer Ratskeller Town-hall cellar in N Germany's commercial town of Bremen, founded in 1405, UNESCO World Heritage Site. Oldest wine is a barrel of 1653 RÜDESHEIMER Apostelwein.

Let's hear it for freckles

How do you tell which are the best berries? By measuring sugar and acidity? By tasting? Do you use a laser sorting table, or just your eyes? KARL FRIEDRICH AUST, grower at Radebeul nr Dresden, can tell the best TRAMINER berries by sight – and it's not just those with the darkest colour. Instead, you should look for brightness of colour, and freckles. Cheaper than buying a fancy sorting table.

Breuer Rhg ★★★→★★★★ Exquisite RIES from RAUENTHAL, RÜDESHEIM and, newly, LORCH. Nonnenberg transforms austerity into age-worthiness, Berg Schlossberg 90' 96' 97' 08' 12' 13' 14 15' 16 17' 18 19 20 has depth at 12% alc, Pfaffenwies (1st vintage 19') full of floral elegance. Exciting experiments with historic grape Gelber Orleans.

Buhl, Reichsrat von Pfz ★★★ Historic PFZ estate at DEIDESHEIM, recently lacking consistency: three different winemakers in 3 yrs.

Germany is 3rd-biggest producer of Pinot N worldwide.

Bürgerspital zum Heiligen Geist Frank ★★★ Ancient charitable estate with great continuity: only six directors in past 180 yrs. Traditionally made whites from best sites in/around WÜRZBURG. SILVANER GG from monopole Stein-Harfe 15' 16' 17' 18' 19 20 is a monument. RIES RANDERSACKER Pfülben (20') has depth and elegance.

Bürklin-Wolf, Dr. Pfz ★★→★★★★ 30 ha of best MITTELHAARDT vyds incl important holdings of FORST's Kirchenstück, Jesuitengarten, Pechstein, bio. Forerunner of VDP classifcation (1994).

Busch, Clemens Mos ★★★★ Steep Pündericher Marienburg farmed by hand, bio, for seven GGS from different parcels. Best usually: Felsterrasse (mineral, deep), Raffes (power, balance), Rothenpfad (silky, balsamic). Now also Res line: 2 yrs barrel ageing.

Castell'sches Fürstliches Domänenamt Frank ★★→★★★ Ferdinand Fürst zu Castell – family making wine since 1057 – is used to thinking in large time frames: SILVANER GG from sole possession Schlossberg 08 09 11 12 15' 16 17' is marketed only 5 yrs after harvest.

Chat Sauvage Rhg ★★★→★★★★ The most Burgundian estate in RHG: full-bodied and aromatic PINOT N from ASSMANNSHAUSEN, JOHANNISBERG, LORCH, RÜDESHEIM; round CHARD.

Christmann Pfz ★★★ VDP President Steffen Christmann, MITTELHAARDT bio pioneer, now joined by daughter Sophie, a PINOT N expert. Promising new Christmann & Kauffmann SEKT project together with Mathieu Kauffmann (ex-Bollinger, *see* France).

Clüsserath, Ansgar Mos ★★★ Tense TRITTENHEIMER Apotheke RIES. KABINETTS delicious. Precise wines aged in old oak FUDER.

Corvers-Kauter Rhg ★★★ Organic estate at Mittelheim, 31 ha: textbook RIES from RAUENTHAL (Baiken), RÜDESHEIM, MARCOBRUNN; PINOT N from ASSMANNSHAUSEN equally gd.

Crusius, Dr. Na ★★→★★★ Family estate at TRAISEN. Young Rebecca C brings fresh air: sensational 20s (eg. Schlossböckelheim Felsenberg), combining finesse, expression.

Deidesheim Pfz ★★→★★★★ Center of MITTELHAARDT. Best vyds: Grainhübel, Hohenmorgen, Kalkofen, Kieselberg, Langenmorgen. Top growers: BASSERMANN-JORDAN, Biffar, BUHL, BÜRKLIN-WOLF, CHRISTMANN, Fusser, MOSBACHER, Seckinger, Siben, Stern, VON WINNING. Gd co-op.

Diel, Schlossgut Na ★★★→★★★★ Caroline D follows her father: exquisite *GG Ries* (best usually Burgberg of Dorsheim). Magnificent SPÄTLESEN, serious *Sekt* (Goldloch RIES and Cuvée Mo).

Doctor M-M Emblematic steep vyd at BERNKASTEL, the place where TBA was invented (1921, THANISCH). Only 3.2 ha, and five owners: both Thanisch estates, WEGELER (1.1 ha), Lauerburgand local Heiligen Geist charity (0.26 ha, leased until 2024 to MOLITOR, SCHLOSS LIESER). RIES of extraordinary depth and richness, but pricey – up to a record-breaking €1437/bottle for Molitor's 19 TROCKEN.

Dönnhoff Na ★★★→★★★★ Cornelius D has found his style: drier and more mineral than father Helmut, but equally delicious. Superb 19s, 20s not far behind, eg EISWEIN and GG (Auction) from monopoly Brücke.

Egon Müller zu Scharzhof Mos ★★★★ 59 71 83 90 03 15 16 17 18 19 20 21 Legendary SAAR family estate at WILTINGEN with a treasury of old vines. Racy SCHARZHOFBERGER RIES among world's greatest wines: sublime, vibrant, immortal. *Kabinetts* feather-light and long-lived. New (from 18 on): TROCKEN version of Scharzhofberg.

Einzellage Individual vyd site. Never to be confused with GROSSLAGE.

Eiswein Made from frozen grapes with the ice (ie. water content) discarded, thus v. concentrated: of BA ripeness or more. Outstanding Eiswein vintages: 98 02 04 08. Less and less produced in past decade: climate change is Eiswein's enemy.

Eller, Juliane Rhh ★★ Ambitious young grower at Alsheim, known for Juwel series and collaboration with German TV stars from *Joko und Klaas* (III Freunde).

Emrich-Schönleber Na ★★★ Werner Schönleber and son Frank make precise RIES from Monzingen's classified Frühlingsplätzchen and Halenberg vyds.

Erden M-M ★★★→★★★★ Village on red slate soils; noble AUSLESEN and TROCKEN RIES with rare delicacy. GROSSE LAGE: Prälat and Treppchen. Growers: BREMER RATSKELLER, Erbes, JJ Christoffel, LOOSEN, MERKELBACH, MOLITOR, Mönchhof, Rebenhof, Schmitges.

Erste Lage Classified vyd, 2nd-from-top level, similar to Burgundy's Premier Cru, only in use with VDP members outside AHR, M RH, MOS, NA, RHH.

Erzeugerabfüllung Analogous to GUTSABFÜLLUNG, but also allowed on the labels of co-ops.

Escherndorf Frank ★★★ Village with steep GROSSE LAGE Lump ("scrap" – as in tiny inherited parcels). Marvellous *Silvaner* and RIES (dr sw). Growers: Fröhlich, H SAUER, R SAUER, Schäffer, zur Schwane.

Feinherb Imprecisely defined traditional term for wines with around 10–25g sugar/litre, not necessarily tasting sweet. More flexible than HALBTROCKEN.

Forst Pfz ★★→★★★★ Outstanding MITTELHAARDT village, famous for GROSSE LAGE vyds Freundstück, Jesuitengarten, Kirchenstück, Pechstein, Ungeheuer. ORTSWEIN usually excellent value. Top growers: Acham-Magin, BASSERMANN-JORDAN, BÜRKLIN-WOLF, H Spindler, Margarethenhof, MOSBACHER, VON BUHL, VON WINNING, WOLF.

Franken / Franconia Region of distinctive dry wines, esp SILVANER, often bottled in round-bellied flasks (BOCKSBEUTEL).

Fricke, Eva Rhg ★★★ Dense, full-bodied, even silky RIES from 17 ha at KIEDRICH, LORCH. Some bottlings auctioned at Sotheby's, 2021.

Fuder Traditional German cask, sizes 600–1800 litres depending on region, traditionally used for fermentation and (formerly long) ageing.

Fürst, Weingut Frank ★★★→★★★★ *Spätburgunders* 05' 09 10 15' 16 17 18' 19 20 of great finesse from red sandstone (most dense, Hundsrück; most powerful, Schlossberg; most typical, Centgrafenberg; best value, Bürgstadter Berg). FRÜHBURGUNDER and whites equally outstanding.

Gallais, Le Mos EGON MÜLLER ZU SCHARZHOF 2nd estate, with 4-ha monopoly Braune Kupp at WILTINGEN. Soil is schist with more clay than in SCHARZHOFBERG; AUSLESEN can be exceptional.

Hermannshöhle vyd takes name from Hermes cult site nr old Roman trade route.

Geisenheim Rhg Town primarily known for Germany's university of oenology and viticulture. GROSSE LAGE Rothenberg (WEGELER) is less famous, but one of RHG's best.

GG (Grosses Gewächs) "Great/top growth". The top dry wine from a VDP-classified GROSSE LAGE.

Goldkapsel / Gold Capsule Mos, Na, Rhg, Rhh Designation (and bottle seal) mainly for AUSLESE and higher; v. strict selection of grapes, which should add finesse and complexity, not primarily weight and sweetness. Lange Goldkapsel (Long Gold Capsule) even better. Not a legal term.

Graach M-M ★★★→★★★★ Small village between BERNKASTEL and WEHLEN. GROSSE LAGE vyds: Domprobst, Himmelreich, JOSEPHSHÖFER. Growers: *JJ Prüm*, Kees-Kieren, KESSELSTATT, LOOSEN, MOLITOR, SA PRÜM, SCHAEFER, *Selbach-Oster*, Studert-Prüm, WEGELER.

Griesel & Compagnie Hess ★★★ SEKT startup at Bensheim, top Prestige series (Rosé Extra Brut, PINOT Brut Nature). Since 2016, also excellent still wine under Schloss Schönberg label.

Grosse Lage Top level of VDP's classification, but only for VDP members. Dry wine from a Grosse Lage site is called GG. *Note* not on any account to be confused with GROSSLAGE. Stay awake, there.

Grosser Ring Mos Group of top (VDP) MOS estates, whose annual Sept auction at TRIER sets world record prices.

Grosslage Term destined, maybe even intended, to confuse: a collection of secondary vyds without identity. Not on any account to be confused with GROSSE LAGE. Newest legislation (2021) stipulates term "region" must precede *Grosslagen* name; mention of village together with *Grosslage* (eg. Piesporter Michelsberg) no longer permitted, but there will be a transition period.

Gunderloch Rhh ★★★→★★★★ Historical NACKENHEIM estate portrayed in Carl Zuckmayer's play *Der fröhliche Weinberg* (1925). Known for nobly sweet RIES and culinary *Kabinett Jean-Baptiste* from prime ROTER HANG sites. Now more emphasis on TROCKEN, eg. Rothenberg GG 15 16 17' 18' 19' 20.

Gut Hermannsberg Na ★★★ Historic Prussian state dom at NIEDERHAUSEN. Some of densest RIES GGs of NA. Kupfergrube from Schlossböckelheim marketed only after 5 yrs. Stylish new guesthouse.

Gutsabfüllung Estate-bottled, and made from own grapes.

Gutswein Wine with no vyd or village designation, but only the producer's name: entry-level category. Ideally, Gutswein should be an ERZEUGERABFÜLLUNG (from own grapes) but is not always the case.

Haag, Fritz Mos ★★★★ BRAUNEBERG's top estate; Oliver H follows footsteps of his late father Wilhelm (died 2020), but wines more modern. *See* SCHLOSS LIESER.

Haag, Willi Mos ★★→★★★ BRAUNEBERG family estate, led by Marcus H. Old-style RIES, mainly sweet, rich but balanced and inexpensive.

Haart, Reinhold M-M ★★★→★★★★ Best estate in PIESPORT, with important holding

---

> **Grosse Lage / Grosslage: spot the difference**
>
> Up to now, German wine labels can say GROSSLAGE or GROSSE LAGE. They are not the same: the former is a mix of usually hundreds of ha of secondary vyds, the latter refers to the exact opposite – to a top single vyd, an EINZELLAGE, a German "Grand Cru" according to the classification set up by growers' association VDP. Until now, a *Grosslage* could disguise itself as a single vyd: who would know if Forster Mariengarten is an *Einzellage* or a *Grosslage*? (It's Gross.) But luckily, this is not the end of the story: after 2026, a *Grosslage* name will not be allowed to be mentioned together with a village name. Instead of Forster Mariengarten, it must be written: region Mariengarten. The term *Bereich* will also be replaced by region. *Bereich* means district within an *Anbaugebiet* (region). *Bereich* on a label should equally be treated as a flashing red light; the wine is a blend from arbitrary sites within that district. Do not buy.

in famous Goldtröpfchen ("gold droplet") vyd. RIES SPÄTLESEN, AUSLESEN and higher PRÄDIKAT wines are *racy, copybook Mosels*.

Haidle Würt ★★★ Family estate now led by young Moritz H, using cool climate of Remstal area for wines of distinctive freshness. Try 20' Pulvermächer RIES GG. LEMBERGERS equally rewarding.

Halbtrocken Medium-dry with 9–18g unfermented sugar/litre, inconsistently distinguished from FEINHERB (which sounds better).

Hattenheim Rhg ★★→★★★★ Town famous for its *Brunnen* ("well") vyds (Nussbrunnen, Wisselbrunnen), which have water-bearing layers underneath, and for legendary STEINBERG high above the village. Estates: Barth, HESSISCHE STAATSWEINGÜTER, Kaufmann, Knyphausen, Ress, Schloss Reinhartshausen and SPREITZER.

Heger, Dr. Bad ★★★ KAISERSTUHL estate known for parcel selections from volcanic soils in Achkarren and IHRINGEN, esp Vorderer Berg (PINOTS N/GR/BL, RIES) from steep Winklerberg terraces. Häusleboden Pinot N from old Clos de Vougeot (*see* France) cuttings planted in 1956. Joachim H now joined by daughter Rebecca.

Heitlinger / Burg Ravensburg Bad ★★→★★★ Two leading estates of KRAICHGAU, under same ownership: Heitlinger more elegant, Burg Ravensburg full-bodied.

Hessische Bergstrasse Hess ★→★★★ Germany's smallest wine region (only 460 ha), n of Heidelberg.

Hessische Staatsweingüter Hess, Rhg ★★→★★★★ State dom with vinotheque in KLOSTER EBERBACH; 238 ha in top sites all along RHG and HESSISCHE BERGSTRASSE, rich stock of mature wines going back to 1706.

Hey, Weingut ★★★ Dynamic young estate at Naumburg, SA-UN, tense RIES GG from Steinmeister vyd. Now VDP.

Heymann-Löwenstein Mos ★★★ Reinhard L pioneered terroir-minded viticulture with spontaneously fermented RIES from steep terraces at WINNINGEN nr Koblenz. Now daughter Sarah ready to take over.

Hochheim Rhg ★★→★★★★ Town e of main RHG, on River Main. Rich, earthy RIES from GROSSE LAGE vyds: Domdechaney, Hölle, Kirchenstück (often best), KÖNIGIN VICTORIABERG, Reichestal. Growers: Domdechant Werner, Flick, HESSISCHE STAATSWEINGÜTER, Himmel, Im Weinegg, KÜNSTLER.

Hock Traditional English term for Rhine wine, derived from HOCHHEIM.

Hövel, Weingut von Mos ★★★ Fine SAAR estate, bio, with vyds at Oberemmel (Hütte is 4.8 ha monopoly), at KANZEM (Hörecker) and in SCHARZHOFBERG.

Huber, Bernhard Bad ★★★→★★★★ Young Julian H has vision of BAD as Germany's Burgundy. Wildenstein SPÄTBURGUNDER is generous, dense, costly; CHARDS AITF REBEN and Hecklingen Schlossberg tight, demanding; v.gd ORTSWEIN (r/w).

Ihringen Bad ★→★★★ Village KAISERSTUHL known for fine PINOTS N/GR/BL (historically also for SILVANER) on steep volcanic Winklerberg. Top growers: DR. HEGER, Konstanzer, Michel, Stigler.

Ingelheim Rhh ★★→★★★★ RHH town across Rhine from RHG, with limestone beds under vyds; historic fame for SPÄTBURGUNDER being reinvigorated by dynamic estates as ADAMS, Werner, Bettenheimer, Dautermann, NEUS, Schloss Westerhaus, Wasem.

Iphofen Frank ★★→★★★ STEIGERWALD village with famous GROSSE LAGE Julius-Echter-Berg. Rich, aromatic, well-ageing SILVANER from gypsum soils. Growers: Arnold, Emmerich, JULIUSSPITAL, Popp, RUCK, Seufert, Vetter, Weigand, WELTNER, *Wirsching*, Zehntkeller.

Jahrgang Year as in "vintage".

Johannisberg Rhg ★★→★★★★ An RHG village. SCHLOSS JOHANNISBERG legendary; other estates incl CHAT SAUVAGE, JOHANNISHOF (ESER), Prinz v. Hessen, Schamari-Mühle. GROSSLAGE (avoid!): Erntebringer.

Johannishof (Eser) Rhg ★★→★★★ Family estate with vyds at JOHANNISBERG, RÜDESHEIM. RIES with perfect balance of ripeness and steely acidity.

Josephshöfer Mos ★★→★★★ GROSSE LAGE vyd at GRAACH, the sole property of KESSELSTATT. Harmonious, berry-flavoured RIES.

Jost, Toni M Rh ★★★ Leading estate in BACHARACH with monopoly Hahn, now led by Cecilia J. Aromatic RIES with nerve, and recently remarkable PINOT N. Family also run estate at WALLUF (RHG).

Jülg Pfz ★★→★★★ Dense PINOT N and sharply mineral CHARD, SAUV BL from limestone soils at SCHWEIGEN. New VDP member.

Juliusspital Frank ★★★ Ancient WÜRZBURG charity with top vyds all over FRANK known for firmly structured *dry Silvaners* that age well.

Kabinett See box, p.165. Germany's unique featherweight contribution, increasingly popular, but (or because) climate change makes it ever more difficult to produce: altitude vyds needed, and early, speedy picking.

Kaiserstuhl Bad Extinct volcano nr Rhine in S BADEN, notably warm climate, black soil. Altitudes up to 400m (1312ft). PINOTS N/GR of class, renown.

Kanzem Mos ★★★ SAAR village with steep GROSSE LAGE vyd Altenberg (slate, weathered red rock). Growers: BISCHÖFLICHE WEINGÜTER TRIER, Cantzheim (plus guesthouse), VAN VOLXEM, VON OTHEGRAVEN.

Karthäuserhof Mos ★★★★Historical RUWER estate, neck-only label stands for refreshing dry wines. After mixed results over last few yrs, a new team should build on the old glory: Mathieu Kauffmann (ex-Bollinger, *see* France), Dominik Völk (ex-VAN VOLXEM).

Keller, Franz Bad *See* SCHWARZER ADLER.

Keller, Klaus Peter Rhh ★★★★ Star of RHH, cultish for ALTE REBEN RIES G-Max from undisclosed (calcareous) parcel, and GGS Hubacker, Morstein. Also Ries from NIERSTEIN (Hipping, Pettenthal), M-M (PIESPORT Schubertslay). Record prices at auction.

Kesseler, August Rhg ★★★★ Outstanding estate at ASSMANNSHAUSEN. August K's 85 Höllenberg PINOT N was start of new German Pinot N miracle. Meanwhile, long-time employees have taken over – and work in same perfectionist manner. Equally outstanding RIES (berry-scented, dense).

Kesselstatt, Reichsgraf von Mos ★★→★★★ Top vyds on Mosel and tributaries, 46 ha. After visionary Annegret Reh-Gartner died 2016, in troubled waters. Long-time winemaker and managing director left just before 21 harvest.

Kiedrich Rhg ★★→★★★★ Top RHG village, almost a monopoly of the WEIL estate; other growers (eg. FRICKE, Knyphausen, PRINZ VON HESSEN) own only small plots. Famous church and choir.

Kloster Eberbach Rhg Atmospheric C12 Cistercian abbey in HATTENHEIM, the place where *The Name of the Rose* was filmed. Domicile of HESSISCHE STAATSWEINGÜTER.

Klumpp, Weingut Bad ★★★ Brothers Markus and Andreas K have led winery to

---

Nihon arigatō!

It is well known that Japanese orchestras can give masterful interpretations of Bach and Beethoven. But do Japanese cellarmasters understand RIES and PINOT N? The answer is yes, as a small number of individuals in German cellars show. Young Chie Sakata produces sublime Pinot N at Bernhard Koch in S PFZ; Masato Nagasawa, balanced Ries at Prinz Salm. Kazuyuki Kaise is responsible for Sekthaus RAUMLAND's exquisite sparkling. In NIERSTEIN, Hideki Asano produces an own line of off-dry Ries at Strub, conceived to match Japanese cuisine. And Fumiko Tokuoka, daughter of a merchant from Osaka, is not only a winemaker but also the owner of the notable Biffar winery at DEIDESHEIM.

top in KRAICHGAU, N BAD, esp fruit-driven PINOT N and refined PINOT GR from Rothenberg vyd at Bruchsal.

Knebel, Weingut Mos ★★★ Family estate in top form at WINNINGEN. Refined TBA, and elegant, aromatic dry wines.

Knewitz, Weingut Rhh ★★★ Young brothers Björn and Tobias K needed less than a decade to turn their parents' unknown estate into one of RHH's leaders, and to join the VDP. Masterpiece Appenheim Hundertgulden RIES 15' 16' 17' 18 19' 20'.

Knipser, Weingut Pfz ★★★→★★★★ Werner K advocated the late marketing of GGS over 10 yrs ago, and now it's fashionable. Some Res are sold only after 5 yrs (eg. iconic PINOT N RdP 15', ★★★★ CHARD 15'). Many specialities, incl rare Gelber Orleans and CAB SAUV-based Clarette Rosé.

Königin Viktoriaberg Rhg ★★→★★★ Historic RIES vyd at HOCHHEIM, 4.5 ha along shores of River Main, today run by Flick estate of Wicker. After 1845 visit, Queen Victoria granted owner right to rename vyd as "Queen-Victoria-mountain".

Kraichgau Bad Small district se of Heidelberg. Top growers: HEITLINGER/BURG RAVENSBURG, Hoensbroech, Hummel, KLUMPP.

Kranz, Weingut Pfz ★★★→★★★★ Top estate at Ilbesheim, S PFZ. Outstanding RIES, SPÄTBURGUNDER, WEISSBURGUNDER from classified Kalmit vyd. Organic.

Krone, Weingut Rhg ★★★ Famous SPÄTBURGUNDER estate with old vyds in ASSMANNSHAUSEN (slate), run by WEGELER. Best: cuvée Juwel 13' 14 15' uses fruit from legendary Höllenberg plus cooler neighbouring plots.

Kühling-Gillot Rhh ★★★★ Top bio estate, run by Caroline Gillot and husband HO Spanier. Best in already outstanding range of ROTER HANG RIES: GG Rothenberg Wurzelecht from ungrafted, 70 yrs+ vines.

Kühn, Peter Jakob Rhg ★★★ Excellent estate in OESTRICH led by PJ Kühn and son. Obsessive bio vyd management and long macerations shape *nonconformist but exciting* RIES. Res RPJK Unikat aged 4 yrs in cask. Doosberg GG 16' 17' 18' 19' reliable at highest level.

Kuhn, Philipp Pfz ★★★ Great versatility; v. reliable. RIES (eg. SAUMAGEN, Schwarzer Herrgott), barrel-aged SPÄTBURGUNDER, specialities (CHARD, FRÜHBURGUNDER, SAUV BL, SEKT).

Künstler Rhg ★★★ Gunter K was able to lease ex-Schönborn vyds in MARCOBRUNN (1st vintage 21), so has GGs from every part of RHG. But also continues to focus on home town of HOCHHEIM: outstanding RIES GGs 20' from Hölle, Kirchenstück.

Kuntz, Sybille Mos ★★★ Progressive organic 12-ha estate at Lieser, esp Niederberg-Helden vyd. Pioneer of MOS TROCKEN; intense wines, one of each ripeness category, intended for gastronomy, listed in many top restaurants.

Laible, Andreas Bad ★★★ Crystalline dry RIES from Durbach's Plauelrain vyd (granite). Andreas Jnr's younger brother Alexander has estate of his own.

Landwein Technically "ggA" (*see* box, p.165), meant to label wines with only broadly defined origin. But now popular among ambitious growers to avoid official quality testing (eg. because of spontaneous fermentations, or low sulphur levels). Best known: Brenneisen, Enderle & Moll, Forgeurac, Greiner, Höfflin, Nieger, Vorgrimmler, WASENHAUS, ZIEREISEN (all BAD); Drei Zeilen, Vetter, Weigand (FRANK); Schmitt (RHH); Konni & Evi (SA-UN).

Lauer Mos ★★★ Fine, precise SAAR RIES: tense, poised. Parcel selections from huge Ayler Kupp vyd. Best: Kern, Schonfels, Stirn.

Leitz, Josef Rhg ★★★ RÜDESHEIM-based family estate, outstanding GGs, best usually Berg Schlossberg 10' 11' 12' 13' 15' 16 17 18 19. Reclaimed some altitude vyds from fallow in Berg Kaisersteinfels. Inexpensive, reliable Eins Zwei-Dry label.

Liebfrauenstift-Kirchenstück Rhh Walled vyd in city of Worms; flowery RIES from gravelly soil: Gutzler, Schembs, Weingut Liebfrauenstift. Don't confuse with Liebfraumilch, a cheap, tasteless imitation.

Loewen, Carl Mos ★★★ RIES of elegance, tension, complexity. Best vyd Longuicher Maximin Herrenberg (planted 1896, ungrafted). Entry-level Ries Varidor excellent *value.*

Loosen, Weingut Dr. M-M ★★→★★★★ Charismatic Ernie L produces fine traditional RIES from old vines in M-M. ERDEN Prälat AUSLESE cultish for decades. Also series of dry Res (2 yrs cask-ageing, 1st vintage 2011), and unique Indotiumarus from WEHLEN Sonnenuhr (vintage 81, 27 yrs in cask, marketed 2021). Dr. L Ries, from bought-in grapes, is reliable. *See also* WOLF (PFZ), Ch Ste Michelle (Washington State, US).

Lorch Rhg ★→★★★ Village in extreme w of RHG, conditions more M RH-like than Rhg-like. Sharply crystalline wines, both RIES and PINOT N, now re-discovered. Best: BREUER (Pfaffenwies 19' 20'), CHAT SAUVAGE, FRICKE, KESSELER, SOLVEIGS, von Kanitz.

Löwenstein, Fürst Frank, Rhg ★★★ Princely estate, classic RIES from HALLGARTEN (RHG), unique SILVANER, Ries from ultra-steep Homburger Kallmuth (FRANK).

Lützkendorf, Weingut Sa-Un ★★→★★★ Pioneer in SA-UN after the wall came down. Uwe L continuing after his father Udo's death in 2020.

Marcobrunn Rhg Historic 7-ha vyd in Erbach, GROSSE LAGE. Potential for rich, long-lasting RIES. Growers: CORVERS-KAUTER, HESSISCHE STAATSWEINGÜTER, Höhn, Knyphausen, KÜNSTLER, PRINZ, Schloss Reinhartshausen, von Oetinger.

Markgräflerland Bad District s of Freiburg, cool climate (breezes from Black Forest), limestone soils. Typical GUTEDEL a pleasant companion for local cuisine. Pinot varieties increasingly successful.

Markgräfler Winzer Bad ★→★★★ Co-op, 940 ha, formerly mediocre, now turned upside down by ex-LVMH manager. Remarkable top range DER CHARD, DER SPÄTBURGUNDER.

Markgraf von Baden Bad ★★→★★★ Important noble estate (112 ha) at Salem Castle (BODENSEE) and Staufenberg Castle (ORTENAU).

Maximin Grünhaus Mos ★★★★ Maximin von Schubert has taken the helm at this supreme RUWER estate; v. traditional winemaking shapes herb-scented, *delicate, long-lived Ries.* To be taken more seriously each yr: creamy WEISSBURGUNDER, elegant PINOT N.

Merkelbach, Weingut M-M ★★→★★★ Tiny estate at ÜRZIG, 2 ha. Brothers Alfred and Rolf (both c.80). Inexpensive MOS made not to sip, but to drink.

Meßmer, Weingut Pfz ★★→★★★ Family estate in S PFZ, best among many specialities Burrweiler Schäwer RIES.

Meyer-Näkel Ahr ★★★→★★★★ Fruit-driven, refined SPÄTBURGUNDER from steep terraces at Walporzheimer Kräuterberg 09 12 13' 14 15' 16' 17' 18 19' and other prime AHR v sites. Except for six barriques, entire 20 vintage lost in flood (*see box*, p.156). In S Africa (Zwalu, with Neil Ellis) and Portugal (Quinta da Carvalhosa).

Mittelhaardt Pfz North-central and best part of PFZ, incl DEIDESHEIM, FORST, RUPPERTSBERG, WACHENHEIM; largely planted with RIES.

Mittelmosel Central and best part of MOS, a RIES Eldorado, incl BERNKASTEL, BRAUNEBERG, GRAACH, PIESPORT, WEHLEN, etc.

Nazi law banned red wine from Mos in 1937 – not lifted until 1987.

Mittelrhein M Rh ★★→★★★ Dramatically scenic Rhine area nr tourist magnet Loreley. Best villages: BACHARACH, BOPPARD, Oberwesel. Delicate yet *steely Ries, underrated* and underpriced. Long-time decline in production now finally halted.

Molitor, Markus M-M, Mos ★★★→★★★★ Perfectionist Markus M farms more than 20 EINZELLAGEN (120 ha), many of them produced at different PRÄDIKAT levels, and in dry (white capsule), off-dry (green) and sweet (golden) versions – resulting in a price list of encyclopaedic proportions.

Germany's quality levels

The official range of qualities and styles in ascending order is (take a deep breath):

1 **Wein** formerly known as Tafelwein. Light wine of no specified character, mostly sweetish.

2 **ggA (geschützte geographische Angabe)** or protected geographical indication, formerly known as LANDWEIN. Dryish Wein with some regional style. Mostly a label to avoid, but some thoughtful estates use the Landwein designation to bypass official constraints.

3 **gU (geschützte Ursprungsbezeichnung)** or protected designation of origin. Replacing QUALITÄTSWEIN. Up to now, only six small-scale appellations in the narrower sense of the word approved by the EU.

4 **Qualitätswein** dry or sweetish wine with sugar added before fermentation to increase its strength, but tested for quality and with distinct local and grape character. Don't despair.

5 **Kabinett** dry/dryish natural (unsugared) wine of distinct personality and distinguishing lightness. Can occasionally be sublime – esp with a few yrs' age.

6 **Spätlese** stronger, sweeter than KABINETT. Full-bodied (but no botrytis). Dry SPÄTLESE (or what could be considered as such) is today mostly sold under Qualitätswein designation (even if not sugared).

7 **Auslese** sweeter, stronger than Spätlese, often with honey-like flavours, intense and long-lived. Occasionally dry and weighty. The lower the alc (read the label), the sweeter the wine.

8 **Beerenauslese (BA)** v. sweet, dense and intense, but seldom strong in terms of alc. Can be superb.

9 **Eiswein** from naturally frozen grapes of BA/TBA quality: concentrated, pungent acidity and v. sweet. Should not display botrytis character (but does sometimes).

10 **Trockenbeerenauslese (TBA)** intensely sweet and aromatic; alc is slight. Extraordinary and everlasting.

Mosbacher Pfz ★★★ Some of best GG RIES of FORST: refined rather than massive. Traditional ageing in big oak casks. Excellent SAUV BL too ("Fumé").

Mosel Wine-growing area formerly known as Mosel-Saar-Ruwer, 8690 ha in total, 62% RIES. Conditions on the RUWER and SAAR tributaries are v. different from those along the Mosel (aka Moselle in French).

Nackenheim Rhh ★→★★★★ NIERSTEIN neighbour with GROSSE LAGE Rothenberg on red shale, famous for *Rhh's richest Ries*, superb TBA. Top growers: *Gunderloch*, KÜHLING-GILLOT.

Nahe Tributary of the Rhine and dynamic region (4230 ha) with a couple of famous (eg. DIEL, DÖNNHOFF, SCHÄFER-FRÖHLICH) and dozens of lesser-known producers, excellent *value*. Great soil variety; best RIES from slate (almost MOS-like raciness).

Neipperg, Graf von Würt ★★★ Arguably most traditional estate of WÜRT, members of N family said to have brought, in C17, LEMBERGER to Germany. SPÄTBURGUNDER of grace and purity, fruit-driven Lemberger, red PINOT M (Schwarzriesling) displays nobility too. Count Karl-Eugen von N's younger brother Stephan makes wine at Canon-la-Gaffelière in St-Émilion (*see* Bordeaux) and elsewhere.

Neus Rhh ★★★ Revived historic estate at INGELHEIM, excellent PINOT N (best: Pares).

Niederhausen Na ★★→★★★★ Village of the middle NA v. Complex RIES from famous GROSSE LAGE Hermannshöhle and neighbouring steep slopes. Growers: CRUSIUS, *Dönnhoff*, GUT HERMANNSBERG, J Schneider, Mathern.

Nierstein Rhh ★→★★★★ Avoid by any means GROSSLAGE Gutes Domtal designation.

Genuin Nierstein RIES is rich, tense, complex, eg. GROSSE LAGE vyds Brudersberg, Hipping, Oelberg, Orbel, Pettenthal. Growers: Bunn, Gehring, Gröhl, GUNDERLOCH, Guntrum, Hofmann, Huff (both), KELLER, KÜHLING-GILLOT, Manz, SCHÄTZEL, ST-ANTONY, Strub.

Ockfen Mos ★★→★★★ SAAR village known for GROSSE LAGE vyd Bockstein. Growers: M MOLITOR, OTHEGRAVEN, SANKT URBANS-HOF, VAN VOLXEM, WAGNER, ZILLIKEN. Historical Geisberg vyd was abandoned in C20 because too cool, but now Van Volxem has re-cultivated 10 ha.

Some German regions have become so dry that vyds need to be irrigated.

Odinstal, Weingut Pfz ★★→★★★ Highest vyd of PFZ, 150m (492ft) above WACHENHEIM. Bio farming and low-tech vinification bring pure GEWÜRZ, RIES, SILVANER. New VDP member (2021).

Oechsle Scale for sugar content of grape juice. Until 90s, more Oechsle meant better wine. But climate change has altered game.

Ökonomierat-Rebholz Pfz ★★★ Top SÜDLICHE WEINSTRASSE estate: bone-dry, zesty and reliable RIES GGS, best usually Kastanienbusch from red schist, also outstanding CHARD, SPÄTBURGUNDER. Father Hansjörg R now supported by twins Hans and Valentin.

Oppenheim Rhh ★→★★★ Neighbour of NIERSTEIN, but with different microclimate (no direct Rhine influence) and different soil (limestone rather than red shale). GROSSE LAGE Kreuz, Sackträger. Growers: Guntrum, Kissinger, KÜHLING-GILLOT, Manz. Spectacular C13 church.

Ortenau Bad ★★→★★★ District around and s of city of Baden-Baden. Mainly Klingelberger (RIES) and SPÄTBURGUNDER from granite soils.

Ortswein The 2nd rank up in VDP's pyramid of qualities: a village wine, many bargains. *See* next entry.

Ortswein aus Ersten Lagen Rhh New designation of VDP RHH indicating a village wine grown in classified vyds. Typically a blend of different ERSTE LAGE sites. Funnily enough, there are no Erste Lage single-vyd wines in Rhh.

Othegraven, von Mos ★★★ Fine SAAR estate, 16 ha, best known for dry and sweet RIES from superb GROSSE LAGE Altenberg at KANZEM. Since 2010 owned by TV star (and von Othegraven family member) Günther Jauch; recent emphasis on (four different) fruity KABINETTS.

Palatinate English for PFALZ.

Pfalz The 2nd-largest German region, 23,700 ha, balmy climate, Lucullian lifestyle. MITTELHAARDT RIES best; S Pfalz (SÜDLICHE WEINSTRASSE) is better suited to PINOT varieties. ZELLERTAL now fashonable: cool climate.

Piesport M-M ★→★★★★ Village in M-M for rich, aromatic RIES. GROSSE LAGE vyds Domherr, Goldtröpfchen. Growers: Grans-Fassian, Hain, Joh Haart, Julian Haart, KESSELSTATT, *Reinhold Haart*, SANKT URBANS-HOF, SCHLOSS LIESER. Avoid GROSSLAGE Michelsberg.

Piwi ★→★★★ Crossings of European and American vines, for fungal resistance ("Pilz-Widerstandsfähigkeit"). Most popular: Regent (r). A new generation of crossings deserves to be followed, eg. Souvignier Gr (w), Satin N (r).

Prädikat Legally defined category of ripeness at harvest. *See* QMP.

Prinz, Weingut ★★★ Distinctly fresh, elegant RIES from Hallgarten's altitude vyds, organic. KABINETT GOLDKAPSEL can be magnificent. Now also MARCOBRUNN.

Prüm, JJ Mos ★★★★ 59 71 76 83 90 03 11 15 18 19 20 Legendary WEHLEN estate; also BERNKASTEL, GRAACH. Delicate but extraordinarily long-lived wines with finesse and distinctive character. Rising prices for SONNENUHR KABINETT.

Prüm, SA Mos ★★→★★★ Quality revolution since Saskia A Prüm took over in 2017. Sublime 20 SONNENUHR SPÄTLESE, 20 WEHLEN ORTSWEIN a bargain.

QbA (Qualitätswein bestimmter Anbaugebiete) "quality wine", controlled as to area, grape(s), vintage. May add sugar before fermentation (chaptalization). Intended as middle category, but now VDP obliges its members to label their best dry wines (GGS) as QbA. New EU name gU scarcely found on labels (*see* box, p.172).

QmP (Qualitätswein mit Prädikat) Top category, six levels according to ripeness of grapes: KABINETT to TBA. No sugaring of must or other forms of enrichment allowed.

Randersacker Frank ★★→★★★ Village s of WÜRZBURG with GROSSE LAGE Pfülben. One of best FRANK villages for RIES. Top growers: Bardorf, BÜRGERSPITAL, Göbel, JULIUSSPITAL, Schmitt's Kinder, Staatlicher Hofkeller, Störrlein & Krenig, Trockene Schmitts.

Rauenthal Rhg ★★→★★★★ *Spicy, austere but complex* RIES from inland slopes. Baiken, Gehrn and Rothenberg vyds contain GROSSE LAGE and ERSTE LAGE parcels, while neighbouring Nonnenberg (monopole of DREUER) is unclassified, despite its equal quality.

Raumland Rhh ★★★ SEKT expert with deep cellar and full range of fine, balanced cuvées; 1st Sekt-only estate to become VDP member. Best. CHARD Brut Nature (disgorged after 10 yrs), Cuvée Triumvirat, MonRose.

Restsüsse Unfermented grape sugar remaining in (or, in cheap wines, added to) wine to give it sweetness. Can range from 1g/l in TROCKEN to 300g in TBA.

Rheingau ★★→★★★★ Birthplace of RIES. Historic s- and sw-facing slopes overlooking Rhine between Wiesbaden and RÜDESHEIM. Classic, substantial Ries, famous for steely backbone, and small amounts of delicate SPÄTBURGUNDER.

Rheinhessen ★→★★★★ Germany's largest region by far (26,940 ha and rising), between Mainz and Worms. Much *Fasswein* (bulk wine), but also treasure trove of well-priced wines from gifted young growers.

Richter, Max Ferd M-M ★★→★★★ Reliable estate, at Mülheim; esp gd, full, aromatic sweet RIES. Magnificent 19 WEHLEN SONNENUHR ★★★ AUSLESE No 42. Round, pretty Brut (EISWEIN dosage).

Riffel Rhh ★★★ Bio family estate, known for purist style, to put Bingen's once-famous Scharlachberg (red soils) back on the map.

Rings, Weingut Pfz ★★★→★★★★ Dense, emphatically fresh RIES and SPÄTBURGUNDER from prime sites in N PFZ. Das Kreuz is popular B'x blend.

Roter Hang Rhh ★★→★★★★ Leading RIES area of RHH (NACKENHEIM, NIERSTEIN). Name ("red slope") refers to red shale soil. At best rich and mineral, but susceptible to drought.

Ruck, Johann Frank ★★★ Spicy, age-worthy RIES, SCHEUREBE, SILVANER, TRAMINER from IPHOFEN.

Rüdesheim Rhg ★★→★★★★ Small town in W RHG, tourist magnet and centre of Rhine romanticism. Unique RIES from Berg (mtn) vyds on slate (Kaisersteinfels,

---

"Not at all acid"

None other than Thomas Jefferson has left us an RHG travel report from C18. On 11 April 1788, the future US president travelled the Rhg in a kind of paddleboat, accompanied by his servant, from Mainz to RÜDESHEIM. Jefferson called the Rüdesheim wines "not at all acid, and to my taste much preferable than to HOCHHEIM". On his way back to Mainz, he was even more enthusiastic about the wines of SCHLOSS JOHANNISBERG: "the best made on the Rhine without comparison". Jefferson also stopped in Erbach at the MARCOBRUNN vyd and judged that it produced "wine of the second quality". Even back then, it seems, wine critics were sometimes overly critical.

Roseneck, Rottland, Schlossberg). Full-bodied but never clumsy, floral, esp gd in off-yrs. Best: BISCHÖFLICHES WEINGÜT RÜDESHEIM (not to be confused with BISCHÖFLICHE WEINGÜTER TRIER), *Breuer*, CHAT SAUVAGE, CORVERS-KAUTER, HESSISCHE STAATSWEINGÜTER, JOHANNISHOF, *Kesseler*, KÜNSTLER, *Leitz*, Ress.

Ruwer Mos ★★→★★★★ Tributary of MOS nr TRIER, cool and late-ripening. Quaffable light dry and intense sweet RIES. Best growers: Beulwitz, Karlsmühle, KARTHÄUSERHOF, KESSELSTATT, MAXIMIN GRÜNHAUS.

Mos is longest walnut-tree alley in world. Planted by Napoleon, scrumped by tourists.

Saale-Unstrut ★→★★★ A ne region around confluence of these two rivers nr Leipzig. Terraced vyds have Cistercian origins. Growers: Böhme, Born, Gussek, Hey (VDP member), Kloster Pforta, Konni & Evi (LANDWEIN), LÜTZKENDORF (VDP), Pawis (VDP).

Saar Mos ★★→★★★★ Tributary of Mosel, bordered by steep slopes. Most austere, steely, *brilliant Ries* of all, consistency favoured by climate change.

Saarburg Mos Small town in SAAR v. Growers: WAGNER, ZILLIKEN. GROSSE LAGE: Rausch.

Sachsen ★→★★★ Region in Elbe V around Meissen and Dresden. Characterful dry whites, TRAMINER a speciality. Best growers: AUST, Drei Herren, F Fourré, Gut Hoflössnitz, Rothes Gut, Schloss Proschwitz, SCHLOSS WACKERBARTH, Schuh, Schwarz, ZIMMERLING.

St-Antony Rhh ★★→★★★ NIERSTEIN organic estate. Exceptional vyds, known for sturdy, ageable ROTER HANG RIES. Since 2019, steep slopes all worked by hand.

Salm, Prinz zu Na, Rhh ★★→★★★ Owner of Schloss Wallhausen in NA and vyds there and at BINGEN (RHH); ex-president of VDP.

Salwey Bad ★★★→★★★★ Leading KAISERSTUHL estate. Konrad S picks early for freshness. Best: GGS Henkenberg and Eichberg GRAUBURGUNDER, Steingrubenberg and Kirchberg SPÄTBURGUNDER and WEISSBURGUNDER.

Sankt Urbans-Hof Mos ★★★ Large family estate (40 ha) led by Nik Weis based in Leiwen, vyds along M-M and SAAR. Limpid RIES, impeccably pure, racy, age well.

Sauer, Horst Frank ★★★ Man who put FRANCONIA on sweet-wine map. Sensational BA and TBA from ESCHERNDORF's steep Lump vyd; v.gd dry wines too. Daughter Sandra now in charge.

Sauer, Rainer Frank ★★★ Top family estate producing seven different dry SILVANERS from ESCHERNDORF's steep slope Lump. Best: GG am Lumpen, and L 99' 03' 07' 18 19'. Highly recommendable ORTSWEIN and ERSTE LAGE Silvaner.

Saumagen Popular local dish of PFZ: stuffed pig's stomach. Also one of best vyds of region: a calcareous site at Kallstadt producing excellent RIES, PINOT N.

Schaefer, Willi Mos ★★★ Willi S and son Christoph finest in GRAACH, 4.2 ha in over 100 parcels. MOS RIES at its best: pure, crystalline, feather-light. Outstanding 20' WEHLEN SONNENUHR KABINETT.

Schäfer-Fröhlich Na ★★★→★★★★ Family estate in NA, spontaneously fermented RIES of great intensity, GGS incl Bockenau Felseneck, Schlossböckelheim Felsenberg and Kupfergrube. Excellent mid-price Schiefergestein, Vulkangestein.

Scharzhofberg Mos ★★→★★★★ Superlative SAAR vyd, 28 ha: a rare coincidence of microclimate, soil and human intelligence to bring about the perfection of RIES. Top: EGON MÜLLER, HÖVEL, KESSELSTATT, VAN VOLXEM.

Schätzel, Weingut Rhh ★★→★★★ Family estate at NIERSTEIN, 15 ha; Kai S has made a name for brilliant Pettenthal KABINETT, and for dry RIES full of extract (and even tannin) at 11.5% abv. Try 18' Fuchs (= Hipping, but classified LANDWEIN).

Schloss Johannisberg Rhg ★★★ Historic RHG estate and Metternich mansion, the place where SPÄTLESE was invented (1775); 100% RIES, owned by Henkell (Oetker group). Currently in top form; even entry-level Gelblack shows nobility of site.

Schloss Lieser M-M ★★★★ Thomas Haag, elder brother of Oliver Haag (FRITZ HAAG) produces painstakingly elaborate RIES both dry and sweet from BRAUNEBERG, Lieser (Niederberg Helden), PIESPORT, WEHLEN. Now also small (leased) plot in BERNKASTEL'S DOCTOR. Hotel Lieser Castle has no ties to wine estate.

Schloss Proschwitz Sachs ★★→★★★ New team brings notable improvements in Prince zur Lippe's 70-ha estate: 18 PINOT NS (both ERSTE LAGE, GG) top choice.

Schloss Vaux Rhg ★★→★★★ Rhg SEKT house, co-owned by Prinz zu Salm, best known for single-vyd RIES Sekt.

Schloss Wackerbarth ★★→★★★ Saxon state dom on outskirts of Dresden, conceived as experience winery, with restaurant, park, events, 190,000 visitors/yr. Also wine-wise in top form; best: Protze RIES TROCKEN, ALTE REBEN, TRAMINER SPÄTLESE (all from Radebeul's Goldener Wagen vyd).

Schnaitmann Würt ★★→★★★★ Excellent barrel-aged reds from Fellbach (nr Stuttgart). Whites (eg. RIES, SAUV BL), SEKT (Evoé!), and wines from lesser grapes (SCHWARZRIESLING, TROLLINGER) tasty too.

Schneider, Markus Pfz ★★ Shooting star in Ellerstadt, PFZ. Full range of soundly produced, trendily labelled wines.

Schoppenwein Café (or bar) wine, ie. wine by the glass.

Schumacher, Paul Ahr ★★★ Perfectionist grower at Marienthal, only 4 ha. Age-worthy SPÄTBURGUNDERS (Kräuterberg 09' 11 12 13 15' 16 17' 18 19'). Was close to giving up after AHR flood (see box, p.156), but luckily will continue.

Schwarzer Adler Bad ★★★→★★★★ Top estate at Oberbergen, KAISERSTUHL, led by Fritz Keller and son Friedrich. Pinots (r/w) to show that France is nr. Family-owned restaurant Schwarzer Adler features one of the most comprehensive wine lists in Germany: a who's who of B'x and Burgundy back to the 50s.

Schwegler, Albrecht Würt ★★★→★★★★ Family estate, 11 ha, led by young Aaron S. Red blends Beryll, Granat, Saphir, powerful CHARD, and RIES ALTE REBEN.

Seckinger, Weingut Pfz ★★→★★★ Three brothers (30 and younger) from Niederkirchen/MITTELHAART making name for RIES from DEIDESHEIM in style of Vin Nature, but without becoming too extreme. Organic.

Sekt ★→★★★★ German sparkling. Avoid cheap offers: bottle fermentation is not mandatory. Serious Sekt producers making spectacular progress, eg. ALDINGER, Bardong, Barth, BATTENFELD-SPANIER, BERGDOLT, BREUER, BUHL, DIEL, F John, GRIESEL, Gut Hermannsberg, H Bamberger, HEYMANN-LÖWENSTEIN, LAUER, Leiner, Melsheimer, MOLITOR, RAUMLAND, Reinecker, Schembs, SCHLOSS VAUX, SCHWARZER ADLER, Solter, S Steinmetz, Strauch, WAGECK, WEGELER, Wilhelmshof, ZILLIKEN. A VDP classification of Sekt vyds under way.

Selbach-Oster M-M ★★★ Scrupulous ZELTINGEN estate with excellent vyd portfolio, known for classical style and focus on sweet PRÄDIKAT wines.

Solveigs Rhg ★★→★★★★ PINOT N from red slate at ASSMANNSHAUSEN and LORCH, only 2 ha, organic viticulture and minimal winemaking. Best: plots Micke 06' 13' 15' 16' 18', Present 95' 99' 03' 04 06 09' 12 13' 15' 16' 18'.

Sonnenuhr M-M Sundial. Name of GROSSE LAGE sites at BRAUNEBERG, Pommern, WEHLEN, ZELTINGEN.

Germany's highest vyds: Hohentwiel (Bad), Kappishäusern (Würt), c.560m (1837ft).

Spätlese Late-harvest. One level riper and sweeter than KABINETT. Needs to be aged for 5 yrs at least.

Spreitzer Rhg ★★★ Brothers Andreas and Bernd S produce deliciously *racy, harmonious* RIES from vyds in HATTENHEIM, Mittelheim, Oestrich. Mid-price range ALTE REBEN a bargain. Breathtaking TBA too (20' Eiserberg).

Staatsweingut / Staatliche Weinbaudomäne State wine estates or doms exist in BAD (IHRINGEN, Meersburg), NA (Bad Kreuznach), PFZ (Neustadt), RHG

(HESSISCHE STAATSWEINGÜTER), RHH (OPPENHEIM), SACHS (Wackerbarth), SA-UN (Kloster Pforta), WÜRT (Weinsberg).

Steigerwald Frank District in E FRANK; vyds at considerable altitude. Best: CASTELL, Hillabrand, Roth, RUCK, VETTER, WELTNER, *Wirsching*.

Steinberg Rhg ★★★ Walled-in vyd above HATTENHEIM, est by Cistercian monks 700 yrs ago: a German Clos de Vougeot. Monopoly of HESSISCHE STAATSWEINGÜTER. Classified parcels (14 ha out of 37) have unique soil (clay with fragments of decomposed schist in various colours). Fascinating old vintages. Not to be confused with Stein-Berg GU of WÜRZBURG.

Steinwein Frank Wine from WÜRZBURG's best vyd, Stein. Goethe's favourite. Only six producers: BURGERSPITAL, JULIUSSPITAL, L Knoll, Meinzinger, Reiss, Staatlicher Hofkeller. Hugh J once tasted the 1540 vintage.

Stodden Ahr ★★★→★★★★ AHR SPÄTBURGUNDER with Burgundian touch, best usually ALTE REBEN (sensational 19'), Rech Herrenberg GOLDKAPSEL. Buildings devastated in flood (*see* box, p.156), but harvest saved.

Südliche Weinstrasse Pfz District known esp for Pinot varieties. Best growers: BECKER, JÜLG, KRANZ, Leiner, Minges, Münzberg, REBHOLZ, Siegrist, WEHRHEIM.

Taubertal Bad, Frank, Würt ★→★★★ Cool-climate river valley, divided by Napoleon into BAD, FRANK, WÜRT sections. SILVANER (limestone soils), local red Tauberschwarz. Frost a problem. Hofmann, Schlör, gd co-op at Beckstein.

TBA (Trockenbeerenauslese) Sweetest, most expensive category of German wine, extremely rare, viscous and concentrated, with dried-fruit flavours. Made from selected dried-out berries affected by noble rot (botrytis). Half-bottles a gd idea.

Thanisch, Weingut Dr. M-M ★★★ BERNKASTEL estate, founded 1636, famous for its share of the DOCTOR vyd. After family split-up in 1988 two homonymous estates with similar qualities: Erben (heirs) Müller-Burggraef (more modern) and Erben Thanisch (VDP, classical style).

Trier Mos The n capital of ancient Rome, on the Mosel, between RUWER and SAAR. Big charitable estates have cellars here among awesome Roman remains.

Trittenheim M-M ★★→★★★ Racy, textbook M-M RIES if from gd plots within extended GROSSE LAGE vyd Apotheke. Growers: A CLÜSSERATH, Clüsserath-Weiler, E Clüsserath, FJ Eifel, Grans-Fassian, Milz.

Trocken Dry. Defined as max 9g/l unfermented sugar. Generally the further s in Germany, the more Trocken wines.

Ürzig M-M ★★★→★★★★ MOS village on red sandstone and red slate, famous for ungrafted old vines and *unique spicy Ries* from Würzgarten vyd. Growers: Berres, Christoffel, Erbes, *Loosen*, MERKELBACH, *Molitor*, Mönchhof, Rebenhof.

Van Volxem Mos ★★★ Historical SAAR estate revived by obsessive Roman Niewodniczanski. Low yields from top sites bring about monumental (mostly

---

German vintage notation

The vintage notes after entries in the German section are mostly given in a different form from those elsewhere in the book. If the vintages of a single wine are rated, or are for red wine regions, the vintage notation is identical with the one used elsewhere (*see* front jacket flap). But for regions, villages or producers, two styles of vintage are indicated:

Bold type (eg. **16**) indicates classic, ripe vintages with a high proportion of SPÄTLESEN and AUSLESEN; or in the case of red wines, gd phenolic ripeness and must weights.

Normal type (eg. 17) indicates a successful but not outstanding vintage. Generally, German white wines, esp RIES, can be drunk young for their intense fruitiness or kept for a decade or even two to develop their potential aromatic subtlety and finesse.

dry) RIES. Spectacular castle-like new cellar building in a Saar loop nr Wiltingen – and capacity to age selected wines 5 yrs+ in tank.

VDP (Verband Deutscher Prädikatsweingüter) Influential association of 200 premium growers setting highest standards. Look for its eagle insignia on wine labels, and for GG logo on bottles. VDP wine is usually a gd bet. President: Steffen CHRISTMANN.

Wachenheim Pfz ★★★ Celebrated village with, according to VDP, NO GROSSE LAGE vyds. See what you think. Top growers: Biffar, BÜRKLIN-WOLF, Karl Schäfer, ODINSTAL, WOLF, Zimmermann (bargain).

Wageck Pfz ★★→★★★ MITTELHAARDT estate for unaffected, brisk CHARD (still, sp), PINOT N of great finesse. Best: Geisberg vyd.

Wagner, Dr. Mos ★★→★★★ Estate with vyds in OCKFEN and Saarstein led by young Christiane W. SAAR RIES with purity, freshness.

Wagner-Stempel Rhh ★★★ Seriously crafted RHH wines from Siefersheim nr NA border. Best usually RIES GGs Heerkretz (porphyry soil).

Walluf Rhg ★★★ Underrated village, 1st with important vyds as one leaves Wiesbaden going w. GROSSE LAGE vyd: Walkenberg. Growers: *JB Becker, Jost*.

Wasenhaus Mos ★★★ Burgundy-inspired PINOT N Bellen and PINOT BL Möhlin from 2-ha limestone sites in MARKGRÄFLERLAND, produced by young Christoph Wolber (who interned at Dom Leflaive and Comte Armand, *see* France) and Alexander Götze (who continues to work at Dom de Montille, *see* France). All vines farmed by hand, all wines labelled as LANDWEIN.

Wegeler M-M, Rhg ★★→★★★★ Important estate in Oestrich and BERNKASTEL (both in top form) plus a stake in the famous KRONE estate of ASSMANNSHAUSEN. Geheimrat J blend maintains high standards, single-vyd RIES and ORTSWEIN usually outstanding value. Old vintages available (Vintage Collection).

Wehlen M-M ★★★→★★★★ Wine village with legendary steep SONNENUHR vyd expressing RIES from slate at v. best: rich, fine, everlasting. Top: JJ PRÜM, Kerpen, KESSELSTATT, LOOSEN, MF RICHTER, MOLITOR, Pauly-Bergweiler, SA PRÜM, Sankt Nikolaus Hospital, SCHLOSS LIESER, SELBACH-OSTER, Studert-Prüm, THANISCH, W SCHÄFER, WEGELER.

Wehrheim, Weingut Dr. Pfz ★★★ Top organic estate of SÜDLICHE WEINSTRASSE; v. dry, culinary style, esp white Pinot varieties.

Weil, Robert Rhg ★★★→★★★★ 01 05 09 12 15 16 17 18 19 20 Outstanding estate in KIEDRICH with GROSSE LAGE vyd Gräfenberg (generous, parcel-selection Monte Vacano available only by subscription) and ERSTE LAGE vyds Klosterberg (balanced) and Turmberg (austere). Superb sweet KABINETT to TBA.

Weingart M Rh ★★★ Outstanding estate at Spay, vyds in BOPPARD (esp Hamm Feuerlay). Refined, taut RIES, low-tech in style, superb value.

Weingut Wine estate.

Weissherbst Pale-pink wine, made from a single variety, often SPÄTBURGUNDER; v. variable quality.

Weltner, Paul Frank ★★★ STEIGERWALD family estate. Densely structured, age-worthy SILVANER from underrated Rödelseer Küchenmeister vyd.

Wiltingen Mos ★★→★★★★ Heartland of the SAAR. SCHARZHOFBERG crowns a series of GROSSE LAGE vyds (Braune Kupp, Braunfels, Gottesfuss, Kupp). ORTSWEIN usually a bargain.

Wind, Katrin Pfz ★★→★★★ Shooting star at Arzheim. Straightforward but nuanced wines, eg. Kalmit FRÜHBURGUNDER and SPÄTBURGUNDER, RIES.

Winning, von Pfz ★★★→★★★★ DEIDESHEIM estate with prime vyds there and at FORST. *Ries of great purity*, terroir expression, fermented in new FUDER casks.

Winningen Mos ★★→★★★ Lower MOS town nr Koblenz; powerful dry RIES. GROSSE LAGES: Röttgen, Uhlen. Top: HEYMANN-LÖWENSTEIN, KNEBEL, Kröber, R Richter.

Wirsching, Hans Frank ★★★ Renowned estate in IPHOFEN known for classically structured dry RIES, *Silvaner*. Andrea W extends range with spontaneously fermented Ries Sister Act, kosher SILVANER. Occasionally BA, TBA of great purity.

Wittmann Rhh ★★★ Leading bio estate, crystal-pure, zesty dry RIES GGS Brunnenhäuschen, Kirchspiel and Morstein 05 07' 08 11 12' 15 16 17 18 19'. Top Morstein ALTE REBEN selection La Borne.

Wöhrle Bad ★★★ Organic pioneer (30 yrs), son Markus a PINOT expert, top GGS.

Wolf JL Pfz ★★→★★★ WACHENHEIM estate, leased by Ernst LOOSEN of BERNKASTEL. Quality sound and consistent rather than dazzling.

Württemberg Dynamic s region nr Stuttgart, many young growers eager to experiment. Best usually LEMBERGER, SPÄTBURGUNDER. Only 30% white varieties. RIES needs altitude vyds.

Würzburg Frank ★★→★★★★ Great baroque city on the Main, famous for its best vyd Stein (STEINWEIN). In 2020, best Stein parcels acknowledged as gU (*see* box, below) under name of Stein-Berg.

Zellertal Pfz ★★→★★★★ Area in N PFZ, high, cool, recent gold-rush: BATTENFELD-SPANIER, KP KELLER, PHILIPP KUHN have bought in Zellertal's best vyd Schwarzer Herrgott or neighbouring RHH plot Zellerweg am Schwarzen Herrgott. Try local estates: Bremer, Full, Janson Bernhard, Schwedhelm, Wick.

Zeltingen M-M ★★→★★★ Top MOS village overshadowed by neighbour WEHLEN despite similiar growing conditions. GROSSE LAGE vyd: SONNENUHR. Top growers: JJ PRÜM, MOLITOR, SELBACH-OSTER.

Ziereisen Bad ★★→★★★ Outstanding estate in MARKGRÄFLERLAND, advocating LANDWEIN, GUTEDEL, Pinots and SYRAH. Best are SPÄTBURGUNDERS from small plots: Rhini, Schulen, Talrain. Jaspis = old-vine selections. Top GUTEDEL 10⁴ (priced at €125) is not just a provocation – it shows what great terroir expression the variety is capable of.

Zilliken, Forstmeister Geltz Mos ★★★→★★★★ 01 03 04 05 07 08 09 10 11 12 15 16 17 18 19 20 SAAR family estate taken over by young Dorothee Z: *Ries from Saarburg Rausch* and OCKFEN Bockstein are racy, savoury, delicate as ever, incl superb long-lasting AUSLESE, TBA (18'); v.gd SEKT (and Ferdinand's gin).

Zimmerling, Klaus Sachs ★★★ Small, perfectionist estate, one of 1st to est after Berlin Wall came down. Best vyd is Königlicher Weinberg (King's v'yd) at Pillnitz nr Dresden. Great 18s (esp RIES, WEISSBURGUNDER).

EU terminology

Germany's part in new EU classification involves, firstly, abolishing the term Tafelwein in favour of plain Wein – this is, up to now, the only visible change on labels. LANDWEIN is still called Landwein, even if its bureaucratic name would be **geschützte geographische Angabe (ggA)**, or protected geographical indication. Brussels generally allows continued use of est designations. **Geschützte Ursprungsbezeichnung (gU)**, or protected designation of origin, should technically be replacing QUALITÄTSWEIN and QUALITÄTSWEIN MIT PRÄDIKAT but is, up to now, mainly in place for large geographical units such as AHR, BADEN, FRANKEN, etc. As it's hard and time-consuming (4–6 yrs) to get recognition for a village- or vyd-specific gU, only six such gUs were in place by end of 2021: Bürgstadter Berg (*see* WEINGUT FÜRST), WINNINGEN Uhlen (parcel-specific Blaufuesser Lay, Laubach, Roth Lay, *see* HEYMANN-LÖWENSTEIN), WÜRZBURGER Stein-Berg, and Monzinger Niederberg (NA). Another gU, IPHOFEN Echter-Berg (*see* WIRSCHING), is in the application process. The existing predicates – SPÄTLESE, AUSLESE and so on (*see* box, p.165) – stay in place; the rule for these styles hasn't changed and isn't going to.

Luxembourg

S mall as the Grand Duchy might be, it is a place of superlatives: Luxembourgers drink a record-breaking 61.3 litres/year, and growers here are the highest-earning in the EU: €92,075 p.a. on average. Vineyards lie upstream of the more famous parts of the Moselle, the soil is limestone, with more in common with Chablis than Piesport. Only 11% is Riesling. Müller-Thurgau (aka Rivaner) prevails. Auxerrois and Crémant fizz are specialities. Most whites have some sweetness sometimes too much, given that climate change makes high acidity levels a thing of the past. Labels don't differentiate between dry and off-dry. A common term (but of little significance) is "Premier Grand Cru". More reliable are groups of ambitious winemakers, eg. Domaine et Tradition.

Alice Hartmann ★★★→★★★★ Top dog in Lux's best RIES vyd, Koeppchen. Also in Burgundy (St-Aubin), Mosel/Saar (Trittenheim, and a plot in Scharzhofberg).

Aly Duhr ★★→★★★ Refined Ahn Nussbaum RIES, Terroir-driven Monsalvat 20' (80% CHARD, 20% AUXERROIS).

Bernard-Massard ★→★★★ Big producer, esp Crémant. Top: Ch de Schengen/Thill's and Clos des Rochers. Sekt in Germany too.

Caves Berna ★★→★★★ Outstanding PINOT N Göllebour.

Gales ★★→★★★ Reliable. Best: Crémant, Domaine et Tradition labels.

Kox, R&L ★★→★★★ Foot-trodden old-vine ELBLING (Rhäifrensch), Crémant with no added sulphur, PINOT BL Orange; gd RIES, classical AUXERROIS.

Pauqué, Ch ★★★→★★★★ Passionate Abi Duhr bridges gap between Burgundy and Germany: superb RIES (dr/sw), barrel-fermented CHARD Clos de la Falaise. Part of vyd (1.2 ha) now farmed organically.

Ruppert, Henri ★★★ RIES TBA-style (*see* p.165), dense PINOT N Ma Tâche (yes, really).

Sunnen-Hoffmann ★★★ Leading family estate, organic. RIES Wintrange Felsbierg VV Domaine et Tradition from vyd planted 1943, Schengen Fels AUXERROIS, Kolteschbierg CHARD *fût de chêne*. Stunning Hommelsbierg PINOT N 20'.

Other good estates: Bentz, Cep d'Or, Desom, Häremillen, Kohll-Leuk, Krier Frères, Mathes, Mathis Bastian, Schmit-Fohl, Schumacher-Lethal. Doms Vinsmoselle is a union of co-ops.

Belgium

A n up-and-coming cool-climate wine country, with ten PDO and PGI denominations. Fresh, food-friendly wines, mainly sparkling or white, with a growing number of notable reds. Most are sold and consumed locally, with wine tourism, tastings and food-pairing events increasingly popular. Wine quality and number of producers continue to rise, as does area under vine, now exceeding 750 ha. Main classic grapes are Auxerrois, Chardonnay, Pinots Blanc/Gris/Noir, alongside disease-resistant varieties Johanniter, Regent and Solaris. Good results with Acolon, Riesling, Souvignier Gris, even Albariño, Grüner Veltliner, Tempranillo.

Try: Aldeneyck, Bon Baron, Chant d'Eole, Chapitre, Clos d'Opleeuw, Crutzberg, des Marnières, d'Hellekapelle, Driesse, Entre-Deux-Monts, Genoels-Elderen, Gloire de Duras, Haksberg, Kitsberg, Neuve-Eglise, Pietershof, Schorpion, Stuyvenberg, Ten Gaerde, Vandersteene, Vandeurzen, Vignoble des Agaises, Vin de Liège, Waes, Wijnfaktorij.

Spain

Abbreviations used in the text:

SPAIN

Alel	Alella	**Pri**	Priorat
Alic	Alicante	**P Vas**	País Vasco
Ara	Aragón	**R Bai**	Rías Baixas
Bier	Bierzo	**Rib del D**	Ribera del Duero
Bul	Bullas	**Rio**	Rioja
Cád	Cádiz	**R Ala**	Rioja Alavesa
Can	Canary Islands	**R Alt**	Rioja Alta
C-La M	Castilla-	**R Or**	Rioja Oriental
	La Mancha	**Rue**	Rueda
C y L	Castilla y León	**Som**	Somontano
Cat	Catalonia	**U-R**	Utiel-Requena
C de Bar	Conca de Barberá	**V'cia**	Valencia
Cos del S	Costers del Segre		
Emp	Empordà		
Ext	Extremadura		
Gal	Galicia		
Jum	Jumilla		
La M	La Mancha	**PORTUGAL**	
Mad	Madeira	**Alen**	Alentejo
Mál	Málaga	**Alg**	Algarve
Mall	Mallorca	**Bair**	Bairrada
Man	Manchuela	**Bei Int**	Beira Interior
Mén	Méntrida	**Dou**	Douro
Mont-M	Montilla-Moriles	**Lis**	Lisboa
Mont	Montsant	**Min**	Minho
Mur	Murcia	**Set**	Setúbal
Nav	Navarra	**Tej**	Tejo
Pen	Penedès	**Vin**	Vinho Verde

Spain is the red-wine producer par excellence, isn't it? Well, no, not entirely. While red dominates, across this diverse country new-wave whites are springing up, nearly all made from local varieties. Not just Garnacha Blanca, and Viura/Macabeo, but also Albillo Real, Malvasía Castellana (also known as Doña Blanca), Malvasía de Sitges, Maturana Blanca. One of the key themes in Rioja is the leap in quality of the whites, as producers seek cooler, higher vineyards and manage the oak regimes with real care. There's plenty on the agenda in head offices in Rioja. There are continuing demands by smaller businesses to break away from the DO, as there are by producers in the Basque region of Rioja Alavesa. Nor has the introduction of a single-vineyards category, *Viñedos Singulares*, been easy. It's a trend across Spain, but the Rioja version is clearly a work in progress, with some loud critics of the definition. In Sherry country, meanwhile, there's been a significant step ahead to recognize the changing environment. Long-forgotten grape varieties are being reintroduced. The names of the *pagos*, or vineyards, can now be printed on labels. Importantly, fortification (typically to eg. 15% for Fino) will not be mandatory; and the term "En Rama", hitherto so vague, is to be regulated.

Spain & Portugal

Recent Rioja vintages

2021 Complex weather yr, but deep-coloured structured reds, gd ageing potential.
2020 Difficult yr. Fresh wines, moderate alc.
2019 Lower yields, overall fine quality, officially *excelente*. Could equal 2001.
2018 Gd yr with generous yields, lower alc, fresh wines.
2017 Dramatic frost. What was left is v.gd.
2016 Largest harvest since 05, well-balanced wines, plenty to like.
2015 Top wines are as gd as 10, rich and full of character. Keep the best.
2014 After two small vintages, return to quality, quantity. Keep or drink.
2013 Cool yr, with rain, later harvest, uneven, with some gd wines. Drink now.
2012 One of lowest yields for two decades, giving fine concentration. Drink up.

Aalto Rib del D ★★★→★★★★ Among RIB DEL D royalty. Two wines: Aalto and
flagship PS (from 200 small plots). Co-founder MARIANO GARCÍA, ex-VEGA SICILIA,
builds wines for cellaring. Best with 10 yrs+. Now co-owns with Masaveu, owners
of Enate (SOM), Fillaboa (R BAI), Murua (RIO); v.gd Asturias cider (Valverán).

Abadía de Poblet C de Bar ★★ Exciting project within Cistercian monastery, burial
place of ARA kings. SCALA DEI winemaker and team work with local varieties, esp
(r) Garrut, Trepat, (w) MACABEO, Parellada. Part of CODORNÍU RAVENTÓS group.

Abadía Retuerta C y L ★★★ Height of luxury hotel – and winery. Just outside RIB DEL D; v.gd white DYA Le Domaine. Serious single-vyd reds, eg. Pago Garduña SYRAH, Pago Valdebellón CAB SAUV, PV PETIT VERDOT. Novartis-owned.

Algueira Rib Sac ★★→★★★ Exceptional producer in RIBEIRA SACRA, expert in its extreme viticulture. Fine selection of elegant wines from local varieties. Outstanding is Merenzao (aka Jura's Trousseau), almost burgundian in style.

Alicante ★→★★★ Take a closer look. It's the spiritual home of MONASTRELL: spicy reds and rare traditional fortified **Fondillón**. But also dry wines from formerly unloved MOSCATEL de Alejandría coming to fore; and local (r) Giró. Top: ARTADI, TELMO RODRÍGUEZ. Also Enrique Mendoza, Finca Collado, Les Freses, Murviedro.

Allende, Finca R Alt ★★→★★★★ Top (in all senses) RIO BODEGA at BRIONES in merchant's house with tower looking over town to vyds, run by irrepressible Miguel Ángel de Gregorio. Outstanding white Rio. New rosé.

Almacenista Man, Sherry Sherry stockholding cellar; ages and provides wines for BODEGAS to increase or refresh stocks. Important in MANZANILLA production. Can be terrific. Few left; many have changed direction and now sell direct to consumers, eg. EL MAESTRO SIERRA, GUTIÉRREZ COLOSÍA.

Alonso, Bodegas Man ★★★→★★★★ Part of revival of SANLÚCAR DE BARRAMEDA BODEGAS, following yrs of decline and bankruptcies. Asencio brothers own Dominio de Urogallo (Asturias). Bought famed cellar of Pedro Romero, incl v. fine SOLERAS of Gaspar Florido. Excellent if super-priced four-bottle collection. More accessibly priced is Velo Flor, 9–10-yr-old MANZANILLA.

Sherry producers making vermouth: Barbadillo, González Byass, Lustau.

Alonso del Yerro Rib del D ★★→★★★ Stéphane Derenoncourt (*see* France) entices elegance from extreme continental climate of RIB DEL D. Transformation since 2016, altogether more delicate. Family business, estate wines. Top: María.

Alta Alella Cat ★★→★★★ Family business with toes in Med and just up coast from Barcelona. Excellent CAVAS, sweet red Dolç Mataró from MONASTRELL. Organic. Celler de les Aus is brand for min-intervention, no-added-sulphur wines.

Alvear Ext, Mont-M ★★→★★★★ Historic Alvear has superb array of PX wines in MONT-M; gd FINO CB and Capataz, lovely sweet SOLERA 1927, unctuous DULCE Viejo; v. fine vintage wines. Also Palacio Quemado BODEGA in Ext.

Añada Vintage.

Aragón Former mighty medieval kingdom, stretching s from Pyrénées, home to Calatayud, CAMPO DE BORJA, CARIÑENA, Som DOS. Once a land of bulk wine from co-ops, now gaining attention for new generation recuperating old vines incl GARNACHA, MACABEO, Moristel (r).

Arizcuren Rio ★★ Javier A trained as architect and designed wineries in RIO, incl his own tiny BODEGA in downtown Logroño (worth a visit). Specializes in GARNACHAS at altitude from R OR, Mazuelo (CARIÑENA; relatively rare as single variety in Rio) and amphora wine. One to watch.

Arrayán, Bodegas Mén ★→★★ Winemaker Maite Sánchez gets best from portfolio of local varieties from rare Mizancho and Moravia via Albillo Real and GARNACHA through to international ones, working in CEBREROS, GREDOS, MÉNTRIDA.

Artadi Ala, Alic, Nav, P Vas ★★→★★★★ Intensely focused Juan Carlos López de Lacalle, now working with son Carlos, left RIO DO end 2015, believing it failed to defend quality; wines now carry Àlava origin. Focus on single vyds: luxuriant La Poza de Ballesteros; dark, stony El Carretil; outstanding El Pisón. Also in ALIC (r El Sequé), NAV (r Artazuri, ROSADO DYA). Izar-Leku TXAKOLÍ (w) from Getaria.

Artuke Rio ★★ Brothers Arturo and Kike de Miguel have transformed family wines. Gloriously elegant, subtle use of large oak. Two top single vyds: El Escolladero, on limestone; La Condenada, on iron-rich sandstone.

Astobiza Ala ★★ Young winery in smallest of TXAKOLÍ DOS, v. fine, advised by Ana Martín (*see* CASTILLO DE CUZCURRITA). Mineral Malkoa; also makes gin.

Atlantic wines Gal, P Vas, Rio Unofficial collective term for bright, often unoaked whites with firm acidity. Increasingly used to describe crisp, delicate reds, esp in R BAI, or TXAKOLÍS. Also used to describe cool-climatic influences, eg. on inland GAL DOS, and specific vintages in R Ala, R Alt.

Ausàs, Bodegas Rib del D ★★ Interpretación is wine of Xavier A, former winemaker at VEGA SICILIA; sources grapes from other growers so he can "interpret" Ribera.

Barbadillo Cád, Man ★→★★★★ Wines from supermarket to superb. Astonishing portfolio of v. fine old Sherries gradually appearing at fine wine auctions. Sherry guru Armando Guerra advises on adventurous new releases, incl new-wave unfortified PALOMINO. Pioneer of MANZANILLA EN RAMA. Top-of-range Reliquía wines unbeatable, esp AMONTILLADO, PALO CORTADO. Cheerful Castillo de San Diego (w IGP Cádiz) from Palomino Fino, Spain's bestseller, ubiquitous in Andalusia's seafood bars. Also Vega Real (RIB DEL D), BODEGA Pirineos (SOM).

Barrio de la Estación Rio The "station quarter" of Haro, from where trains shipped wine to B'x when latter's vines were destroyed by phylloxera. Now home to seven top wineries: BODEGAS Bilbaínas (CODORNÍU), CVNE, GÓMEZ CRUZADO, LA RIOJA ALTA, LÓPEZ DE HEREDIA, MUGA, Roda.

Belondrade C y L, Rue ★★→★★★★ One of leaders in VERDEJO as it should be but rarely is. Didier B was early (1994) exponent of finesse in RUE, and lees-ageing.

Bentomiz, Bodegas Mál ★★→★★★ Dutch by birth, Spanish by adoption: Clara (winemaker) and André (chef) are welcoming hosts in Axarquía, inland from MÁL. Learn how they make sweet MOSCATEL and MERLOT. Rare dry Romé (rosé).

Bhilar, Bodegas R Ala ★→★★ David Sampedro Gil is a native of El Villar and has taken his wines to next level; focus on viticulture and traditional methods, uses horses in vyd. Bio. Phinca Revilla Sexto Año spends 6 yrs in oak, homage to traditional whites of RIO. Phinca Lali is dense, spicy red from a centenarian vyd.

Bierzo ★→★★★★ Looking for a different flavour in Spain? Find it in aromatic, mid-weight, often crunchy fresh reds from MENCÍA in nw. On slate soils they become perfumed, *Pinot-like*. No shot to international fame with RAÚL PÉREZ and Ricardo Pérez Palacios (no relation). Take care: quality is uneven. Look for DESCENDIENTES DE J PALACIOS, Raúl Pérez, plus Dominio de Tares, Losada, Luna Berberide, Mengoba, Veronica Ortega. Also fine GODELLO (w).

Bilbao, Ramón Rib del D, Rio, Rue, R Bai ★→★★ Major producer making strides in quality, driving experimentation. BODEGAS in RIO, R BAI (Mar de Frades), RUE and RIB DEL D (Cruz de Alba). In R OR delivering fresh GARNACHA at altitude. Newest winery in Rio is Lalomba, paradise of tailor-made concrete tanks, with two single-vyd reds and Provençal-style ROSADO.

Bodega A cellar; a wine shop; a business making, blending and/or shipping wine.

Butt Sherry 600-litre barrel of long-matured American oak used for Sherry. Filled 5/6 full, allows space for FLOR to grow. Trend for wineries – and whisky distillers – to use former butts for maturation for Sherry influence – eg. CVNE Monopole Clásico, BARBADILLO Mirabras.

Callejuela Man, Sherry ★★→★★★★ Blanco brothers have vyds in some of Sherry's most famous PAGOS. Part of movement to make terroir Sherries.

Campo de Borja Ara ★→★★ Self-proclaimed "Empire of GARNACHA". Heritage of old vines, plus young vyds = 1st choice for gd-value Garnacha, now starting to show serious quality: Alto Moncayo, Aragonesas, Borsao.

Campo Viejo Rio ★→★★ Campo Viejo is what 1st introduces world (or much of it) to RIO, as no's biggest brand. In addition to value RES, GRAN RES, has varietal GARNACHA, and adds TEMPRANILLO Blanco to white Rio; v.gd top Res Dominio. Part of Pernod Ricard (also owns much-improved YSIOS winery in Rio).

SPAIN

Canary Islands ★→★★★ Seven main islands, nine DOS. Tenerife has five of them. One for vine-hunters: unusual varieties, old vines, distinct microclimates, volcanic soils, unique pruning methods. Dry white LISTÁN (aka PALOMINO) and Listán Negro, Marmajuelo, Negramoll (TINTA NEGRA), Vijariego. MOSCATELS, MALVASÍAS, esp fortified El Grifo from Lanzarote. Top: Borja Pérez, ENVINATE, SUERTES DEL MARQUÉS. In volcano-hit La Palma, Victoria Pecis Torres. Lanzarote Puro Rofe, El Chupadero. Beware: also plenty of dull wine for tourists.

Cangas ★→★★ Isolated DO in wild Asturias just beginning to be known. Appeal is it has unique varieties to enjoy: fresh Albarín Blanco, firm reds from Albarín Negro, Verdejo Negro and, most promising, Carrasquín. Producers: Dominio de Urogallo (owned by BODEGAS ALONSO), Monasterio de Corias, VidAs.

Cariñena Ara ★→★★ The one DO that is also the name of a grape variety. Formerly co-op country, offers value. Jorge Navascués, winemaker at RIO'S CONTINO makes own wines at Navascués Enologia; and at ARA'S VINO DE PAGO Finca Aylés.

Casa Castillo Jum ★★→★★★★ José María Vicente proves JUM can be tiptop. Family business high up in *altiplano*. Outstanding MONASTRELLS, esp PIE FRANCO (plot escaped fairly recent phylloxera); v. fine Las Gravas single vyd.

Castell d'Encús Cos del S ★★→★★★ Shades of a philosopher-king, Raül Bobet (also of PRI Ferrer-Bobet) is constantly planning, sketching ideas. At 1000m (3281ft), he can make *superbly fresh, original wines*. Ancient meets modern here: grapes fermented in C12 granite *lagares*, while winery is up to date. Acusp PINOT, Ekam, RIES, Thalarn SYRAH have become classics.

Castilla y León ★→★★★ Spain's largest wine region. Its diversity makes for exciting discoveries. Its DOS: Arlanza, Arribes, BIER, Cigales, RUE, Sierra de Salamanca (one to watch, with red Rufete grape), Tierra de León, Tierra del Vino de Zamora, TORO, Valles de Benavente, Valtiendas. Catch-all DO Vino de la Tierra de Castilla y León can be source of v. fine wines eg. Barco del Corneta, Máquina y Tabla. Top: ABADÍA RETUERTA, MARQUÉS DE RISCAL (VERDEJO), Mauro, Ossian, Prieto Pariente. Plenty of unique grapes: Cenicienta, Juan García, Puesta en Cruz.

Castillo de Cuzcurrita R Alt ★★ Lovely walled vyd and castle, v. fine RIO.

Castillo Perelada Emp, Nav, Pri ★→★★★ Glamorous estate and tourist destination. Vivacious CAVAS, esp Gran Claustro and Stars; modern red blends. Rare 12-yr-old, SOLERA-aged Garnatxa de l'EMPORDÀ. Casa Gran del Siurana (PRI) v. fine. Owns CHIVITE group.

Catalonia Cat Vast umbrella DO, covers whole of Cat: seashore, mtn, in-between. Top chefs and top BODEGAS. Actual DO too large to have identity.

Cava ★→★★★ Cava is Spain's traditional-method sparkling wine. 90%+ made in PEN. In reaction to loss of market share to Prosecco, Cava is finally bringing in new regulations to improve quality. These extend min ageing, est organic viticulture, identify specific origin and more. New categories: Cava de Guarda, Cava de Guarda Superior (incl Res, Gran Res, CAVA DE PARAJE CALIFICADO). Too late for some producers who left to form DO CLÀSSIC PENEDÈS, Conca del Riu Anoia and CORPINNAT; each with tighter quality regulations.

Cava de Paraje Calificado Cava Launched 2017 by CAVA CONSEJO REGULADOR as top category of single-vyd Cava with stringent rules. Single estate; low or no dosage; min 36 mths age, most exceed that.

Cebreros C y L ★→★★ Young (2017) DO illustrating dynamic development of GREDOS, with distinct zones identifying themselves. As elsewhere GARNACHA (r) dominates; also Albillo Mayor (w). Look for Daniel Ramos, Rico Nuevo, Ruben Díaz, Soto Manrique, TELMO RODRÍGUEZ.

Celler del Roure V'cia ★→★★ Remarkable BODEGA in S v'CIA. Underground cellar of vast amphorae buried up to necks; v.gd, fresh, elegant wines. Cullerot, Parotet, Safrà from local grape varieties.

César Florido Sherry ★→★★★ Master of MOSCATEL, since 1887. Underrated secret grape of Sherry. Explore gloriously scented, succulent: Dorado, Especial, Pasas.

Chacolí *See* TXAKOLÍ.

Chipiona Sherry Sandy coastal zone, source of MOSCATEL. Best are floral delicacies, far less dense than PX. Under new regulations Chipiona can be a town for ageing, not just for production.

Chivite Nav ★★→★★★ Great name of NAV winemaking. Colección 125, incl top CHARD, one of Spain's greatest. Gd late-harvest MOSCATEL. Now CASTILLO PERELADA providing welcome investment.

Clàssic Penedès Pen Category of DO PEN for traditional-method fizz, stricter rules than CAVA; min 15 mths ageing, organically grown grapes. Members incl Albet i Noya, Colet, LOXAREL, Mas Bertran.

Clos Mogador Pri ★★→★★★ René Jnr's father, René Barbier, was one of PRI's founding quintet and mentor to many. One of 1st wineries to gain a Vi de Finca designation. Lovely Manyetes CARIÑENA.

Codorníu Raventós Cos del S, Pen, Pri, Rio ★→★★★★ Historic art nouveau CAVA winery worth a visit. Single-vyd, single-variety CAVAS DE PARAJE CALIFICADO trio v. fine; plus 456 (blend of three vyds) most expensive Cava ever. Mainstream Cavas, eg. Anna, Cuvée Barcelona, also improving. Elsewhere, Legaris in RIB DEL D has v.gd village wines; Raimat in COS DEL S is Europe's largest organic estate, though still to find its way; BODEGA Bilbaínas in RIO has bestseller VIÑA Pomal, plus Vinos Singulares collection – back on form. Jewel is outstanding PRI SCALA DEI, which it part owns. Latest project ABADÍA DE POBLET.

From 1519, ships leaving Cádiz for the Americas all carried vine cuttings and seeds.

Conca de Barberà Cat Small CAT DO once a feeder of fruit to large enterprises, now some excellent wineries, incl ABADÍA DE POBLET, TORRES.

Consejo Regulador Organization that controls a DO – each DO has its own. Quality as inconsistent as wines they represent: some bureaucratic, others enterprising.

Contino R Ala ★★→★★★★ Estate incl one of RIO's great single vyds. Promising developments under winemaker Jorge Navascués. CVNE-owned.

Corpinnat Cat ★★→★★★★ Group of producers of traditional sparkling. More stringent quality than CAVA. Can Descregut, Can Feixes, GRAMONA, Júlia Bernet, Llopart, Mas Candí, Nadal, Pardas, RECAREDO, Sabaté i Coca, Torelló.

Corrales, Viña Jerez ★★★ PETER SISSECK's BODEGA in JEREZ, a FINO from PAGO Balbaína, 8–9-yr-old EN RAMA. His intention is to release single-vyd Sherries, in time from organic vyds. Viña La Cruz is AMONTILLADO, from Pago Macharnudo.

Costers del Segre ★→★★★ Geographically divided DO combines mountainous CASTELL D'ENCÚS and lower-lying Castell del Remei, Raimat.

Cota 45 Sherry ★★ From thoughtful, ever-interesting SANLÚCAR winemaker Ramiro Ibáñez. Ube brand is PALOMINO from different famous PAGOS, eg. Carrascal, Miraflores. Saline, appley, unfortified but briefly matured in Sherry BUTTS. Reveals strong terroir differences. *See also* DE LA RIVA, WILLY PÉREZ.

Crianza Declaration of wine age in RIO. Must be min 2 yr old; reds min 1 yr in oak barrels, whites and ROSADOS min 6 mths. Often used, sometimes with different requrements in other regions.

Cusiné, Tomás C de Bar, Cos del S ★★→★★★ Winemaker leading innovative group: wines incl Finca collection, Tomás Cusiné blends. In MONT, COS DEL S, C DE BAR.

CVNE R Ala, R Alt ★★→★★★★ One of RIO's great names, based in Haro's BARRIO DE LA ESTACIÓN, owns 545 ha vyds. Pronounced *"coo-nee"*, Compañía Vinícola del Norte de España, founded 1879. Four Rio wineries: CONTINO, CVNE, Imperial, VIÑA Real. Most impressive at top end. Also wineries in RIB DEL D and VALDEORRAS.

Daniel Gomez Jiménez-Landi Mén Leader in new generation of GARNACHA producers, making wines at higher altitude across GREDOS (CEBREROS, MÉN, etc.).

De Alberto Rue ★★★ Exceptional demijohn- and solera-aged oxidative VERDEJO: caramel and walnut, vanilla and raisin.

Delgado Zuleta Man ★→★★ Oldest (1744) SANLÚCAR firm. Flagship is 6/7-yr-old *La Goya* MANZANILLA PASADA; also 10-yr-old Goya XL EN RAMA, 40-yr-old Quo Vadis? AMONTILLADO.

Dinastía Vivanco R Alt ★→★★ Vivanco is known for its wines, but even more so for its *outstanding wine museum* in Briones.

DO / DOP (denominación de origen / protegida) DOP has replaced the former DO category.

Dominio del Águila Rib del D Jorge Monzón proves there is another face to RIB DEL D: crisp, delicate, occasionally playful. His pale *clarete* is refreshing. Pícaro del Águila a vivid red blend.

Dulce Sweet. Can be late-harvest, botrytis, or fortifed. Seek out treasures: ALTA ALELLA, BENTOMIZ, GUTIÉRREZ DE LA VEGA, OCHOA, TELMO RODRÍGUEZ, TORRES. Also EMPORDÀ, MÁL, TXAKOLÍ, YECLA.

El Puerto de Santa María Sherry One of three towns forming "Sherry Triangle". Few BODEGAS remain: GUTIÉRREZ COLOSÍA, OSBORNE (gd wine bar), Terry. Puerto FINOS are less weighty than JEREZ, not as "salty" as SANLÚCAR. Lustau's EN RAMA trio show differences of Sherries aged in the three towns.

Empordà Cat ★→★★ One of number of centres-of creativity in CAT. Best: CASTILLO PERELADA, Celler Martí Fabra, Pere Guardiola, Vinyes dels Aspres. Quirky, young Espelt grows 17 varieties: try GARNACHA/CARIGNAN Sauló. Sumptuous natural sweet wine from Celler Espolla: SOLERA GRAN RES.

Sherry is recuperating almost extinct grapes: Beba, Cañocazo, Perruno and more.

En Rama Sherry as if bottled directly from BUTT; v. low filtration, max freshness, more flavour. Understood to refer to MANZANILLA and FINO, but any Sherry bottled this way is En Rama. Soon to be properly regulated.

Envínate Quartet of winemakers casting original light on lesser-known regions, Almansa (Albhara), RIBEIRA SACRA (Lousas), Tenerife (Táganan).

Equipo Navazos Man, Sherry ★★★→★★★★ Academic Jesús Barquín and Sherry winemaker Eduardo Ojeda pioneered négociant approach to Sherry, bottling individual BUTTS. Makes strong case for ageing MANZANILLA in bottle. Recent releases fascinating expressions of same SANLÚCAR DE BARRAMEDA PAGO Miraflores vyd: outstanding anniversary La Bota No 100, Manzanilla Pasada; elegant No 99 (PALOMINO 19 without FLOR or fortificatiion); distinctive I Think Manzanilla.

Escocés Volante, El Ara, Gal ★→★★★ Scot Norrel Robertson MW was a flying winemaker in Spain, hence the brand. Settled in Calatayud, focus on old-vine GARNACHA grown at altitude, often blending in local varieties. Individual, characterful wines, part of movement transforming ARA.

Espumoso Means "sparkling", but confusing: incl cheap, injected-bubble wine as well as traditional method, like CAVA.

Fernando de Castilla Sherry ★★★→★★★★ Gloriously consistent quality. Seek out Antique Sherries; all qualify as VOS or VORS, but label doesn't say so. Antique FINO is complex, fortified to historically correct 17% abv. OLOROSO and PX Singular v.gd. Plus v. fine brandy, vinegar. Favoured supplier to EQUIPO NAVAZOS.

Flor Sherry Spanish for "flower": refers to the layer of *Saccharomyces* yeasts that typically grow and live on top of FINO/MANZANILLA Sherry in a BUTT 5/6 full. Flor consumes oxygen and other compounds ("biological ageing") and protects wine from oxidation. It grows a thicker layer nearer the sea at EL PUERTO and SANLÚCAR, hence finer character of Sherry there. Trend to market unfortifed

Palomino aged for a short time with flor. Growing interest in creating unfortified wines with flor: Spain, Jura, Argentina, NZ.

Fondillón Alic ★→★★★ Fabled unfortified *rancio* semi-sweet wine from overripe MONASTRELL grapes. Now matured in oak for min 10 yrs; some SOLERAS of great age. Unfairly fallen out of fashion, production shrinking too fast: Brotons (v. fine 64' 70'), GUTIÉRREZ DE LA VEGA, Louis XIV Colección de Toneles Centenarios. MG Wines with Bodegas Monovár reissuing v. old wines.

Freixenet Cava, Pen ★→★★★ Biggest CAVA producer. Best known for black-bottled Cordón Negro. Casa Sala is CAVA DE PARAJE CALIFICADO. Other Cava brands: Castellblanch, Segura Viudas. Plus: Morlanda (PRI), Solar Viejo (RIO), Valdubón (RIB DEL D), Vionta (R BAI). Also Finca Ferrer (Argentina), Gloria Ferrer (US), Katnook (Australia). Owned by sparkling giant Henkell.

Frontonio Ara ★→★★ Making waves seeking out old-vine GARNACHA, GARNACHA BLANCA. Also v.gd MACABEO, from old-vine El Jardin de la Iguales vyd.

Fundador Pedro Domecq Sherry Former Domecq BODEGAS were sliced up through multiple mergers. VORS wines owned by OSBORNE; *Botaina*, *La Ina*, *Rio Viejo*, VIÑA 25 by LUSTAU. World's largest brandy company bought remainder, focus on Fundador brandy. Group also incl Terry Centenario brandy, Harvey's, famed for Bristol Cream and v. fine VORS, and Garvey, known for *San Patricio* FINO.

Galicia Isolated nw corner of Spain. Building reputation as source of many of Spain's best whites (*see* MONTERREI, R BAI, RIBEIRA SACRA, RIBEIRO, VALDEORRAS), and light, crunchy reds. Isolation ensures rare varieties.

García, Mariano & Sons Mariano G is a fixture in N and NW Spain. For many yrs his life was VEGA SICILIA, where he was winemaker until 1998. He co-founded AALTO, and launched Mauro (C Y L). He and sons Eduardo and Alberto also run Garmón (RIB DEL D), San Román (TORO), specializing in vyd selection. Arrived in RIO 2020, making 1st vintage in Baños de Ebro.

Genérico Rio If there's no category shown on the RIO bottle – such as RES – then it's a *genérico*. *Genéricos* need not follow all DO rules on ageing (such as barrel size, min ageing). An unattractive and unhelpful name, but can be v.gd or outstanding. A case of needing to know the producer.

Gómez Cruzado R Alt BODEGA tucked in between MUGA and LA RIOJA ALTA in the BARRIO DE LA ESTACIÓN; v.gd Montes Obarenes (w) blend, Pancrudo GARNACHA.

González Byass Cád, Sherry ★→★★★★ Family business (1845). Cellarmaster Antonio Flores is a debonair, poetic, but expert presence. From the *Tío Pepe* SOLERA Flores extracts fine EN RAMA and *glorious Palmas series* (latter celebrated 1st decade 2020). Consistently polished VIÑA AB AMONTILLADO, Matúsalem OLOROSO, Noë PX. Visit boutique hotel on premises (1st hotel in a working BODEGA in JEREZ). Other wineries: Beronia (RIO), Pazos de Lusco (R BAI), Vilarnau (CAVA), VIÑAS del Vero (Som); Dominio Fournier (RIB DEL D); plus (not so gd, but popular) Croft Original Pale Cream Sherry. Finca Moncloa, close to Jerez, produces still reds; also succulent Tintilla de Rota (sweet red fortified).

Gramona Cat, Pen ★★→★★★★ Cousins make impressively long-aged traditional-method sparkling, esp Enoteca, *III Lustros*, *Celler Batlle*. Drove founding of CORPINNAT. Hive of research, bit bio; sweet incl Icewines, experimental wines, table wines. Next-generation Roc Gramona developing L'Enclòs de Peralba.

Gran Reserva In RIO red Gran Res ages min 60 mths, of which min 2 yrs in 225-litre barrique, min 2 yrs in bottle. Whites and ROSADOS age min 4 yrs, of which 6 mths in barrel. Seek out superb old Rio vintages, often great value.

Gredos, Sierra de C y L ★→★★ A mtn region nw of Madrid has built reputation on GARNACHA. Best: pale, ethereal. DOS gradually appearing: CEBREROS, Madrid, MÉNTRIDA. Producers: 4 Monos, Bernabeleva, Canopy, Comando G, DANIEL GOMEZ JIMÉNEZ-LANDI, Marañones, TELMO RODRÍGUEZ.

Guita, La Man ★→★★★ Classic *Manzanilla* distinctive for sourcing fruit from vyds close to maritime SANLÚCAR. Grupo Estévez-owned (also VALDESPINO).

Gutiérrez Colosía Sherry ★→★★★ Rare remaining riverside BODEGA in EL PUERTO. Family business. Former ALMACENISTA. Excellent old PALO CORTADO.

Gutiérrez de la Vega Alic ★→★★★ Remarkable BODEGA specializing in sweet wine. In ALIC, but no longer in DO, after disagreement over regulations. Expert in MOSCATEL, FONDILLÓN. Daughter Violeta now leading business alongside own project Curii, which she runs with partner Alberto Redrado (focuses on Giró r).

Hidalgo, Emilio Sherry ★★★→★★★★ Outstanding family BODEGA. All wines (except PX) start by spending time under FLOR. Excellent unfiltered 15-yr-old La Panesa FINO, thrilling 50-yr-old AMONTILLADO Tresillo 1874, rare Santa Ana PX 1861.

Hidalgo-La Gitana Man ★★→★★★★ Historic (1792) SANLÚCAR firm. MANZANILLA La Gitana a classic. Finest Manzanilla is single-vyd Pastrana Pasada, verging on AMONTILLADO. Outstanding VORS, incl Napoleon Amontillado, Triana PX, Wellington PALO CORTADO.

Jerez de la Frontera Sherry Capital of Sherry region, between Cádiz and Seville. "Sherry" is corruption of C8 "Sherish", Moorish name of city. Pronounced "hereth". In French, Xérès. Hence DO is Jerez-Xérès-Sherry. MANZANILLA has own DO: Manzanilla-SANLÚCAR DE BARRAMEDA.

Joven Young, unoaked wine.

Juan Gil Family Estates Jum ★→★★★ Family BODEGA; has helped transform reputation of JUM. Gd young MONASTRELLS (eg. 4 Meses); long-lived top Clio, El Nido. Other wineries incl Ateca (Calatayud), Can Blau (MONT), Shaya (RUE).

Jumilla Mur ★→★★★ Arid vyds in mtns n of Mur with heritage of old MONASTRELL vines. Top: CASA CASTILLO, JUAN GIL.

Juvé y Camps Cava, Pen ★★→★★★ Consistently gd CAVA. RES de la Familia is stalwart, La Capella is CAVA DE PARAJE CALIFICADO.

La Mancha C-La M ★→★★ Don Quixote country; Spain's least impressive (except for its size) wine region, s of Madrid. Key source of grapes for distillation to brandy, particularly neutral AIRÉN. Too much bulk wine, yet excellence still possible: MARTÍNEZ BUJANDA's Finca Antigua, PESQUERA's El Vínculo and newbie VERUM.

Lively Wines Bier, Rib del D, Rio Company name reflects Germán Blanco's smiling attitude. Try Casa Aurora (BIER), La Bicicleta Voladora (RIO), Quinta Milú (RIB DEL D). Organic.

López de Heredia R Alt ★★→★★★★ Haro's oldest (1877), a family business in the BARRIO DE LA ESTACIÓN with wines that have become a cult. Take a look at "Txoritoki" tower and Zaha Hadid-designed shop. See how RIO was made (as it still is, here). Cubillo is younger range with GARNACHA; darker Bosconia; delicate, ripe *Tondonia*. Whites have seriously long barrel-and-bottle-age; GRAN RES ROSADO is like no other. No shortcuts here.

Loxarel Pen ★★ In the world of traditional-method sparkling, Josep Mitjans is different. Passionately committed to his terroir and his XAREL·LO (Loxarel is anagram). Range incl skin-contact and amphora wines. Cent Nou 109 Brut Nature is quirky treat: traditional fizz, aged 109 mths, never disgorged. Bio.

Lupier, Doms Nav ★→★★ Put NAV and its GARNACHA back on map, rescuing old vines to create two exceptional wines: bold El Terroir, floral La Dama. Bio.

Lustau Sherry ★★★→★★★★ Benchmark Sherries from JEREZ, SANLÚCAR, EL PUERTO. Originators of ALMACENISTA collection. Only BODEGA to produce EN RAMA from three Sherry towns. Emilín is superb MOSCATEL, VORS PX is outstanding, carrying age and sweetness lightly. Consistently excellent.

Maestro Sierra, El Sherry ★★★ Discover how a JEREZ cellar used to be. Run by Mari Carmen Borrego Plá, following on from her mother, the redoubtable Pila. Fine AMONTILLADO 1830 VORS, FINO, OLOROSO 1/14 VORS.

Málaga ★→★★★ MOSCATEL-lovers should explore hills of Málaga. TELMO RODRÍGUEZ began process reviving ancient glories with subtle, sweet **Molino Real**. Barrel-aged No 3 Old Vines Moscatel from Jorge Ordóñez is gloriously succulent. BENTOMIZ has impressive portfolio. Sierras de Málaga DO for dry table wines.

Mallorca ★→★★★ Uneven quality, some v.gd, some simply prestige projects. Can be high-priced and hard to find off island. Incl 4 Kilos, Ánima Negra, Bàrbara Mesquida, Biniagual, Binigrau, Can Ribas, Miquel Gelabert (with a wide array), Son Bordils, Toni Gelabert, Tramuntana. Reds blend traditional varieties (Callet, Fogoneu, Mantonegro) plus CAB, SYRAH, MERLOT. Whites (esp CHARD) improving fast. DOS: Binissalem, Pla i Llevant.

Manchuela ★→★★ Somehow a forgotten region. Take another look: Juan Antonio PONCE, and now JAVIER REVERT at FINCA SANDOVAL, are making waves. BOBAL v.gd.

Marqués de Cáceres R Alt ★→★★ A stalwart. Introduced French techniques in 70s. Fresh (w/rosé). Modern Gaudium; traditional classic GRAN RES. In RUE and R BAI.

Marqués de Murrieta R Alt ★★★→★★★★ Between them, marquesses of RISCAL and Murrieta launched RIO. At Murrieta, the step change in quality continues with new BODEGA on estate. Two styles, classic and modern: Castillo Ygay GRAN RES is one of Rio's traditional greats. Latest release of Gran Res Blanco is **86**, and Gran Res Tinto **75**. Dalmau is impressive contrast, glossy modern Rio, v. well made. *Capellania* is fresh, taut, complex white, one of Rio's v. best; ROSADO, v. pale, unusual Primer Rosé from MAZUELO; v.gd Pazo de Barrantes ALBARIÑO (R BAI).

Marqués de Riscal C y L, R Ala, Rue ★★→★★★★ Riscal is living history of RIO, able to put on a tasting of every vintage going back to its 1st in 1862. Take your pick of styles: reliable RES, modern Finca Torrea, balanced GRAN RES. Powerful *Barón de Chirel Res*. The Marqués discovered and launched RUE (1972) and makes vibrant DYA SAUV BL, VERDEJO and v.gd Barón de Chirel Verdejo, though prefers to put wines in C Y L not Rue. Eye-popping Frank Gehry hotel attached to Rio BODEGA.

Mas Doix Pri ★★→★★★★ Elegant new BODEGA in Poboleda, Pri, enables Mas Doix to expand, funded by new joint owners Lede Family Winery (California). Doix's treasures are superb CARIÑENA (grape), all blueberry and velvet, astonishingly pure, named after yr vyd was planted: *1902*; old-vine GARNACHA (*1903*). Now makes white too: Murmuri, DYA GARNACHA BLANCA, Salix (DYA white blend).

Mas Martinet Pri ★★→★★★ Sara Pérez has pedigree as daughter of one of original PRI quintet. She stays independent of that history. Vociferous, adventurous, ready to try new approaches: amphorae, demijohns, blends.

Mendoza, Abel R Ala ★★→★★★ For knowledge of RIO villages and varieties Abel and Maite M have few equals. Discover no fewer than five varietal whites. Grano a Grano are only-the-best-berry-selected TEMPRANILLO and GRACIANO.

Méntrida C-La M ★→★★ Former co-op country s of Madrid, now being put on map by ARRAYÁN, Canopy, DANIEL GOMEZ JIMÉNEZ-LANDI with Albillo, GARNACHA.

Spain's new fine wines

Some unexpected flavours and some unexpected sources for wines that are not classic but are undeniably brilliant. Try **Bodegas Lanzaga**, Las Beatas, RIO – superb refinement; **Cota 45**, Miraflores, Vino de España – new-wave from Sherry zone; **De la Riva**, OLOROSO Viejísimo, JEREZ-Xérès-Sherry – forgotten BODEGA revived; **Finca Sandoval**, La Rosa, MAN – new winemaker, new approach; **Frontonio**, El Jardín de Las Iguales, Vino de la Tiera Valdejalón – old-vine MACABEO; **José Luis Ripa**, ROSADO, Rio – rosado gets serious; **L'Enclòs de Peralba**, Vi Fi Blanc, CAT next-generation white; **Suertes del Marqués**, Vidonia, Tenerife – textured, mineral, original white; **Viña Meín - Emilio Rojo**, O Gran Meín Tinto, RIBEIRO – purity of indigenous varieties.

Monasterio, Hacienda Rib del D ★★★ PETER SISSECK co-owns/consults, where he 1st started in RIB DEL D. More accessible in price, palate than his DOMINIO DE PINGUS.

Monterrei Gal ★→★★★ Small DO on Portuguese border, with traces of Roman winemaking. MENCÍA and GODELLO are the grapes here. Best: Quinta da Muradella (José Luis Mateo) and gd-value Candea (Mateo with RAÚL PÉREZ).

Montilla-Moriles ★→★★★ Andalusian DO nr Córdoba. Still in JEREZ's shadow. PX makes FINO-style dry wines, and also sweetest wines. Shop nr top end for superbly rich treats, some with long ageing in SOLERA. Top: ALVEAR, PÉREZ BARQUERO, TORO ALBALÁ. Important source of PX for use in Jerez DO.

Olavidia, soft goats' cheese from Sierra Morena, has won "World's Best Cheese".

Montsant Cat ★→★★★ Tucked in around PRI, the gd-value neighbour. Fine GARNACHA BLANCA (Acústic). Characterful reds: Can Blau, Capçanes, Domènech, Espectacle, Joan d'Anguera, Mas Perinet, Masroig, Venus la Universal.

Muga R Alt ★★→★★★★ Muga keeps on improving. Two styles: classical GRAN RES *Prado Enea*; modern, powerful *Torre Muga*. Rare family business where siblings and cousins work successfully refining and building the business. Also pale ROSADO; elegant white; lively traditional-method sparkling.

Mustiguillo V'cia ★★→★★★ Toni Sarrión led renaissance of unloved BOBAL grape, also Merseguera (w); created VINO DE PAGO Finca El Terrerazo with top Quincha Corral. Also at Hacienda Solana (RIB DEL D). President, GRANDES PAGOS network.

Navarra ★→★★★ Gently reviving, led by GARNACHA and new generation. Was in shadow of RIO. Early focus on international varieties confused its identity. Best: DOMS LUPIER, CHIVITE, Nekeas, OCHOA, Tandem, VIÑA ZORZAL. Sweet MOSCATELS.

Numanthia Toro ★★→★★★★ One of TORO's heavyweights. Founded by Egurens of SIERRA CANTABRIA who sold to LVMH. Exceptional, powerful wines. Top Termanthia comes round with 10 yrs of age. Termes is entry-level.

Ochoa Nav ★→★★ Ochoa *padre* led modern growth of NAV, with focus on CAB, MERLOT. Winemaker daughter Adriana calls her range 8a, a play on her surname, developing different styles. Range incl Mil Gracias GRACIANO; fun, sweet, Asti-like sparkling MdO; classic MOSCATEL.

Osborne Sherry ★★→★★★★ Historic BODEGA in EL PUERTO, treasure trove, incl AOS AMONTILLADO, PDP PALO CORTADO. Owns former Domecq VORS incl 51–1a Amontillado. Fino Quinta and mature Coquinero Fino typical of town. Wineries in RIO (Montecillo), RUE, RIB DEL D.

Pago de Carraovejas Rib del D ★★★ Refined, elegant RIB DEL D, from a winery strikingly situated on slopes below Peñafiel's romantic castle. Business is called Alma Carraovejas, owns VIÑA MEÍN - EMILIO ROJO (RIBEIRO), Marañones (GREDOS), Milsetentayseis (Rib del D), Ossian (top-quality VERDEJO producer in c Y L).

Pago de los Capellanes Rib del D ★★→★★★ A v. finc estate, once belonging to church, as name suggests, founded 1996. All TEMPRANILLO. El Nogal has plenty of yrs ahead; top El Picón reveals best of RIB DEL D.

Pago / Vino de Pago / Grandes Pagos *Pago* is a vyd, usually with est name, ie. Sherry's *pago* Miraflores and *pago* Balbaína. *Vino de Pago* is officially the top category of DOP; actually, not always. Typically *Vinos de Pago* are found in less famous zones. Not to be confused with *Grandes Pagos*, network of mainly family-owned estates. Some are *Vinos de Pago* but not all. Current president of MUSTIGUILLO leading promising improvements, recruitment of new members.

Palacio de Fefiñanes Gal ★★→★★★★ Standard DYA R BAI one of finest ALBARIÑOS. Two superior styles: barrel-fermented 1583 (yr winery was founded, oldest of DO); super-fragrant, lees-aged III. Visit palace/winery at Cambados.

Palacios, Álvaro Bier, Pri, Rio ★★★→★★★★ Almost single-handedly built modern reputation of Spanish wine by his obsession with quality. One of quintet that revived PRI. Has driven recent designations from village to Grand Cru: Gran Vi de Vinya. In RIO, at PALACIOS REMONDO, restoring reputation of R OR and its GARNACHAS. In BIER with nephew Ricardo at DESCENDIENTES DE J PALACIOS.

Palacios, Descendientes de J Bier ★★★→★★★★ MENCÍA at its best. Ricardo Pérez P, Álvaro's nephew, grows old vines on steep slate. Floral Pétalos, Villa de Corullón gd value; Las Lamas and Moncerbal are different soil expressions, one clay, one rocky. Exceptional *La Faraona* (only one barrel), grows on tectonic fault. Bio.

Palacios, Rafael Gal ★★★→★★★★ Impossible to fault with Rafael's wines. In VALDEORRAS, focus on GODELLO across 32 rocky parcels (*sortes*). Lovely Louro do Bolo; As Sortes, a step up; *Sorte O Soro*, surely Spain's best white. Sorte Antiga (old vines), v. delicate orange wine, Sorte Souto (tiny production, late harvest).

Palacios Remondo R Baj ★★→★★★★ ÁLVARO P has put deserved spotlight on R OR and its GARNACHAS. Complex Plácet (w) originally created by brother RAFAEL P. Top: Quiñón de Valmira from slopes of Monte Yerga.

Pariente, José Rue ★★→★★★ VERDEJOS with shining clarity. Cuvée Especial is fermented in concrete eggs; silky late-harvest Apasionado. Daughter Martina has joined, also runs Prieto Pariente with brother Ignacio, works in C Y L, GREDOS.

Pazo Señorans Gal ★★★ Consistently excellent ALBARIÑOS from glorious R BAI estate. Outstanding Selección de Añada, min 30 mths on lees, proof v. best Albariños age beautifully.

Penedès Cat ★→★★★★ Region w of Barcelona, most significant and diverse of CAT. Best: Agustí Torelló Mata, Alemany i Corrio, Can Rafols dels Caus, GRAMONA, Jean León, Parés Baltà, TORRES.

Peninsula Vinicultores C Y L, Rio ★→★★ Impressive young project committed to sustainability, authenticity. Badiona (RIO) focus on cool-climate R Ala; Vinos de Montana (CEBREROS, GREDOS), fresh, altitude wines; also fine TXAKOLÍ.

Pepe Mendoza Casa Agrícola Alic ★→★★ After a long career in family wine business (Enrique Mendoza), Pepe launched his personal project in 2016 focusing on elegant Giró (r), MONASTRELL, MOSCATEL. Blends, varietals, experiments, amphorae.

Pérez, Raúl Bier ★★→★★★★ A star, but avoids celebrity. Renowned for finesse, non-intervention. Provides generous house-room for new winemakers in cellar in BIER. Magnet for visiting (eg. Spanish, Argentine) winemakers. Outstanding Vizcaina Mencías; *El Rapolao* exceptionally pure.

Pérez, Willy Jerez With colleague Ramiro Ibáñez of COTA 45 leading return to old ways in JEREZ, researching and reviving practices, traditions, rare varieties, terroirs – and jointly writing the history. Family purchased former estate with old lagares. Interest in unfortified PALOMINO. Projects incl DE LA RIVA.

Pérez Barquero Mont-M ★→★★★ Leading producer of MONT-M. GD Gran Barquero FINO, AMONTILLADO, OLOROSO; La Cañada PX. Supplier to EQUIPO NAVAZOS.

Pie franco Ungrafted vine, on own roots. Typically on sandy soils where phylloxera could not penetrate. Some 100 yrs+ – many in TORO, some in C Y L, RUE.

Wine fairs to visit: lacatadelbarriodelaestacion.com (Rio); vinoble.org (sw/fortified).

Pingus, Dominio de Rib del D ★★★★ One of RIB DEL D's greats. Tiny bio winery of Pingus (PETER SISSECK's childhood name), made with old-vine TINTO FINO, shows refinement of variety in extreme climate. Flor de Pingus from younger vines; Amelia is single barrel named after his wife. PSI uses grapes from growers, long-term social project to encourage them to stay on land. *See also* VIÑA CORRALES.

Ponce Man ★★ Juan Antonio P has single-mindedly transformed family business into pre-eminent producer in DO. One of people building reputation of BOBAL grape, and has a PIE FRANCO bottling PF. Bio.

Priorat ★★→★★★★ Some of Spain's finest. Named after former monastery tucked under craggy cliffs. Key is *llicorella* soil. Best show remarkable purity, finesse, sense of place. Pri has pioneered classification pyramid from village wines through Vi de Vila to Gran Vi de Vinya. After a period when CAB SAUV, SYRAH were thought best, producers have returned to traditional GARNACHA, CARIÑENA, and have toned down new oak. More elegant than of yore.

Raventós i Blanc Cat ★→★★★ Pepe R led historic family business out of CAVA in 2012. Created Conca del Riu Anoia for high-quality sparklings with strict controls. Wines: De Nit ROSADO, Mas del Serral, ringingly pure Textures de Pedra. Can Sumoi estate natural-wine project (2017), incl XAREL·LO, pét-nats. Bio.

Recaredo Pen ★★→★★★★ Outstanding producer of traditional-method sparkling, small family concern. Few wines, all outstanding. Hand-disgorges all bottles. Tops are characterful, mineral *Turó d'en Mota*, from vines planted 1940, ages brilliantly, and RES Particular. Bio. Member of CORPINNAT. Celler Credo are still wines, low in alc, strikingly flinty, pure; also bio.

Remelluri, La Granja Nuestra Señora R Ala ★★→★★★ TELMO RODRÍGUEZ's family property; v. fine white blend; old-vine TEMPRANILLOS; focus on exceptional old GARNACHA single vyds, ethereal wines. Organic.

Reserva (Res) Has actual meaning in RIO. Reds: aged min 3 yrs, of which min 1 yr in oak of 225 litres and min 6 mths in bottle. Whites and ROSADOS: min 2 yrs age, of which min 6 mths barrel. Many now follow own rules. *See* GENÉRICO.

Revert, Javier ★→★★ From family vyds with local grapes ie. Tortosí and Trepadell and many others. Three distinct wines: Micalet, Simeta, Sensal. Pure expression of Med terroir. Also updating wines at FINCA SANDOVAL.

Rías Baixas Gal ★★→★★★★ Atlantic DO in GAL split in five subzones, mostly DYA. Best: Forjas del Salnés, Gerardo Méndez, Martín Códax, PALACIO DE FEFIÑANES, Pazo de Barrantes (MARQUÉS DE MURRIETA), *Pazo Señorans*, Terras Gauda, ZÁRATE. Land of *minifundia*, tiny landholdings. Until recently Spain's premier DO for whites, now at risk of overproduction. ALBARIÑO is variety here; v. best can age, reaching burgundian elegance. A few reds, strikingly fresh, crisp.

Ribeira Sacra Gal ★★→★★★ Magical DO with vyds running dizzyingly down to River Sil. Some impressive, original fresh, light MENCÍA reds: Adegas Moure, ALGUEIRA, Castro Candaz, Dominio do Bibei, Guímaro, Rectoral de Amandi.

Ribeiro Gal ★→★★★ Historic region, famed in Middle Ages for Tostado (sw). Deserving rediscovery, with textured whites made from GODELLO, LOUREIRO, Treixadura. Some reds, fresh, crunchy. Top: Casal de Armán, Coto de Gomariz, Finca Viñoa, VIÑA MEÍN - EMILIO ROJO.

Ribera del Duero ★→★★★★ Ambitious DO with great appeal in Spain, created 1982. Anything that incl AALTO, HACIENDA MONASTERIO, PINGUS, VEGA SICILIA has to be serious. Domestic demand for oaky concentration. At last, elegance breaking through. Wineries in Soria (to e) provide most delicate wines (Dominio de Atauta, Dominio de Es). Try ALONSO DEL YERRO, PAGO DE CARRAOVEJAS, PAGO DE LOS CAPELLANES. Also: Arzuaga, Bohórquez, Cillar de Silos, Garmón, Hacienda Solano, Tomás Postigo, Valduero. *See* C Y L neighbours ABADÍA RETUERTA, Mauro.

Rioja ★→★★★★ Spain's most famous wine region. Three subregions: R Ala, R Alt and R OR. Two key provinces: La Rioja and Álava, or the Basque Country, with NAV to the e. Growing political differences between them also reflected in desire of some Alavesa producers to separate from DO. Introduction of Viñedo Singular category not been a complete success. New generation of producers takes different approach, diversifying traditional image of Rio.

Rioja Alta, La R Ala, R Alt ★★→★★★★ For lovers of classic RIO, a favourite choice. Standard keeps going up. *Gran Res 904* and GRAN RES 890 are stars. But rest of range from *Ardanza*, down to Arana, Alberdi each carry classic house style; all qualify as Gran Res. Also owns R Ala modern-style Torre de Oña, R BAI Lagar de Cervera, RIB DEL D Àster. Hard to fault.

Rioja Oriental Rio New name for Rioja Baja. Renamed to remove any pejorative sense of "baja" as "low", and to identify fact it is most e or oriental (and largest) subregion of RIO. Was poor relation, rapidly gaining attention for its GARNACHA, led by PALACIOS REMONDO. Also RAMÓN BILBAO's Lalomba.

Riva, De La Jerez ★★→★★★ Project from WILLY PÉREZ and Ramiro Ibáñez of COTA 45 based on old, abandoned De La Riva BODEGA. Exceptionally fine wines.

Roda R Alt, Rib del D ★★→★★★ At far tip of BARRIO DE LA ESTACIÓN. TEMPRANILLO specialist: Roda, Roda I, Cirsión, approachable Sela. Also RIB DEL D BODEGAS La Horra, Corimbo (w), Corimbo I. Polished, intense. Fine contrast to neighbours LÓPEZ DE HEREDIA, MUGA: all three at top of game.

Rosado Rosé. NAV dark rosados were defeated by Provence pinks. Spain has fought back with pale hues, esp SCALA DEI's Pla dels Àngels (PRI), MARQUÉS DE MURRIETA's Primer Rosé (RIO), DOMINIO DEL ÁGUILA Pícaro Clarete (RIB DEL D).

Rueda C y L ★→★★★ Spain's response to SAUV BL: zesty VERDEJO. Mostly DYA. Too much poor quality. Best: **Belondrade**, JOSÉ PARIENTE. Pálido is FLOR-aged Verdejo, with 3 yrs in oak. Exceptional SOLERA-aged BODEGAS DE ALBERTO.

Saca A withdrawal of Sherry from the SOLERA (oldest stage of ageing) for bottling. For EN RAMA wines most common *sacas* are in *primavera* (spring) and *otoño* (autumn), when FLOR is richest, most protective.

Sancha, Juan Carlos Rio ★★ Professor of oenology turned winemaker, understands soils, traditions of RIO. Works with lesser-known varieties (Tempranillo Blanco, Maturana Tinta, Maturana Blanca, Monastel aka MOURVÈDRE) and GARNACHA.

Sánchez Romate Sherry ★★→★★★ Old (1781) BODEGA with wide range, also sourcing and bottling rare BUTTS for négociants and retailers. 8-yr-old *Fino Perdido*, nutty AMONTILLADO NPU, PALO CORTADO Regente, excellent VORS AMONTILLADO and OLOROSO La Sacristía de Romate, unctuous Sacristía PX.

Escanya-Vella: grape indigenous to Pri. Means "Drowning an Old Woman". Ahem.

Sandoval, Finca Man ★→★★★ Founded by Victor de la Serna, wine critic. New investors brought in consultant winemaker JAVIER REVERT, who has reshaped and revitalized offering. Top wine: La Rosa.

Sanlúcar de Barrameda Man, Sherry Sherry-triangle town (with JEREZ, EL PUERTO) on River Guadalquivir. Port where Magellan, Columbus, admiral of Armada set sail. Humidity in low lying cellars encourages FLOR. Sea air said to encourage "saltiness". Wines aged here qualify for DO MANZANILLA-Sanlúcar de Barrameda.

Scala Dei Pri ★★→★★★ Tiny vyds of "stairway to heaven" cling to craggy slopes. Part-owner CODORNÍU. Winemaker Ricard Rofes has returned to old ways, eg. fermenting in stone *lagares*. Focus on local varieties. Single vyds Sant'Antoni and *Mas Deu* show terroir. Also rare GARNACHA BLANCA/CHENIN BL blend.

Sierra Cantabria R Ala, Toro ★★★ Exceptional family business. Elegant, single-vyd, low-intervention wines. Organza (w). Reds, all TEMPRANILLO. At Viñedos de Paganos, superb El Puntido; structured La Nieta. Other properties: Señorío de San Vicente in RIO and Teso la Monja in TORO, where Alabaster is star.

Sisseck, Peter Rib del D, Sherry Dane who attracted world interest to RIB DEL D with DOMINIO DE PINGUS starting to work magic on JEREZ. His purchase with a partner of FINO VIÑA CORRALES plus vyd in PAGO Balbaína, has re-energized the sector. His statement "Sherry is the best white wine in Spain" has worked wonders. Also at Ch Rocheyron in B'x.

Solera Sherry System for blending Sherry, less commonly, Madeira (*see* Portugal), plus specialities such as DE ALBERTO Dorado and res wine of GRAMONA. Consists of topping up progressively more mature BUTTS with younger wines of same sort from previous stage, or *criadera*. With FINOS, MANZANILLAS it maintains vigour of FLOR. For all wines gives consistency, refreshes mature wines.

Suertes del Marqués Can ★→★★ Leader of new-wave wineries in Tenerife, smart communicator of uniqueness of island to outside world. Works with LISTÁNS Blanco and Negro, Tintilla, Vijariego (vibrant village, single vyd). Exceptional vyds, unique local *trenzado* (plaited) vines. Worth a visit.

Telmo Rodríguez, Compañía de Vinos Mál, Rio, Toro ★★→★★★★ Groundbreaking winemaker Rodríguez returned to family base REMELLURI in RIO, but continues with Pablo Eguzkiza: in ALIC (Al-Murvedre), Cigales (Pegaso), MÁL (*Molino Real* MOSCATEL), RUE (Basa), TORO (Dehesa Gago), Valdeorras (DYA Gaba do Xil GODELLO, plus three exceptional r single vyds). Return to Rio and BODEGA Lanzaga has led to work on recuperating old vyds: exceptionally pure La Estrada, Las Beatas, Tabuérniga. Latest release Yjar from Rio's Sierra de Toloño: sell-out.

Terra Alta Cat Top GARNACHA country, with 90% of CAT's GARNACHA BLANCA vyds, and 75% of Spain's. Produces complex, textured wines. Producers: Bárbara Forés, Celler Piñol, Edetària, Lafou.

Terroir al Limít Pri ★★→★★★ Dominik Huber is celebrating two decades in PRI. Now impressively elegant, toning down oak, building refinement. Les Manyes fine GARNACHA, Les Tosses pure, blueberry and slate CARIÑENA.

Tinaja aka amphora. Clay pots of all sizes used in revival of traditional winemaking. While focus may be on Georgian *qvevri*, Spain has long tradition, and it's being revived everywhere, incl ALVEAR, CELLER DEL ROURE, LOXAREL, MAS MARTINET. To be found in artisan properties, but also in prestige wineries such as NUMANTHIA.

Toro ★→★★★★ Small DO on the Duero. Famed for bold reds from Tinta del Toro (phenotype of TEMPRANILLO, similar but not identical). Today beer more restrained, but still firm tannic grip. Glamour from VEGA SICILIA-owned Pintia, LVMH-owned NUMANTHIA, and early investor the Eguren family of SIERRA CANTABRIA, now owners of Teso la Monja. San Román, owned by MARIANO GARCÍA and family, also working with GARNACHA and v.gd zesty MALVASÍA Castellana (w). Growing trend for Garnacha, a lighter alternative to traditional reds. Also: Dominio del Bendito, Fariña, Las Tierras de Javier Rodríguez, Matsu.

Toro Albalá Mont-M ★→★★★★ Why is MONT-M not better known? Toro Albalá is highlight with historical treasure trove of wines from PX, incl AMONTILLADO Viejísimo, sumptuous Don PX Convento Selección 31, v.gd GRAN RES 90.

Torres ★★→★★★★ Torres family never stops. After 150 yrs they might ease off but Miguel Jnr runs business, sister Mireia is technical director and runs Jean León, Miguel Snr is busy on many eco fronts. There's a clear pyramid of quality wines. Start at the top with Torres Antología, an outstanding Catalan terroir collection: B'x blend *Res Real*, top PEN CAM *Mas la Plana*; C DE BAR duo (burgundy-like *Milmanda*, one of Spain's finest CHARDS, tiptop *Grans Muralles* blend of local varieties); single-vyd *Mas de la Rosa* will become a Gran Vi de Vinya (top category of PRI) in due course. Also PAGO del Cielo in RIB DEL D, Camino de Magarín (RUE), Pazo Torre Pezelas (R BAI), La Carbonera (RIO) and traditional-method sparkling Vardon Kennett. Also famous, consistent, gd-value portfolio eg. Viña Sol. Pioneer in Chile. Marimar T a star in Sonoma (US).

Tradición Sherry ★★→★★★★ BODEGA assembled by the great José Ignacio Domecq from exceptional selection of SOLERAS. Based on oldest-known Sherry house (1650). Glorious VOS, VORS Sherries, also a 12-yr-old FINO. Outstanding art collection, and archives of Sherry history. Worth a visit for art and archives alone.

Txakolí / Chacolí P Vas ★→★★ Wines from Basque country DOS in Getaria, Bizkaya and Álava. Used to be just acidic wines where DYA Txakolí is poured into tumblers from a height to add to spritz. Drink these in local bars. However, Bizkaya wines, with less exposed vyds, have depth and need not be DYA. Top: Ameztoi, ASTOBIZA, Doniene Gorrondona, Izar-Leku (from ARTADI), Txomín Etxaníz. Also Gorka Izagirre, with Michelin three-star restaurant Azurmendi.

Sherry styles

Manzanilla v. dry, biologically aged nr sea at SANLÚCAR where FLOR grows thickly and wine grows "salty". Usually 15% abv. Serve cool with almost any food, esp crustaceans, eg. I Think (EQUIPO NAVAZOS), LA GUITA, La Gitana.

Manzanilla Pasada with 8 yrs+, where flor is dying, starting to turn into AMONTILLADO; v. dry, complex, eg. LUSTAU'S ALMACENISTA Cuevas Jurado.

Fino dry, biologically aged in JEREZ or EL PUERTO; weightier than MANZANILLA; min age is 2 yrs (as Manzanilla) but don't drink so young. Trend for mature FINOS aged 8 yrs+, eg. FERNANDO DE CASTILLA Antique. Trend to cellar and age Finos and Manzanillas in bottle.

Amontillado started as Fino, after which layer of protective flor has died. Oxygen gives more complexity. Naturally dry. Eg. LUSTAU Amontillado del Castillo. Many brands are sweetened, indicated by "medium" on label.

Oloroso oxidative; not aged under flor. Naturally ultra-dry, superbly savoury, even fierce. May be sweetened and sold as CREAM. Eg. EMILIO HIDALGO Gobernador (dr), Old East India (sw).

Palo Cortado a cult. Traditionally wine between Amontillado and v. delicate OLOROSO. Difficult to identify with certainty. Always refined, complex. Eg. BARBADILLO Reliquía, Fernando de Castilla Antique.

Cream blend sweetened with grape must, PX and/or MOSCATEL for a commercial medium-sweet style.

En Rama another cult. Bottled from BUTT with v. low filtration and no cold stabilization to reveal full character of wine. Typically Manzanilla or Fino, but any Sherry bottled with low/no filtration is En Rama. More flavoursome, said to be less stable. SACA or withdrawal is typically when flor is most abundant, in spring. Differing interpretations; soon to be defined and regulated.

Pedro Ximénez (PX) raisined sweet, from partly sun-dried PX grapes. Unctuous, decadent, bargain. Sip with ice cream. Eg. Emilio Hidalgo Santa Ana 1861, Lustau VORS.

Moscatel aromatic appeal, around half sugar of PX. Eg. Lustau Emilín, Valdespino Toneles.

VOS / VORS age-dated Sherries, some of treasures of Jerez BODEGAS; v. necessary move to raise perceived value of Sherry. Wines assessed by carbon dating to be 20 yrs old+ are VOS (Very Old Sherry/Vinum Optimum Signatum); 30 yrs old+ are VORS (Very Old Rare Sherry/Vinum Optimum Rare Signatum). Also 12-yr-old, 15-yr-old examples. Applies only to Amontillado, Oloroso, PALO CORTADO, PX. Eg. VOS Hidalgo Jerez Cortado Wellington. Some VORS wines are softened with PX; sadly producers can overdo the PX.

Añada / Vintage Sherry with declared vintage. Runs counter to tradition of vintage-blended SOLERA. Formerly private bottlings now winning public accolades. Eg. WILLIAMS & HUMBERT series, Lustau Sweet Oloroso Añada 1997.

Unfortified Palomino strong trend. Some of uneven quality but will improve with experience. Eg. COTA 45, Forlong, Muchada-Leclapart.

Coming soon are significant changes: recognition of new ageing zones for Sherry. Naming of PAGOS on labels. Recuperating almost forgotten grapes. Fortifying wine will no longer be mandatory. New: "Fino Viejo" min 7 yrs. Changes to sweetness: limit for "dry" goes down to 4g/l from 5; max for "medium sweet" goes up to 50g/l.

Valdeorras Gal ★→★★★★ Warmest, most inland of GAL'S DOS, named after the fact Romans found gold in the valleys. Exceptional GODELLO, more interesting than many ALBARIÑOS. RAFAEL PALACIOS is making Godello his life's work; TELMO RODRÍGUEZ's MENCÍAS particularly fine. Also Godeval, Valdesil.

Valdepeñas C-La M ★→★★ Big DO s of LA MANCHA. Historic favourite for cheap reds.

Valdespino Sherry ★★→★★★★ Winemaker Eduardo Ojeda oversees Inocente FINO from top Macharnudo single vyd, rare oak-fermented Sherry (EN RAMA bottled by EQUIPO NAVAZOS). Plus terrific dry AMONTILLADO Tío Diego; outstanding 80-yr-old *Toneles* MOSCATEL, JEREZ's v. best. Owned by Grupo Estévez (owns LA GUITA).

Valencia V'cia ★→★★ Was known for bulk wine, and still supplies supermarket sweet, fortified MOSCATEL. News is it's on the move, with higher-altitude old vines and min-intervention winemaking: eg. Aranleon, Baldovar 923, CELLER DEL ROURE, El Angosto, JAVIER REVERT, Los Frailes, Rafael Cambra.

VDT (Vino de la Tierra) Table wine usually of superior quality made in a demarcated region without DO. Covers immense geographical possibilities; category incl many prestigious producers, non-DO by choice to be freer of inflexible regulation and use varieties they want. (*See* Super Tuscan, Italy.)

Vega Sicilia Rib del D, Rio, Toro ★★★★ Carries a heavy burden of history and reputation carefully, aware that is above fashion. Winery pristine, vyds manicured. Wines built to last. Valbuena has 5 yrs in oak and bottle, Único has almost 10 yrs in oak of different sizes and in bottle. RES Especial, NV blend of three vintages. Plans afoot to release some older vintages. Neighbouring Alión (intended as a modern take on RIB DEL D) is coming into its own. Pintia (TORO) much transformed, fresher, less weighty, in powerful region. Macán (RIO), joint venture with Benjamin de Rothschild, is going in right direction. Oremus in Tokaj (Hungary), in addition to sweet wine, has an electrically fresh dry FURMINT, Petracs.

For fab Sherry and flamenco, visit Madrid's corraldelamoreria.com

Verum C-La M ★ Elias López started with most unpromising beginnings and is having success. In Tomelloso, heart of Spain's AIRÉN-growing, brandy-distilling industry. Old vines, amphorae – it's an excellent combination. One to watch.

Viña Literally, a vyd.

Viña Meín - Emilio Rojo Gal ★★★ Rojo's single, eponymous wine is Treixadura blend. Superb, thrilling freshness. Star of RIBEIRO. Purchased by RIB DEL D PAGO DE CARROVEJAS group. Rojo remains to advise, assisted by winemaker Laura Montero.

Williams & Humbert Sherry ★→★★★★ Winemaker Paola Medina transformed historic BODEGA famed for classic brands eg. Dry Sack, Winter's Tale AMONTILLADO, As You Like It sweet OLOROSO. Now pioneering specialities such as organic Sherry, Vintage Sherries, incl FINO. Dynamic; exceptional quality.

Yecla Mur Traditional bulk-wine country, but changing. Drivers are Castaño family with MONASTRELLS. Castaño Dulce a modern classic.

Ysios Rio ★→★★ Winery in every picture book for its undulating roof by Calatrava. At last wines beginning to get as much worthy attention as building itself.

Yuste Man, Sherry ★★→★★★ BODEGA with a growing collection of SOLERAS. Home to MANZANILLAS Aurora, La Kika. Acquired Herederos de Argüeso, bringing with it v.gd San León, dense, salty San León RES and youthful Las Medallas.

Zárate Gal ★★→★★★ No better way to understand R BAI and ALBARIÑO than with wines of owner/winemaker Eulogio Pomares. El Palomar is from centenarian vyd, one of DO's oldest, on own rootstock, aged in *foudre*. Fontecon is unusual rosé. Fascinating set of local single-variety reds.

Zorzal Nav ★→★★★ Family business transformed by entrepreneurial new generation. Young, gd-value wines (GRACIANO). Restores old-vine NAV GARNACHA (Malayeto); projects with old vines; consultant is Jorge Navascués of CONTINO.

Portugal

Portuguese wines have never been better, Tradition is coming back. Once forsaken, indigenous varieties are now part of most producers' portfolios. Amphora winemaking is having a moment, not just in Alentejo (*see* Vinho de Talha). Lighter styles, often blending white and red grapes (*palhete, clarete*) are now enjoyed again. Ancient vines (Vinhas Velhas, now regulated) and field blends (some with 30+ varieties) are making unique wines while helping to face climate challenges. Old regions are new again (Bairrada, Dão, Portalegre). Innovation is happening too: new regions (Azores, Beira Interior, Madeira table wine) and new styles (lighter Douro, serious Vinho Verde). Even Port, known for its centuries-old rules, is approving new categories (50-yr-old Tawny, Very Very Old Tawny). Back to the future.

Recent Port vintages

A vintage is "declared" when a wine is outstanding by shippers' highest standards. In gd but not quite classic yrs (increasingly in top yrs too by single-estate producers), shippers use the names of their estates for single-quinta wines of real character (and value) but with less structure and longevity. Growing number of limited-production, often single-vyd, Vintage Ports. The vintages to drink now are 63 66 70 77 80 83 85 87 92 94 00 03 04 05 07, though v. young Vintage Port can be delicious

2021 Gd quality; gd weather conditions: mild July after fairly wet winter.
2020 Challenging yr: weather, pandemic.
2019 Balance, freshness but less structure. Best: Niepoort, Noval (incl Nacional), Pintas, Vesúvio.
2018 Gd quality, declaration for some, esp Dou Superior. Best: Ferreira, Noval, Sandeman, Taylor's, Vesuvio.
2017 Superlative yr, widely declared; v. hot, dry, compared to historic 1945.
2016 Classic yr, widely declared. Great structure, finesse.
2015 Controversial yr; v. dry, hot. Declared by many (drink: Niepoort, Noval) but not Fladgate, Sogrape or Symington.
2014 Excellent from vyds that ducked September's rain; production low
2013 Single-quinta yr; mid-harvest rain. Stars: Fonseca Guimaraens, Vesuvio.
2012 Single-quinta yr. Stars: Malvedos, Noval. Elegant, drink now.
2011 Classic yr, widely declared. Considered by most on par with iconic 1963. Inky, outstanding concentration, structure. Stars: Dow's, Fonseca, Noval Nacional, Vargellas Vinha Velha. You can even start on them now.
2010 Single-quinta yr; hot, dry. Stars: Senhora da Ribeira, Vesuvio.
Earlier fine vintages: 07 03 00 97 94 92 91 87 83 80 77 70 66 63 45 35 31 27.

Recent table-wine vintages

2021 Quality, quantity; v.gd whites and fresh, elegant reds.
2020 Gd quantity overall; v.gd whites; pick your red producer.
2019 No rain, cool summer; v.gd quality all around. Keep.
2018 Heavy rains; v. low yields. Aromatic whites, concentrated reds.
2017 Fine vintage, 3rd consecutive; v.gd quality all around. Keep for yrs.
2016 Quality v.gd for those who had patience. Keep for yrs.
2015 Fine yr. Aromatic, balanced reds drinking v. well. Keep.
2014 Fresh whites, bright reds (picked before rain). Drink now.

See Portugal map p.174.

Açores / Azores ★→★★★ Mid-Atlantic archipelago of nine volcanic islands with DOCS Pico, Biscoitos and Graciosa. Pico landscape, incl vine-protecting *currai* (pebble walls), is UNESCO World Heritage Site. Dynamic winemakers, volcanic soil, sea-threatened vines make stony, saline whites and some notable *licoroso* (late-harvest/fortified) from indigenous varieties Arinto dos Açores, Terrantez do Pico, VERDELHO. Try ★★ Adega do Vulcão, ★★★ Azores Wine Company, ★★ Magma, ★★ Pico Wines.

AdegaMãe Lis ★→★★★ Ambitious estate run by codfish group Riberalves. Rising star, winemaker Diogo Lopes makes bright, age-worthy Atlantic-influenced whites. Single-variety range Dory (esp ARINTO, Viosinho) gd-value. Top-notch Terroir is mineral, saline, smoky, oak-aged white blend (Viosinho/ALVARINHO/ARINTO). Estate restaurant serves great cod/wine pairings.

Aldeia de Cima Alen ★★★★ QUINTA NOVA's Luisa Amorim's ambitious personal project blends heritage and modernity in fresher ALEN terroir: terrace-planted vines, native grapes, old vines, traditional ageing (amphora, concrete, oak vats). Top-notch, age-worthy range.

Alentejo ★→★★★ Popular, reliable, hot, dry region covering almost a third of Portugal. Known for spicy, fruit-forward, rich reds, esp from ALICANTE BOUSCHET, SYRAH, TOURIGA N, TRINCADEIRA. Subregional diversity allows for other styles: mineral, seaside Costa Vicentina (CORTES DE CIMA, Vicentino); fresh, high-altitude PORTALEGRE (SUSANA ESTEBAN, ★★★ Terrenus); fresh Vidigueira (ROCIM); earthy, ancient clay amphora VINHO DE TALHA (★★ XXVI Talhas, ROCIM). Classics: CARTUXA, ESPORÃO, ★★★ Herdade dos Grous, JOÃO PORTUGAL RAMOS, JOSÉ DE SOUSA, ★★★ MALHADINHA NOVA, MOUCHÃO. Watch: ★★★ ALDEIA DE CIMA, ★★ Fita Preta, ★★ Fonte Souto (SYMINGTON-owned).

Ameal, Quinta do Vin ★★★ Superior organic VIN. Age-worthy, racy, citrus LOUREIRO. Try gd-value Bico Amarelo, single-vyd Solo Único and oaked Escolha. Charming hotel. ESPORÃO-owned.

Andresen Port ★★→★★★★ Portuguese-owned house with excellent TAWNY Ports, esp 20-yr-old. Outstanding *Colheitas* 1900' 1910' (bottled on demand) 68' 80' 91' 00' 03' 05'. Pioneered age-dated WHITE PORTS. Try 10-, 20-, v.gd 40-yr-old.

Aveleda Vin ★→★★ Well-known, largest VIN producer. Growing range of reliable, crisp whites, esp ALVARINHO, LOUREIRO, Parcela and Solos ranges. Top-notch, oaked Alvarinho/Loureiro *Manoel Pedro Guedes*. Owns top estate VALE D. MARIA (DOU), ★★ D'Aguieira (BAI), ★ Villa Alvor (Alg). Visitor centre 30 mins from Porto.

Bacalhôa Vinhos Alen, Lis, Set ★★→★★★★ Principal brand and HQ of billionaire art-lover José Berardo's group. Age-worthy, v.gd QUINTA da Bacalhôa B's blend incl CAB SAUV 1st planted 1974, also used in iconic Palácio da Bacalhôa. Top MOSCATEL

> **Young guns**
>
> Dynamic, adventurous winemakers are going back in time to make wines from native varieties and (often) with min intervention. Best with ★.
> **Alg** Monte da Casteleja. **AZORES** ★ Adega do Vulcão, ★ Azores Wine Company. **ALEN** Argilla Wines, ★ Cabeças de Reguengo, Miguel Louro, ★ XXVI TALHAS. **BAIR** ★ COZs, FILIPA PATO, ★ Giz, ★ Mira do Ó. **Bei Int** ★ Biaia, Casas Altas, Termos. **COLARES** ★ Viúva Gomes. **DÃO** CASA DE MOURAZ, ★ João Tavares de Pina, ★ Textura. **DOU** Bago de Touriga, Carolina, ★ Costa do Pinhão, MUXAGAT, PORMENOR. **LIS** ★ Baías e Enseadas, ★ Espera Wines, ★ Hugo Mendes, Humus, Marinho, Olival da Murta, Serradinha, ★ Vale da Capucha, ★ Várzea da Pedra. **TEJ** Areias Gordas. **Trás-os-Montes** Arribas Wine Company, ★ Casa do Jóa, ★ Menina d'Uva. **VIN** ★ A&D, Aphros. **Multi-region** ★ ANTÓNIO MAÇANITA, ★ Lés-a-Lés, ★ LUÍS SEABRA, ★ NIEPOORT, ★ VINHOS IMPERFEITOS.

DE SETÚBAL barrels, incl rare Roxo; v.gd fruit-forward ALEN Quinta do Carmo reds; popular brands Catarina, Serras de Azeitão (SET), TINTO da Ânfora (ALEN).

Bairrada ★★→★★★★ Atlantic-influenced DOC and Beira Atlântico VR. Famous for roast suckling pig, age-worthy, structured BAGA reds (many from old vines) and citrus Bical whites. Look for Baga Bair for v.gd sparkling. Try ★★★ Bágeiras, ★★ Casa de Saima, CAVES SÃO JOÃO, *Filipa Pato*, *Foz de Arouce*, ★★ Kompassus, LUÍS PATO, ★★ São Domingos, ★★★ *Sidónio de Sousa*, ★★ Vadio. Watch: ★★★ *Mira do Ó*, NIEPOORT'S ★★ QUINTA de Baixo.

Like Pinot N? Try Bastardo from Dou, Jaen from Dão or Rufete from Bei Int.

Barbeito Mad ★★→★★★★ Highly regarded producer; subtle, elegant range. Unique, single-vyd, single-cask FRASQUEIRAS. Outstanding 20-, 30-, 40-yr-old MALVASIAS and 50-yr-old Bastardo. Historic Series honors MAD'S US popularity in C18 and C19. Growing range of v.gd, mineral, salty table wines.

Barca Velha Dou ★★★★ Portugal's iconic and most expensive red; 1st bottled in 1952, decades ahead of DOU's table-wine revolution. Aged several yrs pre-release, launched only in exceptional yrs incl 91' 95' 99 00 04 08' 11'. Can't afford it? Try v.gd *Res Especial*, released in yrs not declared BV, from CASA FERREIRINHA'S best barrels, 89' 94' 97' 01' 07 09. Both last decades. Luckily (and arguably) 89' 94' 97' 01' 09 could have been BV.

Barros Port ★★→★★★ Founded 1913, maintains substantial stocks of aged TAWNY; v.gd COLHEITAS from every decade since 30s; v.gd 20-, 30-, 40-yr-old Tawny; gd VINTAGE PORT. Sogevinus-owned since 2006.

Blandy's Mad ★★→★★★★ Family-owned for seven generations. Charming, historic *Funchal lodges* hold vast library of FRASQUEIRA (BUAL 1920' 57' 66', MALMSEY 88' 77' 81', SERCIAL 68' 75' 80' 88' 08', Terrantez 75' 80', VERDELHO 76 79'); v.gd 20-yr-old Malmsey, Terrantez and COLHEITAS (Bual 96 08, Malmsey 99 04, Sercial 02, Verdelho 00 08). Superb 50-yr-old Malmsey; Expensive, rare MCDXIX blends 11 yrs between 1863 and 2004, celebrates Mad's 600-yr history. New 1977 Listrão from Porto Santo Island. Also RAINWATER, Atlantis table wine.

Boavista, Quinta da Dou ★★★★ Historic, iconic Cima-Corgo estate, now owned by Sogevinus. Top-notch DOU range from old field blends: fine 80-yr-old vines, single-vyd spicy Oratório and piney Ujo; seductive *Res* blends old and new vines

Bual (or Boal) Mad Classic MAN grape: medium-rich (sweet), tangy, smoky wines; less rich than MALVASIA. Perfect with harder cheeses. Old vintages can be stunning. Try with roast duck for a decadent dinner

Burmester Port ★→★★★ SOGEVINUS-owned Port house making elegant, gd-value TAWNY, esp 20-, 40-yr-old, COLHEITAS. Delectable 30-, 40-yr-old WHITE PORTS; gd VINTAGE PORT, DOU wines. Visitor centre in GAIA.

Cálem Port ★→★★★ Sogevinus-owned Port house. Bestseller: entry-level Velhotes. Try COLHEITA 61', 10-, 40-yr-old TAWNY. Visit lodge in GAIA and restaurant.

Canteiro Mad Natural cask-ageing method for finest MAD (now also TINTA NEGRA). Slow warming in humid lodges for greater complexity than ESTUFAGEM.

Carcavelos Lis ★★★ Hidden gem. Unique, mouthwatering, gripping, off-dry fortified. Villa Oeiras breathed life into v. old, tiny seaside DOC.

Cartuxa, Adega da Alen ★★→★★★★ Historic ALEN estate with old cellars. Restaurant, modern art centre in Évora. Flagship, full-bodied Pêra Manca (r) draws connoisseurs; best-buy Cartuxa RES; single-variety Scala Coeli different each yr.

Carvalhais, Quinta dos Dão ★→★★★ SOGRAPE-owned DÃO estate. Consistent, age-worthy range v.gd, esp oak-aged, flinty ENCRUZADO, sumptuous RES (r/w), earthy Alfrocheiro, floral TOURIGA N, dense, top red Único. Unusual, rich Branco Especial blends Encruzado yrs. Popular brands Duque de Viseu, Grão Vasco.

Chocapalha, Quinta de Lis ★★★ Coastal family-run estate making age-worthy, saline wines. Winemaker Sandra Tavares da Silva (WINE & SOUL). *Among Lisboa's best.* Try vibrant, mineral whites (v.gd CHARD, RES, old-vine ARINTO CH); gd-value reds (QUINTA, CASTELÃO); v.gd, strucutred Vinha Mãe, rich TOURIGA N CH.

Chryseia Dou ★★★★ Prestigious Bruno Prats (B'x) and SYMINGTON FAMILY ESTATES DOU partnership. Polished, structured TOURIGA-driven. *Post Scriptum* v.gd-value.

Churchill Dou, Port ★★★ Family-run Port house, est 1981 by John Graham. Dry WHITE PORT, 20-, 30-yr-old, unfiltered LBV, VINTAGE PORT; QUINTA da Gricha DOU reds, esp gd-value Churchill's Estates and sumptuous TOURIGA N, all v.gd. Visit lodge in GAIA and C19 guesthouse in the Douro.

Like Nebbiolo? Aged Baga is deliciously similar to old Barolo or Barbaresco.

Cockburn's Port ★★→★★★ SYMINGTON-owned house, back on form. Drier, fresher style of VINTAGE PORT in 11' 15' 16' 17' 18'. Extraordinary 08' 27' 34 63 67 70'. Consistently gd Special RES aged longer in wood than others; vibrant LBV aged 1 yr less; v.gd single-QUINTA dos Canais. Appealing GAIA visitor centre incl cooperage tour. New range Tails of the Unexpected.

Colares Lis ★★★ Unique, historic coastal DOC (1908). Europe's most westerly vyds. Windswept, ungrafted vines on sandy soil make age-worthy wines: tannic Ramisco reds and crispy, salty MALVASIA whites. Try ★★★ Adega Regional de Colares and Viúva Gomes (incl old vintages), ★★ Casal Santa Maria.

Colheita Mad, Port Thriving category of drink-on-release nutty, oxidative single-yr TAWNY Port. Cask-aged: min 7 yrs, often 50 yrs+, some superb 100 yrs+ bottled on demand. Bottling date printed on label. Best: ANDRESEN, Dalva, GRAHAM'S, KOPKE, NIEPOORT, NOVAL, POÇAS, TAYLOR'S. Also MAD, cask-aged min 5 yrs.

Cortes de Cima Alen ★★★ ALEN SYRAH pioneer undergoing organic revolution. New seaside, fresh, salty, LOUREIRO/ALVARINHO blend. Consistent fruit-forward range, esp (r/w) Cortes de Cima, RES, varietals (ARAGONEZ, PINOT N, Syrah, TRINCADEIRA), VINHO DE TALHA, top red Incógnito; v.gd whites from coastal vyds.

Cossart Gordon Mad ★★★ Oldest MAD shipper (1745) known for distinct, drier style; v.gd single-yr bottlings. MADEIRA-WINE COMPANY-owned.

Crasto, Quinta do Dou, Port ★★★→★★★★ Prestigious family-run estate; striking hilltop location. Jewels in crown: field-blend, single-vyd, lush, complex reds Vinha da Ponte and Vinha Maria Teresa. Honore, dense "super-blend" of both. Also name of exquisite 100-yr-old+ TAWNY. Seductive old-vyd RES gd-value. Superb varietals TINTA RORIZ, TOURIGA N. Attractive Dou Superior range: supple red, innovative acacia-aged white and SYRAH with VIOGNIER dash.

Croft Port ★★→★★★ FLADGATE-owned shipper with visitor centre in glorious vyds at Pinhão. Sweet, fleshy VINTAGE PORT 75 77 82 85 91 94 00 03' 07 09' 11' 16' 17'; v.gd-value *Quinta da Roêda* Vintage Port 07 08' 09 12' 15' 18' 19'. Superlative old-vine Sērikos 17'. Popular: Indulgence, Pink Rosé Port, Triple Crown.

Crusted (Port) Port Port's affordable hidden gem. Seductive, age-worthy, opulent, rare NV Port style. Blend of two or more vintage-quality yrs, aged up to 4 yrs in casks, 3 yrs in bottle. Unfiltered; forms deposit ("crust"), so decant. Best: CHURCHILL, DOW'S, FONSECA, NIEPOORT.

Dão ★★→★★★★ Historic DOC makes elegant, food-friendy, age-worthy reds and flinty ENCRUZADO-based whites. Modern projects CASA DA PASSARELLA, TABOADELLA add prestige. Quality classics incl ★★ Boas Quintas, ★ Cabriz, ★★ CARVALHAIS, ★★ Casa de Santar, ★ Lusovini, ★★ Ribeiro Santo. Try low-intervention ★★ António Madeira, organic ★★ CASA DE MOURAZ, world-class ★★★ Druida, elegant ★★ Terra Chama, textured ★★ Textura, high-end VINHOS IMPERFEITOS. Dão Nobre ("noble") is top designation. Rare, age-worthy, v.gd-value GARRAFEIRAS. Often called Portugal's Burgundy.

DOC (denominação de origem controlada) Quality wine designation: denomination of controlled origin, cf. AOC in France. DOP (P = protegida) is similar newer denomination, not much used yet. *See* VR.

Dona Maria Alen ★★★ Reputed C19 estate, once a gift by the king to his mistress, now revived. Superb, classic, age-worthy ALEN range: top foot-trodden, old-vine ALICANTE BOUSCHET reds (dense *Júlio B. Bastos*, spicy Grande RES); v.gd single-variety range (TOURIGA N, PETIT VERDOT), elegant Amantis; v.gd-value *estate label*.

Douro ★→★★★★ World's 1st demarcated and regulated wine region (1756), high up the eponymous river. Dramatic UNESCO World Heritage Site. Once inaccessible, now welcomes tourism. Famous for Port, now also for quality table wine (Dou DOC). Three subregions (cooler Baixo Corgo, milder Cima Corgo and warmer, fast-expanding Dou Superior). Over 100 native varieties (incl 80 yr+ field blends) in terraces of unforgiving schist. Powerful, structured reds; fine, high-altitude whites. Best: ALVES DE SOUSA, *Barca Velha*, BOAVISTA, Carvalhas, *Casa Ferreirinha*, *Chryseia*, CHURCHILL, CRASTO, *Maria Izabel*, *Muxagat*, *Niepoort*, ★★★ Poeira, QUINTA NOVA, RAMOS PINTO, *Vale D. Maria*, ★★★ Vale Meão, ★★ Vallado, VESUVIO, *Wine & Soul*. Try ★★ Costa Boal, KRANEMANN, ★★ Murças, ★★★ *Luis Seabra*, MÁRCIO LOPES, Nicolau de Almeida, POÇAS, ★★ *Quanta Terra*.

Dow's Port ★★★★ Historic, reputed SYMINGTON-owned house. Long-lived, drier-style VINTAGE PORT. Legendary yrs: 27, 45, 55, 63, 66, 70, 80, 94; recent: 07' 11' 16' 17'. Single-QUINTAS do Bomfim and *Senhora da Ribeira* (v.gd 15 18 19) in non-declared vintage yrs. *Bomfim visitor centre* in Pinhão.

Duorum Dou, Port ★★ →★★★ DOU Superior partnership between icon winemakers: ALEN'S JOÃO PORTUGAL RAMOS and ex-FERREIRA/BARCA VELHA José Maria Soares Franco. Top-notch, aged several yrs pre-release O. Leucura. Consistent, gd-value range: fine RES, fruity, entry-level *Tons*, COLHEITA. Plus v.gd dense VINTAGE PORT from 100-yr-old vines; gd-value LBV.

Esporão Alen ★★→★★★ Dynamic eco-conscious group with landmark estate. High-quality, fruit-focused, modern. Sophisticated GARRAFEIRA-like wood-aged Private Selection: vibrant, creamy SÉM white and rich, dense native-varieties blend red. Fine, rare Torre de Esporão. New VINHO DE TALHA, native varieties range. Also in DOU (QUINTA dos Murças), VIN (AMEAL). Leading olive oil producer.

Espumante Sparkling. Best are age-worthy, made in BAIR since 1890 (try Aliança, Marquês de Marialva, Poço do Lobo, ★★ São Domingos, ★★ São João). Look out for DADA Dão quality designation. Gd-value Távora-Varosa (esp MURGANHEIRA). DOU'S ★★★ Vértice.

Esteban, Susana Alen ★★→★★★ Focus on v-old, high altitude PORTALEGRE vyds. Flagship Procura (r/w); gd-value second label Aventura. Innovative: Sidecar (with other winemakers) and mineral, high-altitude, 80-yr-old vines Foudre.

Estufagem Mad "Stove" process of heating MAD up to 50°C (122°F) for min 3 mths for faster ageing, characteristic scorched-earth tang. Used mostly on entry-level wines.

Falua Tej ★→★★ French-owned estate undergoing revival. Entry-level Conde de Vimioso (RES a step up) gd-value; Falua Res (r/w) range gd. Stony-soil Vinha do Convento (r/w) v.gd. Barão do Hospital auspicious VIN project.

Ready-to-drink White Port & Tonic? Yes you CAN. Try Cockburn's, Offley, Taylor's.

Ferreira Port ★★★→★★★★ Historic, prestigious SOGRAPE-owned Port house. Stand out 11' 16' 18' vintages, v.gd-value single-QUINTA do Porto 17' 19'. Winemaker Luis Sottomayor (BARCA VELHA) reckons *LBV* now as gd as last decade's VINTAGE PORT; v.gd-value spicy, elegant TAWNY incl Dona Antonia RES, 10 , 20 , new limited-edition 30-yr-old.

Ferreirinha, Casa Dou ★★→★★★★ Large, reputed SOGRAPE-owned DOU range. Home to iconic *Barca Velha*, rare *Res Especial*, age-worthy, classy *Quinta da*

Lêda, v.gd-value Vinha Grande. Growing range incl top-notch *Antónia Adelaide Ferreira* (r/w), single-variety Tinta Francisca, Touriga Fêmea and elegant *Castas Escondidas* (from little-known native varieties).

Fladgate Port Important independent family-owned partnership. Owns leading Port houses (CROFT, FONSECA, KROHN, TAYLOR'S) and growing travel empire with hotels, restaurants in Lisbon, Pinhão, Porto, GAIA. Michelin-starred The Yeatman and six-museum World of Wine (WOW) visitor centre highlights.

The Dou has 48 white grape varieties, 64 red.

Fonseca Port ★★★→★★★★ FLADGATE-owned, founded 1815. Top-notch VINTAGE PORT 63' 70' 85' 94' 00' 03' 11' 16' 17'. Superb Fonseca Guimaraens 13' 15' 18' 19'. Single-QUINTA Panascal; v.gd 20-, 40-yr-old TAWNY; gd-value Bin 27.

Fonseca, José Maria da Alen, Set ★→★★★★ A 200-yr-old, 7th-generation producer. Jewel in crown is fortified *Moscatel de Setúbal*, which mines aged stock to great effect, esp great-value *20-yr-old Alambre* and Moscatel Roxo, stunning *Superior* 55' 66 71. Top-notch, dense Hexagon (r/w). BSE, Lancers, João Pires, PERIQUITA popular brands. Owner of historic, amphora-based ALEN *José de Sousa*.

Foz de Arouce Bei At ★★★ Historic, family-run estate, JOÃO PORTUGAL RAMOS's Beira Atlântico outpost. Noteworthy, age-worthy, BAIR-style range: characterful Cercial, TOURIGA N/BAGA blend and superb, dense, old-BAGA-vines *Vinhas Velhas*.

Frasqueira Mad Highly respected, sought-after MAD category. Also called Vintage. Single-yr, single-noble-variety aged min 20 yrs in CANTEIRO, usually much longer. Date of bottling required. Best: BARBEITO, MADEIRA WINE COMPANY.

Gaia, Vila Nova de Dou, Port Historic home of major Port shippers. Best tourist attractions incl cable car, double-deck bridge, boat tours, hotels (The Yeatman, Michelin-starred fine dining), restaurants/bars (Enoteca 17·56, Vinum), classy cellars (CÁLEM, COCKBURN'S, FERREIRA, GRAHAM'S, POÇAS, SANDEMAN, TAYLOR'S), FLADGATE's ambitious new World of Wine museum.

Garrafeira Often a hint for v.gd, age-worthy value. Reds aged for min 30 mths (often much longer), min 1 yr in bottle. Whites need 12 mths, min 6 in bottle.

Global Wines Bair, Dão ★★→★★★ Formerly Dão Sul. DÃO-based, with estates in many other regions. Great-value brands Cabriz (esp RES) and Casa de Santar (esp Res, superb Nobre). Top-notch, age-worthy Vinha do Contador: dense red and creamy, organic white (rare "Grand Jury" quality label voted by industry). Striking architecture, visitor centre at BAIR's QUINTA do Encontro. Other brands: Encostas do Douro (DOU), Grilos, Monte da Cal (ALEN), Quinta de Lourosa (VIN).

Graham's Port ★★★→★★★★ Highly reputed SYMINGTON-owned Port house. Top age-worthy VINTAGE PORT 27' 63' 66' 85' 91' 94' 97 00' 03' 07' 11' 16' 17', now enjoyable when released. Cellar releases: 94' (also DOW'S, WARRE'S). Sumptuous Stone Terraces 11' 15' 16' 17'; gd-value RES RUBY Six Grapes; v.gd-value Single-QUINTA dos Malvedos 12 15 18' 19', attractive 20-, 30-, 40-yr-old TAWNY, LBV. Fine Single-Harvest (COLHEITAS) collection, esp 40' 50' 52' 61' 63' 69' 72' 82 94' 03. Top-notch, stunning Ne Oublie VERY OLD TAWNY, one of three 1882 casks.

Gran Cruz Port ★→★★★ Port's largest brand (Porto Cruz) owned by French group La Martiniquaise, focused on volume and cocktails incl Port-with-ice programme. GAIA museum, popular rooftop terrace bar, Porto hotel. Quality-focused Dalva PORT brand has outstanding TAWNY stocks: v.gd COLHEITAS, 10-, 20-, -40-yrs-old dry white, *stunning golden white* 52' 63' 71' 89'. Pinhão-based QUINTA de Ventozelo range gd, *charming hotel*.

Grous, Herdade de Alen, Dou ★★→★★★ Prestigious estate/hotel owned by Pohl family. Superior range incl RES (rich, oak-aged w; ripe, fine r); new, top-notch Concrete label aged 2 yrs in cement. Plush Moon Harvested and best-barrels 23 Barricas labels; v.gd, old-vines DOU reds in QUINTA de Valbom.

Henriques & Henriques Mad ★★→★★★★ Superior MAD shipper owned by rum giant La Martiniquaise. Thrilling *20-yr-old Malvasia and Terrantez*, great FRASQUEIRAS (some aged in old bourbon barrels): **1997' 1998'**, BUAL **2000'**, VERDELHO **1957**, Terrantez **1954'**, SERCIAL **1971'**. Riveting TINTA NEGRA *50-yr-old*.

Justino's Mad ★→★★★★ Largest MAD shipper, owned by rum giant La Martiniquaise, makes Broadbent label. Stars: Terrantez Old Res (NV, probably around 50 yrs old), *Terrantez* **1978'** (oldest in cask), MALVASIA **1964' 1968' 1988'**.

Kopke Port ★→★★★★ Sogevinus-owned. Oldest Port house, est 1638. Known for outstanding spicy, structured COLHEITAS **35'** onwards, remarkable WHITE PORT range, esp now-rare **35' 40'** and **20-, 30-, 40-**, new 50-yr-old. DOU range (now called São Luiz) incl v.gd Winemaker's Collection, old-vines DOU red.

Kranemann Wine Estates Dou ★★ →★★★ Historic, C12 estate revival by German surgeon. Consultant Diogo Lopes (ADEGAMÃE) makes DOU and PORT; v.gd, high-altitude, QUINTA do Convento RES, *20-yr-old* TAWNY; Gd-value, fresh Hasso (r/w).

Krohn Port ★→★★★ FLADGATE-owned, reputed shipper (1865); v.gd stocks of cask-aged TAWNY, source of TAYLOR'S 50-yr-old Single Harvest range; v.gd, rich COLHEITAS **10-, 20-yr-old**; gd, elegant VINTAGE PORT.

LBV (Late Bottled Vintage) Port Splendid, affordable alternative to VINTAGE PORT. Single-yr, cask-aged 4–6 yrs (twice as long as VP) for early drinking. Try v.gd, age-worthy, unfiltered versions: FERREIRA, NIEPOORT, NOVAL, RAMOS PINTO, SANDEMAN, TAYLOR'S, WARRE'S. Chocolate's best friend or a delicious dessert on its own.

Lisboa ★→★★★ Undervalued seaside region with large contrasts: small/large brands; native/French varieties. Small, quality-focused, dynamic, often organic producers increasing. Great-value, age-worthy reds, try CHOCAPALHA, SYRAH pioneer MONTE D'OIRO. Historic, unique COLARES microclimate and dry, racy Bucelas (ARINTO-based whites, ROMEIRA is highlight). Try ADEGAMÃE (superb w), QUINTA DE SANT'ANA (v.gd Ramisco), ★★ Pancas (age-worthy range), ★★ Pinto (value blends), ★★ Ramilo (mineral w). Young producers refocus region on Arinto, CASTELÃO, Vital: authentic Casal Figueira, mineral Espera, original Hugo Mendes, natural Serradinha, organic Vale da Capucha, Viuva Gomes.

Lopes, Márcio Dou, Vin ★★→★★★ Innovative VIN-based rising-star winemaker with other interests: gd-value ALVARINHO/LOUREIRO Pequenos Rebentos range, DOU (Anel, Proibido Clarete, *clairet*-style, *see* France), single vyds (Vinha do Pombal: Dou Superior – Anel), RIBEIRA SACRA (70-yr-old vine Telegrafo).

Maçanita, António Alen, Dou ★★→★★★ Unstoppable winemaker/consultant, interests in many regions, ALEN (fashionable Sexy, characterful Fita Preta, esp v.gd Palpite, traditional Chão dos Eremitas range), AZORES (started revival of volcanic Pico Island wines incl v.gd ARINTO do Açores, superb *Terrantez do Pico* and old-vine Vinha dos Utras) and DOU (gd-value r/w range with sister Joana).

Like Chard? Try Encruzado (Dão) or Arinto from around Portugal.

Madeira ★→★★★★ Island and DOC. Only eight firms making famous, thrilling fortified. Stars: BARBEITO, MADEIRA WINE COMPANY. Others: BORGES, HENRIQUES & HENRIQUES, JUSTINO'S, ★★ Pereira d'Oliveira, ★ Faria & Filhos, ★ Madeira Vintners. Growing mineral, saline table-wine range: ★★ Atlantis, ★★★ Barbeito, ★ Terras do Avô, ★ TINTA NEGRA-based Ilha.

Madeira, Rui Roboredo Bei Int, Dou ★★→★★★ Dynamic, high-altitude pioneer, DOU Superior winemaker: gd-value wines, popular Castello d'Alba, mineral *Pedra Escrita*. Top-notch, eponymous, old-vines label. Also in Bei Int with v.gd-value, fresh, high-altitude-vyds *Beyra* range incl elegant *Jaen* and Grande RES.

Madeira Wine Company Mad Family-owned company run by Chris Blandy. Owns BLANDY'S, COSSART GORDON, Leacock, Miles. Makes 50%+ bottled MAD exports.

Malhadinha Nova, Herdade da Alen ★★★ Now fully bio, family-run S ALEN estate and luxury *country house*. Age-worthy range incl *estate label*; rich late-released Marias da Malhadinha; mineral MM; piney, dense ALICANTE BOUSCHET *Menino António* and CAB SAUV Pequeno João. New v.gd old-vine (1949) Vale Travessos (r/w); gd single-variety often single-vyd range. Monte da Peceguina range gd-value. High-altitude, bio, trendy PORTALEGRE vyd is new bet.

Malvasia (Malmsey) Mad Sweetest, richest of MAD's noble grape varieties, yet keeps unique sharp tang. Decadent end to a meal.

Maria Izabel, Quinta Dou ★★★ Rising star. Elegant range made with Dirk NIEPOORT, esp QUINTA, old-vine Vinhas da Princesa (r/w), ARINTO-based Ana. Unusually light, aromatic, complex Sublime red, and rare, expensive Bastardo. Experimental GAMAY (Dou's 1st) in partnership with Marcel Lapierre (*see* France).

Mateus Rosé ★ Popular light, fresh, now off-dry, bubbly rosé and wider range.

Mendes, Anselmo Vin ★★★→★★★★ Known as "Mr. Alvarinho". Benchmark, mineral, age-worthy. Try lees-aged *Contacto*, skin-contact *Curtimenta*, flinty Parcela Única, orange Tempo, Muros de Melgaço. And v.gd-value Muros Antigos range.

Monte d'Oiro, Quinta do Lis ★★→★★★ Reputed, organic family estate started with Chapoutier's Hermitage vines. Age-worthy, fine SYRAH/VIOGNIER *Res*, esp new *11' cellar release*; v.gd, creamy but vibrant Viognier *Res*. Ex-Aequo label: top notch; Syrah/TOURIGA N Bento/Chapoutier partnership; gd entry-level range.

Monte de Ravasqueira Alen ★★→★★★ Family-owned estate, v.gd terroir (high amphitheatre, clay-limestone, granite). Premium range v.gd, esp ALICANTE BOUSCHET; gd-value, single-vyd Vinhas das Romãs (r/w).

Moscatel de Setúbal Set ★★→★★★★ One of Portugal's fortified treasures. Exotic, sweet MOSCATEL, esp exquisite *Roxo* and *Superior*. Best: BACALHÔA VINHOS, ★★ Horácio Simões, JOSÉ MARIA DA FONSECA (holds 100-yr-old+ stocks). Value: ★ Piloto, ★ SIVIPA. Pair with crème brûlée or hard, salty cheeses.

Mouchão, Herdade de Alen ★★★ Oldest estate in ALEN. Family-run, ALICANTE BOUSCHET pioneer; foot-trodden, large-barrel-aged estate red (1st bottled 1949), cellar-release COLHEITAS Antigas label and structured *Tonel 3–4*, unique Alen fortified *Ponte* (r). v.gd-value; gd old-vine Dom Rafael (r/w).

Mouraz, Casa de Dão ★★ Bio pioneer with family-owned vyds. Rebuilding after cellar, some vyds lost in wildfires. Vibrant, authentic range; v.gd 80-yr-old vine *Elfa*; gd-value AIR range. Nina is delicious new *clairet* (*see* France).

Murganheira, Caves ★★ A gd ESPUMANTE producer; owns popular Raposeira. Blends and varietal (native, French grapes) fizz: Vintage, Grande RES, Czar rosé. Rare, expensive Esprit de la Maison PINOT (N/BL/M).

Muxagat Dou ★★★ DOU Superior (Mêda) estate with distinct, fresh, elegant wines (consultant LUIS SEABRA). Benchmark, complex, high-altitude Rabigato-based *Xistos Altos*; v.gd varietal reds, unique *Cisne* blends (r/w) grapes.

Nicolau de Almeida Dou ★★★ Fine, boutique project run by historic Nicolau de Almeida family (grandfather created BARCA VELHA). Top-notch seductive, tense Monte Xisto and v.gd elegant Órbita, Oriente.

Niepoort Bair, Dou, Dão ★★★→★★★★ Estate owned by table-wine pioneer, larger-

---

Splash the cash: these are some top reds
ALEN Alyantiju, Grande Rocim, J de JOSÉ DE SOUSA, Júlio B. Bastos, Marias da Malhadinha, MOUCHÃO Tonel 3–4. **BAIR** FOZ DE AROUCE VINHAS VELHAS, Nossa Missão, Quinta do Ribeirinho. DÃO CASA DA PASSARELLA Vindima. DOU Abandonado, Aeternus, Antónia Adelaide Ferreira, BARCA VELHA, Charme, CHRYSEIA, Manoella VV, MARIA IZABEL Sublime, Maria Teresa, Mirabilis, Oratório, Pintas, QUINTA da Leda, RES Especial, Robustos, Turris, VESUVIO, Vinha da Ponte, Vinha do Moinho, Vinha do Rio. LIS Monte d'Oiro Ex-Aequo.

than-life Dirk Niepoort. Fine DOU range, esp *Redoma incl Res (w)*, Batuta, iconic *Charme*, top-notch *Coche* (w), single-vyd, structured Robustus and Turris. Diálogo/Fabelhaft is gd-value globetrotter. Easy-drinking, natural Nat'Cool range. Dirk's vision incl BAIR's QUINTA de Baixo (esp GARRAFEIRA, Poeirinho, VV), DÃO (esp Conciso), VIN. Port highlights: VINTAGE PORT 15' 17' 19', great-value *Crusted*, unique demijohn-aged *Garrafeira* and single-vyd, organic Bioma. TAWNY v.gd, esp bottle-aged COLHEITAS. New visitor centre in historic GAIA cellar.

Match made in heaven: seafood with Alvarinho, Arinto, Bical or Loureiro.

Noval, Quinta do Dou, Port ★★★→★★★★ Historic estate owned by AXA since 1993. Reputed VINTAGE PORT now released every yr. Jewel in crown *Nacional* 00' 01' 03' 04' 11' 16' 17' 19' from 2.5 ha ungrafted vines; v.gd unfiltered LBV. Superb, fresh COLHEITAS, 20-, 40-yr-olds. Growing table-wine range: Cedro (r/w); age-worthy RES; single-variety PETIT VERDOT, TOURIGA N. Now owns ★★ Passadouro.

Offley Port ★★ 1737 house owned by SOGRAPE; gd, fruit-driven VINTAGE PORT, unfiltered LBV. Attracitve, cocktail-ready Clink range.

Passarella, Casa da Dão ★★→★★★★ Historic estate. Star winemaker Paulo Nunes makes benchmark wines, brings prestige to DÃO. Jewel in the crown is late-release *Vindima 09'* (80-yr-old field blend). Superb, classy Villa Oliveira range, esp oaked ENCRUZADO, dense, 80-yr-old-vine Pedras Altas, unique 2ª Edição (blends five Encruzado vintages); v.gd-value boutique Fugitivo.

Pato, Filipa Bair ★★→★★★ Star bio, BAGA winemaker, daughter of LUIS P. Stands for "wines with no makeup". Superb reds: silky, pre-phylloxera, 130-yr-vine Nossa Missão and earthy, 90-yr-vine Nossa Calcario (also v.gd stony w). Seductive, amphora-aged Post-Quercus range.

Pato, Luís Bair ★★→★★★★ Nonconformist star, father of FILIPA P; *age-worthy, single-vyd Baga* (Vinhas Barrio, Barrosa, Pan) and two Pé Franco (ungrafted) wines (bright, fresh sandy-soil *Ribeirinho*; elegant, chalky-clay Valadas). Ready to drink: v.gd-value VINHAS VELHAS (r/w), BAGA Rebel. Age-worthy whites: creamy, rich, single-vyd Vinha Formal and sharp Parcela Cândido.

Pereira d'Oliveira Mad ★★→★★★★ Family-run producer with vast stocks (1.6 million litres) of bottled-on-demand old FRASQUEIRA, many available to taste at travel-back-in-time 1619 cellar door. Best incl stunning C19 vintages (MOSCATEL 1875, SERCIAL 1875, Terrantez 1880) and rare *Bastardo 1927*.

Periquita Grape, aka CASTELÃO. Trademark of JM DA FONSECA's successful brand.

Pico Wines ★★ Largest, oldest AZORES co-op. Volcanic wines rising in quality; gd-value Terras de Lava, Frei Gigante labels; v.gd, salty, varietal VERDELHO, ARINTO, Terrantez do Pico; 100-yr old-vine Gruta das Torres. Delicious 10-yr-old Licoroso.

Poças Dou, Port ★★→★★★ Family-owned, 100-yr-old+ firm with reputed TAWNY, growing table-wine range. Top-notch Símbolo (with Hubert de Boüard of B'x) and oak-aged Branco da Ribeira (w); v.gd RES (r), v.gd-value Vale de Cavalos (r/w). Unusual, compelling Fora da Série range. Old Tawny stocks allow for fabulous 90-yr-old+ 1918, fine 20-, 30-, 40-yr-old, COLHEITAS; v.gd LBV, VINTAGE PORT.

Popa, Quinta do Dou ★★ Young, dynamic, family-run estate. Carlos Raposo (VINHOS IMPERFEITOS) now a consultant. Top-notch field-blend *Vinhas Velhas* released only in special yrs; v.gd structured Homenagem and seductive, single-variety range (TOURIGA N, *Tinta Roriz*, v.gd-value Popa Black r/w), and Unoaked. New compelling amphora-aged and Curtimenta (skin contact) range.

Pormenor Dou ★★★ Boutique, min-intervention project: fine, age-worthy, old-vines (VINHAS VELHAS) DOU range: classy, elegant, mineral high-altitude *Trilho*; firm, austere A de ARINTO; creamy yet vibrant, high-altitude *Pormenor Res*; v.gd value.

Portalegre Alen ★→★★★ Most n subregion of ALEN undergoing a revival. ESPORÃO, SYMINGTON and SOGRAPE acquisition of cooler, high-altitude (often v. old field-blend) vyds bodes well. Revival pioneers incl ★★ Cabeças do Reguengo, ★★ Monte da Cal, RUI REGUINGA, SUSANA ESTEBAN, ★★ Tapada do Chaves.

Quinta Portuguese for "estate". "Herdade" in ALEN. "Single-quinta" denotes single-estate VINTAGE PORTS (often v.gd value) made in non-declared yrs (increasingly made in top yrs too).

Modern Port dilemma: best can last 50 yrs+ but are hedonistic to drink young.

Quinta Nova Dou ★★→★★★★ Hilltop 250-yr-old estate revamped by Amorim family (of cork fame). *Charming hotel.* Superb Vinha Centenária range blends TINTA RORIZ or TOURIGA N with 100-yr-old vines. Seductive 100-yr vines Aeternus; v.gd-value (r/w) Grainha, Pomares. Luisa Amorim is unstoppable: TABOADELLA estate in DÃO and ALEN personal project ALDEIA DE CIMA.

Rainwater Mad Lighter, drier style of MAD. TINTA NEGRA with dash of VERDELHO. Aged min 5 yrs. Enjoy chilled as an apéritif, with food.

Ramos, João Portugal Alen ★→★★★ ALEN pioneer, also in VIN, DOU (DUORUM), Beira Atlântico (FOZ DE AROUCE). Value brands: Marquês de Borba (r, RES r), VINHAS VELHAS (r/w), Vila Santa. Single vyds Jeremias, São Lázaro, Viçosa. Top Estremus.

Ramos Pinto Dou, Port ★★★ Owned by Champagne Roederer. Complex single-quinta TAWNY 10-yr-old and best-in-class vibrant 20-yr-old. New rich Ervamoira (r) from eponymous QUINTA; gd seductive range, esp Duas Quintas RES (r), Res Especial (mainly TOURIGA N); v.gd age-worthy VINTAGE PORT incl single QUINTAS Bom Retiro, Ervamoira.

Real Companhia Velha Dou, Port ★→★★★ Historic (1756) firm run by Silva Reis family renewing Port (incl Delaforce, Royal Oporto) and DOU portfolio with precision viticulture (500 ha+), winemaking: v.gd old-vine Carvalhas (r/w), Cidrô range, Síbio whites, 20-yr-old TAWNY, VINTAGE PORT. Grandjó is best-in-class late-harvest in Portugal; gd-value Aciprestes, Evel. Port museum, wine bar.

Reguinga, Rui Alen, Tej ★★★ Own projects of star consultant winemaker. PORTALEGRE: v.gd old-vines *Terrenus* range, incl 100-yr-old-vine Vinha da Serra (w), elegant amphora-fermented Vinha da Ammaia. TEJO: Rhône-inspired elegant *Tributo*, creamy Vinha da Talisca.

Reserve / Reserva (Res) Port Higher quality than basic, or aged before being sold. Grande Reserva one step up, approved by regional tasting boards. Rules vary between regions, apply pinch of salt. In Port, bottled without age indication (used in RUBY, TAWNY, often gd value).

Rocim Alen ★★→★★★ Dynamic firm in fresher Vidigueira, VINHO DE TALHA revival pioneer. Top Grande Rocim (dense, piney ALICANTE BOUSCHET r; firm, oaky ARINTO w) and dense, graphite, late-release Crónica #328. Clay-pot/*talha* range, v.gd: foot-trodden, polished, piney Clay Aged (r); elegant Amphora (r/w); thrilling VIN ALVARINHO aged in clay pots. Fresh from Amphora (r/w) Nat'Cool (NIEPOORT) collaboration; gd old-vines, single-vyd Olho de Mocho (r/w), bio Alicante Bouschet Indígena. Delicious old-field-blend DOU red Bela Luz.

Romeira, Quinta da ★★ SOGRAPE-owned, historic Bucelas/LIS estate. ARINTO-focused, v.gd-value: crisp Prova Regia, oaked Morgado Sta Catherina.

Rosa, Quinta de la Dou, Port ★★★ Family-run Pinhão estate (with charming guesthouse), with Port and DOU range ever better under winemaker Jorge Moreira (Poeira); v.gd LBV, 30-yr-old TAWNY, VINTAGE PORT. Rich but elegant wines, esp RES (r/w). Generous gd-value Passagem label.

Rozès Dou, Port ★★★ Owned by Vranken-Pommery (*see* France). LBV from warmer DOU Superior; gd Terras do Grifo table wines incl age-worthy RES, Grande Res.

Ruby Bottle-aged Port after time in wood: 2 yrs for VINTAGE PORT, 4–6 yrs LBV, 6 yrs RES. Also name for most simple, young, cheap sweet Port; can be delicious.

Sandeman Port Port ★★→★★★ Historic house, SOGRAPE-owned, famous for caped man (the don) image; v.gd-value 20-, 30-, 40-yr-old TAWNY, unfiltered LBV. New VINTAGE PORT releases brought back polish. Superb Very Old Tawny Cask 33.

Sant'Ana, Quinta de Lis ★★→★★★ Historic, idyllic, family-run estate; mineral, saline wines. MAÇANITA works native/French varieties. Crisp whites, esp ALVARINHO, ARINTO, v.gd. Age-worthy reds, esp Homenagem, RES. Exquisite NV *Ramisco*.

São João, Caves Bair ★★→★★★ Firm famous for vast stock, 100 yrs old, 50-yrs+ cellar releases. Best: BAIR (Frei João, Poço do Lobo), DÃO (Porta dos Cavaleiros); gd ARINTO/CHARD (w), sparkling blends.

Seabra, Luis Dou, Vin ★★★ New-wave, min-intervention, star winemaker making classy, elegant single-vyd DOU and VIN from indigenous varieties. Top *Xisto Cru* range: stony, high-altitude (w), mineral (r). Consultant at Pormenor, MUXAGAT.

Sercial Mad Racy, driest style of MAD. *Supreme apéritif;* perfect with sushi or seared tuna. Also in mainland table wines. *See* Grapes chapter.

Setúbal, Península de ★→★★ Seaside region s of Lisbon with two DOCS covering same area, different styles: Set (MOSCATEL DE SETÚBAL) and Palmela (CASTELÃO-focused table wines). Try ★★ António Saramago, BACALHÔA VINHOS, ★★ Brejinho da Costa, ★★ Cebolal, JOSÉ MARIA DA FONSECA, ★★ Horácio Simões, ★★ Pegos Claros (esp Grande Escolha), ★★ Piloto, ★★ Portocarro. Value-driven VR wines. Popular: Adega de Pegões, Casa Ermelinda Freitas.

Soalheiro, Quinta de Vin ★★→★★★ Reputed ALVARINHO specialist; v.gd age-worthy range, esp fresh Clássico, mineral *Granit*, subtly barrel-fermented old-vine *Primeiras Vinhas*. New cellar and tasting room.

Sobroso, Herdade do Alen ★★→★★★ Large estate (1600 ha) with boutique country house. Top-notch red: velvety, piney Arché (ALICANTE BOUSCHET, CAB SAUV, TOURIGA N) and dense, rich Grande RES (Alicante Bouschet, Cab Sauv). Also gd consistent Cellar Selection, Res, Sobroso.

Sogrape Alen, Dou, Vin ★→★★★★ Huge firm, global interests. Makes MATEUS ROSÉ. PORT brands: FERREIRA, OFFLEY, SANDEMAN. By region: ALEN (gd-value Herdade do Peso), DÃO (boutique CARVALHAIS), DOU (reputed CASA FERREIRINHA, iconic BARCA VELHA), LIS (gd-value QUINTA DA ROMEIRA), VIN (gd-value Azevedo). Growing fine wine range: *Antónia Adelaide Ferreira* (r/w), 100-yr-old+ vine Legado, Série Ímpar (BAIR *Sercialinho*, old-vine PORTALEGRE *Retorto*).

Sousa, Alves de Dou, Port ★★→★★★ Family-run DOU pioneer located in cooler Baixo Corgo. New-generation Tiago Alves de Sousa is talented winemaker, oenology professor; gd age-worthy *Quinta da Gaivosa*, unique late-released RES Pessoal; v.gd old-vine field blends Abandonado, Vinha de Lordelo. Expanding Port range incl elegant 20-yr-old TAWNY, VINTAGE PORT.

Sousa, José de Alen ★★→★★★ Historic JOSÉ MARIA DA FONSECA-owned estate keeping tradition alive with Portugal's largest TALHA collection (114). Makes superb-value, spicy, ripe, *talha* José de Sousa; elegant J and great-value, dense *Mayor*.

Symington Family Estates Dou, Port ★★→★★★★ DOU's biggest landowner, sustainability pioneer. Traditional but innovative, 5th-generation, family-run; owns clutch of top Port houses incl COCKBURN'S, DOW'S, GRAHAM'S, VESUVIO, WARRE. Benchmark *1890 Lodge* visitor centre/restaurant in GAIA. Top-notch table-wine range: polished CHRYSEIA, fruit-driven Vesuvio. Plus gd-value Altano range. Fonte Souto is fresh, high-altitude PORTALEGRE outpost.

Taboadella Dão ★★★ Amorim's ambitious project brings prestige to DÃO.

Splash the cash: here are some top whites
ALEN Aldeia de Cima GARRAFEIRA, ESPORÃO Private Selection, Grande ROCIM, Retorto. DÃO Villa Oliveira. DOU Branco da Ribeira, Coche, Guru, Mirabilis, NIEPOORT Res, No Millésime, Vinhas da Princesa. LIS ADEGAMÃE Terroir. VIN Curtimenta, Manoel Pedro Guedes, Parcela Única.

Winemaking/viticulture by QUINTA NOVA's star team. Promising range: top-level elegant Grande Villae (r/w); seductive varietal range, esp *Alfrocheiro*; gd-value, unoaked entry-level Villae. Striking, cork-coated cellar.

Talha, Vinho de Alen Clay-amphora Roman tradition, undergoing a revival. Now an ALEN DOC. Vinification: some whole-bunch, 5–6 wk maceration. Decades-old amphorae now hunted treasures: historic JOSÉ DE SOUSA owns 114, makes v.gd range. Also try ★★ Casa Relvas, CORTES DE CIMA, ESPORÃO, ★★ Gerações da Talha, ★★ José Piteira, ROCIM, ★★ XXVI Talhas (esp Mestre Daniel).

Tawny Port Wood-aged Port, drink-on-release. Oxidative, nutty, sweet, racy. Lasts 2 mths after opening. Serve chilled. Enjoy with crème brûlée or dry, salty, hard cheese. Age-dated wines, blend old aged lots, go up in complexity, price: 10-, 20-, 30-, 40-yr-old (*20-yr-old* best balance between ages); new 50-yr-old category; single-year COLHEITAS cask-aged up to 100 yrs, can be expensive; luscious *Very Old Tawny*. Best in class: ANDRESEN, Dalva (GRAN CRUZ), GRAHAM'S, KOPKE, KROHN, NIEPOORT, NOVAL, Otima (WARRE'S), POÇAS, RAMOS PINTO, TAYLOR'S, VASQUES DE CARVALHO. Cheap RES (6 yrs in wood) can be delicious.

Taylor's Port ★★→★★★★ Historic Port shipper, FLADGATE's jewel in the crown. Imposing VINTAGE PORTS incl 63' 66' 70' 77' 94' 97', single QUINTAS (Terra Feita, Vargellas) and rare, splendid Vargellas VINHA VELHA from 70-yr-old+ vines. Outstanding TAWNY esp 50-yr-old COLHEITAS 68' 69' 70' 71', VERY OLD TAWNY 1863' Scion and new Kingsman Edition (average 90 yrs old).

Tejo ★→★★★ Riverside region shifting from quantity to quality. Native varieties CASTELÃO (r), FERNÃO PIRES (w) produce gd results, esp with old vines. Solid: ★★ Alorna, ★★ Casa Cadaval, ★★ Lagoalva de Cima, ★★ Lapa. More ambitious: ★★★ Casal Branco, ★★★ Casal das Aires, FALUA, RUI REGUINGA.

Vale D. Maria, Quinta do Dou, Port ★★→★★★ Reputed DOU table-wine pioneer. Best-in-class, rich reds incl *estate label*, single-vyd Vinha do Rio, *Vinha da Francisca* and new top-notch, late-release Vinha do Moinho; gd smoky, oaky but brisk whites incl VVV, Vinha do Martim; gd-value Dou Superior range; gd LBV, RES, VERY OLD TAWNY, VINTAGE PORT. Owned by AVELEDA.

Vale Meão, Quinta do Dou ★★★ Family-run estate, birthplace of BARCA VELHA. Fine, age-worthy range: elegant top red; v.gd single-vyd, varietal Monte Meão range; gd-value Meandro (r/w); gd VINTAGE PORT.

Vallado, Quinta do Dou ★★→★★★ Family-owned Baixo Corgo estate, modern hotel/ winery; v.gd-value DOU Superior organic (QUINTA do Orgal); v.gd RES field blend, Sousão range, varietal TINTA RORIZ, TOURIGA N; gd 10-, 20-, 30-, 40-yr-old TAWNY. Top 80-yr-old-vine Adelaide red. Thrilling, rare, pre-phylloxera VERY OLD TAWNY.

Vasques de Carvalho Dou, Port ★★★ New producer (rare in Port), est 2012 from inherited family cellars, stock, vyds. Top-notch, expensive, stylish *10-, 20-, 30- and 40-yr-old* TAWNY. New delicious 10-, 20-yr-old WHITE PORTS; v.gd VINTAGE; gd DOU range. Charming boutique in GAIA.

Verdelho Mad Classic, most versatile MAD grape: sweeter than SERCIAL, drier than BUAL. Turns curry or fish and chips into a decadent meal; gd apéritif with hard cheese. Now also in MAD table wine.

Where to enjoy a glass in Portugal

The best for ambience and wine lists; top with a ★: **Alg** ★ Epicur, Rolha Wine Bar; **Braga** ★ Delicatum; **Cascais** ★ Terroso; **Estremoz** Howard's Folly; **Évora** Enoteca Cartuxa, Fita Preta; **LIS** Black Sheep, By the Wine, ★ Comida Independente, ★ JNcQuoi, Jobim, ★ Senhor Uva, ★ Vino Vero, ★ Wines by Heart; **Penafiel** Casa da Viúva; **Porto or nr** A Cave Do Bon Vivant, Capela Incomun, ★ Enoteca 17·56, ★ Garage Wines, Portologia, ★ Prova Wine, ★ Vinum, Wine Quay Bar.

Vértice Dou ★★★→★★★★ Fizz producer in DOU, often considered Portugal's best; v.gd-value Gouveio, Millésime. Remarkable, high-altitude, 84-mth-aged PINOT N.

Very Old Tawny Port ★★★★ Deluxe, luscious TAWNY category for wines aged for decades in wood. Time-travelling experience with two stops: Very Old Tawny (wines aged 40–80 yrs) and Very Very Old Tawny (aged min 80 yrs old). Time capsules: 1900 1910 (ANDRESEN) 1918 (POÇAS), 5G (WINE & SOUL), 1888 ABF (VALLADO), António Vieira de Sousa, CNK (KOPKE), Costa Boal, Honore (CRASTO), Ne Oublie (GRAHAM'S), Scion and Kingsman Edition (TAYLOR'S), VV (NIEPOORT).

Portuguese egg tart (pastel de nata)? Try a Mad (sw), Moscatel or Tawny.

Vesuvio, Quinta do Dou, Port ★★★→★★★★ Magnificent, historic riverside QUINTA making age-worthy range: old-vine estate red; v.gd-value second label *Pombal do Vesuvio*; gd-value entry-level Comboio. Port on par with best, still foot-trodden.

Vilacetinho, Casa de Vin ★★ C18 VIN estate on DOU's border; gd-value Avesso-based range, esp Escolha, blends with ALVARINHO; v.gd Superior, *Res.*

Vinhas Velhas Old vines. Frequently found on labels but different meaning by region, now regulated in DOU: min 40-yr-old vyds, min four-variety field blend, low yield. Other regions under review.

Vinhos Imperfeitos Dão, Vin ★★★ High-end, ambitious project by Carlos Raposo (ex-NIEPOORT, QUINTA DO PÔPA consultant). Exquisite, age-worthy whites: oak-, cement-aged 100-yr-old field-blend saline DÃO; complex, mineral VIN (ARINTO/ LOUREIRO/Avesso); D&V is floral, mineral blend of both regions. Most expensive whites in Portugal. WWW is elegant, sharp, affordable DÃO range (r/w).

Vinho Verde ★→★★★ Portugal's biggest, rainiest, most verdant region, n border with Spain. Step up in quality in recent yrs with gd-value, elegant whites. Best by subregion: high-end ALVARINHO from Monção e Melgaço (ANSELMO MENDES, LUIS SEABRA, MÁRCIO LOPES, ★★★ Regueiro, ★★ Santiago, SOALHEIRO, ★★ Vale dos Ares); LOUREIRO from Lima (eg. AMEAL, bio ★★ Aphros), Avesso from Baião (bio ★★ A&D, ★★ Covela), Amarante (Casa de Cello, CASA DE VILACETINHO, ★★ Sem Igual). Large brands (★ Adega de Monção, ★ AVELEDA, ★ Azevedo) often slightly fizzy (Casal Garcia, Gazela, Muralhas), DYA. Minho is VR covering same area.

Vintage Port Port One of world's great classic wines. Made only in v. finest years (known as "classic" or "declared" vintages), aged v. slowly for decades in bottle after a yrs in wood. Increasingly hedonistic to drink young. Increasingly made back-to-back due to precision viticulture/winemaking. New super-selections incl Capela (VESUVIO), Sērikos (CROFT), Stone Terraces (GRAHAM'S), Vinha da Pisca (NIEPOORT). Unfiltered, throws deposit, always decant. Single-QUINTA Vintage Ports made in non-declared yrs. Pair with dreams.

VR / IGP (vinho regional / indicação geográfica protegida) Same status as French IGP. More leeway for experimentation than DOC.

Warre's Port ★★★→★★★★ The 1st and oldest of British Port shippers (1670), now owned by SYMINGTON. Rich, long-aging VINTAGE and unfiltered LBV. Elegant QUINTA da Cavadinha. Superb-value, rich but fresh *20-yr-old Tawny Otima*.

White Port From white grapes. Mostly off-dry, some dry (mentioned). Growing, high quality, niche: age-dated 10-, 20-, 30-, or 40-yr-old (ANDRESEN, KOPKE, Quevedo, Vieira de Sousa), rare COLHEITAS esp *Dalva*, *Kopke*. Lágrima is cheap, v. sweet. Port and tonic is popular, refreshing apéritif.

Wine & Soul Dou, Port ★★★→★★★★ Reputed family-owned estate. Remarkable age-worthy DOU range from old vines. Top notch: oak-aged *Guru* (No Millésime edition blends 3 yrs); spicy, elegant QUINTA da Manoella VINHAS VELHAS and complex, powerful Pintas. Great value: field-blend *Pintas Character*, second label Manoella (r/w), crisp *Vinha do Altar*; gd Pintas VINTAGE PORT. Oustanding 100-yrs+ VERY OLD TAWNY 5G.

Switzerland

Abbreviations used in the text:

Aar	Aargau
Ber	Bern
Gris	Grisons
Luc	Lucerne
Neu	Neuchâtel
Schaff	Schaffhausen
Thur	Thurgau
Tic	Ticino
Val	Valais
Vd	Vaud
Zür	Zürich

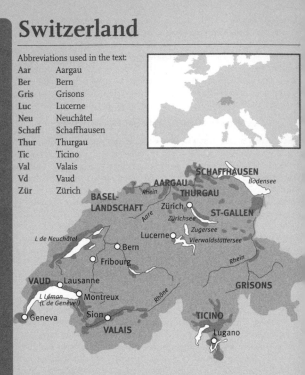

True, Swiss wines are often overlooked. But not by everybody. Take Caroline Frey, co-owner of Château La Lagune in Bordeaux and Paul Jaboulet Aîné in Rhône's Hermitage (*see* France): what does it mean that she is realizing her heart's project in Valais, in a mini-vineyard at Fully? Furthermore, how is it that famous wine locations like the private club 67 Pall Mall in London or the Eleven Madison Park restaurant in NYC have been featuring Swiss wines? And doesn't it say something that Japanese wine connoisseurs are ready, at the Louis Robuchon wine shop, to spend ¥88,000 ($780) on a bottle of Pinot Noir by Martin Donatsch? Maybe it's time to stop praising Swiss wine so loudly...

Recent vintages

2021 Frost, hail, mildew: smallest crop in 20 yrs. Light wines of sound quality.

2020 Early, small, 20–40% less: uneven flowering, drought, v.gd quality.

2019 Rain at harvest time, esp E Switzerland; Vd and Val better.

2018 Powerful, round wines all over the country.

2017 Frost; some cantons only 20% of normal crop; v.gd quality.

2016 Frost in April, rainy summer then sun: mostly mid-weight wines.

Fine vintages: 13 10 (Pinot N) 15 09 05 (all) 99 (Dézaley) 97 (Dézaley) 90 (all).

Aigle Vd ★→★★★ Popular CHASSELAS AOC, but be v. selective; BADOUX, Terroir du Crosex Grillé.

AOC Equivalent of France's appellation contrôlée, 62 AOCs countrywide.

Auvernier, Ch d' Neu ★★→★★★ Important estate (60 ha) with reliable, typical NEU CHASSELAS, plus CHARD, OEIL-DE-PERDRIX, PINOT N (best: single-vyd Les Argiles).

Bachtobel, Schlossgut Thur ★★★ Since 1784 owned by Kesselring family, known for delightful, refined PINOT N labelled 1–4 (higher number = better quality). 2020 No 1 and 2019 No 2 are bargains.

Bad Osterfingen Schaff ★★★ Restaurant and estate in historic spa (est 1472) known for PINOTS BL/N (Badreben, Badreben Abt); Res Privée only in top yrs. Wines made for food, and ageing. Co-producer of ZWAA.

Badoux, Henri Vd ★★ AIGLE les Murailles (classic lizard label) is most popular Swiss brand. Ambitious Lettres de Noblesse series has gd barrel-aged YVORNE.

Baumann, Ruedi Schaff ★★★ Perfectionist family estate at Oberhallau. Beatrice, Ruedi, son Peter best known for delicately fruity PINOT N (eg. Ann Mee, R, ZWAA).

Bern Capital and canton, wine villages on Lake Biel and Lake Thun. Top growers: Andrey, Johanniterkeller, Keller am See, KREBS & STEINER, Schlössli, Schott.

Besse, Gérald Val ★★★ Leading VAL family estate; Gérald and Patricia B, daughter Sarah. Mostly steep terraces up to 600m (1969ft); intense old-vines *Ermitage Les Serpentines* 10' 11' 13' 15 16 17 (MARSANNE on granite soils from 1945).

Bonvin Val ★★→★★★ Old name of VAL, intriguing local grapes: *Nobles Cépages* series (eg. HEIDA, PETITE ARVINE, SYRAH).

Bovard, Louis Vd ★★→★★★★★ Top estate at Cully, LAVAUX. La Médinette 99' 05' 07' 11' 12' 15 16 17 18' 19 20 is textbook DÉZALEY, other AOCs equally reliable. Old vintages available from dom.

Archaeological sites in Val: grape seeds dating back to 790 BC. Should plant them.

Bündner Herrschaft Gris ★★→★★★★ Little Burgundy: four villages nr Chur (FLÄSCH, Jenins, Maienfeld, MALANS), PINOT N with structure, fruit, great capacity to age. Climate balanced between mild s winds and coolness from nearby mtns.

Calamin Vd ★★★ GRAND CRU of LAVAUX, 16 ha of deep calcareous soils on a landslide, CHASSELAS tarter than nearby DÉZALEY. Growers: BOVARD, Dizerens, Duboux.

Chablais Vd ★★→★★★★ Wine region at upper end of Lake Geneva around AIGLE and YVORNE. Name is from Latin *caput lacis*, head of the lake.

Chanton Val ★★★ Noah's Ark for old VAL varieties (Eyholzer Roter, Gwäss, HEIDA, Himbertscha, Lafnetscha, Resi), vyds up to 800m (2625ft).

Chappaz, Marie-Thérèse Val ★★★→★★★★ Small bio estate, famous for sweet Petite ARVINE and Ermitage (MARSANNE), arguably Switzerland's best, but taste the basic FENDANT La Liaudisaz. You can buy everything here.

Colombe, Dom La Vd ★★→★★★ A bio estate at FÉCHY, LA CÔTE, best known for ageable CHASSELAS Brez 12' 15' 19' 20'. Now also terracotta-aged field-blend white (Curzilles) and stunning SAVAGNIN (Amédée).

Cortaillod Neu Village on shores of Lake Neuchâtel renowned for refined PINOT N. Eponymous, low-yielding local clone.

Côte, La Vd ★→★★★ A breadbasket of CHASSELAS: 2000 ha w of Lausanne on Lake Geneva, v. variable in quality. Villages incl FÉCHY, Mont-sur-Rolle, Morges.

Cruchon Vd ★★★ A bio producer of LA CÔTE, now led by young Catherine C, lots of SPÉCIALITÉS (eg. outstanding Altesse). Top PINOT N from local clone Servagnin dating back to C15 (Raissennaz 10' 13' 15' 17', Servagnin 18 19').

Dézaley Vd ★★★ LAVAUX GRAND CRU on steep slopes of Lake Geneva, 54 ha; planted in C12 by Cistercian monks. Potent CHASSELAS develops with age (7 yrs+). Best: Duboux, *Fonjallaz*, LEYVRAZ, *Louis Bovard*, Monachon, VILLE DE LAUSANNE.

Dôle Val ★→★★★ VAL's answer to Burgundy's Passetoutgrains: PINOT N plus GAMAY for light, quaffable red.

Donatsch, Thomas Gris ★★★ Barrique pioneer (1974) at MALANS. Now son Martin in charge: PINOT N Res Privée 13' fetched CHF 1075/bottle at auction.

> **Thank the Franks**
> Switzerland may be small, but it's a PINOT N giant and 7th in worldwide
> Pinot production (3875 ha). Switzerland has been making Pinot N for
> longer than most. In C9, Carolingian Charles III brought Pinot N to the
> shores of Lake Constance. Amadeus VIII of Savoy and his wife Mary of
> Burgundy are said to have given vines to growers at Lake Geneva in C15.
> Finally, the grape reached Graubünden, according to legend, via Henri II,
> Duke of Rohan, during the Thirty Years' War. All of this would not mean
> much if Swiss Pinot was not as excellent as it can be – locally different,
> but always with the Pinot-typical combination of delicacy and depth.

Duboux, Blaise Vd ★★★ Family estate, 5 ha, in LAVAUX. Outstanding DÉZALEY vieilles vignes Haut de Pierre (v. rich, mineral), CALAMIN Cuvée Vincent.

Féchy Vd ★→★★★ Famous though unreliable AOC of LA CÔTE, mainly CHASSELAS.

Fendant Val ★→★★★ Full-bodied VAL CHASSELAS, recognizable by its balsamic scent, ideal for fondue/raclette. Try BESSE, Cornulus, GERMANIER, PROVINS, SIMON MAYE.

Fläsch Gris ★★★→★★★★ Village of BÜNDNER HERRSCHAFT known for PINOT N from schist and limestone. Lots of gd estates, esp members of Adank, Hermann, Marugg families. *Gantenbein* is outstanding.

Flétri / Mi-flétri Late-harvested grapes for sweet/slightly sweet wine.

Fromm, Georg Gris ★★★ Top grower in MALANS (until 2008 also in NZ), known for subtle single-vyd PINOT N (Fidler, Schöpfi, Selfi/Selvenen, Spielmann). Now bio.

Gantenbein Gris ★★★★ 10' 12 13' 15' 16 17 18 19 20 Star growers Daniel and Martha G in FLÄSCH. PINOT N is famous, but CHARD even more intriguing. PINOT N from schist and limestone. Lots of gd estates, esp members of Adank.

Geneva 1400 ha of vines remote from the lake (vyds there belong mainly to canton VD). Growers: Balisiers, Grand'Cour, Les Hutins, Novelle.

Germanier, Jean-René Val ★★→★★★ Big estate (150 ha), reliable FENDANT Les Terrasses, SYRAH Cayas, local AMIGNE from schist at Vétroz a speciality (nobly sw Mitis, dr Balavaud 13 15 16 17' 19).

Glacier, Vin du (Gletscherwein) Val ★★★ A sort of Alpine "Sherry" from rare Rèze grape of Val d'Anniviers, aged in larch casks. Taste at the Grimentz town hall.

Grain Noble ConfidenCiel Val Quality label for authentic sweet wines, eg. CHAPPAZ, DOM DU MONT D'OR, Dorsaz (both estates), GERMANIER, Philippe Darioli, PROVINS.

Grand Cru Val, Vd Inconsistent term, used in VAL and VD, sometimes linked to specific varieties or restricted yields, or to designations like clos, ch, abbaye. Only few are classifications of a vyd site (eg. CALAMIN, DÉZALEY).

Grisons (Graubünden) A mtn canton, German- and Rhaeto-Romanic-speaking. PINOT N king. *See* BÜNDNER HERRSCHAFT. Also Manfred Meier, VON TSCHARNER.

Huber, Daniel ★★→★★★ Pioneer who reclaimed historical sites from fallow in 1981. Partly bio. Son Jonas taken over. Top: ageable MERLOT/CAB FR blend Montagna Magica (01' proved dewy-fresh in 2021).

Johannisberg Val Name for SILVANER in VAL, often off-dry or sweet; great with fondue. Excellent: *Dom du Mont d'Or*.

Joris, Didier Val ★★★→★★★★ Organic mini-estate (3 ha, dozen varieties). Reviving nearly extinct local Diolle (w). Hard to find.

Krebs & Steiner Ber ★★★ Merger of two Lake Bienne family estates. Best: Clos au Comte CHARD under Steiner label.

Lavaux Vd ★★→★★★★ 30 km (19 miles) of steep s-facing terraces e of Lausanne; UNESCO World Heritage Site. Uniquely rich, mineral CHASSELAS. GRANDS CRUS CALAMIN, DÉZALEY, several village AOCs.

Leyvraz, Pierre-Luc Vd ★★★→★★★★ Perfectionist LAVAUX grower; intensely terroir-driven ST-SAPHORIN Les Blassinges and DÉZALEY.

Litwan, Tom Aar ★★★ Career changer (mason to bio grower, 2006) at Schinznach;

5 ha. Substantial CHARDS Büel, Wanne; series of single-vyd Chalofe (generous), PINOT N Auf der Mauer (refined), Rüeget (tight-knit, mineral).

Maison Carrée, La Neu ★★★ Family estate, est 1827, 10 ha, v. traditional winemaking incl use of old wooden press, esp PINOT N (Auvernier, Hauterive, single-vyd, old-vines Le Lerin).

Malans Gris ★★→★★★★ Village in BÜNDNER HERRSCHAFT. Top PINOT N producers: DONATSCH, FROMM, Liesch, Studach, Wegelin. Late-ripening local Completer gives a long-lasting phenolic white. Adolf Boner 01' 05' is keeper of the grail.

Maye, Simon & Fils Val ★★★ Family estate, 11 ha. Dense SYRAH Vieilles Vignes perhaps best in Switzerland; spicy, powerful Païen (HEIDA).

Mémoire des Vins Suisses Union of 57 leading growers in effort to create stock of Swiss icon wines, to prove their ageing capacities. Oldest wines from 1999.

Mercier Val ★★★→★★★★ SIERRE family estate, now young Madeleine M in charge. Meticulous vyd management produces dense, aromatic reds, eg. rare CORNALIN 05' 09' 10' 11 15' 16 17 18 19 and SYRAH.

Mont d'Or, Dom du Val ★★→★★★★ Emblematic VAL estate for nobly sweet wines, esp JOHANNISBERG Saint-Martin. Recently, more emphasis on dry wines.

Morcote, Castello di Tic ★★★ Castle ruin and eponymous 14-ha winery, converted to bio by 3rd-generation owner Gaby Gianini. Excellent Riserva 16' 17 18'.

Neuchâtel ★→★★★ 606 ha around city and lake on calcareous soil. Slightly sparkling CHASSELAS, exquisite PINOT N from local clone (CORTAILLOD). Best: CH D'AUVERNIER, Dom de Chambleau, LA MAISON CARRÉE, PORRET, TATASCIORE.

Oeil de Perdrix "Partridge's eye": PINOT N Rosé, originally from NEU, now elsewhere.

Pircher, Urs Zür ★★★→★★★★ Wines with admirable consistency from steep s-facing slope overlooking Rhine at Eglisau. Crystal-clear whites, complex PINOT N Stadtberger Barrique 15' 16 17 18 19 20 from old Swiss clones. Urs P handing over to Gianmarco Ofner.

Porret Neu ★★→★★★★ Leading family estate at CORTAILLOD with Burgundian approach to CHARD, PINOT N (best: Cuvée Elisa, made to age for decades).

Provins Val ★→★★★ Co-op with 4000+ members, Switzerland's biggest producer, 1500 ha, 34 varieties. Sound entry level, v.gd oak-aged Maître de Chais range.

R3 Zür ★★★ 12' 17' 18 19 20 Räuschling (local grape) collaboration of three leading ZÜRICHSEE growers: Lüthi, Rütihof, SCHWARZENBACH.

Riehen, Weingut Bas ★★→★★★★ New boutique winery (2014) in eponymous town nr Basel, led by Hanspeter Ziereisen (*see* Germany) and merchants Jacqueline and Urs Ullrich. PINOT N (Le Petit, Le Grand): distinction, great potential.

Rouvinez Vins Val ★→★★★ VAL giant at Sierre, est 1947; cuvées La Trémaille (w) and Le Tourmentin (r) founded reputation. Strong expansion through takeover of BONVIN (2009), Caves Orsat (1998), Imesch (2003).

Ruch, Markus Schaff ★★★ Excellent, ultra-rare PINOT N from Hallau (Chölle from 60-yr-old vines, Haalde from steep slope, Buck on sandstone), Gächlingen (Schlemmweg on limestone); 3.5 ha. Also amphora MÜLLER-T and Cidre.

St-Saphorin Vd ★→★★★★ Neighbour AOC of DÉZALEY, lighter, but equally delicate. Best: LEYVRAZ, Monachon.

Sion (Val) has 581mm (23in) annual rainfall, damp Locarno (Tic) 1855mm (73in).

Schaffhausen ★→★★★ Deemed "BLAUBURGUNDERLAND", but a flood of cheap supermarket wines has damaged reputation. Top growers (eg. BAD OSTERFINGEN, BAUMANN, RUCH) among Switzerland's best.

Schenk SA Vd ★→★★★ Wine giant with worldwide activities, based in Rolle, founded 1893. Sound wines (esp VD, VAL); substantial exports.

Schwarzenbach, Hermann Zür ★★★ Family estate on Lake Zürich, 10 ha, a third of vyd devoted to local grape Räuschling, which Hermann S Snr saved from extinction in the 50s/60s. Cultish, age-worthy single-vyd Seehalden 07' 10' 15' 17' 18' 19' 20 21. Many other SPÉCIALITÉS (eg. Completer, Freisamer) too.

Spécialités / Spezialitäten Quantitatively minor grapes producing some of best Swiss wines, eg. Räuschling, GEWURZ or PINOT GR in German Switzerland, or local varieties (and grapes like JOHANNISBERG, MARSANNE, SYRAH) in VAL.

Sprecher von Bernegg Gris ★★★ Historic estate at Jenins, BÜNDNER HERRSCHAFT, esp PINOT N: Lindenwingert, vom Pfaffen/Calander.

St Jodern Kellerei Val ★★→★★★ VISPERTERMINEN CO-op famous for *Heida Veritas* from ungrafted old vines.

Stucky-Hügin Tic ★★★→★★★★ MERLOT pioneer Werner S and son Simon plus Jürg H. Best: Conte di Luna (MERLOT/CAB SAUV), Soma (Merlot/CAB FR), Temenos (Completer/SAUV BL).

Tatasciore, Jacques Neu ★★★★ Refined (and rare) NEU PINOT N.

Ticino ★→★★★★ Italian-speaking. MERLOT (leading grape since 1948) in a taut style. Best: Agriloro, CASTELLO DI MORCOTE, Chiericati, Delea, Gialdi, HUBER, Klausener, Kopp von der Crone Visini, Pelossi, STUCKY, Tamborini, Valsangiacomo, Vinattieri, ZÜNDEL.

Tscharner, von ★★★ Family estate at confluence of Vorderrhein and Hinterrhein. Epic, tannin-laden PINOT N (Churer Gian-Battista, Jeninser Alte Reben), worthy of position as 1st wine estate along River Rhine's 1230-km (764-mile) length.

Twann Ber ★→★★★ Village of Bielersee, famous for GUTEDEL (CHASSELAS), PINOT N, but variable quality. Johanniterkeller, Klötzli, KREBS & STEINER, Schott (bio) all gd.

Valais (Wallis) Largest wine canton, in dry, sunny upper Rhône V, best MARSANNE, SYRAH rival French legends. Many exquisite local varieties. Plenty of gd family estates and gifted youngsters.

Vaud (Waadt) Second-largest wine canton, on shores of Lake Geneva. Family estates known for conservative spirit. Big houses incl Bolle, Hammel, Obrist, SCHENK.

Ville de Lausanne Vd ★★→★★★ Five estates (together 36 ha) owned by commune of Lausanne, esp Clos des Moines and Clos des Abbayes in DÉZALEY, bio.

Visperterminen Val ★→★★★ Upper VAL vyds, esp for HEIDA. One of highest vyds in Europe (at 1000m/3281ft+; called Riben). Try CHANTON, ST JODERN KELLEREI.

Vully Fri, Vd ★→★★★ Small but fine AOC on shores of Lake Murten, split between cantons Fribourg (116 ha) and VD (46 ha); CHASSELAS, PINOT N and TRAMINER. Best: Ch de Praz, Chervet, Cru de l'Hôpital, Javet & Javet, Petit Ch.

Yvorne Vd ★★→★★★ CHABLAIS village with vyds on detritus of 1584 avalanche, eg. BADOUX, Ch Maison Blanche, Commune d'Yvorne, Dom de l'Ovaille.

Zündel, Christian Tic ★★★→★★★★ Bio estate, 4 ha, at Beride. Perfectionist Christian Z now joined by daughter Myra. Wines of purity and finesse, esp MERLOT/CAB SAUV Orizzonte and CHARDS Velabona, Dosso.

Zürich Biggest city, and largest wine-growing canton in German Switzerland, 610 ha. Mainly BLAUBURGUNDER. Best: Besson-Strasser, E Meier, Gehring, Lüthi, PIRCHER, SCHWARZENBACH, Staatskellerei, Zahner.

Zürichsee Zür Dynamic AOC uniting vyds of cantons ZÜR and Schwyz on shores of Lake Zürich. Best: Bachmann, Diederik, E Meier, Höcklistein, Kloster Einsiedeln, Lüthi, Rütihof, Schipf, Schnorf, SCHWARZENBACH.

Zwaa Schaff ★★★ Collaboration: BAUMANN (calcareous, deep soil) and BAD OSTERFINGEN (light, gravelly). PINOT N **94' 09' 13' 15'** 16 17 18 19; PINOT BL/CHARD equally long-lasting.

Wine regions
Switzerland has six major wine regions: VAL, VD, GENEVA, TIC, Trois Lacs (NEU, Bienne/BER, VULLY/Fribourg) and German Switzerland (Aar, GRIS, SCHAFF, St Gallen, Thur, ZÜR and some smaller wine cantons).

Austria

Abbreviations used in the text:

Burgen	Burgenland
Carn	Carnuntum
Kamp	Kamptal
Krems	Kremstal
Nied	Niederösterreich
Stei	Steiermark
S Stei	Südsteiermark
Therm	Thermenregion
Trais	Traisental
V Stei	Vulkanland Steiermark
Wach	Wachau
Wag	Wagram
Wein	Weinviertel
W Stei	Weststeiermark

Austria has quietly risen to the top of the wine tree: it's hard to think of a European country where the average quality is higher. The style is pristine: whites of crystalline clarity, reds of freshness and balance, with plenty of acidity and fruit. The flavours are just unfamiliar enough to lure anyone looking for something different, and they are perfect with contemporary food. Add to that a tendency to biodynamism at the top names and a liking for natural yeasts and experimental techniques, and you get a country where everything is worth trying. Regional identity matters: the change to an appellation system, DAC (districtus austriae controllatus), started in 2002 and is almost finished. After the latest arrivals, just one is missing: Thermenregion.

Recent vintages

2021 Changeable summer, ideal autumn: balanced, ripe wines, gd acidity.
2020 Some disease pressure, some hail. Balanced and classic.
2019 Dream vintage boasting both ripeness and freshness.
2018 The heatwave yr: ripe wines. Gd and plentiful.
2017 Gd juicy, rounded wines.
2016 Lovely fruit expression, fine freshness, but choose growers carefully.
2015 V.gd quality. Full-bodied, ripe, with the stuffing to age.

Alzinger Wach ★★★★ 10 13 15 16 17 Clear-cut, crystalline, long-lived RIES; GRÜNER V.
Ambrositsch, Jutta Vienna ★★ Delicious field blends. Great pop up HEURIGE in VIENNA: Buschenschank in Residence.
Atzberg Wach ★★★ Impressive high-altitude re-cultivation project of abandoned vyds under the auspices of GRITSCH family and a Viennese property investor.

> **Sekt: go for the best**
> Fizz can be cheap and cheerful or really rather gd. The bottom of the
> Sektpyramide is fizz with min 9 mths ageing on lees: basic; don't expect
> wonders. **Sekt Austria Res**: 18 mths on lees, must be traditional method
> and Brut, Extra Brut or Brut Nature. **Grosse Res**: 36 mths on lees; max
> sugar content is 12g/l (Brut) though is often Extra Brut or Brut Nature.
> Grapes are anything allowed for Qualitätswein. GRÜNER V is popular,
> as well as PINOT blends. ZWEIGELT too. Best, look to gd growers in KAMP
> and WEIN: BRÜNDLMAYER, Ebner-Ebenauer, JURTSCHITSCH, LOIMER, MALAT,
> SCHLOSS GOBELSBURG, Szigeti and Zuschmann-Schöfmann.

Ausbruch Quality designation in Rust. Ruster Ausbruch = botrytized, dried grapes. Min must weight 30° KMW or 145.8° Oechsle.

Braunstein, Birgit Burgen ★★★ 15 16 17 19 Individualistic bio star of LEITHABERG. Wonderful BLAUFRÄNKISCH. Try amphora-aged series Magna Mater.

Bründlmayer, Willi Kamp ★★★★ 10 13 15 17 19 20 Iconic producer: GRÜNER V, RIES, esp Heiligenstein Lyra, Alte Reben. Also fine Sekt, francophile PINOT N.

Burgenland Federal state and wine region bordering Hungary. Warmer than NIED, reds like BLAUFRÄNKISCH, ST-LAURENT, ZWEIGELT prevalent. Shallow NEUSIEDLERSEE, eg. at RUST and in SEEWINKEL, creates ideal botrytis conditions.

Carnuntum Nied Previously unsung region e of VIENNA now blossoming with accomplished fresh reds, esp ZWEIGELT marketed as Rubin Carnuntum. Look for BLAUFRÄNKISCH from Spitzerberg. Best: G Markowitsch, GRASSL, MUHR, TRAPL.

Christ Vienna ★★★ Leading light of VIENNA and favourite HEURIGE. Exquisite GEMISCHTER SATZ and unusual red blends.

DAC (districtus austriae controllatus) Provenance- and quality-based appellation system for regional typicity. Creation of 1st DAC 2002, WEIN. Hierarchy inspired by Burgundy: Regional; Ort (village); Ried (cru). This pyramid was widely adopted. Currently 17 DACs: EISENBERG, KAMP, KREMS, LEITHABERG, MITTELBURGENLAND, NEUSIEDLERSEE, Rosalia, S STEI, TRAIS, V STEI, Wein, Wiener GEMISCHTER SATZ, W STEI; CARN, Ruster AUSBRUCH, WACH and WAG as the most recent.

Domäne Wachau Wach ★★★ World-leading co-op with enviable vyds. Clean-cut and expressive, purist style. Whistle-clean GRÜNER V, RIES, esp Achleiten, Kellerberg. Always great value. Try also experimental Backstage.

Ebner-Ebenauer Wein ★★★★ More expressive each yr. Driving force in WEIN. Single-vyd GRÜNER V, old-vine ST-LAURENT, PINOT N. Fine Blanc de Blancs Brut Nature.

Edlmoser Vienna ★★★ Leading winery in s of VIENNA. Refined and mineral white, powerful red. Try eg. RIES Sätzen at superb HEURIGE.

Eichinger, Birgit Kamp ★★★ Consistently outstanding RIES, esp Heiligenstein, unusually spicy and sleek GRÜNER V, esp Hasel.

Eisenberg Burgen Small DAC, BLAUFRÄNKISCH from slate soil. Powerful but elegant.

Erste Lage Single-vyd quality designation in ÖTW wineries in CARN, KAMP, KREMS, TRAIS, VIENNA, WAG: 68 member estates, 81 Erste Lagen, on-going process.

Federspiel Wach VINEA WACHAU middle category of ripeness, min 11.5%, max 12.5% abv. Understated, gastronomic wines as age-worthy as SMARAGD.

Feiler-Artinger Burgen ★★★★ Intuitive bio-winemaker of fine-boned reds and stellar AUSBRUCH in historic Baroque house in RUST town centre.

Fritsch, Karl Wag ★★★ 18 19 20 Passionate for bio: RIES Mordthal; top PINOT N P.

Fritz, Josef Wag ★★ 19 20 Leading producer of ROTER VELTLINER. Try all.

Gemischter Satz Vienna Revived historic concept of co-planted/fermented field blend of white varieties. Complex wines. Prevalent in WEIN, VIENNA: determined producers achieved DAC status called Wiener Gemischter Satz in 2013 for Vienna. No variety to exceed 50%. Try AMBROSITSCH, CHRIST, GROISS, WIENINGER.

Geyerhof Krems ★★★ World-class estate with super-bio credentials. Wonderful GRÜNER V, RIES, notable entry-level Stockwerk.

Grassl, Philipp Carn ★★★ Impressive ZWEIGELT, gorgeous ST-LAURENT, PINOT N.

Gritsch Mauritiushof Wach ★★★ Always vividly fruit-driven, concentrated RIES, esp 1000-Eimerberg vyd. Peppery GRÜNER V. Winemaker for ATZBERG project.

Groiss, Ingrid Wein ★★★ Known for reviving old vyds, specializes in GEMISCHTER SATZ and peppery GRÜNER V. Lovely rosé Hasenhaide.

Gruber-Röschitz Wein ★★★ Unusually for WEIN this sibling-run bio estate also makes racy RIES from granite soils, fine GRÜNER V.

Gut Oggau Burgen ★★ Not only the labels are eye-catching at this bio estate.

Hager, Matthias Kamp ★★ Longstanding bio icon. Try GRÜNER V Mollands, exquisite TBA and Eiswein, unsulphured Urgestein and pét-nats.

Harkamp S Stei ★★★ Best Sekt in STEI; try long-aged Grosse Res, Zero and Solera.

Hartl, Heinrich Therm ★★★ Rising THERM star specializing in sinuous PINOT N and champion of indigenous whites ROTGIPFLER, ZIERFANDLER.

Heinrich, Gernot Burgen ★★★★ Made name with powerful BLAUFRÄNKISCH; new focus on compelling, skin-fermented whites.

Herrenhof Lamprecht V Stei ★★→★★★ Renegade wine-grower in V STEI. Adorable FURMINT, PINOT BL, RIES. Try any from Buchertberg.

Heuriger Wine of most recent harvest. **Heurige:** homely tavern where growers serve own wines with rustic, local food often in the open air – integral to Austrian culture. Called Buschenschank outside Vienna.

Hiedler Kamp ★★★ Beautiful, spicy GRÜNER V, esp Thal and Kittmannsberg.

Hirsch Kamp ★★★★ 15 17 19 20 Consistently overdelivering RIES, GRÜNER V from Heiligenstein, Lamm. Fun entry-level Grüner V Hirschvergnügen.

Hirtzberger, Franz Wach ★★★★ 08 10 13 14 15 19 20 Iconic producer in Spitz, instrumental in defining WACH style. Known for opulence; dialling into elegance now. Outstanding single-vyd RIES, GRÜNER V, esp Honivogl, Singerriedel.

Huber, Markus Trais ★★★ Emblematic of the fine-boned TRAIS style from limestone soils with GRÜNER V, RIES. Citrus brilliance, radiance, slenderness.

Illmitz Burgen Town on NEUSIEDLERSEE, in SW SEEWINKEL famous for BA and TBA (see Germany). Best from Angerhof-Tschida, KRACHER.

Jalits Burgen ★★★ An est EISENBERG estate; concentrated, complex BLAUFRÄNKISCH, single-vyds Szapary and Diabas long-lived.

Jamek, Josef Wach ★★★ Stalwart and WACH institution, run by his descendants, on way back to former glory, esp RIES, GRÜNER V from single-vyds Achleiten, Klaus.

Johanneshof Reinisch Therm ★★★ Champion of local specialities ROTGIPFLER, ZIERFANDLER. Deservedly famous for long-lived PINOT N; world-class ST-LAURENT.

Jurtschitsch Kamp ★★★★ Young German-Austrian couple pushing boundaries in bio; invigorating KAMP with stellar site interpretations of single-vyds Heiligenstein, Käferberg, Loiserberg, as well as with sparkling and pét-nats.

Kamptal Nied Wine region along Danube tributary Kamp n of WACH; rounder style, lower hills, impressive minerality, precision. Top vyds: Heiligenstein, Käferberg, Lamm, Loiserberg, Seeberg. Best: BRÜNDLMAYER, EICHINGER, HIEDLER, HIRSCH, JURTSCHITSCH, LOIMER, SCHLOSS GOBELSBURG. DAC for GRÜNER V, RIES.

From Austria-Hungary with love

FURMINT (Šipon in Slovenia) is rediscovering its place among the white grapes of BURGEN. It was widespread when Burgen was still part of Hungary. After Burgen joined Austria in 1921, Furmint was supplanted by GRÜNER V. But now Furmint is being replanted and used for refreshing, spicy dry whites. Try Hannes Schuster, HEIDI SCHRÖCK, FRANZ WENINGER, MICHAEL WENZEL, and also Herrenhof Lamprecht from V STEI.

Klosterneuburg Wag Wine town in WAG, seat of 1860-founded viticultural college and research institute. *See* next entry.

KMW KLOSTERNEUBURGER Mostwaage ("must level"), Austrian unit of must weight, ie. sugar content of juice. 1° KMW=4.86° Oe (*see* Germany). 20° Bx=83° Oe.

Knoll, Emmerich Wach ★★★★ 05 06 07 08 10 15 19 20 WACH icon perpetuating a tradition of cool and expressive purity in RIES, GRÜNER V. Single-vyd Ried Schütt emblematic of crystalline style. Look for long-lived Vinothekfüllung bottlings.

7242 ha certified bio vyds in Austria. Not a lot, but they make a lot of noise.

Kopfensteiner Burgen ★★★ In front row of EISENBERG producers; go for graceful BLAUFRÄNKISCH from meagre soils of RIED Saybritz.

Kracher Burgen ★★★★ 05 07 09 10 13 15 16 17 18 Towering genius of botrytized wines. Unoaked range is Zwischen den Seen; oak-matured range is Nouvelle Vague. Dazzling array of super-complex, deliriously sweet wines across varieties. Releases at same time with Kollektion, a "ten years later" range.

Kremstal Krems Wine region and DAC for GRÜNER V, RIES. Top: GEYERHOF, MALAT, MOSER, NIGL, PROIDL, SALOMON-UNDHOF, STIFT GÖTTWEIG, WEINGUT STADT KREMS.

Krutzler Burgen ★★★ 11 13 15 16 19 Local icon for highly concentrated, big, bold BLAUFRÄNKISCH, esp cult wine Perwolff.

Lackner Tinnacher S Stei ★★★★ Precise, mineral SAUV BL, PINOT (w); RIED Steinbach.

Leithaberg Burgen Important DAC on n shore of NEUSIEDLERSEE, limestone and schist soils. Red restricted to BLAUFRÄNKISCH, some of Austria's best; whites can be GRÜNER V, PINOT BL, CHARD or NEUBURGER solo or blended.

Lesehof Stagård Krems ★★★ Pure, bold, electric single-vyd RIES. Try Ries Steiner Gaisberg, ravishing entry-level Ries Handwerk.

Lichtenberger González Burgen ★★★ Spanish-Austrian couple showing best side of LEITHABERG (r/w). NEUBURGER Leithaberg DAC, BLAUFRÄNKISCH Vorderberg.

Loimer, Fred Kamp ★★★★ 10 13 15 19 20 Individualistic bio producer of long standing. Famed for RIES, GRÜNER V, esp single-vyds Heiligenstein, Steinmassl, Seeberg. Increasingly elegant PINOT N and *lovely sparkling* (NV and Vintage). Try sparkling Grosse Res Langenlois or new Grosse Res GUMPOLDSKIRCHEN.

Malat Krems ★★★★ Clean-cut, concentrated RIES, GRÜNER V, esp single vyds Gottschelle, Silberbichl. Pioneering PINOT N producer. Three sparklers to try.

Mantlerhof Krems ★★★ Sumptuous GRÜNER V from bio-farmed loess soils.

Mayer am Pfarrplatz Vienna ★★ Heiligenstadt institution. Try RIES Weisser Marmor from Nussberg. HEURIGE where Beethoven wrote 3rd symphony: tourist heaven.

Mittelburgenland Burgen DAC (2005) on Hungarian border: structured, age-worthy BLAUFRÄNKISCH. Producers: GESELLMANN, J Heinrich, Kerschbaum, WENINGER.

Moric Burgen ★★★★ 10 11 13 15 19 Cult producer, unrelenting focus, deservedly famed BLAUFRÄNKISCH; note old-vine single vyds Lutzmannsburg, Neckenmarkt.

Muhr Carn ★★★★ PR-cum-winemaker reviving limestone slopes of Spitzerberg. Unforced yet profound BLAUFRÄNKISCH of rare, scented beauty.

Neumeister V Stei ★★★★ 12 15 18 19 20 World-class address for aromatic whites. Outstanding SAUV BL, esp single vyds Klausen, Moarfeitl. Look for Stradener Alte Reben. Notable GEWÜRZ. PINOT N.

Neusiedlersee (Lake Neusiedl) Burgen Largest European steppe-lake and nature reserve on Hungarian border. Lake mesoclimate and humidity key to botrytis. Eponymous DAC limited to ZWEIGELT and sweet wines, botrytized or not.

Niederösterreich (Lower Austria) Region in ne of three parts: Danube (KAMP, KREM, TRAIS, WACH, WAG), WEIN (ne) and CARN, THERM (s); 59% Austria's vyds.

Nigl Krems ★★★★ Expressive, flavour-laden RIES, GRÜNER V, esp Privat bottlings.

Nikolaihof Wach ★★★★ 08 10 13 15 17 19 20 One of world's 1st bio wine estates. Exemplary, pure, textured RIES, GRÜNER V; look for late-release Vinothek series.

Nittnaus, Anita & Hans Burgen ★★★★ Bio-pioneer, precise single-vyd CHARD from Bergschmallister, Freudshofer; BLAUFRÄNKISCH from Jungenberg, Lange Ohn, Tannenberg. Enticing, entry-level Kalk & Schiefer. Famous for MERLOT blends.

Ott, Bernhard Wag ★★★★ Iconic bio producer, salty, savoury GRÜNER V of increasingly fine-boned elegance, esp Rosenberg, Spiegel, Stein vyds. Fass 4 has cult status.

ÖTW (Österreichische Traditionsweingüter) Private association working on vyd classification. Currently 68 member estates and 90 classified sites in CARN, KAMP, KREMS, TRAIS, VIENNA, WAG. *See* ERSTE LAGE; excludes WACH.

Pichler, Franz X Wach ★★★★ 08 10 13 15 19 20 Storied estate, long-lived RIES, GRÜNER V, from top WACH sites. Cult Ries Unendlich. Has left VINEA WACHAU, after decades. From vintage 21 no more FEDERSPIEL or SMARAGD on labels.

Pichler, Rudi Wach ★★★★ 10 13 15 17 19 20 Magnificent, clear-cut RIES, GRÜNER V from top sites Achleiten, Steinriegl.

Pichler-Krutzler Wach ★★★ Reliably energetic and thrilling *Ries* from WACH single vyds, esp In der Wand, Kellerberg.

Pittnauer Burgen ★★★★ Rare talent for long-lived, elegant ST-LAURENT, fun bottlings of skin-fermented MashPitt, refreshing pét-nat.

Prager, Franz Wach ★★★★ 10 13 14 15 19 Toni Bodenstein is erudite éminence grise of WACH; philosopher/winemaker, spellbinding RIES, GRÜNER V, esp Stockkultur.

Preisinger, Claus Burgen ★★★ Hip but grounded bio winemaker with both serious and fun offerings like crowncapped Puszta Libre.

Prieler Burgen ★★★★ Leading light of quality revolution in 80s. Famed for muscular BLAUFRÄNKISCH requiring bottle-age, esp Goldberg, Marienthal; lesser known for compelling age-worthy, profound PINOT BL.

Proidl, A&F Krems ★★★ Reliably brilliant RIES, GRÜNER V, both from Ehrenfels. Even better since son Patrick joined. Try library Ries releases.

Rebenhof Aubell S Stei ★★★ Exciting, vivid skin-fermented bio whites.

Reserve (Res) Attribute for min 13% abv and prolonged (cask) ageing.

Ried Vyd. As of 2016 compulsory term for single-vyd bottlings.

Rust Burgen Fortified C17 town on NEUSIEDLERSEE; noisy nesting storks. Famous Ruster AUSBRUCH. Top: FEILER-ARTINGER, SCHRÖCK, TRIEBAUMER, WENZEL.

Sabathi, Erwin S Stei ★★★ Best known for elegant, profound SAUV BL and MORILLON from Pössnitzberg limestone and marl soils (*Opok*).

Sabathi, Hannes S Stei ★★★ Leading estate, esp single-vyd SAUV BL Kranachberg. Revived vyds in Styrian capital Graz under the label Falter Ego.

Salomon-Undhof Krems ★★★★ Consistently sleek, elegant, slender GRÜNER V, RIES, single vyds Kögl, Plattenberg, Wachtberg. Fun pét-nat, summery pink fizz.

Sattlerhof S Stei ★★★★ 12 15 18 19 20 Effortlessly brilliant, world-class SAUV BL, MORILLON, esp single vyds Kranachberg, Sernauberg. Top DAC SAUV BL Gamlitz.

Sausal Subregion in SÜDSTEIERMARK with schist and limestone soils; vyds up to 600m (1969ft). Small enclave, extraordinarily elegant RIES, SAUV BL.

Schauer S Stei ★★★ Light-bodied but consistently profound whites from schisty SAUSAL. RIES, SAUV BL and esp notable PINOT BL Höchtemmel.

AUSTRIA (vertical side tab)

Ageing gracefully

Austria is notorious for drinking wines too young: it used to be the case that anything older than a year was reckoned an unsaleable leftover. This started to change a few years ago; top wines are released later now and will age even longer. GRÜNER V ages brilliantly but also makes young everyday wines: don't confuse the two. SAUV BL and PINOT from STEI and BLAUFRÄNKISCH from BURGEN will age too. Among the local heroes, ZIERFANDLER from THERM and ROTER VELTLINER from WAG are worth mentioning. Some producers make releases of older wines: *see* p.336.

Schiefer & Doms Kilger Burgen ★★★ Individualistic BLAUFRÄNKISCH in EISENBERG; expressive, transparent WELSCHRIESLING.

Schilcher W St Racy, peppery rosé of local importance from indigenous Blauer Wildbacher grape, speciality of w STEI. Also try sparkling and/or orange version.

Schloss Gobelsburg Kamp ★★★★ 13 14 15 16 19 20 Cistercian-founded mansion and estate making exquisite RIES and GRÜNER V worth ageing under quality champion Michael Moosbrugger. Notable Tradition series and single vyds Gaisberg, Grub, Heiligenstein, Lamm, Renner. Fine sparkling. Also elegant ZWEIGELT, PINOT N.

Fave vyd helpers: sheep. They crop vegetation, loosen ground, add manure.

Schlumberger Vienna C19 Sekt pioneer. Today high-volume, value producer of traditional-method fizz. Try sparkling GRÜNER V.

Schmelz Wach ★★★ Exquisite, authentic wines strangely below the radar.

Schröck, Heidi Burgen ★★★★ Doyenne of RUST speciality AUSBRUCH on mission to broaden its food-pairing appeal, also notable dry FURMINT. Now with twin sons.

Seewinkel Burgen Region e of NEUSIEDLERSEE; ideal conditions for botrytis.

Smaragd Wach Ripest category of VINEA WACHAU, min 12.5% abv but can exceed 14%, dry, potent, age-worthy. In past often botrytis-influenced but dry.

Spätrot-Rotgipfler Therm Blend of ROTGIPFLER/Spätrot (ZIERFANDLER). Aromatic, weighty, textured. Typical for GUMPOLDSKIRCHEN. *See* Grapes chapter.

Spitzer Graben Wach Side valley and most recent (cool) hotspot in WACH. Steepest terraces, v. dry, meagre soils, excellent RIES, NEUBURGER. Interesting, off-beaten-track wineries: Grabenwerkstatt, Martin Muthenthaler, PETER VEYDER-MALBERG.

Stadlmann Therm ★★★ Exemplary, clear-cut ZIERFANDLER/ROTGIPFLER, esp single vyds Tagelsteiner, Mandelhöh. Subtle, poetic PINOT N.

Steiermark (Styria) Austria's most s region. Aromatic, expressive dry whites: SAUV BL and exciting white Pinots and CHARD (MORILLON). *See* S STEI, V STEI, W STEI.

Steinfeder Wach Lightest VINEA WACHAU category for dry wines of max 11.5% abv. Difficult/impossible to produce in warming conditions, barely exported.

Stift Göttweig ★★★ Prominent hilltop Benedictine abbey surrounded by vyds; quality ethos, crystalline RIES, GRÜNER V from single-vyds Gottschelle, Silberbichl.

Südsteiermark (South Styria) Region of STEI close to Slovenian border, famed for light but highly aromatic MORILLON, MUSKATELLER, SAUV BL from breathtakingly steep slopes. DAC (2018). Best: GROSS, LACKNER TINNACHER, SABATHI (E and H), SATTLERHOF, TEMENT, WOHLMUTH.

Tement S Stei ★★★★ 12 13 15 17 18 19 20 Incredible subtlety and age-worthiness. Esp SAUV BL, MORILLON. Top sites: Grassnitzberg, Zieregg. Try Zieregg RES Sauv Bl made only in exceptional yrs. Today run by Manfred's sons Armin and Stefan.

Thermenregion Nied Spa region e of VIENNA centred on GUMPOLDSKIRCHEN, home to indigenous ZIERFANDLER, ROTGIPFLER; historic PINOT N hotspot. Producers: Alphart, HARTL, JOHANNESHOF REINISCH, STADLMANN.

Tinhof, Erwin Burgen ★★★ Below-the-radar but exquisite bio LEITHABERG estate. BLAUFRÄNKISCH Gloriette from old vines and ST-LAURENT Feuersteig. Also specialist for PINOT BL, NEUBURGER from Golden Erd vyd.

Traisental Nied Tiny district s of KREMS on s bank of Danube. Notable for limestone soils lending finesse, precision. Top: HUBER, Neumayer.

Trapl, Johannes Carn ★★★ Brilliant wunderkind with poetic, expressive, site-specific reds. Floral BLAUFRÄNKISCH, esp Spitzerberg and PINOT-esque ZWEIGELT.

Triebaumer, Ernst Burgen ★★★★ 08 09 10 12 15 17 18 19 20 Iconic RUST producer, now led by founder's progressive children. Try skin-fermented Urwerk series.

Tschida, Christian Burgen Cult star in natural-wine circles. From Illmitz, SEEWINKEL. Notable ZWEIGELT/CAB SAUV Himmel auf Erden.

Umathum, Josef Burgen ★★★★ 13 15 16 17 18 19 20 Bio pioneer, now legend. Exceptionally elegant reds. Probably Austria's best ZWEIGELT: single-vyd Hallebühl. Also BLAUFRÄNKISCH Kirschgarten.

Velich ★★★ Publicity-shy SEEWINKEL producer; cult CHARDS Darscho and Tiglat.

Veyder-Malberg Wach ★★★ Boutique WACH producer, old vines from 4 ha of tiny parcels in SPITZER GRABEN. RIES, GRÜNER V.

Vienna (Wien) Capital boasting 637 ha vyds within city limits. Ancient tradition, reignited quality focus. Local field-blend tradition enshrined as DAC GEMISCHTER SATZ (2013). *Heurigen among vines; must visit.* Best: CHRIST, EDLMOSER, WIENINGER.

Vinea Wachau Wach Pioneering quality WACH growers' association. Ripeness scale for dry wine: FEDERSPIEL, SMARAGD and STEINFEDER.

Vulkanland Steiermark (Southeast Styria) Formerly Südost-Steiermark, DAC (2018), famous for GEWÜRZ from Klöch. Ideal conditions for MORILLON and PINOTS. Best: Frauwallner, NEUMEISTER, Winkler-Hermaden.

Wachau Nied Danube region of world repute for age-worthy RIES, GRÜNER V. Look for top: ALZINGER, DOMÄNE WACHAU, FX PICHLER, HIRTZBERGER, JAMEK, KNOLL, Muthenthaler, NIKOLAIHOF, PICHLER-KRUTZLER, PRAGER, R PICHLER, Tegernseerhof, VEYDER-MALBERG.

Wachter-Wiesler, Weingut Burgen ★★★ Defining producer of sinuous, concentrated but elegant EISENBERG BLAUFRÄNKISCH, also WELSCHRIESLING.

Wagram Nied Region just w of VIENNA, incl KLOSTERNEUBURG. DAC in 2022, based on GRÜNER V, RIES, ROTER VELTLINER. Deep loess soils ideal for Grüner V and increasingly also PINOT N. Best: FRITSCH, JOSEF FRITZ, Nimmervoll, OTT.

Weingut Stadt Krems Krems ★★★ Brilliant municipal wine estate with 31 ha vyds within city limits. Same exacting winemaker as STIFT GÖTTWEIG.

Weinviertel ("Wine Quarter") Austria's largest wine region, eponymous DAC for GRÜNER V in classic and Res version. Region once slaked VIENNA's thirst, now quality counts. Also base wine for Viennese Sekt houses. Try EBNER-EBENAUER, GROISS, GRUBER-RÖSCHITZ, HERBERT ZILLINGER, Schödl Loidesthal.

Weninger, Franz Burgen ★★★★ 10 13 15 16 17 Brilliant, brooding BLAUFRÄNKISCH, esp single-vyds Hochäcker, Kirchholz.

Wenzel, Michael ★★★ 15 17 18 19 20 Quiet man, speaks loudly through excellent FURMINT, esp entry-level Aus dem Quarz, and Prädikatsweine.

Werlitsch S Stei ★★★ Ewald Tscheppe is one of s STEI's natural-wine stars. Try SAUV BL/CHARD blend Ex Vero.

Weststeiermark (West Styria) Small region specializing in SCHILCHER. DAC (2018).

Wieninger, Fritz Vienna ★★★★ 13 14 15 16 18 Driving force behind VIENNA renaissance and bio pioneer. Exemplary GEMISCHTER SATZ from Nussberg, Rosengartl. Great PINOT N. Viennese HEURIGE among Nussberg vines an institution. Experimental playground in his other winery Hajszan-Neumann in Grinzing.

Wohlmuth S Stei ★★★★ Stellar, subtle but dazzling CHARD, RIES, SAUV BL, esp single vyds Edelschuh, Gola, Hochsteinriegl; recultivating steep Dr. Wunsch vyd.

Zillinger, Herbert Wein ★★★★ Brilliant, complex, extraordinary GRÜNER V.

Austria's new fine wines

Well, a few, anyway. Plenty more in the entries, but these growers don't have own entries. **Diwald** GRÜNER-V RIED Goldberg: pineapple in perfect balance, salty finish. **Fuchs und Hase** Pet Nat Rosé: vibrant raspberries. **Grabenwerkstatt** RIES Ried Trenning Smaragd: complex minerality, soft white peach. **Hannes Schuster** ST-LAURENT Zagersdorf undergrowth, dark fruit, stunning balance. **Karl Schnabel** BLAUFRÄNKISCH Hochegg, SAUSAL: elegant, juicy, mineral. **Michael Edlmoser** Wiener GEMISCHTER SATZ Ried Himmel Maurerberg: powerful, warm, quince, pears.

England

It's easy to think that English fizz should be cheaper than the equivalent quality from Champagne. But yields are lower here because planting densities are lower, because humidity is higher and the risk of disease greater. Bear in mind, too, that the cheapest Champagnes tend to come from enormous factory operations. England doesn't have those. Not yet, anyway. What it does have is a top end reaching for ever-higher quality, and a style of freshness and depth, tension and ripeness. Drink the 16s and older now. All the wines listed here are sparkling unless specified otherwise. Abbreviations: Buckinghamshire (Bucks), Cornwall (Corn), East/West Sussex (E/W S'x), Hampshire (Hants), Herefordshire (Her).

Ambriel W S'x Delicate, poised sparkling, no malo, esp Cloud Ten in magnums, 10 yrs age: lovely depth.

Black Chalk Hants ★★★ Library wines tops. Pure, lovely acidity, taut, depth, all from chalk soils. You might guess that from the name.

Breaky Bottom E S'x ★★★★ Beautiful wines, all precision, depth, from tiny vyd. CHARD/PINOT N, but revelatory SEYVAL BL too.

Bride Valley Dorset ★★ Better, more interesting the day after 1st opening, if you can wait that long. Vibrant fruit, super-crisp.

Camel Valley Corn ★★→★★★ Interesting rosé PINOT N Brut, vibrant Cornwall Brut; plenty of flavour, gd balance, gd value.

Chapel Down Kent Commercial, made in quantity. ★★ Kit's Coty range (still, sp).

Coates & Seely Hants ★★★ Gets better and better. La Perfide vintages superb, tense, rich and delicate.

Cottonworth Hants Length and depth gd here. Classic Cuvée is best.

Court Garden E S'x Rich style with bit of power. Subtle, taut Blanc de Blancs; elegant, biscuity Classic Cuvée; rich Blanc de Noirs. Rosé is red-fruited, crunchy.

Denbies Surrey Large, commercial, marked toastiness. Well-organized tourism, with vyd hotel.

Lots of Bacchus still white around: a bit like the shouty end of NZ Sauv Bl.

Digby Hants, Kent, W S'x ★★★ Elegant 13, powerful 14 rosé. made by Dermot SUGRUE under contract at WISTON, so can't go wrong.

Exton Park Hants ★★★ Brilliant new RB (Res Blend) takes it to another level: lovely maturity and grace.

Grange, The Hants ★★★ Savoury, classic lemon shortbread, well aged. Drink at the opera, or anywhere.

Greyfriars Surrey Crisp, savoury fizz, esp pure Rosé Res. Structured Blanc de Noirs.

Gusbourne Kent, W S'x ★★★ More and more precise. Look out for new late-released 12; gd still red Pinot N too, crunchy, juicy.

Hambledon Vineyard Hants ★★★ Deep, sleek wines, gd complexity, poise. Rosé has weight, power. Meonhill is cheaper label.

Harrow & Hope Bucks ★★ Well made, classic, v. pleasing. Try rich, deep Blanc de Noirs with food.

Hart of Gold Her ★★ Rounded and biscuity; v. appealing Brut.

Hattingley Valley Hants ★★★ Crisp, taut wines, always impeccably made; Kings Cuvée 14 toasty, gd depth.

Henners E S'x ★★ Nice biscuity, reliable Brut and cherry-spice Rosé. Always elegant, delicate, structured.

Herbert Hall Kent ★★→★★★ Lovely precision, tension and ripeness, with v.gd length. A treat.

Hoffmann & Rathbone E S'x ★★ Firm, well-made wines from bought-in fruit.

Hundred Hills Oxon New, with tasty, round wines. v. fresh. Esp Blanc de Blancs, Preamble, Rosé.

Hush Heath Estate Kent ★★★ Flavoursome sparkling, still CHARD a speciality. Flagship Balfour Brut Rosé as gd as ever. Well-organized tourism, gd restaurant.

Jenkyn Place Hants ★★ Dermot SUGRUE makes these, so they're going to be gd. Grown-up, poised.

Essex is a place to watch for still wines: slightly warmer climate, more ripeness.

Langham Wine Estate Dorset ★★ Expressive fizz, esp vigorous Blanc de Blancs, poised PINOT M; gd-value Corallian.

Leckford Estate Hants ★ Waitrose's own estate, vinified by RIDGEVIEW. Nicely mature, gd fruit, straightforward.

Litmus Made at DENBIES, outpaces host. Still. Graceful orange BACCHUS and saline Element.

Nyetimber W S'x ★★★★ 1086 for luxury market. Otherwise depth, complexity in NV, excellent Classic Cuvée, Rosé with flesh and bones.

Plumpton College E S'x ★★ UK's only wine college; makes pretty wines with fruit; gd still too.

Pommery England Hants Crisp, poised, slightly better than eponymous Champagne? Made at HATTINGLEY V for now.

Raimes Hants Tense, fresh style; gd fruit. Made at HATTINGLEY V.

Rathfinny E S'x ★★★ Elegant, refined, precision from chalk downland. Cradle Valley still is sappy, pretty. Look for gin and vermouth too.

Ridgeview E S'x ★★ Toasty, rich, reliable. Umpteen cuvées, well-aged Blanc de Blancs 09. Oak Res a bit Marmite.

Roebuck W S'x Crisp style with depth. Savoury, salty Classic Cuvée; food-friendly Blanc de Noirs.

Simpsons Kent ★★ Elegant fizz, esp Chalklands; gd still too, incl savoury Rabbit Hole PINOT N. On the up.

Squerryes Kent North Downs fruit, well-aged wines: juicy, savoury Vintage Brut 15; taut, deep, late-disgorged 11.

Sugrue E S'x ★★★★ Superb. Expensive and worth it. Try the glorious The Trouble with Dreams and tense, deep ZODO (zero dosage). Dermot S is winemaker at WISTON.

Trotton W S'x Spectacular Sparkling (that's the brand, not a description) elegant, fine, well made; gd still BACCHUS/PINOT GR.

Westwell Kent ★★★ Pure, taut, fine Pelegrim fizz; v.gd red Field PINOT N/CHARD with indigenous yeasts. Amphora Ortega pleasant.

Wiston W S'x ★★★★ Made by Dermot SUGRUE; superb elegance, finesse. Steely, tense style that shouldn't be opened too young. It needs bottle age. Blanc de Noirs 14 a treat.

Wyfold Oxon ★★★ Brut has lovely taut richness; pale Rosé is full-flavoured and ripe. Lovely wines.

Pink, pinker, pinkest
Fancy some still rosé – crisp and Provençal-pale – that you probably have to buy from the vyd, because quantities are low? Try these: Artelium, Ashling Park, BALFOUR, Brenley, BRIDE VALLEY, Crouch Valley, Dillions, Folc, Gorsley, Heppington, SIMPSONS, Nutbourne, Whitehall, Woolton Farm.

Central & Southeast Europe

More heavily shaded areas are the wine-growing regions.

Abbreviations used in the text:

Bal	Balaton	N/S Pann	North/South Pannonia
Cri & Mar	Crișana & Maramureș	Pod	Podravje
Cro Up	Croatian Uplands	Pos	Posavje
Dalm	Dalmatia	Prim	Primorje
Dan P	Danubian Plain	Sl & CD	Slavonia & Croatian Danube
Dob	Dobrogea	Thr L	Thracian Lowlands
Is & Kv	Istria & Kvarner	Tok	Tokaj
Mold	Moldovan Hills	Trnsyl	Transylvania
Mun	Muntenia & Oltenia Hills	U Hun	Upper Hungary

HUNGARY

Extensive volcanic soils, fabulous grapes like Furmint, Hárslevelű, Kadarka and Kékfrankos and a dynamic winemaking scene all stand Hungary in great stead as a powerhouse for the region. These are grapes that can offer freshness, elegance and originality.

Aszú Tok Botrytis-shrivelled grapes and the resulting sweet wine from TOK. Legal min for Aszú is 120g/l residual sugar, equivalent to 5 PUTTONYOS (similar to Sauternes); 6-Puttonyos category is even richer, as is the no-longer-permitted Aszú Eszencia. Aszú is always balanced by amazing acidity. Most producers only make in certain yrs.

Badacsony Bal Distinctive volcano n of Lake Balaton for rich, mineral, structured whites, esp revived, rare Kéknyelű. Look for Borbély, Gilvesy, Laposa, Patzay, Sabar, Szászi, SZEREMLEY, ValiBor, Villa Sandahl.

Balassa Tok ★★★ Personal project of viticulturist for GRAND TOKAJ. No surprise that vyd FURMINT selections are beautifully crafted, esp Mézes-Mály, Szent Tamás plus gorgeous, luscious Betsek Kvarc SZAMORODNI.

Balaton Wine region around Lake Balaton (Central Europe's largest lake), incl districts of BADASCONY; Balatonboglár (gd: Budjosó, GARAMVÁRI, IKON, Kislaki, KONYÁRI, bio Kristinus, Légli plus giant TÖRLEY); Balatonfüred-Csopak (n of lake, protected status for vyd-selected OLASZRIZLING, try Béla És Bandi, Dobosi, Figula, Homola, Jasdi, Liszkay, St Donát; Balaton Uplands (SOMLÓ and Zala, try Bussay, Pálffy). Mineral-rich whites, esp Olaszrizling dominate, also FURMINT, RIES and increasingly gd reds from KÉKFRANKOS, MERLOT, CAB FR.

Barta Tok ★★★→★★★★ Stunning winery in restored Rákóczi mansion, owns highest vyd in TOK. Vivien Ujvári crafts intense yet elegant wines, esp Öreg Király FURMINT, HÁRSLEVELŰ. Glorious SZAMORODNI, ASZÚ from pure Furmint.

Béres Tok ★★→★★★ Famous for a popular medicine in Hungary. Dramatic estate that reliably delivers superb, honeyed ASZÚ and gd dry wines, esp Lőcse FURMINT.

Bikavér ★→★★★ Means "Bull's Blood". PDO only for EGER and SZEKSZÁRD. Always a blend, min four varieties. In Szekszárd, at least 5% KADARKA is compulsory, with min 45% KÉKFRANKOS, specified oak ageing. Look for: Eszterbauer (esp Tüke), HEIMANN, Markvart, Meszáros, Sebestyén, TAKLER, Tuske, Vestergombi, VIDA. Egri Bikavér is 30–65% Kékfrankos, min four varieties, oak-aged for 6 mths. Superior and Grand Superior restricted yield, 12 mths in barrel. Best Egri Bikavér: BOLYKI, Bukolyi, Csutorás, GÁL TIBOR, Grof Buttler, Kovács Nimród, ST ANDREA, Thummerer.

Hungary has over 1500 thermal springs – spas are a national habit.

Bock, József S Pann ★★→★★★ Quality wine pioneer in VILLÁNY, making ripe, full-bodied, oaked reds: Bock CAB FR Fekete-Hegy, SYRAH, Capella, new 70+ blend.

Bolyki N Hun ★★ Stunning winery in a quarry in EGER, great labels, appealing juicy wines: v.gd Egri Csillag, Meta Tema, rosé and BIKAVÉR.

Csányi S Pann ★→★★ Largest winery in VILLÁNY. Much-improved entry-level varietals plus premium Ch Teleki, serious, structured Kővilla.

Degenfeld, Gróf Tok ★★→★★★ Stunning castle hotel with organic vyds. Luscious sweet wines impress: Andante botrytis FURMINT, Fortissimo, ASZÚ.

Demeter, Zoltán Tok ★★★★ Benchmark cellar for elegant, intense dry wines, always vyd selections, esp Boda, Veres FURMINTS; excellent Szerelmi HÁRSLEVELŰ. PEZSGŐ (sp) pioneer in region. Fine Anett and Eszter SZAMORODNIS, superb ASZÚ.

Dereszla, Ch Tok ★★→★★★ Excellent ASZÚ. Gd dry FURMINT, Kabar. Also reliable PEZSGŐ. Rare flor-aged dry SZAMORODNI Experience.

Disznókő Tok ★★★→★★★★ Dramatic estate with gd restaurant. Superb, silken sweet wines that keep beautifully – one of v. few to make ASZÚ every yr. Wonderful *Kapi* cru in top yrs; v.gd 1413 SZAMORODNI. Inspiration dry white also gd.

Dobogó Tok ★★★ Pretty family winery in TOKAJ town. Benchmark ASZÚ 6-PUTTONYOS and late-harvest Mylitta, excellent long-lived, dry FURMINT and unusual but gd PINOT N Izabella Utca.

Dűlő Named single vyd or cru. Top *dűlő* in TOKAJ incl Betsek, Bomboly, Király, Mézes-Mály, Nyúlászó, Szent Tamás, Úrágya.

Duna (Danube) Duna Great Plain, largest region for light-bodied wines. Districts: Csongrád, Hajós-Baja (Koch, Sümegi), Kunság (Frittmann, Font, Gedeon).

Eger N Hun The s-facing slopes of Bükk Mtns, giving finer, cooler-climate, red blends and elegant fresh whites, esp Egri BIKAVÉR and Egri Csillag ("Star of Eger"), dry white blend of Carpathian grapes. Try Bukolyi, Csutorás, Gróf Buttler, Kaló Imre (natural), Ostorosbor, Petrény, Thummerer, Tóth Ferenc.

Essencia / Eszencia Tok Legendary, syrupy free-run trickle from ASZÚ grapes, min 450g/l residual sugar, barely ferments Reputed to raise the dead.

Etyek-Buda N Pann Rolling limestone hills producing expressive, crisp whites (esp SAUV BL), gd sparklers, promising PINOT N. Leading producers: ETYEKI KÚRIA, HARASZTHY, Kertész, Nyakas, Rókusfalvy, TÖRLEY Sparkling Cellar.

HUNGARY

Etyeki Kúria N Pann ★★ Leading winery in ETYEK-BUDA, producing v.gd SAUV BL, elegant reds. Winemaker Meresz Sandor also has natural project, gd Zenit.

Figula Bal ★★→★★★ Family winery nr Csopak, notable vyd selections of OLASZRIZLING, esp Öreghegy, Sáfránkert, Szákas. Excellent Köves (w blend).

Gál Tibor N Hun ★★ Appealing Egri Csillag, fine KADARKA, modern TiTi BIKAVÉR.

Garamvári Bal ★★ Leading bottle-fermented PEZSGŐ specialist, esp FURMINT Brut, Evolution Rosé; gd Garamvári range, esp SAUV BL, IRSAI OLIVÉR. Lellei label consistent, great-value varietals.

Gere, Attila S Pann ★★★→★★★★ Some of country's best reds from pioneering family winery in VILLÁNY: Attila Cuvée, Kopar Cuvée, Solus MERLOT, VILLÁNYI FRANC (CAB FR). Try elegant Fekete-Járdovány (rare historic grape from organic plot), plus gd TEMPRANILLO.

Gizella Tok ★★★ Superb small family winery. Inviting pure FURMINT, delightful Barát HÁRSLEVELŰ, delicious SZAMORODNI.

Grand Tokaj Tok ★→★★ Huge investments at TOK's largest winery have improved top wines under guidance of highly regarded winemaker Karoly Áts and viticulturist István BALASSA. Enjoyable honey-and-lemon Arany Késői Late Harvest, stylish dry FURMINT Kővágó DŰLŐ, v.gd ASZÚ.

Haraszthy N Hun ★★ Beautiful estate at ETYEK-BUDA for expressive, precise SAUV BL, crisp, aromatic Sir Irsai (w).

Heimann S Pann ★★→★★★ Family winery in SZEKSZÁRD, esp intense Barbár, Franciscus. Fine KADARKA, Alte Reben KÉKFRANKOS. New Heimann & Fiai terroir range with local grapes, natural fermentation.

Hétszőlő Tok ★★★ Historic cellar and stunning organic vyd, owned by Michel Reybier of Cos d'Estournel (*see* B'x). Try elegant Kis-Garai FURMINT, fine ASZÚ.

Heumann S Pann ★★★ German/Swiss-owned estate in Siklós making great KÉKFRANKOS Res, CAB FR Trinitás, delicious rosé, classy SYRAH.

Hilltop Winery N Pann ★★ In Neszmély. Meticulous, value DYA varietals, Hilltop, Moonriver export labels. Kamocsay Premium range (CHARD, Ihlet Cuvée) v.gd.

Holdvölgy Tok ★★→★★★ Super-modern winery in MÁD, noted for complex dry wines (esp Vision, Expression), plus v.gd Eloquence SZAMORODNI.

Ikon Bal ★★ Well-made wines from KONYÁRI and former Tihany abbey vyds. Try Evangelista CAB FR.

Juliet Victor Tok ★★★ Ambitious investment by founder of Wizzair. Impressing with estate and vyd-selection dry FURMINTS (notable *Bomboly* and *Király*) and superb rich SZAMORODNI and ASZÚ.

Kikelet ★★★ Beautifully elegant wines (try vyd selections of HÁRSLEVELŰ and FURMINT) from small family estate owned by French winemaker and her Hungarian husband.

Királyudvar Tok ★★★ Bio producer in old royal cellars at Tarcal. Top FURMINT Sec, Henye PEZSGŐ, Cuvée Ilona (late-harvest), flagship 6-PUTTONYOS Lapis ASZÚ.

Konyári Bal ★★→★★★★ Next-generation family estate nr BAL. Try DYA rosé; Loliense (r/w); v.gd Jánoshegy KÉKFRANKOS, plush MERLOT Sessio, impressive Páva.

Kovács Nimród Winery N Hun ★★→★★★ In heart of EGER. Try Battonage CHARD, Sky FURMINT, Rhapsody BIKAVÉR, Blues KÉKFRANKOS, 777 PINOT N, NJK.

Somló wines used to be sold by chemists for their mineral richness.

Kreinbacher Bal, Somló ★★→★★★ Leading producer of PEZSGŐ, always FURMINT in blend; v.gd Classic Brut, superb Prestige Brut. Also gd still wines: Juhfark, Selection range for Furmint and HÁRSLEVELŰ.

Mád Tok Historic wine-trading town with superb vyds. Mád Circle of leading producers: Árvay, Áts (family project of Karoly Áts also of GRAND TOKAJ), BARTA, Budaházy, Demetervin (gd Mád FURMINT, Úrágya DŰLŐ), HOLDVÖLGY, JULIET

VICTOR, Lenkey (unique long-aged complex wines), Mád Hill, Orosz Gabor, Pelle, ROYAL TOKAJI, SZEPSY, Tokaji Classic.

Mad Wine Tok ★★ Sizeable winery on edge of MÁD (with handy café), previously Szent Tamás, now under Mad label; gd Dry FURMINT, late-harvest, v.gd single-vyd Furmints and SZAMORODNI. Mad One for flagship wines.

Malatinszky S Pann ★★★ Certified organic VILLÁNY cellar. Top long-lived Kúria **Cab Fr**, Kövesföld (r). Noblesse labels gd.

Tokaj is the 2nd-oldest delimited wine region in the world – 1737.

Mátra N Hun ★→★★ Region for fresh lively whites, rosé and lighter reds. Better producers: Balint, Benedek, Centurio, Gábor Karner, NAG, Nagygombos (rosé specialist), Szöke Mátyás.

Mór N Pann Small region, famous for fiery local Ezerjó. Try ★★ **Czetvei Winery**.

Oremus Tok ★★★→★★★★ Part of the Vega Sicilia (*see* Spain) stable: dry FURMINT Mandolás continues to be a benchmark. Also fine, elegant ASZÚ, SZAMORODNI.

Pajzos-Megyer Tok ★★→★★★ Two properties under one (French) ownership. Try Megyer for inviting, gd-value late-harvest varietals and remarkable, complex flor-aged dry SZAMORODNI. Pajzos for more serious dry but fruity Furmint T, tok ASZÚ.

Pannonhalma N Pann ★★→★★★ Winery beneath stunning 800-yr-old Abbey. Look for top Hemina blends plus vgd SAUV BL. Tricollis blends gd entry-point.

Patricius Tok ★★→★★★ Steely, vibrant dry FURMINTS incl organic range. PEZSGŐ is a new focus. Amazing ASZÚ (esp 17).

Pezsgő Hungarian for sparkling; growing trend. PDO TOK must be bottle-fermented.

Puttonyos (Putts) Sweetness/style guide on labels for TOK ASZÚ, based on residual sugar. Traditionally number of 25-kg puttonyos or hods of Aszú grapes added to each 136-litre barrel (*gönci*) of base wine.

Royal Tokaji Wine Co Tok ★★★→★★★★ MÁD winery that led renaissance of TOK (Hugh Johnson was co-founder in 1990). Excellent 6-PUTTONYOS single-vyd bottlings: esp Betsek, **Mézes-Mály**, Nyulászó, **Szent Tamás**. Blue label is benchmark 5-Puttonyos. Also gd-value Late Harvest and dry wines.

St Andrea N Hun ★★★→★★★★ Leading EGER father-and-son winery for exciting, new-wave BIKAVÉR (fruity Áldás, single-vyd Hangács, Axios made by son, barrel-selection Merengő, stunning flagship Nagy-Eged-Hegy). Also gd Egri Csillag (w): Boldogságos, Napbor, Örökké. FURMINT-based Mária is white flagship.

Sauska S Pann, Tok ★★→★★★★ Returning expat founded stunning wineries in VILLÁNY, TOK, with new PEZSGŐ cellar in progress – a winery focus. Refined KADARKA, KÉKFRANKOS, CAB FR and impressive red blends, esp Cuvée 7 and Cuvée 5. Consistent, mineral FURMINT from Tok and complex vyd selections Medve and Birsalmás.

Somló Bal Dramatic extinct volcano famous for firm, mineral whites, esp Juhfark (sheep's tail). Small wineries dominate, esp Fekete, Györgykovács, Kolonics, Royal Somló, Somlói Apátsági, Somlói Vándor, Spiegelberg.

Sopron N Pann An extension of Burgenland into Hungary – almost burgundian reds, esp from most-grown KÉKFRANKOS. Bio WENINGER is standard-setter. Also look for Luka, Pfneiszl, Taschner, Vinceller, Wetzer.

Szamorodni Tok Name of Polish origin for TOK made from whole bunches, with partial botrytis. Gaining popularity since 3- and 4-PUTTONYOS ASZÚ stopped – seen as more authentic than late-harvest. *Édes* (sweet) style is min 45g/l sugar (usually sweeter), 6 mths oak ageing. Try BALASSA, BARTA, Bott, DEMETER, Demetervin, GIZELLA, HOLDVÖLGY, JULIET VICTOR, KIKELET, Kvaszinger, MAD WINE, Nobilis, OREMUS, Pelle, SZEPSY. Best dry (*szaraz*) versions flor-aged like Sherry, try Breitenbach, CH DERESZLA, Karádi-Berger, **Megyer**, **Tinon**.

Szekszárd S Pann Famous for rich reds, now increasing focus on BIKAVÉR,

KÉKFRANKOS, reviving lighter KADARKA. Try Dúzsi (rosé), Eszterbauer (Nagyapám Kadarka, Tüke Bikavér), HEIMANN, Lajver, Markvárt (Kadarka, Kékfrankos), Pósta (Kadarka), Sebestyén (Ivan-Volgyi Bikavér), Szent Gaál, TAKLER, Tüske (Kadarka, Menek), VESZTERGOMBI, VIDA.

Szepsy Tok ★★★★ Standard-setting, no-compromise 17th- and 18th-generation family winery in MÁD, now guided by István S Jnr. Focus on complex, terroir-selected dry FURMINT (esp Bányász, Percze, Szent Tamás, Urbán), excellent sweet SZAMORODNI. Estate Furmint also v.gd. Superb ASZÚ incl new single-vyd releases.

Szeremley Bal ★★ Pioneering estate in BADACSONY. Volcanic slopes give intensely mineral, long-ageing whites esp RIES, Szürkebarát (aka PINOT GR), rare KÉKNYELŰ.

Takler S Pann ★★ Family estate for ripe, supple SZEKSZÁRD reds. Best: Res selections of BIKAVÉR, CAB FR, KÉKFRANKOS.

Tinon, Samuel Tok ★★★ Frenchman rooted in TOK since 1991. Long-ageing, complex ASZÚ in more traditional style with long maceration, oak-ageing; gd vyd-selected FURMINT. Flor-aged dry *Szamorodni* wonderful.

Tokaj-Nobilis Tok ★★★ Fine small bio producer run by Sarolta Bárdos, one of TOK's inspirational women. Excellent dry Rány & Barakonyi FURMINT, v.gd ASZÚ, pretty 3 Gracia (sw). Benchmark for Furmint PEZSGŐ.

A vehicle invented in 1500 took name from town of Kocs – or "coach".

Tokaj / Tokaji Tokaj is the town and wine region; Tokaji the wine. Recommended producers without individual entries: Árvay, Áts, Bardon, Basilicus, Bodrog Bormühely, Bott Pince, Breitenbach, Budahazy, Carpinus, Demetervin, Erzsébet, Espák, Füleky, Harsányi, Hommona Atilla, Karádi-Berger, Kvaszinger, Lenkey, Orosz Gábor, Pelle, Pendits, Peter, Sanzon, Szarka, Szóló, TR, Zombory, Zsadányi, Zsirai.

Törley ★→★★ Bright, gd-value DYA international and local varieties (labels incl Chapel Hill, St Stephen's Crown), György-Villa for top selections. Major fizz producer (esp Törley, Gala, Hungaria labels), v.gd classic method, esp *François President Rosé Brut*, CHARD Brut.

Tornai Bal ★★ Signficant SOMLÓ estate; gd-value entry-level varietals, excellent, complex mineral Top Selection range FURMINT, Juhfark, TRAMINI.

Tűzkő S Pann ★★ Antinori-owned estate in Tolna; gd CAB FR, KÉKFRANKOS, MERLOT and TRAMINI.

Vesztergombi S Pann ★★→★★★ Family SZEKSZÁRD estate impressing with next generation at helm. Look for Bikavér, Kétvölgy Kékfrankos, Vintage.

Vida S Pann ★★→★★★ Next-generation family winery in SZEKSZÁRD. Appealing entry-point Tündértánc. Wonderful Bonsai (old-vine) KADARKA, Hidaspetre KÉKFRANKOS, La Vida.

Villány S Pann Most s wine region. Noted for serious ripe B'x varieties, esp top-performing CAB FR – labelled Villányi Franc with its own rules for premium and super-premium categories. Juicy examples of KÉKFRANKOS, PORTUGIESER. High quality without own entry: Gere Tamás & Zsolt (Aureus Cuvée), Günzer Tamás (Mátyás Cuvée, Bocor), Hummel, Jackfall, Janus, Kiss Gabor, Lelovits (CAB FR), Maul Zsolt (Creátor Kékfrankos, Dávid), Polgar, Riczu (Symbol Cuvée), Stier (MERLOT, Villányi Cuvée), Ruppert, Tiffán, Wassmann (bio).

Vylyan S Pann ★★→★★★ Red specialist making v.gd vyd selections, esp Gombás PINOT N, Mandolás CAB FR, Montenuovo, Pillangó MERLOT. *Duennium Cuvée* is flagship red. Also delicious rare Csókaszőlő.

Weninger N Hun ★★★ Standard-setting bio winery in SOPRON run by Austrian Franz Weninger Jnr. Single-vyd *Steiner Kékfrankos* is superb. SYRAH, CAB FR and red Frettner blend also impressive. Try Rózsa too.

BULGARIA

New projects, new wineries and old grapes revived are the story in this small but dynamic wine country. Good-value international varietals for export still exist, but there's so much more to discover – you will have to seek out specialist importers or, better still, visit Bulgaria (where you can be sure of a warm welcome) to find them.

Around half of Bulgaria's winemakers are women: an equality world leader.

Alexandra Estate Thr L ★★ Res continues to impress, as does v.gd acacia-aged VERMENTINO, plus elegant rosé.

Angel's Estate Thr L ★★ Stallion range for smart, sleek reds and impressive Deneb labels, esp SYRAH.

Bessa Valley Thr L ★★★ One of 1st proper estates of new era, now with mature vines for real depth. Try Enira, v.gd SYRAH and Enira Res, excellent **Grande Cuvée**.

Better Half Thr L ★★★ Family garage winery, small batches. Try Dalakov Kvevri range and v.gd Res (r).

Black Sea Gold Thr L ★ 600 ha nr Black Sea. Better labels: Salty Hills, Vera Terra, gd-value Golden Rhythm.

Bononia Dan P ★★ Renovated brewery-turned-winery, close to Danube; v.gd CHARD, **Gomotartzi Gamza**, Istar SAUV BL, Ooh La La rosé.

Borovitsa Dan P ★★★ Handcrafted terroir wines in tiny parcels. Dux is long-lived flagship. Also try Cuvée Bella Rada (RKATSITELI), rare local Evmolpia, GAMZA (Black Pack), MRV, Sensum.

Boyar, Dom Thr L ★→★★★ Large modern winery, dramatic painted tanks. Reliable entry-point ranges Bolgaré, Deer Point; mid-range Elements, Quantum. Top: single-vyd Solitaire (MERLOT). Boutique Korten cellar: v.gd Grand Vintage, Natura.

Rose V grows c.85% of world's rose oil: should be able to smell it from space.

Bratanov Thr L ★★ Family estate in Sakar using wild yeast. Try CHARD, SYRAH, 3-Blend, Symbiose, Tamianka.

Burgozone Dan P ★★ Family estate overlooking Danube. Gd whites, esp Eva, SAUV BL, VIOGNIER. Refined reds, esp Iris Creation, Esperanto.

Damianitza Thr L ★★→★★★ Holistic producer in STRUMA v. Try Redark, Dzindzifkite CAB FR, Uniqato and flagship Kometa.

Dragomir Thr L ★★→★★★ Intense long-lived reds, esp Pitos, flagship RUBIN Res, CAB FR. Try fruit-focused Sarva range (esp MAVRUD, rosé).

Eolis Thr L ★★→★★★ Tiny bio estate; v.gd VIOGNIER, SYRAH, Inspiration (r blend).

Katarzyna Thr L ★★ Supple, ripe, well-made wines. Flagship Res impresses, also MAVRUD, MERLOT, Question Mark, Seven Grapes.

Logodaj Thr L ★★ Struma winery for super-ripe reds; esp Nobile Early MELNIK, v.gd rosé, fine bottle-fermented **Satin**.

Maryan Dan P ★★ Family-run; v.gd Res (r), Ivan Alexander (r), Kera Tamara CHARD.

Medi Valley Thr L ★★→★★★ Own vyds at 550m (1684ft) nr Rila Monastery; v.gd reds, esp Great Bulgarian, Incanto Black, MELNIK 55; fine VIOGNIER.

Projects and more

Bulgaria continues to be a hotbed of experimentation, new wines and new names. Keep an eye on: Abdyika, Augeo, Balar, Bendida, Ch Copsa, Damyanov, Four Friends, Georgiev/Milkov, Glushnik, Haralambievi, Ivo Varbanov, Kapatovo, Levent Wine House, Libera Estate, Melvino, Pink Pelican, Red Church (from Alex Kanev), Roxs, Rupel, Seewines, Staro Oryahovo, Stratsin, Uva Nestum, Varna Winery, Via Verde, Via Vinera Karabunar, Villa Bassarea, Villa Yustina, Yalovo, Zaara Estate.

Menada, Dom Thr L ★ Large producer, cheerful Tcherga blends.

Midalidare Estate Thr L ★★→★★★ Immaculate boutique winery. Precise whites v.gd reds (esp Grand Vintage), country's best sparkling.

Minkov Brothers Thr L ★→★★ Large boutique winery nr Karnobat. Try CAB FR, Cycle range, Enoteca Rubin, Le Photografie PINOT N, SYRAH.

Miroglio, Edoardo Thr L ★★★ Immaculate estate noted for superb bottle-fermented sparkling. Exciting PINOT N, esp Res; v.gd Elenovo range, esp CAB FR, MAVRUD Rubin and flagship Soli Invicto.

Neragora Thr L ★★ Organic estate, gd MAVRUD, CAB SAUV Selection.

Orbelia Thr L ★★ Family winery in STRUMA V; fine CAB FR, v.gd Via Aristotelis range new estate res.

Orbelus Thr L ★★ Organic STRUMA winery. Vibrant Orelek whites, gd MELNIK 55.

Rossidi Thr L ★★→★★★ Pioneering winery nr Sliven. Top concrete egg-fermented CHARD; v.gd MAVRUD, RUBIN, excellent SYRAH. Intriguing orange GEWURZ.

Rumelia Thr L ★★ MAVRUD specialist, v.gd, try Rumelia Res, Erelia, unoaked Merul.

Salla Estate Dan P ★★ Precise whites, esp Barrel CHARD, Vrachanski MISKET, RIES. Elegant CAB FR.

Santa Sarah Thr L ★★★ Quality pioneer, now with own estate. Bin reds are v.gd; Privat is flagship. Appealing No Saints rosé.

Stefan Pirev Wines ★★→★★★ Personal project of respected winemaker; v.gd Eager blends (r/w), CHARD Kosara.

Yoghurt too. Bacteria that make yoghurt 1st identified in Bulgaria, 1905.

Struma Valley Thr L Stunning scenery in Bulgaria's warmest region, well organized for tourism. Focus on local grapes: MELNIK 55, Sandanski MISKET, Shiroka Melnik (aka Broadleafed Melnik). Names to watch: Abdyika, Augeo (Ruen), Kapatovo (GRENACHE/MOURVÈDRE/SHIRAZ), Libera (Hotovo, Orange Keratsuda), Rupel, Seewines (esp Colorito w, Disegno), Via Verde, Zlaten Rozhen and newcomer Damyanov (Broadleafed Melnik).

Terra Tangra Thr L ★★ Large estate in Sakar, certified organic red vyds; gd MALBEC, MAVRUD (r/rosé), serious Roto.

Tohun Dan P ★★ Precise, refreshing whites and rosé, esp Barrique CHARD, Tohun rosé, promising Tohun CAB SAUV/MERLOT.

Tsarev Brod Dan P ★★ Innovative estate in n. Try pét-nat RIES, rare local Gergana, Amber CHARD, Evmolpia rosé, complex SAUV BL Res; v.gd Ries Icewine.

Villa Melnik Thr L ★★ Family winery, focus on local grapes, esp MAVRUD, MELNIK; gd orange SAUV BL, impressive Res and Hailstorm labels.

Villa Yambol Thr L ★ Fair value, varietal wines and blends in Kabile range.

Vinex Slavyantsi Thr L ★→★★ "Fair for Life"-certified for work with local Roma. Reliable budget varietals and blends, esp Leva brand.

Yamantiev's Thr L ★→★★ Sound, commercial Kaba Gayda, SHIRAZ. Excellent *Marble Land (r)* and Yamantiev's Grand Res CAB SAUV.

Zagreus Thr L ★★ Organic vyd, MAVRUD in all styles: white via rosé to complex Amarone-style Vinica from semi-dried grapes. St Dimitar label gd value.

Zelanos Thr L ★★ Pristine winery/wines. Try fresh Red MISKET, PINOT GR, elegant Z series PINOT N and CAB FR.

SLOVENIA

Winemaking in Slovenia can require heroic viticulture on steep, often rocky hillsides. Hand labour is pretty much essential, so costs are high, but quality is so good that value is still strong. This is a green, forested landscape with strong environmental values, so sustainable, often organic winemaking is widespread, and the result is some of the most exciting wines in the region. It continues to be a hotspot for skin-contact, orange and low-intervention wines, but don't overlook great producers with more classic approaches.

You can ride a Lipizzaner at Slovenia's Lipica stud. Mind those high kicks, though.

Albiana Pos ★★ Small family estate with stunning vyds in DOLENJSKA. Bright whites, rosé and gd Modra Frankinja Alto.

Batič Prim ★★ Highly regarded for bio/natural wines in VIPAVA: Angel blends, MALVAZIJA, PINELA, REBULA, Zaria (w).

Bjana Prim ★★★ PENINA specialist from BRDA, always refined, elegant traditional-method sparklers using local REBULA as key ingredient.

Brda Prim Top-quality district. Try (without own entries) Benedetič, Blažič, Dobuje, Emeran Reya, Klinec, Medot (v.gd PENINA), Moro (v.gd Margherita r/w), Mulit, Zanut (Brjač, REBULA, SAUV BL).

Burja Redd ★★★★ Low-intervention, organic VIPAVA estate. Focus on local grapes, spontaneous ferments. Try Burja Bela, Burja Noir (PINOT N), Burja Reddo.

Dolenjska Pos Overlooked region, once reliant on traditional sharp red Cviček. Look for much-improved wines from ALBIANA, Dular, FRELIH, Klet Krško (Izbor LAŠKI RIZLING SW, Turn MODRA FRANKINJA), KOBAL (PINOT GR, Luna Modra Frankinja), Kozinc (Joker sp, SAUV BL). Local Žametovka proving great for sparkling, esp Dom Slapšak, Frelih.

Dolfo Prim ★★→★★★ Family winery in BRDA; gd Spirito PENINA, appetizing Gredic (w), MALVAZIJA, REBULA.

Dveri-Pax Pod ★★→★★★ Historic Benedictine owned estate nr Maribor (name means Gate of Peace). Crisp, bright, gd value whites; v.gd old-vine selections, esp FURMINT Ilovci, also excellent *Furmint Penina*.

Erzetič Prim ★★ Family winery, bio, next-generation winemaker: v.gd amphora wines: PINOT GR, REBULA.

Ferdinand Prim ★★★ Hilltop estate with new winery. Delicious cross-border Sinefinis (sp) with Gradis'ciutta (Italy), complex Brutus, excellent Epoca (r/w).

Frelih Pos ★★ Female-run family winery: Echo, PENINA, MODRA FRANKINJA, SILVANER.

Gašper Prim ★★★ Impressive brand of Gašper Čarman with KLET BRDA; v.gd MALVAZIJA, REBULA Selekcija, new Markisa rosé, excellent CAB FR.

Gross Pod ★★★ Husband and wife making amazing terroir wines. Superb Gorca and Iglič FURMINTS, Colles SAUV BL, RIES and impressive Furmint Brut Natur.

Guerila Prim ★★ Estate in VIPAVA, bio, v.gd PINELA, MALVASIA, BARBERA, CAB FR.

Istenič Pos ★★ Pioneering fizz specialist. Try Prestige Extra Brut, Gourmet Rosé, No 1 Brut and new Rare Brut Natur.

Istria (Slovenska Istra) The Slovenian part of the peninsula; main grapes: MALVAZIJA, REFOŠK. Best: Bordon, Brič, Korenika & Moškon (bio), MonteMoro, Pucer z Vrha, Rodica (organic), SANTOMAS, Steras (superb Epulon, Saurin Malvazija), VINAKOPER, Zaro.

Jakončič Prim ★★★ BRDA producer, v.gd Carolina range, skin-contact Uvaia PINOT GR.

Joannes Pod ★★ Consistent long-lived RIES from bio estate.

Slovenia is Europe's 3rd most forested country, and 53% of its land is protected.

Kabaj Prim ★★★ Highly regarded for complex skin-contact whites and bold reds from French-led BRDA estate.

Klet Brda Prim ★★→★★★ Slovenia's largest co-op, forward-thinking and dynamic; v.gd fresh Quercus range, unoaked Krasno, Colliano for US. Bagueri vyd selections excellent, esp REBULA. Refined De Baguer range and superb flagship A+ (r/w). Also produces v.gd Schumacher wines.

Kobal Pod ★★ Personal vision of Bojan K; v.gd FURMINT, Black Label SAUV BL.

Kogl Pod ★★ Historic estate nr Ormož (1542). Vibrant, precise whites, vivid rosé

Krapež, Vina Prim ★★→★★★ VIPAVA family producer. Lapor Belo, Rdece always v.gd, fantastic but tiny production of MALVAZIJA.

Kras Prim Best known for controversial TERAN denomination. Try Vinakras, Čotar for gd long-lived low-intervention wines.

Kristančič Prim ★★ Family producer in BRDA. Try Pavó from old vines.

Kupljen Pod ★★ Dry white pioneer nr Jeruzalem: Aldebaran RIES, Loona, ŠIPON, White Star.

Marof Pod ★★→★★★ Pioneering estate: Breg, Bodonci SAUV BL, Kramarovci CHARD.

Movia Prim ★★★→★★★★ Bio winery led by charismatic Aleš Kristančič, now joined by son Lan. Excellent Veliko range, esp *Belo (w)*, *Rdeče* (r), showstopping *Puro Rosé* (sp). New Kapovolto (sp). Also admired for long-macerated orange Lunar (CHARD, REBULA versions).

Pasji Rep Prim ★★ Next-generation organic VIPAVA estate; v. well-made wines, esp lovely MERLOT Breg, plus Jebatschin blends, PINOT N.

Penina Name for quality sparkling wine (Charmat or traditional method). Trendy.

Podravje Largest region covering Štajerska Slovenija and Prekmurje in e. Best for vibrant, dry whites, impressive sweet, improving reds typically from MODRA FRANKINJA, PINOT N.

Posavje Region in se covering DOLENJSKA, Bizeljsko-Sremič (ISTENIČ, Klet Krško, Dular, Tajfl) and Bela Krajina (look for Gaube, Metlika, Prus, Šturm, ŠUKLJE). Increasing focus on v.gd fizz and elegant MODRA FRANKINJA. Izbor (sw) can be excellent: great acidity.

PRA-VinO Pod A 70s pioneer of private production. Best for sweet, incl Icewine (*ledeno vino*), botrytis.

Primorje Region in w covering Slovenian ISTRIA, BRDA, VIPAVA, KRAS. Aka Primorska.

Puklavec Family Wines Pod ★★→★★★ Large family winery, consistent expressive whites in Puklavec & Friends and Jeruzalem Ormož ranges; v.gd Seven Numbers label, amazing archive wines from 70s. Also in N Macedonia (gd r).

Pullus (Ptujska Klet) Pod ★★ Reliable fresh bright Pullus label; G range and sweet wines impress.

Radgonske Gorice Pod ★→★★ Historic sparkling cellar making top-selling Srebrna (silver) PENINA; v.gd classic-method Rosé Brut, elegant Untouched by Light picked/made in dark.

Santomas Prim ★★→★★★ Benchmark producer of REFOŠK, esp Antonius from 60-yr-old vines. Also v.gd CAB SAUV, Mezzo Forte (r).

Ščurek Prim ★★★ Family estate in BRDA, five sons; gd varietal entry-point wines. Superb *Rebula Up*, attractive Stara Brajda (r/w).

Simčič, Edi Prim ★★★★ Beautiful family winery and standard-setter in BRDA, now with excellent Fojana and Kozana vyd selections and some of country's top reds, esp barrel-selection Kolos, Kozana MERLOT. Plus v.gd Classic range, Lex blends.

Pinpointing single vineyards

The next mtn to climb for Slovenian growers is single-vyd wines. Slovenska Velika Lega is a group of top producers (EDI SIMČIČ, MAROF, ŠUKLJE, Vino Gross) raising awareness of Slovenia's special and historic terroirs, and they plan to lobby to update legislation – current national vyd list is 40 yrs old. All four producers in group already produce amazing single-vyd wines, and there are others like BURJA (Stranice), KLET BRDA (De Baguer MERLOT, Motnik), MARJAN SIMČIČ (Opoka), PASJI REP (Merlot Breg), Zanut (Brjač, Jama) with vyd releases. It's only fair to mention that many small producers may only be working with a single vyd anyway, but it's a welcome move to build story of Slovenia's terroir.

Šimčič, Marjan Prim ★★★★ Single-vyd old-vine Opoka range world-class (new PINOT N Breg and Sauv Vert). Superb UnicoM (r blend). Also gd Selekcija, Teodor blends. Leonardo (sw) consistently great.

Štajerska Slovenija Pod Major region in e incl important districts of Haloze, Ljutomer-Ormož, Maribor. Crisp, vibrant whites and top sweet. Try Conrad Furst & Sohne, Dom Ciringa, Doppler, Frešer, Gaube, Heaps Good Wine, Horvat, Krainz, M-vina, Roka, Sanctum, SiSi, Šumenjak, Valdhuber, Zlati Grič.

Steyer Pod ★★ TRAMINER specialist in ŠTAJERSKA, esp Vaneja.

Šuklje ★★→★★★ Small family estate setting new standards with vyd selections in Bela Krajina: Lozice SAUV BL, Lodorna and excellent Vrbanjka MODRA FRANKINJA.

Sutor Prim ★★★→★★★★ Excellent small producer from VIPAVA. Lovely Sutor White, fine CHARD, elegant MERLOT-based red.

Tilia Prim ★★→★★★ Consistent fresh expressive Sunshine whites, owner's passion for PINOT N shown in range from easy-drinking estate to more serious Merljaki and black label.

Verus Pod ★★★ Fine, beautifully made, vibrant whites: v.gd FURMINT, PINOT GR, refined RIES, aromatic SAUV BL. Great value too.

Vinakoper Prim ★★ Large Istrian winery, continues to improve. Look for Capo d'Istria, Capris, Rex Fuscus labels, esp MALVAZIJA, REFOŠK.

Vipava Prim Dramatic valley noted for *Burja* wind in PRIM. Try Benčina, Bizjak, Fedora, Ferjančič, Guštin, JNK, Lepa Vida, Lisjak, Marc, Miška, Mlečnik, Saksida, Štokelj, Vina Ušaj Ussai.

Vipava 1894 Prim ★→★★ Much improved. Try Lanthieri range, Terase MALVAZIJA.

CROATIA

Nearly 20 million tourists to Croatia drank a lot of wine before lockdown, and while numbers partially recovered during 2021, wineries in tourist regions are still struggling. Improving quality and a host of indigenous grapes (as many as 120), spread across diverse regions, make for a lifetime's fascination. They are best explored in person.

Antunović Sl & CD ★★ Croatia's 1st female-owned winery. Mother and son make lovely GRAŠEVINA, complex Jubilea Res.

Arman, Franc Is & Kv ★★ Family estate founded in 1850 for v.gd reds, fresh MALVAZIJA, skin-contact Malvazija Classic.

Arman, Marijan Is & Kv ★★ Owner Marijan A sadly passed away In early 2022 leaving a great legacy of impressive MALVAZIJA, esp G Cru and v.gd TERAN.

Badel 1862 ★→★★ Owns Benkovac, Duravar, Ivan Dolac wineries. Try smooth ripe Korlat reds. Also sound DINGAČ 50°, PLAVAC MALI.

Benvenuti Is & Kv ★★★ Standard-setting family winery run by two brothers. TERAN, esp Santa Elisabetta, Caldierosso blend, excellent. Reliable fresh MALVAZIJA, gorgeous San Salvatore MUŠKAT (sw).

Bibich Dalm ★★→★★★ Family estate from C15, focus on local grapes, esp Debit (w), plus SYRAH, blends. Try Lučica single-vyd and sweet Ambra.

Bire Dalm ★★→★★★ Specialist on Korcula; v.gd Grk (all styles), esp complex Defora.

Boškinac Dalm ★★→★★★ Michelin-starred restaurant on Pag Island plus impressive long-lived reds, esp Cuvée. Also try rare Gegić Ocu.

Bura-Mrgudić Dalm ★★ Family winery renowned for weighty, traditional Bura PLAVAC MALI. Also modern Benmosche DINGAČ, ZIN.

Cattunar Is & Kv ★★ Hilltop winery for MALVAZIJA from Istria's four different soils. Late-harvest Collina always impresses.

Clai Is & Kv ★★ Natural-wine pioneer, admired for skin-contact orange: Sveti Jakov MALVAZIJA, Ottocento blends. Red Bombonero ages well.

> **Istria & Kvarner**
> Dynamic Adriatic peninsula and nearby islands, also great olive oil and truffles. Versatile MALVAZIJA is main grape and controversial TERAN (must be labelled Hrvatska Istra – Teran). Also gd spot for MERLOT. Try Banko Mario, Capo, Cossetto, Degrassi, Deklić, Dom Koquelicot, Dubrovac, Franković, Ipša, Medea, Meneghetti, Misal Peršurić (sp), Novacco, Piquentum, Radovan, Rossi, Sirotić, Tercolo, Trapan, Veralda, Zigante. On Kvarner: KATUNAR WINE ESTATE; Ivan Katunar and Sipun for Žlahtina and rare Sansigot.

Coronica Is & Kv ★★→★★★ A 3rd-generation winemaker on *terra rossa*. Highly regarded for barrel-aged Gran MALVAZIJA and Gran TERAN.

Dalmatia Rocky coastline, lovely islands, many exciting wineries.

Damjanić Is & Kv ★★→★★★ Increasingly impressive family winery; v.gd Borgonja (aka BLAUFRÄNKISCH), Clemente (r/w), MALVAZIJA.

Dingač Dalm Historic PDO for weighty reds from PLAVAC MALI, on Pelješac peninsula. Try Benmosche, BURA-MRGUDIĆ, Crna Ovca, KIRIDŽIJA, Lučić, Madirazza, Matuško, Miličić, SAINTS HILLS, SKARAMUČA, Vinarija Dingač.

Enjingi, Ivan Sl & CD ★★ Legendary natural-wine maker in SLAVONIJA. Noted for GRAŠEVINA and long-lived Venje.

Enosophia Sl & CD ★→★★ Improving. DYA GRAŠEVINA, Miraz Frankovka, CAB FR.

Fakin Is & Kv ★★★ Exciting young winemaker continues to impress. Try fresh MALVAZIJA and top label La Prima, plus Il Primo TERAN.

Galić Sl & CD ★★★ Precise, refined wines: v.gd Brut Natur (sp), Crno 9, Leon GRAŠEVINA, CHARD, PINOT N.

Grabovac Dalm ★★★ Family winery impressing with Trnjak, Modro Jezero Ris, skin-contact Kujundžuša.

Gracin Dalm ★★→★★★ Professor Leo G pioneered revival of BABIĆ making a complex yet elegant version. Also v.gd Kontra in JV with KIRIDŽIJA.

Grgić Dalm ★★ California legend Mike Grgich, of Judgement of Paris fame, returned to Croatian roots to make super-rich PLAVAC MALI, elegant POŠIP on Pelješac peninsula with daughter and nephew.

Hvar Dalm Island with UNESCO listing for Stari Grad Plain and its vyd *chora*, dating to C4 BC. Noted for PLAVAC MALI; Carić, Duboković, PZ Svirče, TOMIĆ, ZLATAN OTOK all gd

Iločki Podrumi Sl & CD ★→★★ Oldest winery in Europe, continuous production, cellar from 1450. Try Premium GRAŠEVINA, TRAMINAC, Principovac range.

Kabola Is & Kv ★★→★★★ Immaculate organic estate, watchword for quality: v.gd MALVAZIJA as fizz, young wine, cask-aged Unica, Amfora. Also gd Amfora TERAN.

Katunar Wine Estate Is & Kv ★★ Leading producer of Žlahtina found only on island of Krk. Try Sveta Lucija. Also gd PLAVAC MALI.

Kiridžija Dalm ★★ PLAVAC MALI gd, weighty but fruity DINGAČ v.gd.

Komarna Dalm Newest wine region: seven producers, all organic. Volarević impresses with gd PLAVAC MALI as rosé, red and complex Gold Edition. Rizman for Nonno POŠIP, SYRAH; Terra Madre for *sur lie* Pošip; SAINTS HILLS for Sv Roko.

Korta Katarina Dalm ★★ Croatian wine with US roots and luxury hotel on Korcula; gd POŠIP, weighty PLAVAC MALI.

Kozlović Is & Kv ★★★→★★★★ Beautiful winery, superb wines: v.gd MALVAZIJA, TERAN, esp Santa Lucia selections, delicious Sorbus MUŠKAT Momjanski.

Krajančić Dalm ★★→★★★ Cheerleader for revival of exciting POŠIP: Intrada, Sur Lie, Statut, orange.

Krauthaker Sl & CD ★★★ Pioneering producer and GRAŠEVINA specialist from Kutjevo, esp Mitrovac, sweet TBA. Also try Zelenac Rosenberg.

Kutjevo Cellars Sl & CD ★★ Large producer with cellar from 1232; gd-value, consistent GRAŠEVINA main focus. Also premium De Gotho, lovely Icewine.

Laguna, Vina Is & Kv ★★→★★★ Signficant winery on *terra rossa*. Reliable gd-value MALVAZIJA and top, age-worthy Vižinada label. Premium Festigia and Riserva ranges increasingly impressive.

Matošević Is & Kv ★★★ Benchmark MALVAZIJA, esp Alba, Alba Antiqua, Alba Robinia (in acacia). Also v.gd Grimalda (r/w), SAUV BL.

Međimurje Cro Up Coolest wine county in n, focus on Pušipel (aka FURMINT), fresh SAUV BL. Try Cmrečnjak, DK Vina, Horvat, Jakopić, Kocijan, Lovrec, Štampar.

Miloš, Frano Dalm ★★ Admired for traditional structured Stagnum, also easier PLAVAC MALI and rosé.

Pilato Is & Kv ★★ Consistent family winery; v.gd MALVAZIJA, PINOT BL and reds, esp Grande Cuvée.

Prošek Dalm Controversial sweet wine made from sun-dried local grapes in DALM; 1st mention 1556. Has applied for PDO status from 2021: Italy's Prosecco producers object.

Roxanich Is & Kv ★★→★★★ Wine hotel, natural producer noted for long-macerated whites (Antica, Ines, Milva); complex reds, esp TERAN Ré, Superistrian Cuvée.

Saints Hills Dalm, Is & Kv ★★→★★★ Two estates in Istria and DALM, consultant Michel Rolland; v.gd Frenchie (w), Nevina (w), Posh (POŠIP), structured DINGAČ, richly fruity PLAVAC MALI St Roko.

Skaramuča Dalm ★★ One of largest private wineries in Croatia, family run; gd fruit-driven PLAVAC MALI, POŠIP. Step up to Elegance label for v.gd DINGAČ, Plavac.

Croatia's kuna currency is named after pine marten: pelts used to be currency.

Slavonija / Slavonia Inland region, also famous for oak. GRAŠEVINA and whites most planted; much-improved reds. Try Adzic, ***Antunović***, Bartolović, Belje, ENJINGI, ENOSOPHIA, GALIĆ, KRAUTHAKER, KUTJEVO, Mihalj, Orahovica, Sontacchi, Zdjelarević.

Stina Dalm ★★→★★★ Dramatic steep vyds on Brač Island; v.gd PLAVAC MALI (esp Majstor label, top Remek Djelo), POŠIP, Vugava; gd Tribidrag (aka ZIN).

Testament Dalm ★★→★★★ Young winery in historic area at Sibenik, organic vyds; v.gd POŠIP, Tribidrag (aka ZIN), BABIĆ as red and Opolo (rosé). Fun Dalmatian Dog (r/w). Sister winery Merga Victa on Korcula.

Tomac Cro Up ★★→★★★ Amphora pioneer and natural-wine maker nr Zagreb. Notable sparkling, pét-nat and orange TRAMINEC.

Tomaz Is & Kv ★★★ Family winery pushing the boundaries in Istria. Seriously impressive Barbarossa TERAN, complex MALVAZIJA Sesto Senso.

Tomić Dalm ★★ Bold wines from leading personality on HVAR; organic PLAVAC MALI; gd reds (esp Plavac Barrique); sw PROŠEK Hectorovich.

Zlatan Otok Dalm ★★ Family winery from HVAR with vyds also at Makarska, Šibenik. Famous for ripe reds, gd DYA POŠIP.

BOSNIA & HERZEGOVINA, KOSOVO, MONTENEGRO, NORTH MACEDONIA, SERBIA

The wines of the western Balkans are endlessly fascinating, evolving quickly as new-wave wineries pop up and established names move with the times. Local drinkers still love the glamour of international grapes; for outsiders the raft of indigenous grapes is what grabs attention.

Bosnia & Herzegovina White Žilavka is star, esp from sun-drenched karst vyds around Mostar. Juicy supple red Blatina, darker, more structured Trnjak are gaining attention too. Look out for: Andrija (Arhivsko, Selekcija Žilavka, Res) Begić (PLAVAC MALI r/rosé), bio *Brkić (Mjesečar)*, Carska Vina (David, Žilavka) Hercegovina Produkt (Zlatna Dolina, Charisma labels), Jungić (Premium CAB SAUV, Šikar), Keža (Ž range), Marjanović (Blatina Barrique), Nuić (Blatina, Trnjak), Rubis (Veteribus Blatina, Žilavka), *Škegro (Krš Orange, Krš Trnjak, Carsus Blatina)*, Tolj (Kavalkada Žilavka), Tvrdos Monastery (VRANAC), Vilinka (Žilavka, X-Line), Vinarija Čitluk (Teuta range), Vino Milas (Blatina Res Žilavka), *Vukoje (Carsko-Vino, Selekcija, Vranac).*

Kosovo Declared independence in 2008 but still has two wine laws – its own and Serbia's. Largest region is Dukagjini (Metohija in Serbian rules) with Rahovec the largest zone; 37 licensed wineries, Stonecastle and Old Cellar (Bodrum i Vjeter) biggest. Smaller names like Kosova, Sefa improving, esp VRANAC. Most planted grapes: Vranac, Prokupac, Smederevka/DIMIAT, GAMAY, WELSCHRIESLING. Exports mainly to neighbours.

Montenegro Inky-dark, full-bodied VRANAC dominates, as does 13 Jul Plantaže with 2310-ha single vyd, one of Europe's largest. Wines are pretty gd, esp Vranac: try Premijer, Pro Corde, Stari Podrum, and best can age. Small wineries getting better: Lipovac, Rupice, Savina, Sjekloča, Vukicevic. Research shows Montenegro as viticultural hotspot: home of Vranac, ZIN (local name Kratošija).

North Macedonia Continues to move away from cheap bulk towards quality, esp with flagship Vranec grape (local spelling), which produces inky-dark wines with ageing potential, capable of terroir expression. Giant *Tikveš* has French-trained winemaker, intensive research programme; impresses with *Barovo, Bela Voda* single vyds; gd Special Selection and rich, oaked Dom Lepovo, reliable Alexandra Cuvée. Also gd: Bovin (Alexandar, super-rich A'gupka, Dissan Barrique), Dalvina (esp Armageddon, Dionis, Hermes, Synthesis labels); Ezimit (gd-value varietals), Imako (Black Diamond), Lazar (Erigon r, Kratošija), Puklavec Family Vyds (Instinct Vranec); *Stobi (v.gd Vranec Veritas, also gd Aminta r)*, Cuvée, RKATSITELI Vranec classic). Ch Kamnik nr Skopje is leading boutique winery, v.gd 10 Barrels, Winemaker's selection (organic) Cuvée Prestige, Vranec Terroir.

Serbia A hotspot for new wineries and experimentation – hard to keep up. Over 20,000 ha and 353 wineries. Mix of international grapes and revived local grapes, esp Prokupac, plus obscurities like Bagrina, Morava, Neoplanta, Probus, Seduša. Try Aleksandrović (Regent Res, Rodoslov, Trijumf range), Aleksić (Amanet, Tamjanika), Bikicki (Uncensored TRAMINER), Botunjac, BT Winery (Marselan Limited Edition, President VRANAC), Budimir (Boje Lila, Svb Rosa, Triada), Chichateau, Cilić, Čokot, Despotika, Deurić (Aksiom Probus, Brut Nature, Morava), *Doja (Breg Prokupac), Dukay-Sagmeister (Kadarka, esp cru selections)*, Erdevik (Stifler's Mom SHIRAZ, CHARD), Grabak (Vivat Prokupac), Ivanović (organic from 2020; try Prokupac, No 1/2), Janko (Bifora, Zavet Stari), Kovačević, Lastar (Chard, Tamjanika, Triangl PINOT N), Matalj (Kremen Kamen, Terasa), *Maurer* (Fodor, *Kadarka 1880*), Pusula, Radovanović, *Rajković(Rskavac)*, Rubin (Rubinov Prokupac), Šijački (Seduša), Temet (Beli Kamen, Tri Morave), Tonković, Virtus (Credo, Credo Beli, Ergo), Zvonko Bogdan (Cuvée No 1, Icon Campana Rubimus, MERLOT).

CZECHIA

With all the recent restrictions affecting oeno-gastronomy, Czech winemakers have needed much agility to stay afloat; e-shops with free deliveries and tastings conducted on Zoom or Skype have become the norm. Consumers are also moving up the quality scale, demanding wines with a clear declaration of all added substances and not just the sulphites required by law. Many more are choosing artisanal production with a clear origin and specific qualities rather than run-of-the-mill products. Two regions: Bohemia (Boh) and Moravia (Mor).

Baloun, Radomil Mor ★★→★★★ Wide range of highly quaffable wines, all dry. Award-winning MERLOT; white Pinot N Bl and Blaufränkisch Bl are curiosities.

Cibulka, Vino Mor ★★ Impressive CAB SAUV and MERLOT, intriguing Blauer Silvaner Brut, all organic.

Dobrá Vinice Mor ★★★ Excellent RIES, SAUV BL, PINOT N, WELSCHRIESLING in *qvevri* (clay amphorae) from Georgia. Top blends: Cuvée Kambrium, Quatre and long-lived VDB, VDČ.

Dva Duby Mor ★★★ Focus on BLAUFRÄNKISCH/ST-LAURENT grapes grown on granodiorite subsoil typical of Dolní Kounice. Bio principles. Flagship labels: Ex Opere Operato, Impera, Mille e Tre, Rosa Inferni, Vox Silentium.

Lobkowicz, Bettina Boh ★★→★★★ Admirable PINOT N Barrique Selection, classic-method RIES (sp). Saphíra and v.gd-value entry-level Lady Lobkowicz (r/w/rosé).

Mádl Mor ★★ Nicknamed "Malý vinař" (small vintner), family-run, top labels BetOn, Mlask, Cuvée 1+1, also v.gd PINOT GR, SAUV BL.

Naturvini Mor ★★ Originally from Slovakia, Patrik Staško est in 2010. Distinctive Pálava, WELSCHRIESLING from foot of Pálava Hills where both excel; 90% bio.

Porta Bohemica Boh ★★ Marlstone bedrock for vyds n of Prague. RIES, PINOT N, MÜLLER-T. Outstanding Frühroter Veltliner, interesting blends Charpin, MüVé.

Springer, Jaroslav Mor ★★★ Father with son Tomáš make remarkable single vyd Záhřebenské PINOT N. Rouči blend (Jaroslav label), Čtvrtě (Tomáš label).

Stávek, Richard Mor ★★→★★★ Dedicated terroirist, best v'yd sites: Kolberg, Špigle-Bočky, Veselý. Orange, pét-nat hit in NYC, Japan.

Vican Mor ★★★ Est 2015 by film producer. White-grape oriented, accent on PINOT GR and Pálava grapes matured in Moravian oak. Also v.gd Quevri House range.

Znovín Mor ★→★★ Important producer and wine centre nr Austrian border. Emphasis on SAUV BL, RIES (esp Robinia matured in acacia wood), PINOT N Duel.

SLOVAKIA

Growing conditions no longer seen as marginal. Six wine regions: Central Slovakia (C Slo), Eastern Slovakia (E Slo), Lesser Carpathians (L Car), Nitra (Nit), Southern Slovakia (S Slo), Tokaj (Tok). New-wave and authentic wineries are on the increase.

Belá, Ch S Slo ★★★ Fine Mosel-style RIES by Egon Müller (*see* Germany) and Miroslav Petrech joint venture.

Bott Frigyes S Slo ★★★ Impressive collection of bio wines. Traditional varieties, many like FURMINT, Lipovina (HÁRSLEVELŰ), JUHFARK or KADARKA from nearby Hungary, plus engaging PINOT N, RIES.

Dubovský & Grančič L Car ★★★ Small young winery. St George Edition, two Sekts.

Dudo, Miroslav L Car ★★ Prize-winning CAB SAUV, also unusual cultivars and crosses: FETEASCĂ ALBĂ, Devín or Dunaj.

Elesko L Car ★★→★★★ Large super-modern facility unrivalled in Central Europe with art gallery featuring Warhol originals.

J&J Ostrožovič Tok ★★★ Leading producer, extension of larger Hungarian namesake.

Karpatská Perla L Car ★★→★★★ Important vintner specializing in RIES from single vyds (Kramáre, Suchý Vrch) and GRÜNER V (Ingle, Noviny).

Pivnica Brhlovce Nit ★★★ Photographer Ján Záborský est 2011. Artisanal made "volcanic" wines in troglodyte dwellings.

Rúbaň, Ch S Slo ★★ Ambitious enterprise. Interesting full-bodied Alibernet, Milia, Noria, Svojsen varietals.

ROMANIA

New wineries keep popping up in Romania, helped by EU funds and a burgeoning food-and-wine scene in urban areas. Romanians are big wine drinkers, so exports tend to be limited to a few larger wineries. A good excuse for a visit.

Averești, Domeniile Mold ★→★★ Much-improved winery, esp Diamond, Nativus labels for local Busuioacă, FETEASCĂ NEAGRĂ, Zghihară.

Avincis Mun ★★ Dramatic hilltop winery in DRĂGĂȘANI; v.gd Negru de Drăgășani, Cuvée Grandiflora blend. Attractive Crâmpoșie Selecționată and sparkling.

Balla Géza Cri & Mar ★★→★★★ Miniș estate run by professor of horticulture, v.gd. Excellent Stone Wine range from red soils at 400m (1312ft), plus v.gd selection Cadarca, FETEASCĂ NEAGRĂ.

Banat Dynamic region in w. Try Agape Artă & Natură, Crama Aramic, Thesaurus.

Bauer Winery Mun ★★→★★★ Excellent low-intervention winery of Oliver and Raluca Bauer (also at PRINCE ȘTIRBEY). Small batches from old vines eg. PETIT VERDOT, Sauvignonasse, sweet CRÂMPOȘIE. Orange-wine pioneer.

Budureasca Mun ★→★★ DEALU MARE estate with longstanding British winemaker. Noble 5 is gd, plus Origini, Zenovius, Vine-in-Flames export range.

Catleya Mun ★★ Personal project of French winemaker Laurent Pfeffer; gd Freamăt, excellent Epopée selection.

Corcova Roy & Dâmboviceanu Mun ★★ Superb vyds and C19 royal cellar. Try FETEASCĂ NEAGRĂ, SYRAH, appealing SAUV BL, rosé. PINOT N Res gd with age.

Cotnari, Casa de Vinuri Mold ★★ Next-generation Cotnari estate growing local varieties. Try Valdoianu for FETEASCĂ NEAGRĂ, Colocviu for GRASĂ de Cotnari, Busuioacă de Bohotin, Vibe sparkling range.

Cotnari Winery Mold ★ Former state winery, same name as DOC. Mostly dry and semi-dry whites from local grapes. Aged sweet Collection can impress.

Crișana & Maramureș Region in nw. Look for Carastelec (v.gd Carassia bottle-fermented sp and RIES), organic Nachbil (BLAUFRÄNKISCH, Ries, Grandpa), Weingut Edgar Brutler (roșu).

Dagon Clan Mun ★★→★★★ Small, impressive vyd: Sandridge and Clearstone.

Davino Winery Mun ★★★★ Excellent producer in DEALU MARE. Focus on blends for v.gd, age-worthy Dom Ceptura, Flamboyant, Rezerva, Revelatio (w); v.gd FETEASCĂ NEAGRĂ under Purpura Valahica label.

Dealu Mare / Dealul Mare Mun Means "Big Hill", possibly Romania's top region, esp for reds. Planning Romania's 1st DOCG (DOC Garantat). New, promising producers: Ferdi, Gramofon, La Migdali, Velvet. Also long-est bio Domeniile Franco-Române for burgundian PINOT N.

Dobrogea Warm region moderated by Black Sea. Improving wineries: Alcovin-Macin (Curtea Regala for fresh, clean varietals), Alira (rosé, Grand Vin), Bogdan (bio), Histria (v.gd Nikolaos CAB SAUV, rosé), La Sapata (bio, focus on BĂBEASCĂ NEAGRĂ), Rasova (esp Tortuga), Trantu, Vladoi.

DOC Romanian term for PDO. Sub-categories incl DOC-CMD: harvest at full maturity; DOC-CT: late-harvest; DOC-CIB: noble-harvest. PGI is vin cu indicatie geografică, or simply IG.

Domeniul Coroanei Segarcea Mun ★→★★ Historic royal estate. Famous for TĂMÂIOASĂ Roze. Try Minima Moralia, Principesa Margareta, Simfonia red blend.

Drăgășani Mun Dynamic region with long history and unique grapes incl Crâmpoșie Selecționată, Negru de Drăgășani, Novac.

Iconic Estate Mun ★→★★ Consistent gd-value commercial La Umbra range, Colina Pietra blends. Try Theia CHARD, Kronos PINOT N and top Hyperion label: FETEASCĂ NEAGRĂ.

Jidvei Trnsyl ★→★★ Romania's largest single vyd, 2500 ha+. Best: Owner's Choice (with Marc Dworkin of Bulgaria's Bessa V); Eiswein and Extra Brut (sp).

LacertA Mun ★★ Quality estate in DEALU MARE. Cuvée IX (r), Cuvée X (w), SHIRAZ.

Licorna Wine House Mun ★★ In DEALU MARE. Impressing with Serafim for local grapes, Bon Viveur for international blends. Anno, limited top selection.

Liliac Trnsyl ★★★ Impeccable Austrian-owned estate. Crisp fine whites, delicious sweet Nectar, Icewine with Kracher (Austria); v.gd super-premium Titan.

Metamorfosis, Viile Mun ★★★ Part Antinori-owned (*see* Italy) estate in DEALU MARE. Top: Cantvs Primvs, esp FETEASCĂ NEAGRĂ; v.gd Coltul Pietrei SAUV BL, Via Marchizului Negru de DRĂGĂŞANI, PINOT N, fruit-driven Metamorfosis range.

Moldovan Hills Mold Largest wine region in ne. Fresh whites and rosé, refined reds incl Gramma (vibrant, fresh w/rosé), Hermeziu (esp Busuioacă de Bohotin), promising newcomer Strunga (v.gd FETEASCĂ ALBĂ, FETEASCĂ NEAGRĂ). Gîrboiu for Tectonic (try Şarba), Epicentrum and top Constantin (r).

Muntenia & Oltenia Hills Mun Major region in s. DOC areas: DEALU MARE, Dealurile Olteniei, DRĂGĂŞANI, Pietroasa, Sâmbureşti, Stefaneşti, Vanju Mare.

Oprişor, Crama Mun ★★→★★★ Consistent German-owned winery; v.gd La Cetate range; Crama Oprişor CAB SAUV, Jiana Rosé, Rusalca Alba, excellent Smerenie (r).

Petro Vaselo Banat ★★ Organic vyd in BANAT; gd Bendis (sp), Melgris FETEASCĂ NEAGRĂ, Ovas (r). Fruit-driven entry-level range, PV label for top wines.

Romanian language is 1700 yrs old, the only Latin language in E Europe.

Prince Ştirbey Mun ★★★ Pioneering estate in DRĂGĂŞANI. Fine, vibrant dry whites, esp Crûmpoşie Selecţionată (still and sp), FETEASCĂ REGALĂ, TĂMÂIOASĂ, Genius Loci SAUV BL and local reds (Novac, Negru de Drăgăşani).

Recaş, Cramele Banat ★★→★★★ Romania's most successful exporter; progressive, consistent wines with longstanding Australian and Spanish winemakers; v.gd-value varietals, multiple labels incl Calusari, Paparuda, Schwaben Wein, Wildflower. Mid-range: Regno Recas, Sole, Solo Quinta. Excellent premium wines, esp Cuvée Uberland, La Stejari, Selene.

Sarica Niculiţel, Via Viticola Dob ★→★★ Signficant investment. Caii de la Letea for gd ALIGOTÉ, FETEASCĂ NEAGRĂ, rosé. Owns Domeniile Prince Matei (DEALU MARE).

SERVE Mun ★★→★★★ The 1st private winery in Romania, founded by late Count Guy de Poix; v.gd Terra Romana (Cuvée Amaury w), Cuvée Sissi rosé, PINOT N; v.gd Guy de Poix FETEASCĂ NEAGRĂ. *Cuvée Charlotte* quality red benchmark.

Transylvania Cool mtn plateau encircled by Carpathians. Noted for crisp whites. New producers in Lechinţa zone incl organic Gorgandin, Jelna, Lechberg.

Valahorum Mun ★★→★★★ Premium project founded by Tohani winery in 2018. Now several wineries and brands incl flagship super-rich Apogeum FETEASCĂ NEAGRĂ, Mennini, Mierla Alba, Tohani. La Salina from TRNSYL joined 2021.

Villa Vinèa Trnsyl ★★ Italian-owned. Diamant, GEWURZ, KERNER; red blend Rubin.

Vişinescu, Aurelia Mun ★★ DEALU MARE estate. Artisan uses local grapes. Top label is Anima, esp Fete Negre from selected FETEASCĂ NEAGRĂ.

MALTA

Malta has produced wine for hundreds of years, but quality only began to revive this century. The north makes white, the south red, roughly speaking, and Gozo makes both. Meridiana is a widely found label, Marsovin the biggest. Both have some good wines. Mar Casar makes an interesting amphora Chardonnay. On Gozo, try Ta'Mena. Local grapes are red Gellewza, relatively light in style, and white Girgentina, soft and delicate. Sirakusan is Nero d'Avola. Otherwise it's a roster of international grapes, and often a bit too much oak.

Greece

The first thing to note about Greece is its extraordinary wealth of native grape varieties – which are being exploited to the full now that Greek producers are more eager than ever to produce wines packed with individuality, character and sheer quality. Single-vineyard wines are increasingly in vogue, while growers are getting more experimental with rare yet ancient grape varieties. Grapes for these styles of wine are usually picked earlier rather than later, pushing alcohol levels down and highlighting varietal definition. Abbreviations: Aegean Islands (Aeg), Attica (Att), Central Greece (C Gr), Ionian Islands (Ion), Macedonia (Mac), Peloponnese (Pelop), Thessaloniki (Thess).

Aivalis Pelop ★★★ Top-quality producer in NEMEA, with cult followers in Greece and abroad, esp for 4 and Armakas reds. Full, new oak style, demands ageing.

Alpha Estate Mac ★★★ Superb KTIMA in AMYNTEO with largest acreage. Ktima Alpha (r/w) are classics. Ecosystem range a terroir study, Barba Yiannis XINOMAVRO from century-old vines. Meticulous work in both cellar and vyd.

Amynteo Mac (POP) Captivating XINOMAVRO (r/rosé/sp) from coolest Greek POP. Several new wineries in the works.

Argyros Aeg ★★★★ Top SANTORINI producer; VINSANTOS (older the better). Evdemon and Nyhteri are top of the tree. Largest vyd holder on the island.

Avantis C Gr ★★★ Boutique winery in Evia and SANTORINI (called Anhydrous). Exquisite Agios Chronos (SYRAH/VIOGNIER) and Icon Santorini.

Biblia Chora Mac ★★★ Popular KTIMA and justifiably so, as shown by exemplary SAUV BL/ASSYRTIKO. Ovilos range (r/w) rivals B'x at triple the price. Sister wineries in Pelop (Dyo Ipsi), SANTORINI (Mikra Thira) and GOUMENISSA (Mikro Ktima).

Boutari, J & Son ★→★★★★ Historic brand. Excellent value, esp **Grande Res Naoussa** to age 40 yrs+. Top: 1879 Legacy NAOUSSA, from v. old vyd.

Carras, Dom Mac ★★ Historic estate at Halkidiki, recently changed hands. Ch Carras a timeless classic.

Cephalonia Ion Island famous for mineral, floral ROBOLA (w) and getting famous for dry MAVRODAPHNE (r). Many excellent growers, from ultra-natural Sclavos to experimental-yet-reserved GENTILINI.

Dalamaras ★★★→★★★★ Prodigious producer in NAOUSSA. Delectable range; Palaiokalias is world-class. Try to find Vieilles Vignes.

Douloufakis Crete ★★★ Nikos D broke the news of the VIDIANO variety to the world. Some of the most elegant wines on the island, reds and whites.

Dougos C Gr ★★★ On foothills of Mt Olympus, an ambassador for RAPSANI, like Old Vines. All reds top-class: MAVROTRAGANO, Opsimo.

Economou, Ktima Crete ★★★★ One of great artisans of Greece; in Sitia; v. rare.

Gaia Aeg, Pelop ★★★ Exquisite SANTORINI range all along. Sister winery in NEMEA, **Gaia Estate** one of most sophisticated in POP.

Gentilini Ion ★★★ Historic CEPHALONIA winery, incl **steely Robola**, esp Wild Paths. Dry MAVRODAPHNE Eclipse (r) is benchmark.

If it is all Greek to you...

If you love NEBBIOLO, try XINOMAVRO, esp from NAOUSSA and AMYNTEO.
If you enjoy MERLOT, try AGIORGITIKO from MEMEA but also from Mac.
If you like GRENACHE N, try Liatiko from Sitia and Daphnes in Crete.
If you are fond of MENCIA, try Limniona from Thessaly. If MOURVÈDRE fascinates you, try MAVROTRAGANO from Tinos and SANTORINI.

> **Tinos is the next big thing**
> The island of Tinos, if you're a tourist, is a bit like Mykonos and,
> if you're a wine-lover, approaches SANTORINI. With varieties like
> ASSYRTIKO and MAVROTRAGANO and top-quality wineries like Dom de
> Kalathas, T-Oinos, Vaptistis, Volacus and more on the way, make
> sure you plan a visit.

Gerovassiliou Mac ★★★ Trendsetter. Captivating ASSYRTIKO/MALAGOUSIA, top Malagousia (he's the specialist). Best range: Museum. Linked with BIBLIA CHORA.

Goumenissa Mac ★★→★★★ (POP) Excellent XINOMAVRO/Negoska (r). Try Chatzyvaritis, Mikro KTIMA, TATSIS.

Hatzidakis Aeg ★★★ Top producer of SANTORINI; children of late Haridimos now in charge, less adventurous. Try Skytali (and oaked version), esp 19 and thereafter; also Louros, if you can find any. Top cuvées back on track in terms of quality.

Karydas Mac ★★★ Tiny family KTIMA and amazing vyd in NAOUSSA, crafting complex, age-worthy XINOMAVRO.

Katogi Averoff Epirus, Pelop ★★→★★★ Historic name; Katogi a popular red. Rossiu di Munte range from plots at 1000m (3281ft)+.

Katsaros Thess ★★★ Tiny winery on Mt Olympus. KTIMA red can age for two decades. Try XINOMAVRO Valos.

Kechris ★★→★★★ ASSYRTIKO-based The Tear of the Pine, possibly *world's best Retsina*: fantastic wine. No kidding.

Kir-Yianni Mac ★★→★★★ Fantasic NAOUSSA and AMYNTEO producer, owns SIGALAS in SANTORINI too. Moves also into GOUMENISSA, Paros and Tinos.

Ktima Estate in Greek. Increasingly seen on labels.

Lazaridi, Ktima Costa Att, Mac ★★★ KTIMA in Drama and Att (Oenotria Land label). Top: Cava Amethystos CAB FR. Julia MERLOT (p) convincing PROVENCE lookalike. Plantings in high Drama v. promising: try MALAGOUSIA.

Only ten Greek vyds are as big as an average French vyd.

Lazaridi, Nico Mac ★→★★★ Originally from Drama. Several large-volume, gd-value ranges. Top: Magiko Vouno (r/w).

Lyrarakis Crete ★★→★★★ *Single-vyd versions* extraordinary. Karnari Kotsifali is Cretan tradition re-imagined. Exciting range throughout.

Malvasia Group of four POPS recreating famous medieval "Malmsey". Not from MALVASIA, but local varieties. POPs are Monemvassia-M in Laconia, M of Paros, M Chandakas-Candia and M. of Sitia, both from Crete.

Manoussakis, Ktima Crete ★★★ Initially Rhône-inspired, but Greek varieties here to stay: ASSYRTIKO, full Muscat of Spinas, Vidiano and even Romeiko (r). Opulent styles.

Mantinia Pelop (POP) High-altitude, cool region. Crisp, almost Germanic whites from MUSCAT-like *Moschofilero*. Excellent sparklers from Spiropulos and TSELEPOS. Bosinakis, Troupis rising stars.

Mercouri Pelop ★★★ Beautiful KTIMA on w coast; v.gd dry MAVRODAPHNE/REFOSCO, delicious Foloi RODITIS (w).

Naoussa Mac ★★★→★★★★ (POP) Amazing region for breathtaking XINOMAVRO. Best on par in quality, style (but not price) with Barolo. Top: DALAMARAS, KARYDAS, KIR-YIANNI, THIMIOPOULOS. Even standard bottlings age for 10 yrs++.

Nemea Pelop ★★→★★★ (POP) AGIORGITIKO reds. Always charming; styles from fresh to classic to exotic. Try Driopi from Aivalis (New World style), GAIA, Ieropoulos, Mitravelas, PALYVOS, PAPAÏOANNOU, SKOURAS, TSELEPOS.

Palyvos Pelop ★★→★★★ Fine Ktima in NEMEA making big-framed reds with AGIORGITIKO and French varieties.

> **POP go the bubbles**
> Sparkling wines are booming everywhere, incl Greece. There are suitably
> cool climates in POPS like MANTINIA or AMYNTEO and great grapes, like
> ASSYRTIKO, Athiri or XINOMAVRO vinified as Blanc de Blancs. Karanikas
> is leader, followed by Cair, Douloufakis, Edenia, KIR-YIANNI, Milia Riza,
> Spiropoulos, TSELEPOS, Tsililis, Zoinos,

Papaioannou Ktima Pelop ★★★ Put NEMEA on map. Top wines: Palea Klimata (old vines), Microclima, Terroir. Keep for ages.

Pavlidis Mac ★★★ Premium KTIMA in Drama. Thema (w) ASSYRTIKO/SAUV BL, rosé TEMPRANILLO v. popular. Emphasis: top range incl AGIORGITIKO, Assyrtiko.

PGE Regional wines. Greek term for PGI (protected geographical indication).

POP Greek equivalent of AOP (appellation d'origine protégée). Many amazing wines not incl.

Rapsani C Gr POP on Mt Olympus. Made famous by TSANTALIS (try Grande Res); now DOUGOS and others add excitement. XINOMAVRO-based.

Retsina New Age Retsinas (GAIA, KECHRIS, natural-style Kamara) have freshness, character – great alternative to Fino Sherry. The perfect food wine if you eat family-style like Greeks.

Samos Aeg ★★→★★★ (POP) Island famed for sweet MUSCAT BL, esp fortified Anthemis, sun-dried Nectar. Rare old bottlings are steals. New producers emerging, eg. Nopera, as well as dry, non-POP wines.

Santorini Aeg ★★★→★★★★ Dramatic volcanic island with POP white (dr/sw) wines to match. Luscious VINSANTO, salty, *bone-dry Assyrtiko*. Top: ARGYROS, GAIA, HATZIDAKIS, SANTO, SIGALAS, Vassaltis. Cheapest ★★★★ whites around, age 20 yrs.

Santo Wines Aeg ★★→★★★ Successful SANTORINI co-op. Try rich Grande Res, complex VINSANTOS. Great value by Santorinian standards.

Semeli C Gr, Pelop ★★ Vast range, gd value. Main focus on NEMEA (try Grand Res) and MANTINIA.

Sigalas Aeg ★★★★ Leading light of SANTORINI. Kavaliero and Nychteri out of this world. Paris S, the founder, is no longer involved but the KIR-YIANNI team is doing wonders.

Skouras Pelop ★★★ Innovative, modern. Lean, wild-yeast Salto MOSCHOFILERO. Top reds: high-altitude Grande Cuvée NEMEA, Megas Oenos. Try solera-aged Labyrinth, complex Peplo rosé. *Recioto*-like Titanas is a rare Mavrostyfo.

Tatsis Mac ★★★ Natural producer in GOUMENISSA yet clean, beautiful in style. Whites more challenging but pure.

Thimiopoulos Mac ★★★★ Superstar producer in NAOUSSA and RAPSANI (Terra Petra). Kaiafas Naoussa best, but Earth and Sky conquers markets.

T-Oinos Aeg ★★★→★★★★ Putting Tinos on the global wine map, with impressive ASSYRTIKO and MAVROTRAGANO. Prices to match the quality.

Tsantalis Mac ★→★★★★ Long-est producer. Huge range; gd RAPSANI reds. Monastery wines from *Mount Athos* noteworthy.

Nemea Lions: may sound like a football team, but it's the new premium category of this POP.

Tselepos Pelop ★★★ Leader in MANTINIA, NEMEA (as Driopi) and SANTORINI (Canava Chrysou). Greece's best MERLOT (★★★★ Kokkinomylos). Avlotopi CAB SAUV not far behind. Great Driopi Res and delectable Santorinis.

Vassaltis Aeg ★★★ Relative newcomer in SANTORINI. Strong winemaking team, creating wines of great precision.

Vinsanto Aeg ★★★★ Sun-dried, cask-aged luscious ASSYRTIKO and Aidani from SANTORINI that can age forever. Insanely low yields.

Eastern Mediterranean & North Africa

EASTERN MEDITERRANEAN

The cradle of wine culture, totally rejuvenated in the last 20 years. The East Med countries share hot temperatures, high elevation, mountainous areas, lots of limestone, volcanic and stony soils, old vines and an on-going new interest in indigenous varieties. It is a winemaking paradise, with great deal of variety too.

CYPRUS

Cyprus is taking the next step in learning about its terroir and its leading grape Xynisteri. Many single-vyd versions, often from serious altitudes. Historic Commandaria is getting a fresh face too, with new approaches to drying grapes and ageing for brighter sweet wines.

Anthony gifted Cyprus to Cleopatra. Mind those snakes.

Aes Ambelis ★★ CAB SAUV, Promara, XYNISTERI, rosé gd; fortified COMMANDARIA v.gd.

Anama Concept ★★ Tiny producer of super-rich complex sweet wine from sun-dried old-vine MAVRO in the style of COMMANDARIA.

Argyrides Vineyards ★★ Beautiful 4th-generation winery in pretty village of Vasa. Excellent MARATHEFTIKO, MOURVÈDRE; v.gd CHARD, VIOGNIER.

Commandaria Rich, sweet PDO wine from sun-dried XYNISTERI and/or MAVRO grapes; 1st mentioned by Hesiod 800 BC, may be fortified. Now fresher. Gerolemo, KYPEROUNDA, Oenou Yi-Vassiliades, TSIAKKAS. Traditional AES AMBELIS, Alasia (Loel), Centurion (ETKO), St Barnabas (KAMANTERENA), St John (KEO).

ETKO & Olympus ★ Oldest Cyprus winery (1844). Traditional, long-aged COMMANDARIA, Haggipavlou label for better dry wines.

Ezousa ★→★★ Boutique winery. Try Gris, MARATHEFTIKO, VIOGNIER, XYNISTERI.

Kamanterena (SODAP) ★→★★ Large co-op in Pafos hills. Kamanterena label for better wines, esp gd-value XYNISTERI, single-vyd Project X, rosé, Stroumbeli MARATHEFTIKO. Excellent St Barnabas COMMANDARIA.

KEO ★→★★ Large drinks group; winery at Mallia for better wines. Notable traditional-style St John COMMANDARIA.

Kyperounda ★★→★★★ Highest winery on Cyprus: vyds up to 1390m (4560ft). *Petritis* remains standard-setting XYNISTERI; plus look for new single-vyd releases and excellent limited-edition Epos. Also gd reds and notable modern COMMANDARIA.

Makarounas ★★ New boutique family winery; gd results so far for Spourtiko, XYNISTERI, Yiannoudi and En Arhi (r).

Tsiakkas ★★→★★★ Standard-setting winery, highest vyds 1460m (4790ft). Beautiful SAUV BL, Promara, XYNISTERI. Excellent COMMANDARIA; v.gd Vamvakada (aka MARATHEFTIKO), Yiannoudi, organic Rodinos rosé. New single-vyd releases.

Vasilikon, K&K Winery ★★ Family winery, v. consistent. Try Vasilissa, XYNISTERI. Ayios Onoufrios is juicy and appealing, gd varietal LEFKADA, long-lived Methy.

Vlassides ★★→★★★ Stunning family winery, pioneer of excellent CAB SAUV, SHIRAZ, but now also v.gd local grapes, esp Alátes vyd XYNISTERI, Óroman based on Yiannoudi, top Opus Artis. Also gd fruit-driven Grifos range.

Vouni Panayia ★★→★★★ Early new-era private winery on Cyprus, passion for local grapes. Try Alina XYNISTERI, MAVRO/MARATHEFTIKO, Spourtiko, Plakota, Promara, and experimental micro-vinifications.

Zambartas ★★→★★★ Australia-trained winemaker making v.gd single-vyd range incl Margelina from centenarian vines, XYNISTERI; v.gd Zambartas range, esp MARATHEFTIKO, SHIRAZ/LEFKADA, fruity Koukouvagia range.

ISRAEL

The trends are for more blends, more Med varieties, ever-improving whites and high-elevation vyds. There is more focus on vyd and regional expressions than before. Upper Galilee, Golan Heights and Judean Hills (Jud) may produce the best quality, but Negev Desert may be the most interesting. R&D in varieties and viticulture of great interest.

Largest Byzantine winery, 1500 yrs old, excavated at Yavne, s of Tel Aviv.

1848 ★→★★ Best of Shor family, winemakers since 1848. Classic CAB FR, elegant Orient red blend, single-vyd Argaman, monster PETITE SIRAH.

Abaya ★★ Rehabilitates sick vyds. Gluggable Proletair and more refined Snunit, both CARIGNAN. Low or no added sulphites.

Agur ★→★★ Characterful winemaker reinforced with new partner, Kessem (r).

Ahat ★★ Quality *garagiste*. Superb CHENIN BL, complex ROUSSANNE/VIOGNIER, rosé.

Amit Toledo ★★ Wine troubadour. Bright, refreshing wines. Precise, chic SYRAH.

Ashkar ★→★★ Israeli-Arab domestic winery; gd SAUV BL, SHIRAZ.

Barkan-Segal ★→★★★ Largest winery; winemaker MW. Barkan: international style. Cherry-berry Argaman, new orange wine; Segal: innovative, wild style. Whole Cluster SYRAH recommended. Marawi and other native varieties.

Bat Shlomo ★→★★ Reviving traditions in village. Crisp, citrus SAUV BL.

Carmel ★→★★ Historic, est 1882. Carmel Signature label for top wines.

Castel, Dom du ★★★★ Jud pioneer. Set standards for style, quality in Israel. Beautiful, advanced winery. Grand Vin, complex B'x-style blend. Plush Petit Castel, great-value second label. Fresh rosé. Well-balanced "C" Blanc du Castel (CHARD). La Vie (entry-level r/w). Next generation becoming more involved.

Clos de Gat ★★★ Estate with style, individuality. Powerful Sycra SYRAH; rare, rich MERLOT, gd CHARD. Great-value Harel Syrah, entry-level Chanson (w).

Cremisan ★→★★ Central Mtns. Palestinian wine in a monastery est 1885. Pioneer of indigenous grapes. Hamdani Jandali (w) best.

Dalton ★★→★★★★ Family winery, creative winemaker. Lively Levantina (r), refreshing CARIGNAN, Mineral SEM.

Feldstein ★★→★★★ Individualist. Argaman pioneer, Dabouki specialist. Pale, zingy GRENACHE rosé.

Flam ★★★→★★★★ Brothers run this family winery. Elegant B'x blend Noble, fruit-forward SYRAH, deep MERLOT. Classico always great value. Fresh, fragrant SAUV BL/CHARD, crisp rosé. Prestige Chard Camellia: silky, elegant.

Galil Mountain ★→★★ Sustainability pioneer. Lively PETIT VERDOT. Refreshing GRENACHE; gd-value Yiron.

Golan, Ch ★★★ Innovative, expressive winemaker; v.gd Geshem (r/w) Med blends. New kosher label.

Gvaot ★★ Researching local varieties; gd PINOT N; light, fruity Bittuni.

Jezreel Valley ★→★★★ Flavourful CARIGNAN v.gd, Big, oaky prestige Icon.

Kishor ★→★★ Estate. Village for special-needs adults. Bright, fresh GSM.

Lahat ★★★ Rhône specialist. Enchanting GSM, peppery SYRAH. ROUSSANNE/VIOGNIER, can age.

Latroun Trappist Monastery since 1890. Best: PINOT GR, SYRAH. Winemaker also vigneron of promising Palestinian winery Dom Kassis.

Lewinsohn ★★★ Garage de Papa was born in a garage; v.gd CHARD, graceful, spicy SYRAH. Rare whole-cluster red. New rosé.

Margalit ★★★ Father and son. Israel's 1st cult wine. Rich CAB FR. No wine in 2021.

Mia Luce ★★→★★★ *Garagiste*. Rhôney SYRAH, superb MARSELAN, fine COLOMBARD.

Nana ★→★★ Desert pioneer. Fruity barrel-fermented CHENIN BL, robust SYRAH.

Pelter-Matar ★★ A v. popular brand. Light, fragrant SAUV BL. Matar kosher label.

Psagot ★→★★ Successful, fast growing. Peak is big, succulent Med blend.

Razi'el ★★★ Ben Zaken family (CASTEL). Barrel-aged rosé; fine-textured SYRAH/
CARIGNAN. Exquisite new handcrafted traditional-method fizz: NV, Rosé.

Recanati ★★→★★★ Vivid, wild CARIGNAN, Opulent, bold Special Res. New winery.

Sea Horse ★★→★★★ Quirky Counoise. Intriguing Oz, mainly CINSAULT/GRENACHE.

Shiloh ★★ CAB S: robust, rich, oaky. Regular award winner. Prestige Mosaic.

Shvo ★★★ Non-interventionist winemaker, a true vigneron. Super-rustic chewy red,
rare Gershon SAUV BL, fresh BARBERA, characterful rosé.

Sphera ★★★→★★★★ Cool-climate white; fizz due. Crisp White Concept varietals.
Harmonious First Page (SEM/ROUSSANNE/CHENIN BL). Complex, rare White
Signature (Sem), gd for ageing.

Tabor ★→★★★ Ecological vyds; v.gd whites, esp SAUV BL, ROUSSANNE. Flavourful Eco.

Teperberg ★→★★★ Largest family winery, since 1870. Full-bodied CAB FR; complex
Ramato PINOT GR; gd value at every price point.

Tulip-Maia ★★ Works with adults with special needs. Opulent Black Tulip, bracing
SAUV BL. Maia: Med style. Greek consultants. Refreshing wines.

Tzora ★★★★ Terroir-led, precision winemaking. Beautiful Shoresh vyd. Certified
sustainable by Fair'n Green.Talented winemaker (MW). Wines show intensity,
balance, elegance. Crisp, complex Shoresh Bl; Jud (r/w) always superb value.
Graceful Misty Hills (CAB SAUV/SYRAH) with finesse from "fossil" plot.

Vitkin ★★→★★★ ABC icebreaker. Quality CARIGNAN pioneer, esp old-vine Carignan,
PETITE SIRAH. Floral PINOT N. Complex GRENACHE BL.

Vortman ★→★★ Passionate vigneron. Racy SEM; gd-value Shefaya (r). Fun pét-nat.

Yaacov Oryah ★★ Creative artisan. Wonderful, luscious The Old Musketeer (sw).

Yarden ★★→★★★★ The pioneering winery of Israel. Rare, prestige Katzrin. Kings
of CAB SAUV at every price point. New dry GEWURZ. Superb Blanc de Blancs (sp).
Second label: Gamla. SANGIOVESE of interest. Mt Hermon (r) big-selling brand.

Yatir ★★→★★★ Desert winery, forest vyds. Velvety Yatir Forest (r), fine but rare
GSM; gd-value Mt Amasa. New Darom by Yatir.

LEBANON

Wineries weather political, economic and financial crises as Lebanon finally
discovers its DNA. Native varieties Aswad Karesh, Asmi N and Meksassi
are suddenly cool. They join Obeideh and Merwah, which – alongside the
adopted "heritage" grapes, CINSAULT, GRENACHE, CARIGNAN, MOURVÈDRE, as well
as regional "hired guns" ASSYRTIKO and SAPERAVI – are challenging the butch
B'x/Rhône classics for top billing. Meanwhile, high-altitude (1000m/3281ft+)
CHARD, SAUV BL, VIOGNIER continue to impress with quality and diversity.

Find of yr: scruffy old jug turns out to be C8 BC Phoenician, from Tyre. Yes, really.

Aurora ★★ Small family-owned/-run Batroun winery. Med climate gives different
expression from Bekaa V to gravelly, plush CAB FR; also PINOT N, CHARD.

Belle-Vue, Ch ★★★ Sea-facing slopes of Mt Lebanon. International fanbase. Le
Château and Le Renaissance, plush blends of B'x grapes/SYRAH. Also Petit Geste
(SAUV BL/VIOGNIER).

Clos St Thomas ★→★★★ Hitting right notes with Les Gourmets "cadet" range of
red (aromatic CINSAULT), white, rosé. Textured Obaidy (sic) and muscular, high-
altitude PINOT N.

Coteaux du Liban ★★ CINSAULT blend, Obeideh, CHARD/VIOGNIER. One to watch.

Ixsir ★★→★★★ High vyds, stony SYRAH-based blends, floral whites and prestige El. Altitudes range and Grande Res Rosé excellent.

Kefraya, Ch ★★→★★★★ Innovative; French winemaker. Les Exceptions varietals: SAPARAVI in amphorae; complex *Comte de M*, oaky Comtesse de M, fruity Les Breteches (CINSAULT-based).

Ksara, Ch ★★★ Est 1857; innovative, esp old-vine CARIGNAN, Merwah. Consistent, excellent value. Blanc de Blancs, CHARD outstanding. Sunset iconic rosé.

Marsyas, Ch ★★ Newish Bekaa winery in village of Kefraya. Powerful CAB/SYRAH; B-Qa de Marsyas more Rhône-like; thrilling CHARD/SAUV BL white. Owner of complex ★★★ Bargylus (Syria), miracle wines made in impossible conditions.

Massaya ★★ Terraces de Baalbeck: refined GSM. Entry-level Les Colombiers v.gd value. Also Cap Est (r) from E Bekaa vyds on Anti-Lebanon Mtns. Punchy rosé.

Musar, Ch ★★★→★★★★ (r) 02 03 05' 07' 08 09 10 11 12 13. *Unique recognizable style.* Best after 15–20 yrs in bottle. Indigenous, aged Obaideh, Merweh age indefinitely. Second label: Hochar (r) now higher profile. Musar Jeune is softer, easy-drinking, but white and rosé equally interesting.

Najm, Dom ★★★ Tiny winery in Batroun run by husband and wife Salim and Hiba Najm. One red: seriously earthy, red fruit, MUSAR-esque CAB SAUV/GRENACHE/MOURVÈDRE. Lebanon's unicorn wine.

Sept ★★★ *Garagiste enfant terrible.* Skin-contact Obeideh, other min-contact varietals, especially haunting SYRAH.

Terre Joie ★★ Small W Bekaa winery. Only reds and a crunchy rosé. CAB FR, CINSAULT and rare (in Lebanon) high-quality MERLOT.

Tourelles, Dom des ★★→★★★★ C19 winery now run by dynamic winemaker Fouzi Issa. Blockbuster SYRAH, gd Marquis des Beys (r/w). Outstanding old-vine CARIGNAN, CINSAULT, Obaideh/Merwah blend; equally gd classic red. Classy rosé.

Vertical 33 ★→★★★ Organic CARIGNAN, CINSAULT, Obeideh, PINOT N. Neo-MUSAR!

Wardy, Dom ★★ New owners combine varietals with traditional French blends – Obeideh, SAUV BL, Cinsault – and an outstanding easy-drinking red.

TURKEY

Fun for blind tastings and trying to pronounce those grape varieties. Still devilishly difficult to be a Turkish winery, but a fascinating wine country.

Buzbag ★ Main brand since 40s. Rustic ÖKÜZGÖZÜ/BOĞAZKERE (Kayra). Better now.

Chamlija ★★→★★★ CABS SAUV/FR, different terroirs; gd PINOT N. Prestige: Django.

Corvus ★★ Bozcaada island. Intense, oaky New World style.

Doluca ★→★★★ Family of three generations. Tugra gd-value local varieties.

Kalpak ★★ One to watch. Classic B'x blend (CAB SAUV/MERLOT/CAB FR/PETIT VERDOT).

Kavaklidere ★→★★★ Largest winery, modern wines: fruity Yakut, gd Pendore SYRAH.

Kayra ★→★★ Plush ÖKÜZGÖZÜ, tannic BOĞAZKERE, fresh NARINCE. Owned by Diageo.

Pasaeli ★→★★ Fresh, vibrant B'x-style blends from single vyd.

Sevilen ★→★★★ A gd international style, spicy SYRAH. Aromatic FUMÉ BL.

Shiluh Unique from SE Anatolia. Wines made in buried clay jars, no filtration or fining. Owned by members of ancient Syriac community.

Suvla ★→★★★ Full-bodied B'x blend Sur, and fruity SYRAH backed by oak.

Urla ★★→★★★ Tempus (r) has complexity. NERO D'AVOLA/Karasi is firm, spicy.

Notable names in North Africa
Morocco Baccari ★★ (Première de Baccari), Castel Frères (Boulaouane Vin Gris), Celliers de Meknès (Ch Roslane), Ouled Thaleb ★→★★ (Tandem/Syrocco), Val d'Argan ★→★★ (Orian r), Volubilia (Epicuria SYRAH). **Tunisia** Neferis (Selian CARIGNAN).

Asia & Black Sea & Caucasus

ASIA

China World capital not only of fake wine, but also faux design: Ch Changyu-Moser XV is B'x plus fairytale Loire V. Changyu is China's oldest winery, est 1892 from a fortune made in Batavia, Penang and Singapore. Austrian Laurenz Moser is winemaker of Changyu-Moser XV; top reds inspired by new oak. And gd CAB SAUV rosé: pale, gossamer.

The ideal red in China is CAB SAUV. Lafite's Dom de Long Dai exudes poise and confidence. Moët-Hennessy's Ao Yun competes for eye-watering prices. Chinese Cabs from Chinese wineries: the oakier, the pricier. Wineries have frequent mood swings on oak and length of time in barrel. No water-tight guide but try Ch Rongzi, Grace Vyd, Jiabeilan, Kanaan Winery, Li Family, Silver Heights, Tiansai. Ningxia's Fei Tswei is not afraid of fruit, esp MARSELAN. Xinjiang's Ch Zhongfei and Tiansia, and Shanxi's Grace Vyd produce gd Marselan. Best have violet/floral/dark-cherry fruit.

In China, red wine is served with everything. Doesn't have to match.

PINOT N remains a holy grail. Women winemakers or/and owners gd, incl at Dom du 1er Juin, Jiabeilan, Silver Heights. Ch Chanson makes gd pure CAB FR (they thought they'd planted Cab Sauv). So too Tiansai. Ch Guofei: mouth-smacking dry and off-dry RIES; Taila Winery of Shandong: finest sweet, a PETIT MANSENG. Xinjiang's Puchang Vyd Clovine from Chinese-bred hybrid Beichun is like Ruby Port.

Northern coastal China makes Icewine, but mostly oversulphured; gd ones incl Ch Changyu Black Diamond Golden Icewine Valley, Ji'an Baite Manor, Sanhe's Cailonglin. All are pure VIDAL.

India Sub-tropical India's wines offer ripe, plush New World fruit but aspire to Old World elegance, balance. Oak generous on reds. Best traditional-method sparklers: Chandon and York, both mostly CHENIN BL. Sula and Grover Zampa have gd SAUV BL and CAB/SHIRAZ. Fragrant VIOGNIER and RIES by Vallonné. Fleshy SANGIOVESE by Fratelli. Classic Cab Sauv by boutique KRSMA from World Heritage Hampi Hills. Varieties to watch: CHARD (revival), GRENACHE and TEMPRANILLO.

Japan Yamanashi is the birthplace of Japanese wine, and Grace Wine is the benchmark for KOSHU, Japan's virtually indigenous grape; it arrived in Japan with a returning traveller and has DNA from Chinese wild vines. Styles are evolving and generally improving, and the most sensational is Grace Wine's single-vyd Cuvée Misawa, the 1st subregional Koshu planted in Akeno, which has the longest growing sunshine hours in Japan. Examples of gd Koshu are also made by Aruga Branca, Ch Mercian, Dom Hide, Haramo, Huggy Wine, Kurambon, L'Orient, Lumiere, Manns Wines, Marquis, Soryu and Suntory. Best are without oak. Grace Blanc de Blancs is Japan's finest sparkling: all CHARD, min 5 yrs on lees. Muscat Bailey A, a hybrid created in Japan in the 20s, is most planted red. You either like or are put off by the candy-floss aroma. Dom Hide's has concentration, intensity. Chitose Kimura Vyd PINOT N esp gd. International-variety reds tend to be woody. Ripeness already a challenge; it beggars belief that what delicate fruit there is becomes neutralized by oak. Best come from Grace Cuvée Misawa (CAB SAUV/MERLOT/CAB FR/PETIT VERDO).

BLACK SEA & CAUCASUS

"Ancient world" wines are Georgia's and Armenia's distinctive contribution to the international wine scene. Ancient means the tradition of skin-macerated, amphora-fermented wines produced for millennia and now hugely popular among wine geeks and adepts of natural winemaking. Azerbaijan, known for brandy, is starting to produce quality wine. In Ukraine, locally produced quality dry wines are on the rise, while Moldova, with its value-driven production, offers new top wines. Curiosities can be found in Uzbekistan and Kazakhstan.

Armenia Although Armenia is embracing modern business practices, its winemaking is as old as Georgia's, with a shared use of amphorae, known here as *karas*. Armenia's indigenous stars are (w) Voskeat and (r) Areni; (r) Akhtanak, Karmrahyut also give gd quality. Try larger Armenia Wines and Hin Areni or boutique ArmAs, Old Bridge, Tushpa, Van Ardi, Voskeni, Voskevaz, v.gd Zorah.

Georgia Ancient winemaking methods going back 8000 yrs have been continuously used to produce Georgia's signature whites and reds with skin-macerated fermentation and ageing in clay *qvevris*. Indigenous varieties are another unique heritage: about 40 are in commercial production. Red star SAPERAVI gives inky-black wines in many styles from light, semi-sweet to powerful, dry, tannic and age-worthy. Acidic white RKATSITELI, grown mainly in Kakheti and traditionally fermented on the skins, has prompted the international phenomenon of orange/amber wines. White Chinuri, Kisi, Mtsvane of note. Wine is made by almost every household. For reliable to superb quality, seek wines from Badagoni, Ch Mukhrani, Jakeli, Khareba, Marani, Papari V, Shumi Winery, Tbilvino, Teliani V.

Moldova Neighbouring Ukraine and Romania, Moldova has more vyds than S Africa and boasts the highest density of plantings and largest cellars in the world. Backed by long history and fame in tsarist Russia, its modern production is offering value. International grapes dominate, but worth seeking local: (w) FETEASCĂ ALBĂ, FETEASCĂ REGALĂ, Viorica; (r) FETEASCĂ NEAGRĂ, Rară Neagră. Try unusual red blend Negru de Purcari (CAB SAUV/SAPERAVI/Rară Neagră) and Icewine. Leading producers: Cricova (sp), Milestii Mici, Vinăria Purcari. Plus gd to excellent quality: Asconi, Castel Mimi, Ch Vartely, Et Cetera, Lion Gri, Vinăria Bostavan, Vinăria din Vale.

Russia Sanctioned Russia recognizes geographic and controlled origin only for its own wines, even if they are meaningless for local consumers and the wider world. Natural conditions for production best by Black Sea and River Kuban, but some vyds planted as far as Caspian Sea. Harsh climate in Don V known for indigenous grapes (r Krasnostop, Tsimliansky), requires vines to be buried in winter. Elsewhere, winemaking is centred around international grapes. At the time of going to press, Russia was under international sanctions, so we see little point in recommending individual producers.

Ukraine Until Putin's bloodbath, Ukraine's wine-growing stretched from mild Black Sea to continental Carpathian Mtns. Grapes mainly international, with local curiosities (w) Telti Kuruk, (r) Odessa Black. Premium boutique wines came from Crimea (Oleg Repin, Uppa Winery); est producers for dry still wines incl Beykush Winery, Esse, Guliev Wines, Kolonist, Prince Trubetskoy Winery, Satera, Shabo, Veles, Villa Tinta. Wines modelled on Champagne have important and popular heritage: ArtWinery, Novy Svet, Odessavinprom, Zolotaya Balka. There are also historically proven – now niche production – fortified styles: Ch Chizay, Koktebel, Massandra, Solnechnaya Dolina. But by now, God knows.

United States

WASHINGTON
OREGON
IDAHO
NEVADA
CALIFORNIA
COLORADO
ARIZONA
NEW MEXICO
TEXAS
OKLAHOMA
MISSOURI
WISCONSIN
MICHIGAN
OHIO
MASSACHUSETTS
PENNSYLVANIA
NEW YORK/
NEW JERSEY
MARYLAND
VIRGINIA
NORTH CAROLINA
GEORGIA

NORTH COAST
Mendocino
Redwood Valley
Sierra Foothills
Anderson
Valley
Clear Lake
Clear Lake
Sonoma Coast
Northern
Sonoma
Napa Valley
Sonoma
Valley
Carneros
Coombsville/Oak Knoll
Lodi
Clarksburg
Sacramento
Sacramento
El Dorado
Shenandoah Valley
Amador
Calaveras
San Francisco
Livermore Valley
Santa Clara Valley
Santa Cruz
Mountains
Monterey
Salinas
CENTRAL
VALLEY
San Joaquin
Fresno
Carmel Valley/Arroyo Seco
Santa Lucia
Highlands
San Lucas
CENTRAL COAST
Pacific Ocean
Paso Robles
CALIFORNIA
Lake Tahoe
NEVADA
San Luis Obispo
Edna Valley/Arroyo Grande Valley
Santa Maria Valley
Santa Barbara
Sta Rita Hills
Santa Ynez Valley
Santa Barbara
Los Angeles

Abbreviations used in the text
(*see also* principal viticultural
areas pp.245, 263, 269):

Arroyo GV	Arroyo Grande Valley, CA
Clark	Clarksburg, CA
Coomb	Coombsville, CA
Mad	Madera, CA
Oak K	Oak Knoll, CA
PNW	Pacific Northwest
San LO	San Luis Obispo, CA
Santa Cz Mts	Santa Cruz Mountains, CA
Son	Sonoma, CA

UNITED STATES

How to sum up such a huge and varied country? California is the United States' most famous wine region, and Napa the most famous part of California; and in 20 years' time Napa might look very different to now. Growers wearily replanting vineyards destroyed by wildfires might replant with Cabernet; or they might branch out. (Refosco? Tempranillo? The less expensive parts of the state – Lodi, or the Sierra Foothills, or Mendocino – are the experimental spots.) They're aware that a new generation might want different things from wine, but how to second-guess that? And how, as the climate changes, do you find the ideal grape for your vineyards for 20 years' time? All regions evolve; nothing stands still. At the moment there is some usefully clear diffferentiation across the US: Texas specializes in Rhône grapes, and Viognier thrives in Virginia. Oregon is not entirely about Pinot, but you'd be forgiven for thinking it was. On the East Coast there is serious Riesling, and even in California, styles from old-vine Zinfandel to Chardonnay of purity and tension. But underneath these headline styles there are experiments everywhere. Some will be headlines in the future.

American Viticultural Areas

With no production rules or traditions to protect, AVAs are only loosely comparable to appellations contrôlées. Administered by the US government's TTB, they are instead guides to location – and climate, soil, market – and are a wine-minded alternative to state or county labels. Whether a region within a state or a traits-based overlap such as high-toned Columbia Gorge shared by WA and OR, or NY, PA, OH's cool, water-tempered Lake Erie, there are 251 est AVAs – and a steady queue of applied and pending. Most (140) are in CA, which boasts a wealth of nested AVA subregions, some hyperfocused, like Napa V's Stags Leap District. While an AVA label indicates a min higher standard at the federal level (state and county labels mean 75% provenance, AVA promises 85%) some states have stricter rules: OR famously demands 100% for the former, 95% for the latter. AVA approval standards are rigorous: petitions must show distinguishing features verifiable on US Geological Survey maps, and how they affect viticulture inside vs outside the petitioned zone – whose proposed name must be one historically applied to the area, as shown by "newspapers, magazines, historical or modern books". As wine-growing quality and know-how rise across the US, the TTB says AVAs allow "producers to better describe the origin of their wines and... consumers to better identify wines". Some even translate to higher prices.

Arizona (AZ)

Surge in quality and innovation driven by hip vibe. High-profile names are now est with serious cred. The 2021 addition of Verde Valley gives AZ three AVAs of high-desert terroir: volcanic rock and limestone, gd ripening weather. Spain, Portugal, Italy and B'x varieties excel. Wineries incl **Alcantara Vyds** elegant, earthy Confluence (r) line, on-trend GRENACHE rosé. **Arizona Stronghold ★★★** flagship SANGIOVESE-focused blend Mangus and excellent Tazi (w blend). **Bodega Pierce** estate-grown SAUV BL. **Burning Tree Cellars** artisanal, small-batch, intense red blends. **Caduceus Cellars ★★★** owned, along with sister winery Merkin Vyds, by alt-rocker Maynard James Keenan: excellent Dos Ladrones (w blend), Nagual del Marzo (r) and Kitsuné Sangiovese.

Callaghan Vyds ★★→★★★ Cornerstone of AZ winemaking with quality red blends, esp signature Padres (r) blend of mostly Spanish grapes, try Barrett's sparkling (w/rosé). **Ch Tumbleweed** relative newcomer: focused wines with hip labels, heritage plots; Grenache-driven Willy blend is flagship. **Dos Cabezas WineWorks** blends across AVAs; El Signature El Campo (r/w). **Javelina Leap Vyd & Winery** awarded ZIN. **Page Springs Cellars** GSM, Rhône white specialist esp El Serrano blend (r) and ripe, round MARSANNE. **Pillsbury Wine Company** ★★ estate-grown Rhône varieties from filmmaker Sam Pillsbury. SHIRAZ-led Guns and Kisses, co-fermented Côte-Rôtie, anchors the portfolio, which also incl mineral-inflected VIOGNIER and dry aromatic blend WildChild (w).

California (CA)

California has been a formidable economic and agricultural engine since before the 1849 gold rush. It grows hundreds of wine grape varieties and stellar produce from arugula to zucchini. It offers cool, foggy Pacific coastline, mtns, warm inland valleys and everything in between. Drought and fall forest fires are serious concerns, but 2021 saw fewer fires in wine country and a post-harvest storm that brought much rain to the n half of CA. The harvest was a joyous respite for many, if not all. Natural wines like pét-nats and orange wines are trending (on a small scale), as is cultural inclusiveness. I firmly believe that Americans will pay for great wines made with social awareness and responsible environmental intentions. This discussion should incl not just the new and trendy but also legendary brands that have delivered on these priorities for decades. The wine biz in America is cutthroat. Land prices are insane, distribution is consolidating and wine prices are uncompetitive. And climate change looms large, as it does in many wine regions. If your winery doesn't make incredible wine, have loyal distribution partners, or tell a great story, it's a gd time to consider selling, as many have.

Recent vintages

CA is too diverse for simple summaries. There can certainly be differences between the N, Central and S thirds of the state, but no "bad" vintages in over a decade. Recent wildfires prove challenging, but only for latest-picked grapes.
2021 Small crop but high quality. Fires hurt El Dor Co, parts of Lake County.
2020 Small crop: great whites. N Coast reds: smoke issues.
2019 Solid harvest. Minor late losses in Alex V to fires, smoke taint.
2018 Bumper crop of great quality, but smoke issues in Lake County.
2017 Wildfires in Napa, Son after most grapes picked; quality mostly v.gd.
2016 Gd quality: reds/whites show great freshness, charm.
2015 Dry yr, low yields, but quality surprisingly gd, concentrated.
2014 Despite 3rd year of drought, quality high.
2013 Another large harvest with excellent quality prospects.

Principal viticultural areas

There are well over 100 AVAs in CA. Below are the key players.
Alexander Valley (Alex V) Son. Warm region in upper Son. Best known for gd Zin, Cab Sauv on hillsides.
Amador County (Am Co) Warm Sierra County with wealth of old-vine Zin; Rhône grapes also flourish.
Anderson Valley (And V) Mend. Pacific fog and winds follow Navarro River inland. Superb Pinot N, Chard, sparkling, v.gd Ries, Gewurz, some stellar Syrah.
Atlas Peak E Napa. Exceptional Cab Sauv, Merlot.
Calistoga (Cal) Warmer n end of Napa V. Red wine territory esp Cab Sauv.

Carneros (Car) Napa, Son. Cool AVA at n tip of SF Bay; gd Pinot N, Chard; Merlot, Syrah, Cab Sauv on warmer sites, v.gd sparkling.

Coombsville (Coomb) Napa. Cool region nr SF Bay; top Cab Sauv in B'x style.

Diamond Mountain Napa. High-elevation vines, outstanding Cab Sauv.

Dry Creek Valley (Dry CV) Son. Top Zin, gd Sauv Bl; gd hillside Cab Sauv, Zin.

Edna Valley (Edna V) San LO. Cool Pacific winds; v.gd Chard.

El Dorado County (El Dor Co) High-altitude inland area surrounding Placerville. Some real talent emerging with Rhône grapes, Zin, Cab and more.

Howell Mountain Napa. Briary Napa Cab Sauv from steep, volcanic hillsides.

Livermore Valley (Liv V) Suburban, gravelly, warm region e of SF, gd potential.

Mendocino County (Mend) Large county n of Son County, incl warm Red V and cool And V.

Mendocino Ridge (Mend Rdg). Emerging region in Mend, dictated by elevation over 365m (1198ft). Cool, above fog, lean soils.

Monterey County (Mont) Big ranches in Salinas V provide affordable Chard and Pinot N in cool, windy conditions. Carmel V bit warmer, Arroyo Seco moderate.

Mount Veeder Napa. High mtn vyds for gd Chard, Cab Sauv.

Napa Valley (Napa V) Cab Sauv, Merlot, Cab Fr. Look to sub-AVAs for meaningful terroir-based wines, and mtn areas for most complex, age-worthy.

Oakville (Oak) Napa. Prime Cab Sauv territory on gravelly bench.

Paso Robles (P Rob) San LO. Popular with visitors. Reds: Rhône, B'x varieties.

Pritchard Hill (P Hill) E Napa. Elevated, woodsy, prime terrritory for Cab Sauv.

Red Hills of Lake County (R Hills) N extension of Mayacama range; great Cab Sauv country.

Redwood Valley (Red V) Mend. Warmer inland region; gd Zin, Cab Sauv, Sauv Bl.

Russian River Valley (RRV) Son. Pacific fog lingers; Pinot N, Chard, gd Zin on benchland.

Rutherford (Ruth) Napa. Outstanding Cab Sauv, esp hillside vyds.

Saint Helena (St H) Napa. Lovely balanced Cab Sauv.

Santa Barbara County (Santa B) County n of LA; transverse valleys, several notable subzones, cool and warm.

Santa Lucia Highlands (Santa LH) Mont. Higher elevation, s-facing hillsides, great Pinot N, Syrah, Rhônes.

Santa Maria Valley (Santa MV) Santa B. Coastal cool; gd Pinot N, Chard, Viognier.

Sta Rita Hills (Sta RH) Santa B. Excellent Pinot N.

Santa Ynez (Santa Ynz) Santa B. Rhônes (r/w), Chard, Sauv Bl best bet.

Sierra Foothills (Sierra F'hills) El Dor Co, Am Co, Calaveras County. All improving.

Sonoma Coast (Son Coast) A v. cool climate; edgy Pinot N, Chard, Syrah.

Sonoma Valley (Son V) Note Son V is area within Son County; gd Chard, v.gd Zin, excellent Cab Sauv from Son Mtn sub-AVA.

Spring Mountain Napa. Elevated Cab Sauv, complex soil mixes and exposures.

Stags Leap (Stags L) Napa. Classic red, black fruited Cab Sauv; v.gd Merlot.

Acorn RRV ★★→★★★ Preserving CA heritage making lively co-fermented field blends from historic Alegria vyd featuring ZIN plus 17 other mixed black grapes.

Alban Vineyards Edna V ★★★ Old-guard Rhône Ranger, still making great wine in EDNA V climate sweet spot. Top VIOGNIER, GRENACHE too. His vision has paid off. Always exciting.

Albatross Ridge Mont ★★★ Soaring estate 11 km (7 miles) from Pacific nr Carmel produces nervy, delicious CHARD, PINOT N with wings.

Alma de Cattleya RRV ★★★ Colombian-born Bibiana González Ravé, with mad CV making wine in Europe and US at 1st-tier wineries, stakes her own claim in CA. Resonant, unpretentious and affordable PINOT N; CHARD sublime.

Alma Rosa Sta RH ★★★ Pioneer Dick Sanford's 2nd act produces *au courant* PINOT N, CHARD from five non-contiguous sites. Tasting room in Solvang.

Andrew Murray Vineyards Santa Y ★★★ SYRAH leads Rhône pack, but white VIOGNIER, ROUSSANNE, fresh GRENACHE BL hits too.

Anthill Farms Son Coast ★★★ Three hard-working WILLIAMS-SELYEM alumni produce consistently lively, ethereal cool-climate PINOT N, SYRAH, old-vine, head-trained CHARD from coastal SON COAST and AND V. And V Pinot N great value, worth seeking out. Serious up-and-comer.

Antica Napa Valley Napa V ★★★ Piero Antinori's ATLAS PEAK property showing great potential for CAB SAUV. How did it take this long? Whole AVA rebounding.

Au Bon Climat Santa B ★★★ Jim Clendenen made restrained PINOT N, crisp CHARD before it was hip, and advocated elegant style now trending. Relevant as ever.

Baileyana Edna V ★★★ Sustainable Paragon vyd now owned by WX Brands delivers reliably superb CHARD, balanced PINOT N and peppery, lively SYRAH. Sister brands Tangent and Zocker bang out outstanding whites: peachy ALBARIÑO, pungent SAUV BL from Tangent, edgy GRÜNER V, RIES from Zocker rock solid.

Balletto RRV ★★→★★★ Refreshingly from-the-heart, pure CAB across board at loveable prices, most notably classically-proportioned RRV CHARD.

Banshee Wines Son Coast ★★★ Growing, scrappy PINOT N-driven brand with no vyds but gd connections. Well-made single-vyd wines.

Barnett Spring Mtn ★★★ Under-the-radar mtn-top gem managed by David Tate, who also makes CHARD, PINOT N from SON V. Screaming gd wines across board, towering views, plus 1st-rate NAPA V CAB SAUV – all well worth drive up mtn.

Baxter And V ★★★ Phil B, 2nd-generation winemaker, makes subtly earthy, burgundian PINOT N (Oppenlander or Valenti vyds) using no new oak. Every bottle exudes passion, confidence, competence. Soulful, v. well made.

Beaulieu Vineyard (BV) Napa V ★★→★★★ Iconic Georges de Latour Private Res CAB SAUV continues to be iconic; spin-off wines are serviceable down the line.

Beckmen Vineyards Santa B ★★★ Tom B founded Roland synthesizers, then founded bio Purisma Mtn estate with son Steve. Formidable SYRAH, GRENACHE, GRENACHE BL. Affordable, excellent Cuvée le Bec red blend is popular nationwide.

Bedrock Wine Co Son V ★★★ Morgan Peterson's label is a paean to historic ZIN vyds, techniques. Wisdom of ages seen through clear young eyes.

Beringer Napa ★★→★★★★ Private Res CAB SAUV, single-vyd Cabs serious, age-worthy. ST H historic location well worth a visit. CHARD gotten quite gd.

Berryessa Gap Vineyards Central V ★★ Under-the-radar Yolo County project nr Sacramento making fresh, lightly oaked, Iberian-inspired wines and more. TEMPRANILLO dazzling, VERDEJO and DURIF also delicious. Great value.

Bevan Cellars Napa V, Son ★★★→★★★★ Outsized personality with great taste, Russell B sources from prime single vyds making superb B'x varieties; CABS SAUV, FR most desirable. Bit of PINOT N and heavenly Dry Stack vyd SAUV BL from Bennett V also superb. Expensive, but outstanding.

Blackbird Napa V ★★★→★★★★ Estate has shifted from MERLOT-driven to a B'x-blend strategy under the eye of formidable winemaker Aaron Pott. Arise red blend a steal at $50, CAB FR-focused Paramour also excels.

California Chardonnay renaissance

California CHARD is better than ever. More refined, coastal, cooler-climate wines are focused, mouthwatering and defy cliché. To wit: Ramey Hyde, Ritchie vyds (CAR, RRV), Lioco Demuth vyd (AND V), MACROSTIE (SON COAST), TONGUE DANCER Bacigalupi vyd (RRV) and FORT ROSS VYDS Sea Slopes (Fort Ross-Seaview). All stunning Chards from cool, coastal regions. That MacRostie is c.$25 retail, and *hella* gd!

Boeger Sierra F'hills ★★ Stalwart gold-country winery since 70s known for BARBERA, CHARBONO, ZIN. Surprising CHARD among best from SIERRA F'HILLS.

Bogle Central V, Lodi ★★ Dependable under-$15 grocery-store family-owned brand delivers ever-reliable varietal wines from LODI, Clarksburg and now more coastal zones, all aged in real barrels. Respect.

Boisset Collection Napa V, Son ★★★ Ambitious portfolio of sustainably farmed estates and brands assembled by boulevardier Jean-Charles Boisset incl BUENA VISTA, DELOACH, Lyeth, Wattle Creek (SON COUNTY); JCB and Raymond (NAPA V).

Bokisch Vineyards Lodi ★★→★★★ Lodi star with enthusiastic followers champions Spanish varieties; v.gd TEMPRANILLO heads list backed by superb ALBARIÑO, GRACIANO, flirty rosado.

Bonny Doon Mont ★★★ Ever adventurous and clever, this brand has been a major artistic voice in CA wine. *Vin gris* is superb, juicy Clos de Gilroy GRENACHE, Le Cigare Volant blend a CA Rhône classic. Sold 2020, but founder Grahm stays on for now.

Brewer-Clifton Santa B ★★★ Estate STA RH PINOT N, CHARD producer now owned by JACKSON FAMILY WINES. Old-guard Pinot N brand still in fine form, zesty Chard matured in neutral oak.

Buena Vista Son V ★★→★★★ Historic winery est 1857, bedazzled by owner Jean-Charles BOISSET (DE LOACH, Raymond) with period-costume tours, lights and animatronics straight outta Disneyland. Try The Count's Selection SYRAH.

Cade Howell Mtn ★★★ Superb mtn wines: CAB SAUV, SAUV BL; stunning, ultra-modern winery. Partnership: Getty family, CA Governor Newsom, GM J Conover.

Cakebread Napa V ★★★ Comforting CAB SAUV still has massive cachet with long-time fans. SAUV BL also popular, CHARD v.gd; and an easy stop by on a cruise down Rte 29.

Calera ★★★→★★★★ PINOT N, CHARD on limestone, at altitude. Sold to DUCKHORN (2017), wines still outstanding. Everyone should try epic Jensen Pinot N once in their life – CA wine history in a bottle.

Carlisle Son V ★★★ Best way to save historic vyds is to make extraordinary wines from them. Mike Officer crafts brilliant ZIN-based field blends from N CA, preserving history with updated growing tech. Rhône reds also notable.

Carneros, Dom Car ★★★→★★★★ Taittinger outpost led by women for decades makes one of best Blanc de Blancs sparkling in CA, Le Rêve. Other bubblies and PINOT N also formidable.

Caymus Napa V ★★★ One of NAPA's revered status brands. Special Selection CAB SAUV esp iconic, but on rich, sappy end of style spectrum.

Cedarville Sierra F'hills ★★★ Quality leader in granite-rich Fairplay District of EL DOR CO. Superb red wines across board with red-fruited GRENACHE, SYRAH; fine CAB SAUV, ZIN.

Chandon, Dom Napa V ★★→★★★ LVMH owns luxury bubbly specialist in Yountville. Flagship Étoile Brut NV and tangy Res rosé of PINOT N NV delicious. Al fresco restaurant offers exciting wine/food pairings.

Chappellet Napa V ★★★★ Rugged P HILL property has delivered savoury, age-worthy

---

Tired of toast?
Why so much new oak? There's a profound misconception that it's done around the world. Many great wines in Europe are made without new oak barrels, which impart a lot of flavour. Some CA winemakers have had it with oaky wines. Explore this fresh idea with BERRYESSA GAP (Yolo County), BAXTER PINOT N (all neutral oak from beginning), Foursight and Bee Hunter (AND V), LANG & REED N Coast CAB FR (NAPA V), MAGGY HAWK Pinot N, Saracina (MEND), Shane Wines (SON).

madrone scrub-laced CAB SAUV for decades. Signature-series Cab Sauv superb, dry CHENIN BL a rare treat. Family also owns PINOT N, CHARD-themed SONOMA-LOEB brand.

Chimney Rock Stags L ★★★→★★★★ Underrated Terlato-owned brand making best wines ever under steady hand of winemaker Elizabeth Vianna. Tomahawk Vyd CAB SAUV top-notch.

Cliff Lede Stags L ★★★ Luxe estate flies the Canadian flag and bangs out 1st-rate, well-structured CAB SAUV and rich, leesy SAUV BL that some will adore. Also owns Fel brand in AND V, making elegant PINOT N, exotic PINOT GR.

Constellation Brands ★→★★★ Publicly traded major wine/beer/spirits company owns famed ROBERT MONDAVI brand, Meiomi, THE PRISONER, Woodbridge. Lately re-focusing on beer and cannabis products.

Continuum Napa V, St H ★★★★ Scion Tim Mondavi's P HILL estate spares no expense making one of NAPA V's greatest, most complex B'x blends. Second label Novicium from younger vines.

Copain Cellars And V ★★★ Old World-influenced, classically proportioned wines; recently sold to JACKSON FAMILY. PINOT N is strong suit, esp bright, spicy Kiser vyd versions. Tous Ensemble line easy-going, friendly.

Corison Napa V ★★★ While many in NAPA V follow $iren call of bloated wines for big scores, diminishing pleasure, Cathy C consistently makes elegant, fresh Cabs, esp focused age-worthy Kronos vyd CAB SAUV.

Côte, Dom de la Sta RH ★★★ Raj Parr and Sashi Moorman make critically acclaimed, elegant PINOT N and CHARD. Burgundy-inspired, but with differences in latitude, soils, making for distinctly CA wines steered by great taste.

Cuvaison Car ★★★ Quiet historic property, making great wine yr after yr. Top marks to CHARD, PINOT N from CAR estate; gd CAB SAUV, SYRAH, from MT VEEDER. Single Block bottlings incl lovely rosé, SAUV BL.

Dalla Valle Oak ★★★★ A 1st-rate hillside estate transitioning to 2nd generation. Maya CAB SAUV is legendary, eponymous Cab Sauv a cult wine, Collina label best affordable introduction to luxury NAPA Cab.

Daou P Rob ★★★ Elevated estate in Adelaida District is driving CAB SAUV in P ROB to new heights in altitude and price.

Dashe Cellars Dry CV, N Coast ★★★ RIDGE veteran Mike D makes tasteful, affordable and balanced DRY CV and ALEX V ZIN from urban winery in Alameda. Terrific old-vine CARIGNANE, zesty GRENACHE rosé.

Dehlinger RRV ★★★ PINOT N specialist still on par after more than four decades. Also v.gd CHARD, SYRAH, balanced CAB SAUV.

DeLoach Winery Son ★★★ Flamboyant maestro Jean-Charles BOISSET saw gd value in this progressive organic, bio-oriented winery making great CHARD, PINOT N. Solid down-to-earth investment, if not his sexiest.

Diamond Creek Napa V ★★★★ The extraordinary longevity of CAB SAUV from iconic sites like Volcanic Hill and Gravelly Meadow makes some of most coveted and collectable wines of NAPA V.

Dominus Estate Napa V ★★★★ Moueix-owned (*see* France). Winery is dazzling but not open to public. Wines from gravelly bench soils consistently elegant, impressive. Second label: Napanook, v.gd. Important vision of S NAPA V.

Donum Estate N Coast ★★★→★★★★ Anne Moller-Racke has passionately worked CAR soils since 1981. PINOT N from four sites is focused, generous, complex. Adding vyds in AND V, SON COAST.

Drew Family And V, Mend ★★★→★★★★ MEND RDG visionary making minimalist, savage PINOT N from AND V and higher up hills. Look for estate Field Selections Pinot N from Mend Rdg, SYRAH from coastal Valenti vyd. Hunt these down.

Dry Creek Vineyard Dry CV ★★★ Standard-bearer on its A-game. Trustworthy, Loireinspired, grassy FUMÉ BL and other SAUV BL always delicious, CHENIN BL and all reds better than ever; great stop nr Healdsburg. What's not to love? ZIN/B'x blend Mariner also better than ever.

Duckhorn Vineyards Napa V ★★★→★★★★ Crowd-pleasing, ultra-consistent CAB SAUV, MERLOT, incl famed Three Palms vyd, gd SAUV BL. Second label Decoy hugely successful, excellent value. Parent company Duckhorn Wine Co also owns Migration brand: N Coast CHARD and PINOT N, GOLDENEYE (AND V), CALERA and KOSTA BROWNE and WA state brands. Publicly traded company as of 2021.

More CA winemakers viewing new oak barrels as a flavour additive, which they are.

Dutton-Goldfield RRV ★★★ Classical cool-climate CA CHARD, PINOT N from RRV-based powerhouse grower, not super-edgy or risky, maybe a gd thing.

Edwards, Merry RRV ★★★ PINOT N pioneer has retired, but quality should stay high. Single vyds from SON always wildly popular. Ripe, rounded reds sometimes a tad sweet by today's standards. Slightly sweet musqué SAUV BL also popular.

Emeritus RRV ★★★→★★★★ Emergent estate, three home dry-farmed (!) vyds making focused, structured PINOT N under supervision of gifted winemaker Dave Lattin. Hallberg Ranch bottlings exquisite, singular in style.

Etude Car ★★★ Ever-trustworthy brand that always succeeded at making great CAB SAUV, PINOT N under same roof, using same attentive techniques. Now owned by Treasury Wine Estates, but legacy stays true. Pinot rosé to die for.

Failla Napa V ★★★ Adept Ehren Jordan crafts superb CHARD, PINOT N, SYRAH from scattered prime vyds in N CA, even OR. SON COAST Pinot N shows great blending.

Far Niente Napa V ★★★→★★★★ Pioneer of CAB SAUV, CHARD in big, generous, NAPA style. Hedonism with soul. Dolce: celebrated dessert wine. Also makes Nickel & Nickel single-vyd Cabs.

Farrell, Gary RRV ★★★ Namesake founder long gone, but wines still terrific despite a few ownership changes, much thanks to winemaker Theresa Heredia, Farrell's hand-picked successor. Basic RRV CHARD beams brightly, Hallberg and Fort Ross single-vyd PINOT N among top offerings.

Fetzer Vineyards N Coast ★★★ Early champion of organic/bio viticulture in MEND, still gd, best under Bonterra brand. Owned by Concha y Toro (Chile).

Field Recordings P Rob ★★★ Impressively subtle, perceptive wines from P ROB's Andrew Jones. Best are blends Neverland and Barter & Trade, but don't miss Alloy and Fiction, delicious in 500ml *cans*.

Flowers Vineyard & Winery Son Coast ★★★→★★★★ Extreme SON COAST pioneer 3 km (2 miles) from Pacific; CHARD, PINOT N remain great illustrations of that climate and elevation.

Foppiano Son ★★→★★★ Honest RRV wines loaded with sunny fruit and little pretence. PETITE SIRAH, SAUV BL notable.

Fort Ross Vineyard Son Coast ★★★ Dazzling high-elevation estate a stone's throw from Pacific; zesty CHARD, terrific, savoury PINOT N, surprisingly gd PINOTAGE (!).

Freemark Abbey Napa V ★★★ Classic name claimed by JACKSON FAMILY WINES in 2006, improved. Great values: single-vyd Sycamore, Bosché CAB SAUV bottlings.

Frog's Leap Ruth ★★★ John Williams is a pioneer of bio viticulture in CA. He coaxes best out of NAPA V floor with elegant CHARD, refreshing SAUV BL, supple, juicy CAB SAUV, MERLOT and brambly ZIN.

Gallo, E&J ★→★★★ Biggest wine company in world, titan in under-$20 sector with dozens of major CA brands incl Apothic, Barefoot, Louis Martini. Recent buys: Black Box, Clos du Bois, Jayson, RAVENSWOOD. *See also* GALLO OF SONOMA.

Gallo of Sonoma Son ★★★ Formidable wines from great SON sources and broader lands, unfussy as founders would have wanted. Fruit quality speaks loudly.

Gloria Ferrer Car ★★★ Exceptional CA bubbly. Toast to decades-long team of owners, growers, winemakers that made this Freixenet-owned venture extraordinary. All wines v.gd, Vintage Royal Cuvée best of all.

Goldeneye And V ★★★→★★★★ Well-conceived and managed estate winery with three principal vyds. PINOT N on ripe side, but undeniably delicious, esp The Narrows vyd. Bubbly also special, and visitors welcome.

Graziano Family Redwood V ★★→★★★ Flavourful, reliably delicious, mostly Italian-inspired, with deep roots in MEND. Brands incl Enotria, Graziano, Monte Volpe, Saint Gregory. MONTEPULCIANO, PINOT GR, SANGIOVESE delish.

Gundlach Bundschu Son V ★★★ Terrific wines, welcoming vibe, popular tasting destination with adventurous cool Huichica Fest music concerts for hipster set. Best bets GEWURZ, CAB SAUV, MERLOT.

Hall Napa V ★★★→★★★★ Glitzy ST H winery makes great NAPA CAB SAUV, but bewildering variety of selections. Signature offering best, velvety SAUV BL, MERLOT among best in CA. Also owns coastal CHARD, PINOT N brand Walt.

Halter Ranch P Rob ★★→★★★ Boasting 200 acres+ of sustainably farmed vyds on P ROB's west side, Halter reckons large as premium grower and winery in AVA. Solid CAB SAUV, SYRAH. PICPOUL a sprightly surprise.

Hanzell Son V ★★★ Pinot pioneer of 50s still making CHARD, PINOT N from estate vines. Both reward cellar time. Mineral Chard still thrilling. Sebella Chard (young vines) all bright, crisp fruit.

Harlan Estate Napa V ★★★★ Concentrated, robust CAB SAUV – one of original cult wines only available via mailing list at luxury prices. Still all those things today. Son Will makes The Mascot from younger vines, plus Promontory.

Harney Lane Lodi ★★ Family-owned with century of grape-growing under its belt. Old-vine ZIN from Home Ranch and Lizzy James vyds the stars, but ALBARINO, TEMPRANILLO also impress as Iberian grapes gain momentum in Lodi.

Hartford Family Son Coast ★★★ Winemaker Jeff Stewart is quietly killing it at this JACKSON FAMILY property, with small batches of CHARD, PINOT N, ZIN from sites up and down coast. Don't sleep on the taut Seascape Chard and the subtle Fog Dance Pinot N from Green V.

HdV Wines Car ★★★ A CAR gem: fine complex CHARD with honed edge, v.gd PINOT N from grower Larry Hyde in conjunction with Aubert de Villaine of DRC (*300 Franco*); v.gd CAB SAUV, SYRAH

Heitz Cellar Napa V ★★→★★★ Once-iconic brand sold 2018; steady source of gd CAB SAUV at fair price; gd SAUV BL, medium-bodied GRIGNOLINO a rare treat.

Hendry Oak K ★★★ Classic, soulful, minimalist wines est 1939. Of note: brambly, distinctive CAB SAUV, ZIN (try Block 28) from cool pocket of valley nr Napa town. Never disappointing.

Hess Collection, The Napa V ★★★ Great mtn-top visit with world-class art gallery, also makes gd wine. CAB SAUV from MOUNT VEEDER speciality, esp exceptional 19 Block Cuvée, blockbuster with gd manners.

Honig Napa V ★★★ Sustainably grown NAPA V CAB SAUV, SAUV BL nationwide benchmarks thanks to consistent quality, hard-working family and team. Top Cab from Bartolucci vyd in ST H.

It's rare to share a bottle at a restaurant these days. Everyone orders by the glass.

Hope Family P Rob ★★→★★★ Veteran winemaker Austin H steadily delivers quality reds, excellent Austin Hope CAB SAUV, SYRAH at fair prices. Also makes Treana; solid Liberty School is value label.

Inglenook Oak ★★★ FF Coppola's reclaimed Inglenook brand. Rubicon CAB SAUV is flagship proprietary red. Also v.gd CHARD, MERLOT. Showpiece Victorian estate popular with international set.

Iron Horse Vineyards Son ★★★ Amazing selection of 12 vintage bubblies, all wonderfully made. Ocean Res Blanc de Blancs v.gd, Wedding Cuvée a winner; v.gd CHARD, PINOT N.

Jackson Family Wines ★★→★★★★ Visionary, massive vyd owner in CA with a constellation of stellar properties, brands, esp elevated sites; owns popular Kendall-Jackson brand plus ritzier Cardinale, COPAIN, Edmeades, FREEMARK ABBEY, HARTFORD FAMILY, La Crema, La Jota, Lokoya, MAGGY HAWK, MATANZAS CREEK, Murphy Goode, Siduri, Verité. Jackson Estate series great for N Coast mtn CAB SAUV.

Jessie's Grove Lodi ★★→★★★ Deep roots in Lodi, est 1868, Royal Tee vyd among CA's oldest ZIN plantings. Boss Greg Burns knows Zin inside out, and it shows in fine, generous wines. Try *Westwind bottling*, or ALBARIÑO, VERMENTINO if you're in mood for white.

Jordan Alex V ★★★ Showcase estate generates elegant, balanced, B'x-faithful CAB SAUV homage to B'x for a devoted following. CHARD, now sourced from vyds in RRV, delivers lemony, mouthwatering panache.

Joseph Swan Vineyards Son ★★★ Long-time RRV producer of intense old-vine ZIN and single-vyd PINOT N. Sleeper for overlooked Rhône varieties also v.gd, esp SYRAH, ROUSSANNE/MARSANNE blend.

Josh N Coast ★★ Shooting-star brand from Joseph Carr. Solid varietal bulk brand successfully competing with affordable E&J GALLO offerings.

Keller Estate Son Coast ★★★ One more example of balanced, elegant CA wine from the cool coastal regions. CHARD, PINOT N thrilling.

Kistler Vineyards RRV ★★★ Style of CHARD, PINOT N adapted over yrs, wines only improved. Still from a dozen designated vyds in any given yr. Highly sought.

Korbel ★★ Affordable bubbly that's hard to beat for the price, even vs like-priced Italian or Spanish competition. Lovely tasting room nr Russian River.

Kosta Browne Son Coast ★★★→★★★★ Hyped brand now under DUCKHORN umbrella: technically impeccable PINOT N from prime vyds, more restrained now. I'll leave the value proposition to consumers, but Keefer Ranch Pinot N wowed me.

Krug, Charles Napa V ★★→★★★ Historically important winery made recent comeback, demanding recognition for role in modern NAPA V. Late owner Peter Mondavi was Robert's estranged brother. Supple CAB SAUV, crisp, pure SAUV BL.

Ladera Napa V ★★★→★★★★ Stotesbery clan sold HOWELL MTN winery and set up shop in ST H. Hillside CAB SAUV, MALBEC great; don't miss superb SAUV BL from NZ winemaker.

Lang & Reed Mend, Napa ★★★ Noone in CA has flown CAB FR banner more passionately than L&R's John Skupny. Wines capture perfume, litheness with NAPA generosity. Also delicious MEND CHENIN BL.

Larkmead Napa V ★★★★ Historic gravel-laced NAPA V estate revived; *outstanding Cab Sauv*, supple, balanced; bright, delicious SAUV BL. Rare Tocai FRIULANO a delight.

Lindquist Family – Verdad Arroyo GV, Arroyo Seco ★★★ Rhône Ranger Bob L's 2nd act(s) after trailblazing QUPÉ. More outstanding Rhône varieties under Lindquist brand; Verdad label reserved for Spanish varieties, incl *auténtico* ALBARIÑO, TEMPRANILLO from bio Sawyer Lindquist vyd in nearby EDNA V.

Lioco N Coast ★★★ Influential minimalist brand champions elegant, subtle CHARD, CARIGNAN, PINOT N. Dependable, ethereal, satisfying wines.

Littorai And V, Son Coast ★★★★ Burgundy-trained Ted Lemon's N Coast CHARD, PINOT N are pure, inspiring, with sense of place. A lot of wines, but they're all sleek, exciting. Cerise, Savoy, Wendling vyd offerings thrilling.

Lohr, J ★★→★★★ Prolific producer of Central Coast makes CAB SAUV, PINOT N, CHARD for balance and gd value. Cuvée Pau and Cuvée St E pay homage to B'x. Don't miss seductive, floral Beaujolais-like Wildflower Valdiguié.

Long Meadow Ranch Napa V ★★★→★★★★ Smart, holistic vision incl destination winery with restaurant, cattle on organic farm. Supple, age-worthy, fresh CAB SAUV has reached ★★★★ status; lively Graves-style SAUV BL.

Louis M Martini Napa, Napa V ★★→★★★ Since buying the Martini brand and epic Monte Rosso vyd, E&J GALLO has restored latter to greatness. Martini brand is solid for prosaic CAB, ZIN.

Lucia Santa LH ★★★ Access to great vyds and 2nd-generation knowledge has positioned Jeff Pisoni well. Impressive CHARD, PINOT N. Central Coast, SANTA LH brand to watch.

Macchia Lodi ★★★ One of Lodi's most accomplished winemakers, Tim Holdener vacuums up medals in blind tastings yr after yr. Speciality is balanced old-vine ZIN, but also interesting SANGIOVESE, TEROLDEGO; v.gd PETITE SIRAH.

MacPhail Son Coast ★★★ Now owned by HESS COLLECTION, making mostly PINOT N from cool sites in SON and MEND. Highlights are Gap's Crown, Sundawg Ridge and Toulouse vyd bottlings.

MacRostie Son Coast ★★★ New tasting room is a modern beauty; screwcapped wines steadily improving. Lovely PINOT N, SYRAH; SON COAST CHARD absolute delight. Excellent consistency, value. Buy these wines!

Maggy Hawk And V ★★★ COPAIN alumna Sarah Wuethrich heads team crafting great CHARD, PINOT N in deep end (cold part) of AND V. Don't just drink it, go there.

Marston Spring Mtn ★★★★ Not-quite-cultish small-production mtn CAB SAUV arguably among most complex Cabs of NAPA V, with great structure, foresty flavours. Wines currently made by Sierra Leone-born Marbue Marke; Albion SAUV BL also terrific; gd stories and great wine.

Americans uncomfortable using corkscrews in public; screwcaps are sooo easy.

Masút Mend ★★★ Newish elevated Eagle Peak property, run by Ben and Jake FETZER, shines. Ethereal estate PINOT N. Will inspire others to explore area.

Matanzas Creek Son ★★★ Exceptional JACKSON FAMILY WINES property in cool Bennett V focuses on excellent MERLOT, SAUV BL from lavender-perfumed estate.

Matthiasson Napa V ★★★ Experimental wines become cult hits. Elegant CAB SAUV, racy CHARD, epic white blend, plus esoterica like RIBOLLA GIALLA, SCHIOPPETTINO.

Mauritson Dry CV ★★★ Clay M, 6th-generation grower, captains extraordinary holdings in elevated Rockpile district; wines only got better under his two decades. ZIN flagship, as with many DRY CV wineries; CAB SAUV, SAUV BL excellent.

Mayacamas Vineyards Mt Veeder ★★★ Now owned by Charles Banks, former partner in SCREAMING EAGLE; CA classic has not changed big-boned style, only improved. Age-worthy CAB SAUV, CHARD recall great bottles of 70s, 80s.

Miraflores Sierra F'hills ★★★ Marco Cappelli left NAPA V to set up in Sierra Mtns; vinifies sublime, broad array from estate and region he rightly believes in.

Mi Sueno Napa V ★★★ Rolando started dishwashing at a NAPA resort, worked his way up at wineries. Lorena's family worked vyds, bought property. Now wedded, they make excellent wine, every bit as gd as story. *¡Felicidades y salud!*

Montelena, Ch Napa V ★★★ Tons of history, great continuity of ownership, style. Serious, if slightly funky CAB SAUV cellar-worthy; CHARD holds up well too. Castle-like winery, heavenly setting.

Mount Eden Vineyards Santa Cz Mts ★★★→★★★★ Gorgeous vistas from high vyd, one of CA's 1st boutique wineries with Burgundian clones dating back to Martin Ray days. Taut, mineral CAB SAUV, PINOT N, stunning CHARD since 1945. Inspired by Burgundy, but pure rugged CA character.

Mount Veeder Winery Mt Veeder ★★★ Classic CA mtn CABS SAUV, FR grown at 500m (1640ft) on rugged, steep hillsides. Dense wines: ripe, integrated tannins.

Mumm Napa Valley Napa V ★★★ At RUTH since 1970. Quality bubbly, notably Blanc de Noirs and pricier, complex DVX single vyd left on lees for a few yrs.

Nalle Dry CV ★★★ Refined craftsmanship-level ZIN, impeccable, elegant claret-style reds. Great family-owned stop nr Healdsburg.

Niner Edna V, P Rob ★★→★★★ Young, ambitious family estate with excellent CAB SAUV from P ROB, great EDNA V ALBARIÑO, CHARD. Restaurant (CA cuisine) gd for lunch in P Rob countryside.

Obsidian Ridge Lake ★★★ Star of Lake County extension of Mayacamas mtn range. Super CAB SAUV, SYRAH from hillside vyds, volcanic soils scattered with glassy obsidian. Half Mile Cab 1st-rate. Also owns Poseidon brand from CAR.

Somm elbow: sommelier tendon injury, caused by pouring from 1kg (2lb) bottles.

Ojai Santa B ★★★ In a change of style from big, super-ripe to leaner, finer, former AU BON CLIMAT partner Adam Tolmach making best wines of his career: v.gd CHARD, PINOT N, Rhône styles. SYRAH-based rosé delicious.

Opus One Oak ★★★★ Mouton Rothschild family-controlled standard bearer for fine NAPA CAB SAUV; popular luxury export. Wines designed to cellar 10 yrs+.

Pahlmeyer Napa V ★★★ Jammy, pricey, well-made NAPA V wines: B'x blend, MERLOT, lavish CHARD notable. Popular volume-label Jayson. Now in E&J GALLO portfolio.

Patz & Hall N Coast ★★★★ James Hall is one of CA's most thoughtful winemakers; culls fruit from top vyds from Central Coast to MEND. Style is generous, tasteful, super-reliable. Zio Tony *Chard* v. special, lemony, electric, opulent.

Paul Hobbs Wines N Coast ★★★→★★★★ Globetrotting winemaker Paul H still a local hotshot. Bottlings of single-vyd CHARD, CAB SAUV, PINOT N, SYRAH top; v.gd-value second label: Crossbarn.

Peay Vineyards Son Coast ★★★→★★★★ Standout brand from one of coast's coldest zones. Finesse-driven CHARD, PINOT N, SYRAH superb. Second label, Cep, also v.gd, esp rosé. Weightless, impeccably made wines.

Pedroncelli Son ★★ Old-school DRY CV winery updated vyds, winery; still makes bright, elbow-bending CAB SAUV, ZIN, solid CHARD. Refreshingly unpretentious.

Peter Michael Winery Son ★★★★ Elevated Knight's V making legendarily age-worthy CAB SAUV blends, notably Les Pavots. Sells mostly to restaurants, mailing list. Also offers cool-climate CHARD, PINOT N selections from SON COAST and Coeur à Coeur SAUV BL blend from Les Pavots site.

Phelps, Joseph Napa V ★★★→★★★★ Expensive NAPA "First Growth" Insignia, one of CA's 1st ambitious B'x blends, still dependably great, as is Napa CAB SAUV. Most offerings excellent quality, esp SYRAH.

Philip Togni Vineyards Spring Mtn ★★★★ 10 12 14 15 Legendarily age-worthy SPRING MTN CAB SAUV. Stiff mtn terroir generally needs time, rewards patience. All class.

Pine Ridge Napa V ★★★ Outstanding CAB SAUV from several NAPA V vyds. Estate STAGS L bottling, silky, graceful. Lively CHENIN BL/VIOGNIER innovative classic.

Pisoni Vineyards Santa LH ★★★ Family winery in SANTA LH became synonymous with PINOT N explosion and big, jammy wines. Still, Pinot N is and always was well made and remains popular.

Presqu'ile Santa MV ★★★ New Central Coast winery, elegantly styled PINOT N, SYRAH. Concrete egg-fermented SAUV BL of note, as is estate Pinot N.

Pride Mountain Napa V, Spring Mtn ★★★→★★★★ Epic Mayacamas mtn-top estate straddles NAPA V and SON border; superb, bold CAB SAUV; amazing MERLOT.

Quintessa Ruth ★★★★ Magnificent estate at heart of NAPA V, owned by Chilean international player Augustin Huneeus, makes single wine: superb, refined B'x blend justifies triple-digit price.

Qupé Santa B ★★★ One of original SYRAH champions, brilliant range of Rhônes, esp X Block, from one of CA's oldest vyds. Hillside Estate also epic; don't miss impeccable MARSANNE, ROUSSANNE. Central Coast SYRAH unbeatable for $.

Radio-Coteau Son Coast ★★★ Notable new-wave SON COAST PINOT N, serious coastal

SYRAH and old-vine, dry-farmed ZIN. CHARD and Zin bulletproof, but Pinot N steals show. Veg gardens, cider orchard, goats, chickens, honeybees and cats too.

Ramey Son ★★★→★★★★ Influential David R delivers flinty, reductive burgundian-style whites in his Hyde, Ritchie and RRV CHARD. Reds gd, but Chard should be *de rigueur* tasting at UC Davis winemaking school.

Rancho Sisquoc Santa MV ★★→★★★ Rustic tasting room and historic chapel deliver satisfying spectrum of B'x-style wines incl outstanding CHARD, CAB FR, PINOT N.

Ravenswood ★★★ CONSTELLATION-owned, but single-vyd ZIN still from remarkable sites like Bedrock, Old Hill, Teldeschi. "No wimpy wines" motto still applies.

Red Car Son Coast ★★★ Hip brand with colourful label making precise CHARD, lacy, fruit-forward PINOT N and killer rosé.

Renwood Sierra F'hills ★★→★★★ Historic Sierra Nevada brand on rebound after purchase by international New Frontier Wine Co. Joe Shebl assembles lovely, robust ZIN from ancient, head-pruned, dry-farmed vyds.

Ridge N Coast, Santa Cz Mts ★★★★ Saintly founder Paul Draper semi-retired, but his spirit lives on. Majestic, legendary, age-worthy estate Montebello CAB SAUV always superb. Lasts forever. Outstanding single-vyd field-blend ZIN special. Don't overlook top-rank, minerally CHARD.

Robert Mondavi Winery ★★→★★★ Owned by CONSTELLATION since 2004, many wines could be better; changing of winemaking guard appears at hand. To Kalon vyd still a great site; potential there.

Robert Sinskey Vineyards Car ★★★ Great, idiosyncratic NAPA estate favouring balance, restraint. Impressive CAB SAUV and CAR PINOT N. Racy Abraxas white blend and Pinot rosé excellent.

Rodney Strong Vineyards Son ★★★ Strong indeed, across board; 14 vyds producing sinewy coastal CHARD, PINOT N, super ALEX V CAB SAUV from Alexander's Crown, Rockaway vyds. Also owns Davis Bynum in RRV, try River West vyd CHARD.

Roederer Estate And V ★★★★ Adventurous Champagne Roederer venture brought glamour to AND V. Finesse, class off charts, esp luxury cuvée L'Ermitage. Also makes Scharffenberger fizz. Dom Anderson PINOT N also excellent.

Rombauer Napa V ★★★ Buttery CHARD is calling card; boomers adore this prestige brand. Solid CAB SAUV, MERLOT, ZIN also sunny, flavourful.

St Jean, Ch Son V ★★★ Rock of SON V, solid on all fronts, but consensus flagship for decades has been Cinq Cépages: five B'x varieties. Nectarine-rich Robert Young vyd CHARD also a classic.

Saintsbury Car ★★★ Regional pioneer and benchmark still making v.gd, highly relevant PINOT N, CHARD, yummy Vincent Vin Gris rosé. Lee vyd Swan clone Pinot is vibrant but velvety.

St-Supéry Napa ★★★ Bought by owners of Chanel (and Ch Rausan-Ségla, B'x), but some continuity of talent. Tasteful, balanced Virtú (w) and Élu (r) B'x blends and SAUV BL, esp *Dollarhide Ranch*, thrilling.

Sanford Santa B, Sta RH ★★★ Now owned by Terlato family, wines still exceptional, esp La Rinconada, Sanford & Benedict PINOT N, SANTA B CHARD.

CA needs more heat-tolerant whites inland: Colombard, Ugni Bl, Vermentino.

Schramsberg Napa V ★★★★ Best bubbles in CA? Exacting quality in every bottling, esp luxurious J Schram and Blanc de Noirs kept pace with French competition. Memorable tours of historic caves by reservation.

Screaming Eagle Napa V, Oak ★★★★ Original "cult" CAB SAUV, famously outstanding, rare and four-figures *cher* from primo OAK benchland; v. limited SAUV BL. Sister winery is Jonata.

Scribe Son ★★★ Hipster gentleman-farmer aesthetic a hit with younger set. Tasting room pours well-made esoterica like SYLVANER, ST-LAURENT and PINOT N rosé.

Sea Smoke Sta RH ★★★ Cultish, high-end, opulent CHARD, PINOT N and sparkling made in a strictly estate-driven model with incredible continuity of leadership and talent. Great wines, but drink while young (them, and you).

Seghesio Son ★★★ No longer family-owned, but still classic ZIN from ALEX V, DRY CV. Muscular but graceful reds. Old Vine bottling benchmark for price. Rockpile Zin is dynamite.

Shafer Vineyards Napa V, Stags L ★★★→★★★★ Solid brand, widely respected by critics, sommeliers alike. Hillside Select CAB SAUV a lavish CA classic; Relentless SYRAH/PETITE SIRAH blend powerful, artful. One Point Five, absolute beauty for money. Fine CHARD, MERLOT from nearby CAR.

Shannon Ridge Lake ★★→★★★ Grand, undulating, high-elevation estate overlooking Clear Lake. Terrific reds and whites from PETITE SIRAH to SAUV BL, esp Res series. Great value. Second label Vigilance a big seller.

Silverado Vineyards Stags L ★★★ Walt Disney descendants have owned this Silverado Trail gem since 1976. Always well maintained. Single-vyd Solo CAB SAUV; new release of Geo B'x blend from COOMB. CAB FR excellent.

Silver Oak Alex V, Napa V ★★★ Plush CAB SAUV in consistent, fruit-driven style.

Sonoma-Cutrer Vineyards Son ★★★ Flagship CHARD, classic, by-the-glass pour at restaurants all over country, rich but zesty. Owsley PINOT N from RRV lush.

Sonoma-Loeb Car, RRV ★★★ CHAPPELLET'S SON PINOT N, CHARD label. Sangiacomo vyd CHARD stands out.

Spottswoode St H ★★★★ Crown jewel of ST H. Sublime estate always chasing perfection with every sustainable bona fide in the universe. Estate Cabs thrilling with modest alc. Second wine Lyndenhurst CAB SAUV value, Spottswoode SAUV BL always zesty, delightful.

Spring Mountain Vineyard Spring Mtn ★★★★ Top-notch estate delivers site-driven, age-worthy mtn CAB SAUV. Signature Elivette B'x blend always layered and sturdy, Estate Cab terrific, estate SAUV BL a tropical delight.

Staglin Family Vineyard Ruth ★★★★ Perennial 1st-class, potent CAB SAUV from family-owned estate. Formidable, complex Salus CHARD.

Stag's Leap Wine Cellars Stags L ★★★ Since founder sold to large corporation, gd to see quality maintained. Flagship CAB SAUV (top-of-line Cask 23, Fay, SLV).

Stags' Leap Winery Napa V, Stags L ★★★→★★★★ Important, beautiful estate recently restored, exceptional spot to visit (by appt only), wines great now. CAB SAUV leads, but PETITE SIRAH and field-blend Ne Cede Malis always been special.

Sterling Napa V ★★ Great spot to visit; take aerial tram to tasting room with 90m-(295ft-) high view of valley.

Stony Hill Vineyard Spring Mtn ★★★ Revered NAPA V estate mostly famous for whites, esp mineral, ageable CHARD, plus GEWURZ, RIES. Sold to LONG MEADOW RANCH (2018). Expect to hold steady.

Tablas Creek P Rob ★★★→★★★★ When it comes to Rhône and Med varieties, what doesn't this do perfectly? Killer red and white blends Esprit and Côtes de Tablas plus varietals from MOURVÈDRE to VERMENTINO and GRENACHE BL – don't miss. Impeccable CA wine with min pretence.

California's new fine wines – a handful

What's bubbling under? Interesting varieties and styles that aren't yet mainstream but are finding an audience, eg. tropical, racy St Amant VERDELHO; Saracina COLOMBARD, razor-edge white capable of age; TABLAS CREEK's lucid, mouthwatering VERMENTINO; DASHE Les Enfants Terribles GRENACHE, Beaujolais-like study in wild berries; WIND GAP SYRAH, excitedly blue-fruited and peppery; Donkey and Goat Stonecrusher ROUSSANNE, CA take on hearty white Rhônes.

Terra Valentine Napa V, Spring Mtn ★★★ Wurtele family lovingly rehabbed this winery in early 2000s, then handed off to winemaker Sam Baxter; mtn CAB SAUV is focus, romanticizes SANGIOVESE with some success. Don't underestimate.

Terre Rouge / Easton Sierra F'hills ★★★ Single company with two sides: traditional old-vine ZIN (Easton) and Rhône varieties (Terre Rouge). Mostly reds. Affordable Tête-à-Tête red blend a steal, Ascent SYRAH reliably special.

The Prisoner Napa V ★★ It would be criminal to omit this red-blend renegade created by Dave Phinney, now in custody of CONSTELLATION BRANDS. A brutish offering from mysterious sources with captive audience.

Tongue Dancer And V, Son Coast ★★★ James MacPhail's solo venture after exiting his eponymous brand follows similar plan: robust CHARD, PINOT N from stellar N Coast vyds. Bacigalupi vyd Chard can wow.

Trefethen Family Vineyards Oak K ★★★ Underappreciated winery in cool OAK K deserves more credit, elegant CAB SAUV, MERLOT, CHARD and delicious dry RIES.

Trinchero Family Estates Napa V ★ ★★★ Bewildering slew of labels, incl mass-market Sutter Home; esp pleasing CAB SAUV under Napa Wine Company label.

Turley Wine Cellars P Rob ★★★★ Sells mostly to mailing list. Brilliant brambly old-vine ZINS from vyds scattered across state. True CA treasures.

Unti Dry CV ★★★ Wines start in vyds; grower delivers soulful, luscious, tasty BARBERA, GRENACHE, SYRAH, ZIN.

Villa Creek P Rob ★★★ Former restaurateur Chris Cherry sees P ROB's climate through a Spanish lens; makes brilliant, focused GARNACHA from 60-acre bio Maha estate, plus solid CAB SAUV, Rhône-style blends.

Vineyard 29 Napa V ★★★ ·★★★★ Top winemaker Philippe Melka's fingerprints all over gorgeous CAB SAUV at maturing estate venture; gd but oaky SAUV BL.

Volker Eisele Family Estate Napa V ★★★ Special site tucked way back in Chiles V continues to overdeliver with CAB SAUV and more. Looking for an adventure? Try a twisty Chiles V road trip.

Wente Vineyards ★★ Oldest continuing family winery in CA makes decent whites/reds. Outstanding gravel-grown SAUV BL leads. Murietta's Well: gd blends (r/w).

Wilkes, J Santa B ★★→★★★ Work in progress with passionate, brainiac oenologist Wes Hagen (formerly of Clos Pepe) at reins. Can't fail if tireless Wes is on board. P ROB, SANTA B (r/w).

Williams-Selyem RRV ★★★ Benchmark SON PINOT N since '70s, inspired by Burgundy; put RRV on map as international Pinot centre. Rochioli Riverblock Pinot N legendary, priced accordingly. Wines are min processed might encounter a funky bottle here and there.

Wind Gap Son Coast ★★★ Pax Mahle is one of CA's most talented winemakers, esp in cool climates; CHARD, PINOT N excellent, in best vintages; SON COAST SYRAH displays revelatory new-wave terroir.

Wine Group, The Central V ★ By volume, world's 2nd-largest wine producer; budget brands like Almaden, Big House, Concannon, Cupcake, Glen Ellen.

Colorado (CO)

More than 1000 acres of vines – some of highest-altitude vyds in US, up to 2134m (7000ft); 160 wineries. Extreme continental climate similar to Spain's La Mancha. Cool-climate varieties thrive, and new experimentation with hybrids. Grand V and West Elks AVAs home to 90% of grapes. **Bookcliff** ★★ winning GRACIANO; excellent CAB FR Res, CAB SAUV, MALBEC, SYRAH, VIOGNIER. **Carboy** winery, négociant and tasting room, four locations. Rotating inventory. **Carlson** ★ heritage winery, dry GEWURZ, RIES, T-Red LEMBERGER (r) and full-bodied reds. **Colterris** premium, B'x style. Coral White Cab Sauv, stylish signature wine. **Jack Rabbit Hill Farm** ★ only certified bio winery in state.

Eye-catching orange wine, Lone Eagle Proprietary Blend of Ries and Hungarian Bianca; trendy grower ciders. **Sauvage Spectrum** new label, fresh sparkling. **Snowy Peaks (Grande V)** v. high-altitude, 100% CO grapes, Rhône varieties; Oso (r) blend uses hybrid grapes. **Sutcliff Vyds** solid showing of B'x styles in the Four Corners frontier. **Stone Cottage Cellars** organic, moody MERLOT, high elevation. **The Storm Cellar** relative newcomer focusing on high-elevation whites and rosé. Rosé of St Vincent; gd Ries range.

Georgia (GA)

Blue Ridge Mtns tie GA to VA, NC. Dahlonega Plateau, 1st all-GA AVA, rocky hills; B'x, CHARD, PETIT MANSENG, PINOT N. Best: **Crane Creek**, **Engelheim** (PINOT GR); **Frogtown** (SANGIOVESE); **Habersham**; **Sharp Mtn** (Sangiovese, GEWURZ); **Stonewall Creek** (NORTON); **Three Sisters** (oldest on plateau; PINOT BL, AVA CAB FR); **Tiger Mtn**; **Wolf Mtn** (traditional sp); **Yonah Mtn**.

Idaho (ID)

Young, emerging region, still deciding what to plant. Promise is clear. A subplot? Majority of wines made by women.

Cinder Wines Snake RV ★★ Melanie Krause (ex-Ch Ste Michelle, WA) makes velvety SYRAH, off-dry RIES in Snake River V AVA. VIOGNIER also v.gd.

Colter's Creek ★★ Lewis-Clark V pioneers Mike Pearson (vines), Melissa Sanborn (wines) craft herby CAB, silky SYRAH, helping to put ID on wine map.

Huston Vineyards Snake RV ★ Chicken Dinner RIES blend a fruit-forward delight.

Rivaura ★★ New Lewis-Clark V estate. One to watch, esp SYRAH.

Ste Chapelle Snake RV ★ ID's 1st and largest winery focuses on sweet whites, soft reds. Well-priced quaffers.

Sawtooth Winery Snake RV ★ Valley stalwart founded 1987 makes tasty estate RIES, SYRAH, TEMPRANILLO.

Maryland (MD)

East Shore sandy soils, hills of Garrett and Allegheny mtns, blue-crab-rich Chesapeake Bay checks freezing winters, stifling summers; B'x grapes incl reliable ripeners PETIT VERDOT, SAUV BL; also ALBARIÑO, MERLOT. **Big Cork**; **Black Ankle** 1st post-Prohibition winery; **Boordy**; **Bordeleau**; **Catoctin Breeze**; **Crow**; **Dodon**; **Elk Run**; **Linganore**; **Old Westminster**; **Philosophy** 1st African-American woman winemaker in MD, sourcing local CAB FR, VIOGNER for premium wines raised, poured, in Baltimore's **The Wine Collective**; **Sugarloaf Mtn**.

Massachusetts (MA)

Only AVA: Martha's Vyd, within SE New England, shared with CT and RI. Cool Atlantic climate moderated by Gulf Stream. Like rest of New England, many fruit wines, occasionally v.gd. CHARD, GEWURZ, PINOTS N/BL/GR, RIES, some Cayuga; Concord developed here 1849; 25+ small producers. **Truro Vyds** (CAB FR, MERLOT in sea-breezed vyd). **Turtle Creek** (1–6 barrels of single-varietal, incl Cab Fr). **Westport** (PINOT M: traditional-method, still).

Michigan (MI)

The "Third Coast", on huge Lake Michigan, cold wine frontier, Pinots do well, also CHARD, GEWURZ, RIES; CAB FR, MERLOT, occasional TEROLDEGO hybrids) and cherries – as in some of its best bottles (**Ch Chantal**'s well-regarded Cerise Noir is 80% PINOT N, the rest Montmorency cherry). Versatility like **Karma Vista**'s peppery SYRAH and **Lemon Creek**'s spicy dark-fruit SHIRAZ. **Left Foot Charly**

BLAUFRÄNKISCH, **Mawby** Blanc de Blancs, de Noirs, **Veterra's** PINOT BL – all MI classics. **Tabonne Vyds** Estate Red is lesson in hybrid worthiness (Baco Noir, Chancellor, Leon Millot, Marechal Foch). New; few wineries but growing. Note cold-hardy Itasca for dry whites.

Missouri (MO)

New ambitious plan by Hoffman Family of Cos: $100 million, 700 acres, to create Napa-style destination, golf course, hotel, etc. Until then, try Chambourcin, SEYVAL BL, VIDAL, Vignoles (dr/sw). **Hermannhof**: Chardonel, Norton, Vignoles. **Stone Hill** in Hermann: v.gd Chardonel (frost-hardy hybrid, Seyval Bl x CHARD), Norton and gd Seyval Bl, Vidal. Also **Adam Puchta** for fortifieds and Norton, Vidal, Vignoles. **Augusta Winery** for Chambourcin, Chardonel, Icewine. **Les Bourgeois** for Chardonel, Montelle, Norton, SYRAH; v.gd Chambourcin, Cynthiana. **Mount Pleasant** in Augusta for rich fortified and Norton. **St James** for Norton, Seyval, Vignoles.

Nevada (NV)

Few commercial wineries. **Churchill Vyds** in high-desert region: gd SEM/CHARD, all NV-grown grapes. **Pahrump Valley** oldest winery here, three ranges with irreverent names, of which award-winning Nevada Ridge is most serious.

New Jersey (NJ)

Some of 60+ wineries among best in E US; 150 growers, need more. B'x varieties in S NJ's flat gravelly maritime Outer Coast Plain (incl Cape May); limestone, granite Warren Hills in n for elegant BLAUFRÄNKISCH, GEWURZ, GRÜNER V, PINOT N, RIES, SYRAH. Also quality NJ produce at some wineries.

Blaufränkisch versatile, reliable in NJ, NY; Grüner V growing in NJ, PA.

Alba ★★★ Limestone, granite in Warren Hills AVA. One of largest PINOT N plantings on E Coast. Burgundy aspiration, incl earthy 30-mths Grand Res; excellent CHARD, gd GEWURZ, RIES, 30-days-macerated CAB FR.

Beneduce Vineyards ★★★ Family estate; BLAUFRÄNKISCH, PINOT N, open-fermented, barrel-raised, sweet-spice-emboldened, ripe, cooler-climate fruit. Intermezzo: GEWURZ 10 day open-top maceration for dry, firm aromatics, bitter-cleansing finish. Chambrusco: native-ferment Chambourcin pét-nat.

Mount Salem ★★★ Austrian varieties, Burgundy methods to match terroir: Pattenburg gravelly loam slope 215m (705ft), rich aromas, bold acidities in BLAUFRÄNKISCH, ST-LAURENT, ZWEIGELT (varietal; masterful Matthias blend flagship); barrel-fermented GRÜNER, CHARD. Also CAB FR, three bottlings: Chester, Hunterdon, Monmouth counties. Wild ferment. Planting San Marco.

Saperavi grape: red flesh, mutates easily, reflects terroir. Perfect tourist?

Unionville ★★★ Single-vyd Burgundy takes: Pheasant Hill for lasting herbal, currant, rosé, 12% abv PINOT N, rich-fruited 13% abv CHARD; firmer citrus Chard Home Vyd. Carbonic-maceration, no-oak CAB FR, cool-climate quaff. All else now Rhône varieties: MARSANNE, ROUSSANNE; PICPOUL for 2024.

William Heritage ★★★ 17 19 Cool maritime Outer Coastal Plain AVA; B'x-inspired Norman's Vyd CAB FR, 10 yrs+. BDX (five B'x varieties): deep, red berries. Grounded Estate Res CHARD among best in e. Top Blanc de Blancs 17 18, creamy, salty, taut citrus.

Working Dog ★★★ Barrel-fermented CHARD, VIOGNER (ripeness, lift). MERLOT, SYRAH; several oak-aged CAB FR, incl flagship Retriever (creamy, rustic, lasting).

Chambourcin grape: serious in MD; Lambrusco tribute in NJ; Loire-style in Philly.

New Mexico (NM)

The oldest wine region in the US. High-altitude vyds in three AVAs, largest of which is Mimbres V. Historic focus on fizz, but still wines also gd.
Black Mesa ★★ local grapes MERLOT, PETITE SIRAH. **DH Lescombes Family Vyds** Burgundy heritage in pioneering winery. **Gruet** ★★★ benchmark for traditional-method fizz; mostly sources W Coast grapes. **La Chiripada** ★ early pioneer, 20 varieties+, top-notch Res CAB SAUV, convincing RIES. **Luna Rossa Winery** Italian-focused estate vyds; v.gd Res AGLIANICO, introduced RIBOLLA GIALLA. **Noisy Water** ★★ low-intervention, cool-climate varieties in historic Engle Vyd; gd Res CAB SAUV, Wild Ferment Old Vine PINOT N. **Vivác** ★★ excellent red blends Divino (Italian grapes), Diavolo (French), v.gd Port-style Amante.

New York (NY)

US's 3rd-largest producer. Several wineries on old dairy farms in once cattle-focused state. Knowledge, experimentation keep quality high and rising. Climate like N Europe's: winter freeze; lakes, rivers, ocean influence give BLAUFRÄNKISCH, CAB FR, GEWURZ, RIES, B'X – esp maritime Long Island (Long I) – and give crucial relief for some of nation's most compelling, lower-alc and sparkling; hybrids, increasingly made seriously, dry, esp in colder Finger Lakes (Finger L; sunlight hours same as Napa's over fewer days) and Hudson V (most complex soils). Niagara Escarpment (Niag; offers NY-rare limestone).

21 Brix ★★ Estate on Lake Erie with 1st-rate CHARD, GEWURZ, GRÜNER V, RIES; aromatic BLAUFRÄNKISCH, CAB SAUV; v.gd PINOT N. Serious Noiret. VIDAL Icewine.

Arrowhead Spring Vineyards ★★★ Estate on Niag, starring PINOT N; B'x blends. CAB FR, SYRAH: 13% abv and cool-climate acidity. Focused CHARD.

Bedell Long I ★★★ Pre-eminent estate. Native yeasts, maritime climate shows in powerful, saline wines: Musée (MERLOT/PETIT VERDOT/MALBEC) is top label; v.gd CAB FR, SYRAH. Plus CHARD, SAUV BL, VIOGNIER. Artist labels, eg. April Gornik, Chuck Close. Experimental small batches: Carbonic Malbec; AUXERROIS and VERDEJO (unusual in e) harvested 2021 for 1st time.

Bloomer Creek Finger L ★★★ 16 18 20 Three bio-minded sites. Cleverly austere CAB FR/MERLOT White Horse 16. Barrow vyd RIES 18 macerated for herbal-honey weight. Recent releases: skin-contact GRÜNER V/CHARD 20; library CAB FR 16.

Boundary Breaks Finger L ★★★ Top-notch dry to dessert RIES, all lush, acid-driven; serious GEWURZ; gd cool-climate CAB FR, MERLOT.

Channing Daughters Long I ★★★ Estate famed for beachy terroir via experiments: natural fermentation, Italian varieties. Textured, ageable FRIULANO; uncommon NY cool-maritime LAGREIN. Single-site floral PETIT VERDOT, spicy BLAUFRÄNKISCH/DORNFELDER and bright CAB FR. Field blends: Mosaico 16, Meditazione 17. Co-fermented Research Cab 16 from vyd of eight reds (Italian, French), plus a white.

Element Winery Finger L ★★★ Tenacious, terroir-driven CHARD, RIES; CAB FR, LEMBERGER, PINOT N, SYRAH; belief in local MERLOT (ripe, herbal), heady GAMAY. Colloquial Wines is experimental estate programme: GRENACHE, PINOT N; Blanc de Blancs 21, out in 31.

Fjord Hudson V ★★★ In-the-know celebrated ALBARIÑO and spontaneous-fermented CAB FR; v.gd CHARD Icewine 15. Estate GAMAY, MERLOT planted 2020. Recently planted SAPERAVI in Field Blend 2019 with BLAUFRÄNKISH, CAB FR; Cab Fr 19.

Floral Terranes Long I ★★★ Fruit from N Fork, ambient-ferment in suburban garage into concentrated, wild, moody wines. Hold MERLOT 20 until its ferocious if delicious youth softens. CAB SAUV 20 is earthy Amerena, striking acidity. Cider from apples foraged in untended orchards.

Fox Run Finger L ★★★ RIES range v.gd, incl single vyd and barrel-fermented; rare

Res CAB FR 19, LEMBERGER; herbal CAB SAUV; ageable MERLOT. Winemaker Peter Bell mentored many of Finger L's greatest.

Frank, Dr. Konstantin Finger L ★★★ 17 19 Founder est vinifera in Finger L; winery on Keuka Lake. Lena 19: five B'x grapes. US-rare Siberian/N China Amur is lively, revisit 5–10 yrs. Amber RKATSITELI 19 tribute to Georgian winemaking, precise. Old Vines PINOT N 19 incl 1958 vyd. SAPERAVI 19. Blanc de Blanc 17 (Keuka Lake CHARD) and Blanc de Noirs (Keuka Pinot N) sparkling masterclasses.

Heart & Hands Finger L ★★★ Small production, excellent, and just three grapes: CHARD, RIES, PINOT N for exploring limestone (Devonian, and rare in Finger L) vyd on shores of Cayuga Lake. Classic cool-climate white, rosé; delicate red.

Hermann J Wiemer Vineyard Finger L ★★★ Top US RIES name; incl original bio vyds, some single bottlings. Fine CHARD (also farmed bio), GEWURZ (vines among NY's oldest), CAB FR, PINOT N and superlative fizz. Owns fantastic Standing Stone (SAPERAVI; Blanc de Blancs from 1974 vyd). Planting NEBBIOLO.

Hosmer Finger L ★★★ On Cayuga Lake, vyd est 1972. Awarded RIES, incl limited bottlings; also CAB FR, CHARD; 80s PINOT N vines, some for Blanc de Noirs.

Keuka Lake Vineyards Finger L ★★★ Vivacious RIES, incl Falling Man from steep slopes; v.gd CAB FR. Hybrids incl Vignoles and cult Alsatian Leon Millot.

Keuka Spring Vineyards Finger L ★★★ World-class GEWURZ lineup incl site blends, single sites; Alto Adige-style; 19 esp age-worthy. CAB FR, LEMBERGER, MERLOT.

Lakewood Vineyards Finger L ★★★ A 3rd-generation estate. High-quality Res CAB FR; everyday bottle too. Impressive GEWURZ, PINOTS GR/N, multiple RIES.

Lamoreaux Landing Finger L ★★★ Wildly varied sites, excellent RIES (many bottlings, norm in Ries-happy, varied-terrain Finger L), CHARD, GEWURZ, Icewine, plus CAB FR, MERLOT, PINOT N. Creamy Blanc de Blancs, 4 yrs lees. Library wines.

Liten Buffel ★★★ Estate in Niag, two PINOT N, PINOT GR Ramato, RIES (vinified whole-cluster; also skin-contact). Co-ferment BLAUFRÄNKISCH/SAUV BL. Wild yeasts in neutral oak, no filtering, no sulphur. Noble rot some yrs.

Macari Long I, North F ★★★ Clifftop estate, elegant, bio-minded. Katherine's Field SAUV BL, MERLOT-led Alexandra: NY-plush with racy acidity. CAB FR loves NY: Macari's is light-filled, savoury red candy; also in gleeful Horses pét-nat.

McCall Long I, North F ★★★ Known for top PINOT N, incl single-vyd, Res, rosé; gd CAB FR, SAUV BL; red B'x blends. Also home to French-origin Charolais cattle.

Millbrook Hudson V ★★★ Was 1st to grow vinifera in Hudson V; estate RIES, CHARD, PINOT N. Single-vyd Tocai (FRIULANO), CAB FR. Acidity lets reds age a few yrs.

Paumanok Long I ★★★ Racy, ageable CHENIN BL (still/sp), new PINOT N. Excellent B'x blends, CAB SAUV, RIES. Fine CHARD, single-vyd MERLOT, CAB FR; occasional Grand Vintage. Spontaneous ferments, low sulphur. Heads Palmer Vyds: saline PINOT BL, elegant Merlot, excited about ALBARIÑO.

Ravines Finger L ★★★ 17 18 19 RIES single-vyd (incl renowned steep limestone Argetsinger) and blended; concentrated GEWURZ, among top in e. PINOT N (focus), lovely exuberant CAB FR 19, drink in 5 yrs+.

Red Newt Finger L ★★★ Terroir- and RIES-focused, incl Seneca Lake crus, top US quality. Elegant GEWURZ, Pinot Gr; gd CAB FR, MERLOT, PINOT N. Tierce Ries is collaboration with Anthony Road, ARROWHEAD SPRING. Bistro, local produce.

Delaware makes gd pét-nat: the grape, not the state. Channing D, Chëpika.

Red Tail Ridge Finger L ★★★ Seneca. Super CHARD, RIES (incl one-block, wild-ferment), BLAUFRÄNKISCH, PINOT N. Lean, fruity TEROLDEGO, sells out. Sparkling incl Blanc de Noirs, pét-nats, Sekt.

Shaw Vineyard Finger L ★★★ On Seneca Lake, quieter w side. Res focus: full-bodied reds like CAB SAUV, MERLOT, PINOT N; focused whites like GEWURZ, RIES on fine lees; barrel-aged Ries too. Orange, blends, incl Gewurz, SAUV BL, PINOT GR.

> **Climate cool: NY's new fine wines**
> With naturally moderate alc levels, NY is bastion of both classics and
> new fine wines: new in style or raw material, experimental and an entire
> vocabulary of tart, bright, fresh. BEDELL AUXERROIS 21 "first for the East
> Coast". CHANNING DAUGHTERS' Tocai friulano (Long I), LAGREIN. DR. K
> FRANK's Amber RKATSETELI, 1983 vines, clay-raised, Lena (red blend).
> FLORAL TERRANES' MERLOT, Long I grape naturally fermented in suburbs
> (surprise edgy tannins). BLOOMER CREEK's elegant White Horse, glacial-
> lake play on Right Bank B'x. Yes, that one.

Sheldrake Point Finger L ★★ Cayuga Lake. Exuberant cool-climate GAMAY,
fresh earthy B'x blends, multiple RIES, single-plot PINOT GR, MUSCAT Ottonel.
Unwooded whites. Experiments incl vyd pine staves for fuller-bodied CHARD.

Silver Thread Finger L ★★★ CHARD, RIES; CAB FR, PINOT N all v.gd, bio. Terroir-
convinced: regenerating vyd soil to match surrounding forest's; solar-powered.

Sparkling Pointe Long I ★★★ Convincing *fizz*; French winemaker, Champagne
grapes, loam soil. Cuvée Carnaval range (r/w/rosé) lets MERLOT into mix.

Suhru Long I ★★★ Founded 2008 by Australian. Maritime, glacial-soil SHIRAZ; v.gd
age-worthy B'x blend. SAUV BL, TEROLDEGO. Provence-style rosé (MERLOT/CAB FR).

Weis Vineyards Finger L ★★ Keuka Lake. Winemaker Weis born, trained in Mosel.
Ries from dry to botrytized. Red mainstays PINOT N to SAPERAVI. Estate vines 1st
planted 2020: CHARD, Ries; CAB FR, SAPERAVI.

VineBalance sustainability standards now incl NY: look out for VB logo.

Whitecliff Hudson V ★★ Site-, soils-driven, incl ex-cherry orchard, quartz-rich
historic Olana slope for v.gd barrel-aged CAB FR, GAMAY. PINOT N on limestone
ridge. Peachy, stony CHARD; robust Res RIES.

Wölffer Estate Long I ★★★ Premier S Fork estate and destination; classical
approach. Quality CAB SAUV, MERLOT, PINOT N. CHARD made burgundy-style; new
maritime-minded SAUV BL; gd rosé set off Hamptons craze for the stuff.

North Carolina (NC)
From Blue Ridge Mtns, Piedmont hills in w (**Raffaldini**'s deft Italian-style
appassimento MONTEPULCIANO; umami SAGRANTINO; SANGIOVESE; Tuscan-inspired
VERMENTINO) to coastal e (Outer Banks **Sanctuary Vyds** complex yellow-fruit
Pearl ALBARIÑO; serious CAB SAUV; ageable black-olive Petit Verdot/SYRAH/
TEMPRANILLO Shipwreck; wind-, sand-subdued TANNAT/PETIT VERDOT Double
Barrel). More terroir, incl Yadkin V: **Jones Von Drehle** Tempranillo; MALBEC
Res; **Junius Lindsay** Syrah; **McRitchie** dry MUSCAT; **RayLen**; **Shelton**.

Ohio (OH)
Lake Erie moderates winters. Five AVAs; CHARD, MÜLLER-T, PINOT GR, RIES; B'x
varieties, DOLCETTO, PINOT N. **Debonné** since 70s; family-run **Ferrante** GEWURZ,
GRÜNER V; **Firelands**; **Harpersfield** KERNER/RIES/MUSCAT Ottonel; **Laurentia**
concrete-tank whites; **Markko** Lake Erie CAB SAUV, Chard, Pinot N; learned from
NY's Dr. Frank to plant Ohioan vinifera in 1968; **M Cellars** RKATSITEI;
St Joseph Vyd experiments with CORVINA, SANGIOVESE.

Oklahoma (OK)
Two AVAs: Ozark Mtn and Texoma, c.40 wineries. Mostly reds, esp CAB SAUV.
Clauren Ridge gd Meritage. **Sparks Vyd & Winery** sweet wine specialist.
Stableridge v.gd Bedlam CHARD. **The Range Winery** wide range of blends and
fruit wines; Montage red blend current favourite.

Oregon (OR)

Many small, quality-oriented producers, but outside investments have sizeably increased the production of some. Quality continues to be high, and experimentation abounds in lesser-known regions such as S OR and Col G. Dry-farming remains common; a push towards organic/bio evident, even among larger producers. Best bet for high-quality wines is Will V, generally from PINOT N and CHARD; OR wines remain an outstanding value. Most recent vintages are gd to v.gd. Beware smoke taint in 20, and 18 (S OR), 17 (Col G).

Principal viticultural areas

Columbia Gorge (Col G) is split between WA and OR. Experimentation, variety, sustainable viticulture.

Rocks District of Milton-Freewater (Walla Walla V [Walla]) entirely in OR, producing dense, age-worthy Syrahs.

Southern Oregon (S OR) warmest growing region, encompassing much of W OR, s of Will V. The s sub-AVA Rogue V (Rog V) incl Applegate (App V). The n sub-AVA Umpqua V (Um V) incl Elkton OR and Red Hill Douglas County. Rhône and Spanish varieties best. Quality is spottier than Will V but future looks promising.

Willamette Valley (Will V) sub-AVAs Chehalem Mts (Ch Mts), Dundee Hills (Dun H), Eola-Amity Hills (E-A Hills), McMinnville (McM), Ribbon Ridge (Rib R), Van Duzer Corridor (Van DC), Yamhill-Carlton (Y-Car), Laurelwood District (LD) and Tualatin Hills (Tual H). Coming soon: Lower Long Tom, Mt Pisgah, Polk County. Pinots Bl/Gr, Ries, and Gamay excel; beautiful Chard; Pinot N remains star.

oo (Double Zero) Will V ★★★ CHARD, PINOT N specialist sourcing fruit from prestigious vyds, multiple AVAs. Seek out Hermann cuvée Pinot N and VGR.

Abacela Um V ★★→★★★ The 1ST TEMPRANILLO in US. Barrel Select v.gd. Fiesta for value. Beautiful, ripe ALBARIÑO with classic stonefruit character.

Abbott Claim Y Car ★★★→★★★★ Historic property now owned by Antony Beck (see Graham Beck, S Africa). Organic, dry-farmed. Due North, Orientate PINOT N.

Adelsheim Chehalem Mts ★★→★★★ PINOT N and CHARD cuvées reliably gd. Staking Claim Chard, Breaking Ground Pinot N v.gd value

Alloro Chehalem Mts ★★★ Beautiful site with elegant CHARD, RIES; PINOT N. Estate Riservata, Justina age v. well.

Analemma Col G ★★★ Shifted to estate model 2018. Bio farming; esp Spanish varieties but known for vibrant, fresh sparkling. Seek out floral GODELLO and red-fruited, lavender-scented GRENACHE.

Most structured Pinots come from volcanic Jory soils, lightest from loess, generally.

Antica Terra Will V ★★★ Unique expressions of classic varieties. Spicy, microbial PINOT N. Opulent, golden, tannic CHARD. Tasting experiences here lauded.

Archery Summit Dun H ★★★ Summit, Looney, Arcus vyds excel; also Whole Cluster Cuvée and exotic Ab Ovo PINOT GR fermented in concrete egg.

Argyle Will V ★★ Decent Vintage bubbly, esp Brut Rosé.

Arlyn Chehalem Mts ★★★ Dry-farmed, organic grower. Wines made by LINGUA FRANCA team. All beautifully balanced.

A to Z Wineworks S OR ★ Value-priced, soundly made.

Audeant Will V ★★★ Promising new project, fruit from multiple AVAs. Single-vyd PINOT NS Nysa and Luminous Hills best.

Ayoub Dun H ★★★★ Excellent vyd sources, balanced handling of oak. Outstanding estate PINOT N, both whole-cluster and regular (try comparative tasting).

Ayres Rib R ★★★ Planted 2001; 38-acre, family-owned parcel makes dark-fruited spicy PINOT N; look for Pioneer and One.

Beaux Frères Rib R ★★★ Majority-owned by Champagne Henriot (*see* France). New vyd-designate series highlighting other AVAs; stars remain bio-farmed estate bottlings, esp elegant, earthy Belles Soeurs PINOT N.

Bergström Will V ★★★★ 100% estate fruit since 2020. Family-owned. Organic farming and elegant, powerful PINOT N and CHARD made to go the distance. Sigrid Chard among best in US.

Bethel Heights E-A Hills ★★★ Family-owned. Organic. Long-lived PINOT N. West Block (planted 1977) Pinot N incredible value. CHARD can be skipped but Estate PINOT GR (last vintage 19) outstanding.

Will V now recognized by EU as geographical indication. Only other one is Napa.

Big Table Farm Will V ★★★ Quirky, complex, esp Elusive Queen CHARD, Laughing Pig Rosé and all single-vyd PINOT N.

Brick House Rib R ★★★ All bio farming; all native ferments. Hands-on, family winemaking team. Excellent Cascadia CHARD, Les Dijonnais PINOT N, and older vintages of Cuvée du Tonnelier (replanted 2018).

Brooks E-A Hills ★★★ Family-owned/operated, bio. Exceptional RIES (up to 20 cuvées; try Ara, Bois Joli, Estate). Also v.gd Rastaban PINOT N.

Christopher, J Will V ★★→★★★ Collaboration between Ernst Loosen (*see* Germany) and Jay Somers. Bright, red-fruited, floral, easy-drinking PINOT N, consistently pleasing. Fresh, intense Über (SAUV BL).

Cristom E-A Hills ★★★ 90 acres of estate vyds going bio. Long-lived, spicy, red-fruited PINOT N, esp gd Jessie and Eileen.

David Hill Vineyards & Winery Tua ★★★ Originally planted 1966 by Charles Coury. Alsace varieties; v.gd Blackjack bottling from original-vine PINOT N. Discovery Series experimental, natural.

Divio, Dom Rib R ★★★★ Bottlings of most WILL V AVAs showcase different terroirs. Clos Gallia parcel bio-farmed, best PINOT N. Excellent CHARD. ALIGOTE from 2021.

Drouhin Oregon, Dom Dun H ★★★ Burgundy-inspired; CHARD, Édition Limitée, Louise PINOT N best. Sister label Drouhin Oregon Roserock in E-A HILLS equally fine, esp Zéphirine Pinot N.

EIEIO Y-Car ★★→★★★ Jay McDonald is a one-man operation, leasing vyds (mostly Y-CAR), minimal intervention. Yates Conwill CHARD and PINOT N shine.

Elk Cove Will V ★★★ Family-owned, 400 acres, dry-farmed, some own-rooted, 1st vintage 1977. Single-vyd PINOT N v.gd, Clay Court, outstanding Five Mtn and Mt Richmond; v.gd PINOTS BL/GR. Sister label Pike Road from purchased grapes.

Et Fille Will V ★★→★★★★ Family-owned/operated; small estate plot, six sustainably farmed vyds in WILL V. Outstanding Gabriella PINOT N, Père Honneur (sp).

Eyrie Vineyards, The Dun H ★★★★ Dry-farmed, organic, no-till, min-intervention winemaking. Oldest producer in WILL V, original vines planted 1965. Elegant, age-worthy, low-alc. Incredible library releases. CHARD, Daphne, PINOT GR, Sisters, original South Block PINOT N bottlings textural wonders.

Haden Fig Will V ★★★ Sister brand to Evesham Wood. Single-vyd PINOT N and Juliette CHARD: aromatic, intriguing, gd value.

Idiot's Grace Col G ★★→★★★★ Family-owned, dry-farmed, organic estate. Outstanding CHENIN BL, fresh strawberry GRENACHE; gd GAMAY, SANGIOVESE.

Johan Van D ★★★ Dry-farmed, bio, no-till estate vyd. Bought by Mini Banks, also of Cowhorn, 2021. Unique expressions of CHARD, PINOTS GR/N; v.gd Nils Pinot N.

Ken Wright Cellars Will V ★★→★★★ Long history informs PINOT N vyd selections. Overall quality high. WILL V cuvée overdelivers. Wines better with age.

King Estate Will V ★★★ Estate now largest bio producer in US. PINOT GR (esp

Backbone, Domaine, Johnson School, Steiner) core of portfolio with a dozen PINOT N plus CHARD, GEWURZ, SAUV BL, new sparkling.

Lange Estate Dun H, Will V ★★★ Fine PINOT GR and Assemblage, estate, Freedom Hill PINOT N; CHARD still tops here. Classiqué range v.gd value.

Lingua Franca E-A Hills ★★★ Larry Stone MS co-founded estate, Dominique Lafon (Burgundy, *see* France) consults. Stylish CHARD (Bunker Hill, Estate, Sisters) and balanced PINOT N (Mimi's Mind, The Plow).

Loop de Loop Col G ★★→★★★★ Organic, dry-farmed, no-till. PINOT N ethereal, red-fruited, spicy. Four Winds esp excellent. Second label Wallflower created during 2020 wildfires, now for experimental projects.

Morgen Long Will V ★★★★ Seth ML focuses on CHARD from vyds whose farming he values. Precise laser-focused acid backbone, mineral, intense citrus and well-integrated oak. Incredibly ageable. Seven Springs, X Omni standouts.

Ovum OR ★★★ Artisanal RIES, GEWURZ from both N/S OR. Big Salt blend value.

Patricia Green Cellars Will V ★★★ Excellent single-vyd PINOT N. Founder deceased; value and quality remain intact. Bonshaw Block, Etzel Block, Mysterious, Notorious superb. Rare OR SAUV BL.

Ponzi Lau ★★★ Bought by Bollinger 2021. Luisa P stays. Abetina, Aurora PINOT N v.gd; Aurora, Avellana CHARD also. Tavola, Classico Pinot N v.gd value.

Quady North App V, Rog V, S OR ★★★ Herb Q excels with Rhône varieties; v.gd fresh Pistoleta blend; vibrant Counoise rosé; Mae's SYRAH outstanding.

Résonance Will V ★★★ Jadot's (*see* France) OR project; age-worthy CHARD, PINOT N from winemaker Guillaume Large. Estate wines tops; WILL V cuvée best value.

Rex Hill Will V ★★ Jacob-Hart and fine barrel-fermented CHARD standouts.

Roco Will V ★★→★★★ Crafts superb vintage RMS Brut bubbly and layered, age-worthy CHARD, PINOT N. Private Stash is Res, Gravel Rd for value.

Roy & Fils, Dom Dun H ★★★ Organic estate vyd, 39 acres. Iron Filbert and Quartz Acorn PINOT N best. Excellent Iron Filbert CHARD.

Shea Wine Cellars Will V, Y-Car ★★★ Top-tier winemakers clamour for Shea fruit; in-house wines gd too. Block selections, Homer PINOT N tops; excellent CHARD.

Sokol Blosser Dun H, Will V ★★ Value Evolution series v.gd. PINOT N Peach Tree and Orchard tops; the rest ordinary.

Soter Will V ★★★→★★★★ Tony S, CA legend, shines with CHARD, PINOT N, bubbly. Two labels: estate grown Mineral Springs Ranch, Planet Oregon for value. Estate is bio (with integrated animals), shows best quality.

Stoller Family Estate Dun H ★★→★★★ Steadily improving CHARD (Elsie's Res), PINOT N. Group incl Canned Oregon, Chehalem, Chemistry and History brands.

Trisaetum Rib R, Will V ★★★★ Meticulous estate wines; all styles of RIES, v.gd CHARD, PINOT N, impressive sparkling under Pashey label.

Troon App V ★★★ The 2nd regenerative organic vyd in the US. Excellent Estate SYRAH (Siskiyou best), VERMENTINO. Rest of portfolio tends towards funky, fresh, light reds and orange wines (Glou Glou GRENACHE, Kubli Bench Amber).

Vincent Will V ★★→★★★ Works with dry-farming growers in E-A HILLS and RIB R. No new oak, to highlight fruit. Excels with CHARD, esp Tardive.

Walter Scott E-A Hills ★★★★ Family-owned, focused on vyd sources. Precise, linear, buzz-worthy CHARD. Earthy, balanced PINOT N with well-integrated oak. GAMAY also excellent. X-Novo, Freedom Hill, Sojourner vyds top list. Age-worthy.

Oregon's new fine wines

All notable for gd farming practices and pure wines, often from unusual varieties. **Golden Cluster** (WILL V) Müller?. **Idiot's Grace** (COL G) CHENIN BL. **Johan Drueskall** (VAN DC) PINOT GR. **Niew** (Will V) CHARD. **Teutonic Red Blend** (Will V) GEWURZ/PINOT N. **Troon** (APP V) VERMENTINO.

OREGON

Willamette Valley Vineyards Will V ★★ Many shareholder/owners; extensive vyds, mostly CHARD, PINOT N. Rocks District now home to Maison Bleue, Pambrun.

Winderlea Will V ★★★→★★★★ Bio producer, vibrant single-vyd PINOT N: Crawford Beck, Shea, Weber, Winderlea Legacy. Plus v.gd, age-worthy CHARD.

Pennsylvania (PA)

ALBARIÑO: esp **Galen Glen, Galer, Maple Springs**. GRÜNER V new stars; PINOT N, B'x varieties. Many Italians, BARBERA to FIANO. Continental climate. Lake Erie-softened nw. Gentler temperatures in se: **Va La** cult Avondale field blends incl CORVINA, nine NEBBIOLO clones; **Vox Vineti** Nebbiolo, barrel-fermented rosé, B'x blends, PETIT VERDOT. In central Leigh V: **Galen Glen**, windy, 305m (1000ft) up, Stone Cellar range from oldest vines: Grüner V planted 2003; **Stony Run** v.gd Brut 18, gd plush Albariño. Also **Allegro**, reliable since 70s; **Armstrong Valley** CAB FR, MERLOT; **Fero Vyds** bright LEMBERGER, celebrated SAPERAVI; **Karamoor**; **Mazza** Lake Erie TEROLDELGO; **Mural City Cellars** Philly urban winery, serious natural style; **Penns Woods** SAUV BL, CAB SAUV; **Presque Isle**; **Vynecrest** Lemberger blend; **Waltz** Cab Fr; **Wayvine** gd barrique-aged native Carmine.

Rhode Island (RI)

Smallest US state, just e of NY's North Fork across cold-tempering Sound; B'x reds, PINOT N, CHARD. **Diamond Hill** chemical-free farming, barrel-aged Pinot N; **Greenvale** min Chard, MALBEC, CAB FR; **Mulberry Vyds** PINOT GR, SYRAH; **Newport** GEWURZ; **Sakonnet** Gewurz, red blends; **Verde** biology prof. turned small farmer; BLAUFRÄNKISCH; St Croix, other hybrids.

Texas (TX)

Every yr more producers hit quality markers, more infusion of winemakers from est regions and now, thanks to recent legislation, more (ie. all) TX grapes in wines labelled with a stated county, AVA or vyd. About 400 wineries mostly in two enormous AVAs – Texas Hill Country and Texas High Plains – with producers having more reason to leverage regionality and hyper-local focus. The state is truly a laboratory of successful experiments, with Portuguese and Med varieties as the stars; pét-nat shines too. Vintage 17 one of recent best.

Ab Astris Recent newcomer, family-owned boutique winery with attention to local terroir, varieties that best thrive there. PICPOUL and TANNAT of particular note.

Becker Vineyards ★★★ Big, ripe B'x, Rhône styles: Prairie Rotie signature. Wilmeth Vyd CAB SAUV recent winner. ROUSSANNE Res from Farmhouse Vyd too.

Bending Branch Winery ★★★ Premier TANNAT house; signature bold reds incl PETITE SIRAH, PETIT VERDOT, Sagratino; v.gd PICPOUL, ROUSSANNE, SOUZÃO, Newsom Vyds CAB SAUV. Also 16 acres of experimental varieties.

Brennan Vineyards ★★★ Set in historic homestead featuring dry VIOGNIER, white Rhône blend Lily; v.gd NERO D'AVOLA Super Nero. Deeper MOURVÈDRE and MALBEC dry rosé. Winemaker's Choice changes yearly.

Calais Winery ★★ Hip winery in quality pocket of Hill Country with Frenchman Ben Calais making B'x styles here and for French Connection Wines, showcase for Rhône varieties. High-elevation vyds give complexity, concentration.

Crowson Wines Single-variety, low-intervention natural wines from recent entrant. Known for dry MALVASIA Bianca; rich, intense rosé.

Duchman Family Winery ★★★★ The 1st into wine-on-tap trend with Italian varieties. Known for AGLIANICO and dry VERMENTINO, but also fresh Trebbiano. Grapes from heritage cool-climate High Plains vyds.

English Newsome Cellars Rejuvenated. Rhône styles: Res ROUSSANNE, VIOGNIER.

Fall Creek Vineyards ★★★ Look for GSM Terroir Reflection and super-premium ExTerra TEMPRANILLO from Salt Lick vyds. Early pioneer of CHENIN BL. B'x blend Meritus is delicious homage to André Tchelistcheff.

Haak Winery ★ Madeira-style Blanc du Bois, making inroads with MALBEC. Recent change in ownership.

Kuhlman Cellars Young winery, Burgundy-trained winemaker, Rhône and B'x red blends. TEMPRANILLO/CAB SAUV Ignis. Estate MARSANNE/ROUSSANNE blends.

Lewis Wines ★★★ Focus on single-vyd Portuguese varieties, esp ALICANTE BOUSCHET, Tinta Cão, TOURIGA N. Impressive estate rosé and CHENIN BL.

Llano Estacado ★★★ Excellent MALBEC, 1836 (r/w); v.gd Viviana (w), Viviano (r) mimics Super Tuscan. Outstanding THP TEMPRANILLO from all-TX fruit.

McPherson Cellars **★★★** One of TX's fathers of wine; now hanging his hat on warmer varieties, esp SANGIOVESE and Rhône; v.gd CHENIN BL.

Messina Hof Winery ★★★ Bonarrigo (dry w blend) taps family's Sicilian roots. SAGRANTINO pioneer. Range of desserts.

Pedernales Cellars ★★★ Sourced from Kuhlken family vyds; Spanish, Rhône grapes with benchmark TEMPRANILLO, VIOGNIER and recent single-vyd Signature Series.

Perissos Vineyard and Winery ★★ Family-run winery; deep, exotic reds, esp five-grape Racker's Blend incl AGLIANCIO/TEMPRANILLO. Strong in Italian blends.

Southold Farm and Cellar ★★★ Run by e coasters, tapping trend for freshness, natural wines, neutral ageing vessels. Apera, fortified skin-fermented PICPOUL Bl, is inventive. Names, labels change yearly; try Robert Clay vyd, field blends.

Hill Country, High Plains are 531 km (330 miles) apart: more than Bx, Champagne.

Spicewood Vineyards **★★★** Estate-grown, TX ranch family, sister winery to RON YATES. Outstanding tropical-infused SAUV BL; sparkling SEM, GRENACHE pét-nat. Named after grandfather, TEMPRANILLO-driven Good Guy field blend is signature.

Wedding Oak Balanced, ripe ROUSSANNE, best-in-class AGLIANICO, gd Texedo Red DOLCETTO/TOURIGA N; gd SANGIOVESE; experimenting with hop-infused whites.

William Chris Vineyards **★★★** Low-intervention, vyd-focused; look for MOURVÈDRE, incl crisp rosé and pét-nat blend. Flagship Enchante (r). Small-batch winery Lost Draw Cellars with impressive SANGIOVESE.

Yates, Ron ★★★ Young law student. TEMPRANILLO and focus on Rhône, Spanish, Italian. SANGIOVESE from Newsom Vyd. Pét-nat from GRENACHE and Mencía.

Vermont (VT)

With mtns, harsh winters, brief sunny summers, frost, hail, humidity, a few hardy souls plant vyds anyway, rely on hybrids like Frontenac Noir, La Crescent plus BLAUFRÄNKISCH, RIES in extreme n terroirs. **Lincoln Peak** incl nouveau Marquette. Bio-farmed, natural-thinking **La Garagista** (new Brianna plantings); fizz-focused **Shelburne Vyds** and **Zafa Wines** pioneers to look for. **Iapetus** is experimental bio side of Shelburne winemaker; try the Marquette.

Virginia (VA)

Continental climate; challenge is to beat humidity, winter freeze, harvest-time hurricanes. More vyds being planted on heights, esp in Shenandoah V for some protection plus mtn quality. Elegant outcomes state-wide in classical (CAB FR, MERLOT, PETIT VERDOT) and experimental (hardy, high-acid PETIT MANSENG sings; NEBBIOLO, TANNAT ever more present). CAB SAUV; VIOGNIER favoured too.

Ankida Ridge ★★★ Top, low-alc, ageable PINOT N poss best in VA, for a lucky few: <1000 cases. Steep granite slopes, 518m (1700ft) up in Blue Ridge Mtns. CHARD, gd Blanc de Blancs. GAMAY promising.

OREGON–VA

Barboursville ★★★ 14 17 18 19 In Monticello; B'x mainstays incl PETIT VERDOT (Octagon). New Nascent, refined blend: VERMENTINO (hints of estate's masterful Vermentino Riserva), floral VIOGNER, FALANGHINA's oilness. NEBBIOLO, ageable 10 yrs+. CAB FR Res 05, Monticello ideal. Paxxito, luscious VIDAL/MUSCAT Ottonel.

Boxwood ★★ Founder of Middleburg AVA. Blends based on CAB FR, MERLOT, plus CAB SAUV, PETIT VERDOT in both classic B'x style and drink now; rosé of same grapes; SAUV BL. New: Sauv Gr. Short drive from Washington DC.

Early Mountain **★★★** 15 17 19 20 Rich, earthy Eluvium; Rise in best yrs; 2nd-bottling 17 (12% TANNAT), complex. Five single-site CAB FR, two regional, three site-specific for VA-nuanced quality: Quaker Run is lush. Elegant PETIT MANSENG 19 is state's top, age-worthy.

New VA Peninsula AVA: 50 x 5–15 miles, varies from sub-tropical to continental.

Glen Manor Vineyards ★★ Historic farm, 5th-generation. Vines on steep rocky slopes in Blue Ridge Mtns 305m (1000ft)+ up. Began with SAUV BL, now joined by rich CAB FR from 20–30-yr-old vines, off-dry PETIT MANSENG, PETIT VERDOT.

King Family Vineyards ★★★ French winemaker, age-worthy Meritage; top tiny-production *vin de paille*-style PETIT MANSENG; experimental Small Batch Series, in recent yrs lively no-sulphur CHARD, whole-cluster CAB FR, skin-contact VIOGNER. SAVAGNIN planted 2020.

Lightwell Survey **★★★** 19 Bold blends from single sites. The Weird Ones Are Wolves is CAB FR co-fermented with 6% PETIT MANSENG: earthy, highly floral. Same-method Los Idiots and Dos Idiots (plus PINOT GR): sweet apple-blossom, jasmine, acidity, powerful core. Hintermen is RIES-assisted Petit Manseng, astounding, co-fermented for 1 yr.

Linden ★★★ Estate founded in 80s by early believer in site over fruit; VA wine mentor ever since. Notable high-altitude wines from three sites: rich, mineral CHARD, vivacious SAUV BL, savoury PETIT VERDOT, elegant, complex B'x-style reds often require ageing. Some library wines; vertical pours in tasting room.

Michael Shaps Wineworks ★★ Long-time VA producer with solid CHARD, VIOGNER, luscious PETIT MANSENG; tasty TANNAT, PETIT VERDOT, Raisin d'Être (w Petit Manseng and r blend) from grapes dried in old tobacco barns.

Midland Construction ★★★ Old family farm, limestone soils 400m (1312ft) up. BLAUFRÄNKISCH, CAB FR, CHARD. Also PETIT MANSENG, RIES.

Pollak ★★ Estate since 2003, wines on international side: heftier CABS FR/SAUV, MERLOT, Meritage; creamy PINOT GR; lush, spicy VIOGNIER.

Ramiiisol **★★★** New, no-expense-spared CAB FR from Blue Ridge Mtns, iron-rich granite gneiss; v. elegant, concentrated; bio. NEBBIOLO planted too.

Rausse, Gabriele ★★★ Small, quality estate nr Monticello, Italian viticulturist who planted BARBOURSVILLLE with Gianni Zonin (*see* Italy). Varietal CHARD, PINOT GR; MALBEC, NEBBIOLO; PINOT N as *vin gris*.

RdV Vineyards **★★★★** Red blends (B'x-inspired), from granite soil hillside. Elegance, complexity, power: MERLOT-led Rendevous; CAB SAUV-driven Lost Mountain (released 4 yrs after vintage; 1st, 2010) was VA's 1st $100 wine.

Veritas ★★★ Solid estate, founded 1995, incl steep 20-yr-old forest vyds. Concentrated, floral CAB FR can age 10 yrs+. Paul Schaffer PETIT VERDOT is flagship, for tannin lovers. Restrained SAUV BL, richer VIOGNIER; gd CHARD, MERLOT. Traditional-method sparkling to come.

Washington (WA)

Washington is the 2nd-largest wine-producing state in the US, but most of the 1000+ wineries have limited production, meaning many fly under the radar. With land relatively inexpensive, grapes are too. Meanwhile, summers are always warm and irrigation is a requirement to grow grapes. The end result? Consistently high-quality wines and terrific value – if you can find the wines.

Principal viticultural areas

Columbia Valley (Col V) Huge AVA in central and e WA with a touch in OR. High-quality Cab Sauv, Merlot, Ries, Chard, Syrah. Key sub-divisions incl Yakima Valley (Yak V), Red Mtn, Walla AVAs.

Red Mountain (Red Mtn) Sub-AVA of Col V and Yak V. Hot region known for Cabs and B'x blends.

Walla Walla Valley (Walla) Sub-AVA of Col V with own identity and vines in WA and OR. Home of important boutique brands and prestige labels. Syrah, Cab Sauv and Merlot.

Yakima Valley (Yak V) Sub-AVA of Col V. Focus on Merlot, Syrah, Ries.

Abeja Col V, Walla ★★★ Production now in hands of Dan Wampfler (ex-Dunham Cellars) and wife Amy Alvarez-Wampfler. High-quality COL V CAB SAUV, CHARD.

Andrew Will Col V, Red Mtn ★★★★ 10' 12' 14' 16 One of oldest, most heralded wineries crafting age-worthy B'x blends; reserved style. Will Carmada, 2nd-generation winemaker, now in place. Involuntary Commitment, v.gd value.

Avennia Col V, Yak V ★★★ 10 12' 14' 16' 18' 10-yr-old winery has placed itself among state's best. Distinctive, classy; old vines, top vyds. Sestina B'x blend and Arnaut SYRAH tops. SAUV BL v.gd. Lydian value label.

Betz Family Winery Col V ★★★→★★★★ 10 12' 14' 16' 18' Woodinville stalwart has been making high-quality B'x, Rhône styles for 20 yrs+. Père de Famille CAB SAUV flagship. Untold Story gd value. All noteworthy.

Cadence Red Mtn ★★★ 10' 12' 14 16' 17' Seattle winery (20 yo+) making structured, single-vyd B'x blends from RED MTN in reserved style. Patience required. Bel Canto and Cara Mia tops. Coda from declassified barrels exceptional value.

Cayuse Walla ★★★★ 10 11 12' 14 16' Not only some of best SYRAH in US; some of best on planet. All estate vyd. Stratospheric scores, but mailing list only, with yrs-long wait. Steep prices on secondary market but worth it. Sister wineries Hors Categorie, Horsepower, No Girls also top quality.

Charles Smith Wines Col V ★★ Eponymous winemaker spun off this brand to wine giant Constellation. Focus remains on value CAB SAUV, MERLOT, RIES.

Col Solare Red Mtn ★★★→★★★★ 10 12' 14 18 CH STE MICHELLE and Tuscany's Antinori partner to create estate CAB SAUV with complexity, longevity.

Columbia Crest Col V ★★→★★★ WA's largest producer is all about quality, value; v.gd, well-priced Grand Estates label; Res wines cut above, esp CAB SAUV, Walter Clore.

Columbia Winery Col V ★★ One of WA's founding wineries; value; higher end.

No soggy Seattle here; almost all grapes are grown in the desert to the east.

Corliss Estates Col V ★★★ 08' 10 12' 14 Cult WALLA producer, extended time in barrel/bottle. Sister winery Tranche focuses on Blue Mtn fruit, v.gd value.

Côte Bonneville Yak V ★★★ All estate wines from DuBrul, highly regarded vyd. Extended bottle-age before release. *Carriage House* v.gd value.

DeLille Cellars Col V, Red Mtn ★★★ 10' 12' 14' 16 18 One of Woodinville's founders, known for high-end B'x, Rhône styles. D2 & Four Flags v.gd value. Chaleur Bl one of state's best whites. No misses in lineup.

Devison Walla ★★★ Peter D spent decades making wine for others, now himself. State's best rosé, SAUV BL. Attention-getting reds with voice.

Doubleback Walla ★★★ 10' 12 16 18 Ex-footballer Drew Bledsoe's winery isn't a vanity project; classy, elegant CAB SAUV. Bledsoe Family sister winery. Bledsoe-McDaniels new OR PINOT and WA SYRAH project.

Dunham Cellars Walla ★★ Long-time producer of v.gd CAB SAUV, SYRAH. Trutina & Three-Legged Red gd value.

Dusted Valley Vintners Walla ★★ Focus on COL V, WALLA, from value Boomtown to high-end single-vyd offerings. Stained Tooth SYRAH consistent standout.

Fielding Hills Col V ★★ Producer of estate-vyd wines from Wahluke Slope made in rich, ripe style; v.gd value. Concentric Wine Project winery's playground.

Figgins Walla ★★★ 10 12 14 16 18 Founded by 2nd-generation winemaker Chris F (LEONETTI); focus on single vyd in Upper Mill Creek. Structured B'x blends. Patience/decanting reward.

Force Majeure Red Mtn ★★★ All estate wines: big, bold style, age-worthy.

Gorman Red Mtn ★★★ Hedonism the driver: rich, ripe wines. Evil Twin CAB SAUV/SYRAH blend calling card. Ashan CHARD project.

Gramercy Cellars Walla ★★★ 10 12' 13 16 18 Master Sommelier Greg Harrington produces lower-alc/oak, higher-acid, food-focused wines. Speciality earthy SYRAH, herby CAB SAUV. Lower East value label.

H3 Col V ★ Recent spin-off from COLUMBIA CREST. Value CAB, MERLOT, Red Blend.

Hedges Family Estate Red Mtn ★★ In appellation known for ripe reds; long-time bio producer focuses on more reserved, savoury styles. CMS blend gd value.

Januik Col V ★★★ 10 12' 18 Mike J cut teeth at CH STE MICHELLE before launching this Woodinville luminary 20 yrs+ ago. Consistent-quality, value B'x styles, some of state's best CHARD. Novelty Hill sister winery. Son Andrew has eponymous label.

Kevin White Winery Yak V ★★★ Micro-producer of high quality, outrageous value. The trick? Getting them before they're gone.

Kiona Red Mtn ★★ Founding RED MTN winery; v.gd estate wines, old-vine LEMBERGER.

K Vintners Col V, Walla ★★★ Outsized personality, former rock-band manager CHARLES SMITH makes single-vyd SYRAH, Syrah/CAB SAUV blends with iconic label. Sixto CHARD-focused sister winery. Also CasaSmith, Substance, ViNo.

Latta Wines Col V ★★★ Ex-K VINTNERS winemaker Andrew L; stunning single-vyd GRENACHE, MALBEC, MOURVÈDRE, SYRAH. Latta Latta v.gd value. Disruption, Kind Stranger side projects, value.

L'Ecole No 41 Walla ★★★ 10 12' 14 16 Founding WALLA winery; superb-value Walla, COL V ranges. Ferguson flagship B'x blend. CHENIN BL, SEM v.gd value.

Leonetti Cellar Walla ★★★★ 08 10' 12' 14 18 WALLA's founding winery with well-deserved cult status, steep prices for cellar-worthy CAB SAUV, MERLOT, SANGIOVESE. Res B'x-blend flagship. Single-vyd wines v. limited but knee-buckling.

Liminal Red Mtn ★★★→★★★★ New cult producer of high-elevation reds and whites that focus on purity, demand attention.

Long Shadows Walla ★★★→★★★★ Brings globally famous winemakers to WA to make one wine each. Pedestal Michel Rolland MERLOT. Poet's Leap RIES one of best in state. All high quality, worth seeking.

Luke Col V ★★ Producer of well-priced Wahluke reds that way overdeliver.

Mark Ryan Winery Red Mtn, Yak V ★★★ Original member of Woodinville's "grape killers" known for big, bold style. But there's refinement too. Dissident v.gd value. MERLOT-based Long Haul and Dead Horse CAB SAUV stand out. Board Track Racer second label gd value.

Maryhill Col V ★ Variety prices/styles: quaffable Winemaker's Select to single vyd.

Milbrandt Vineyards Col V ★★ Wahluke Slope winery focusing on value plus smaller-production single-vyd offerings. Look for PINOT GR, RIES.

Northstar Walla ★★★ When MERLOT was WA's guiding star, this producer helped lead way. Decades later, it still does.

Owen Roe Yak V ★★★ Long-time producer of YAK V SYRAH, CAB SAUV and B'x blends emphasizing restraint. Recently purchased by CA giant Vintage Wine Estates.

Pacific Rim Col V ★★ Founded by Randall Grahm, RIES specialist now owned by

Banfi. Oceans of tasty, inexpensive, eloquent Dry to Sweet and Organic. For more depth, single-vyd releases.

Passing Time Col V ★★★→★★★★ Former pro quarterbacks Dan Marino, Damon Huard focus on appellation-specific CAB SAUV. Winemaker Chris Peterson (AVENNIA). Horse Heaven Hills tops. Quickly earning cult status.

Pepper Bridge Walla ★★★ Estate wines, B'x style, from top sites Pepper Bridge, Seven Hills. Structured, classy. Time in cellar required.

Quilceda Creek Col V ★★★★ 04' 07 10 12' 14' 16' 18' Flagship producer of WA, cult CAB SAUV known for richness, layering, ageing potential. One of most lauded in world. Sold by allocation. Buy if it you can find it – and if you can afford to.

Reynvaan Family Vineyards Walla ★★★ 10' 11 12' 14 16 18 Wait-list winery focusing on estate vyds in Rocks District, Blue Mtn foothills. Reds get raves – deservedly so – but don't miss whites.

Rôtie Cellars Walla ★★★ Rhône-style specialist once best known for reds but now increasingly for whites too. Northern Blend consistent standout. House of Bones CHARD project.

Ste Michelle, Ch Col V ★★→★★★ State's founding winery offers gd-value (r/w), plus estate offerings and higher-end Res. World's largest RIES producer, dry and off-dry COL V exceptional value.

Savage Grace Yak V ★★ Producer of low-oak, low-alc, low-intervention, single-vyd wines with something to say.

Saviah Cellars Walla ★★★ Long-time producer of high quality/value-ratio estate WALLA wines. The Jack label gd value.

Unlike OR, don't come for the Pinot (less than 1% production); Cab king in WA.

Seven Hills Winery Walla ★★★ 10 12' 14 16 18 One of WALLA's founding wineries. Age-worthy B'x reds made in restrained, sophisticated style. MERLOT v.gd value.

Sleight of Hand Walla ★★★ Audiophile Trey Busch makes dazzling B'x blends and Rhône styles. Seek Funkadelic SYRAH from Rocks District. Renegade value label.

Sparkman Cellars Red Mtn, Yak V ★★★ Woodinville "grape killer" producer focuses on power, diversity, making two-dozen-plus wines. Ruby Leigh, Stella Mae B'x blends consistent standouts. Kingpin top CAB SAUV.

Spring Valley Vineyard Walla ★★★ CH STE MICHELLE property focusing on estate reds. Uriah MERLOT B'x blend the headliner.

Syncline Cellars Col V ★★★ Rhône-dedicated Columbia Gorge producer with distinct, fresh style. Subduction Red v.gd value. Sparkling GRÜNER V insider wine. PICPOUL consistent standout.

Tamarack Cellars Col V ★★ Long-time WALLA producer of CAB SAUV, MERLOT and CAB FR. Firehouse Red gd value.

Walla Walla Vintners Walla ★★ Long-time valley producer of luscious reds. Estate wines tops.

Waterbrook Walla ★→★★ One of state's oldest wineries, now owned by wine giant Precept, focuses on value. Browne Family sister winery at higher end.

Woodward Canyon Walla ★★★★ 07 10 12' 14 16 18 Founding WALLA producer now run by 2nd generation. Focus on B'x styles: Old Vines CAB SAUV. CHARD always best in state. Nelms Road value label.

WT Vintners Yak V ★★★ Sommelier-winemaker Jeff Lindsay-Thorsen picks earlier, pulls back oak on single-vyd GRENACHE, SYRAH, GRÜNER V.

Wisconsin (WI)

Wollersheim Winery (est 1840s) is one of best estates in midwest, with hybrid and Wisconsin-native American hybrid grapes. Look for Prairie Fumé (SEYVAL BL), Prairie Blush (Marechal Foch).

Mexico

Five valleys comprise the Baja wine country, and Valle de Guadalupe produces 75% of the total. Once the realm of a few pioneers, it now attracts organic and bio winemakers; grapes are mostly Mediterranean, styles full-bodied and fruit-forward. Check out the wine resorts and innovative dining too: the Ruta del Vino rivals Napa Valley for liveliness.

Adobe Guadalupe ★★★★ Hugo d'Acosta helped est destination winery and inn. Wines named after archangels or gardens. Serafiel blend (CAB SAUV/SYRAH) top. Jardín Mágico SAUV BL fresh, tropical; Uriel rosé blends TEMPRANILLO/Syrah.

Bichi ★★ Champion of natural wine (*bichi* means "naked" in Sonoran Yaqui dialect), ex-lawyer Noel Téllez runs Tecate winery with Beaujolais-trained winemaker. Focus on old vines; try pét-nat Pét-Mex, and Listán from MISSION grapes (r).

Bruma Valle de Guadalupe Chic winery attached to eco-luxury hotel; young B'x-trained, rising-star winemaker Lulú Martinez Ojeda. Try Ocho Blanc de Noirs (w) from CARIGNAN Noir.

Camou, Ch ★★★ Pioneering, French-inspired, serious B'x credentials. Award-winning Gran Vinos, complex, elegant CAB SAUV-driven, worthy of ageing.

Carrodilla, Finca La Organic, Mondavi-trained vintner. CAB SAUV, SHIRAZ among best, but gd use of NEBBIOLO in blends; Canto de Luna blend (r).

Casa de Piedra ★★★ Modern winery in historic stone house; 1st project of Hugo d'Acosta. CAB SAUV/TEMPRANILLO v.gd; invested in sparkling.

Casa Vieja, La ★ Natural wines, ungrafted PALOMINO, MISSION, some 120 yrs+.

Corona del Valle Winery/restaurant destination. Napa-trained winemaker, focus on TEMPRANILLO/NEBBIOLO blend; also new GRENACHE rosé.

Henri Lurton, Bodegas ★★ Venture of B'x's Lurton family; new winemaker. Champion of CHENIN BL. Centenario is signature; promising SAUV BL, NEBBIOLO.

Lechuza Small but acclaimed portfolio in winery est by expat couple, now a two-generation operation. Amantes (CAB SAUV/NEBBIOLO/MERLOT) on list at Napa's French Laundry. Surprising burgundian-style CHARD.

Lomita, Hacienda La Another hip destination, with surrealist art-adorned walls, young entrepreneur (also owns FINCA LA CARRODILLA) and a Napa- and Tuscan-trained winemaker. Organically farmed; v.gd CHENIN BL/SAUV BL shines.

Mina Penelope Winemaker Verónica Santiago at helm, one of growing number of women making wine here. Julio 14 (SYRAH-driven GMS blend) is signature. Others have NEBBIOLO focus, but Santiago also produces on-trend amber, rosé.

Monte Xanic ★★★ The 1st modern premium winery; excellent CAB SAUV, v.gd MERLOT. Try SAUV BL, unoaked CHARD, fresh CHENIN BL. Calixa blend (r) and NEBBIOLO Limitada top. High-elevation PINOT N breaks new ground.

Paralelo ★★ Pioneering eco-constructed winery by Hugo d'Acosta, ultra-modern. Small production. Intense Emblema SAUV BL, B'x-style Ensamble (r).

Rubio, Bodegas F Whimsical elephant-themed labels belie serious intent of wines. Henencia Blanco blends PALOMINO/CHENIN BL; gd Mezcla Italiana blend MONTEPULCIANO/NEBBIOLO/SANGIOVESE.

Tres Valles ★★ Taste on site among the sculptures. Powerful reds in Kiliwa language. TEMPRANILLO drives v.gd Kuwal blend. Top-rated single varieties: Maat (GRENACHE), Kojaa (PETITE SIRAH) and new CHARD/CHENIN BL.

Vena Cava ★★★ Now branded as "the hippest winery in Mexico", est by husband/wife former music-industry execs and sailors; built from reclaimed fishing boats with architect Alejandro d'Acosta; farm-to-table dining. Well priced, modern, organic. Top CAB SAUV, SAUV BL, TEMPRANILLO.

Canada

Canadian wine is in the ascendant. Aided by climate change, technology and a generation of environmentally conscious growers, the country's 12,500 ha+ of vineyards spread 9300 km (5779 miles) from west to east, offering a kaleidoscope of grapes as diverse as the people making wine. Pure, savoury and electric, Canadian wine has become the number-one choice of locals, leaving little for the rest of the world. As the legacy of Icewine fades to the background, re-imagined versions of Chardonnay, Cabernet Franc, Pinot Noir, Riesling and red blends have taken the lead.

Ontario

Prime appellations of origin: Niagara Peninsula (Niag), Lake Erie North Shore (LENS) and Prince Edward County (P Ed). There are two regional appellations within the Niagara Peninsula – Niagara Escarpment and Niagara-on-the-Lake – and ten sub-appellations. New LENS sub-appellation South Islands.

Bachelder ★★★★ Thomas B is the conscience of Niag terroir with encyclopedic knowledge of peninsula's vyds; elegant, age-worthy CHARD, GAMAY, PINOT N.

Cave Spring ★★★★ Penachetti family are Niag pioneers tending old-vine vyds. Must-stop for dry, late-harvest, Icewine RIES freaks.

Fielding ★★★Visiting Fielding comes with an inviting family vibe complementing its v.gd small-lot Beamsville Bench CHARD, GAMAY, CAB FR and sparkling.

Henry of Pelham ★★★ Speck family, 6th generation; sustainable growers offering wide range incl Family Res and outstanding Cuvée Catherine Brut, RIES Icewine.

Hidden Bench ★★★★ Harold Thiel and team are reference producers of serious CHARD, RIES, PINOT N; top picks Felseck Ries, Nuit Blanche, Tête de Cuvée Chard.

Inniskillin ★★★ Icon Icewine pioneer since 1984 still innovating with red CAB FR, Cab Fr/VIDAL (sp), Icewines.

Leaning Post ★★★ Warm family vibe greets visitors; wines focus on subregion nuances, v.gd CHARD, PINOT N, eclectic Freaks & Geek series.

Malivoire ★★★★ Canadian winery of the year 2021; v.gd CHARD, GAMAY, PINOT N; and people who make visits a rare combination of welcoming and memorable.

Prince Edward County A holiday playland and home to 40 producers at e end of Lake Ontario on limestone-rich soils. Top CHARD, PINOT N. Closson Chase, Hinterland, Huff, Rosehall Run, Trail Estate, Stanners.

P Ed wine country is home to largest freshwater dune system on earth.

Ravine Vineyard ★★★ Organic 14-ha St David's Bench vyd set on ancient riverbed. Elegant Res CAB FR, CHARD; delicious farm-to-table lunch or dinner.

Stratus ★★★★ JL Groux is Niag legend, making serious CAB FR/GAMAY blends; Charles Baker completes lineup with age-worthy RIES. Must-visit.

Tawse ★★★ Four times Winery of the Year led by dynamic owner Moray Tawse. Top PINOT N, CHARD fashioned after burgundy.

Thirty Bench ★★★ Small-lot, highly specialized experience with electric single-block old-vine RIES. Steel Post, Wild Cask, Wood Post on Beamsville Bench.

Trius / Andrew Peller ★★★ Rising star; shared winemaking under Australian Craig MacDonald. Sparkling, RIES, CAB FR, Icewine. Top restaurants too.

Two Sisters ★★★ Top estate reds CAB FR, CAB SAUV, MERLOT; cult sparkler Blanc de Franc, rustic Italian dining at Kitchen 76.

British Columbia

Geographical indications for BC wines of distinction and BCVQA are BC, Fraser Valley, Gulf Islands, Kootenays, Lillooet, Okanagan Valley (Ok V) – Golden Mile Bench, Naramata Bench, Okanagan Falls, Skaha Bench (subregions of Ok V) – Shuswap, Similkameen Valley (Sim V), Thompson Valley, Vancouver Island (Van I) and its newest subregion, Cowichan Valley.

New Sovereign Opal vine: 4 *Vitis* species, tastes like Torrontés for BC cold winters.

Blue Mountain ★★★ Legendary traditional-method fizz programme incl complex RD versions; age-worthy CHARD, PINOT N, lauded by the trade.

CedarCreek ★★★★ Serious, organic, focused on Platinum single-vyd blocks CHARD, PINOT N. Must-visit Home Block restaurant, visitor centre.

Checkmate ★★★★ Multiple offerings; Next World CHARD and MERLOT from S Ok micro-blocks. High-end visitor/tasting facilities with panoramic valley views.

Clos du Soleil ★★★ Old World-inspired blends grown in the wild, windswept Sim V, organic capital of Canada. Top: Capella (w), Signature (r).

Cowichan Valley PINOT GR, PINOT N hotbed, ancient volcanic soils on Van I. Top: Averill Creek, Blue Grouse, Emandare, Unsworth, Rathjen, Venturi Schulze.

Haywire ★★★ Ok V legend Chris Coletta offers consumer-friendly visits advocating organics and less-is-more winemaking. Pure PINOT GR, PINOT N, sparkling.

La Frenz ★★★★ Twice Canadian small winery of the year. Sublime MUSCAT Liqueur and Tawny, equally exceptional dry wines overseen by co-owner Jeff Martin.

Martin's Lane ★★★★ Outstanding single-vyd PINOT N, RIES from Naramata and Kelowna. World-class Pinot facility and architecture led by Kiwi Shane Munn.

Mission Hill ★★★★ Certified organic vyds, 500 ha, now under rising-star winemaker Corrie Krehbiel. Top (r/w): Legacy, Terroir series. Re-imagined visitor experience, spectacular al fresco dining.

Osoyoos Larose ★★★ Groupe Taillan blends B'x culture with 33-ha, 26-yr-old single vyd planted to all five B'x varieties. Age-worthy Le Grand Vin; solid Petalos.

Painted Rock ★★★ Skaha Bench, 24-ha vyd below 500-yr-old native pictographs. Mix B'x and Ok V style to gd effect: CAB FR, CHARD, SYRAH, signature Red Icon.

Phantom Creek ★★★★ Stunning $100-million facility, panoramic views, outdoor amphitheatre. Certified organic vyds; v.gd CAB SAUV, PINOT GR, SYRAH, red blends.

Quails' Gate ★★★ CHARD, PINOT N pioneer still on point. Busy visitor centre, lauded wine-country restaurant. New 200-acre, E Kelowna vyd with CHARD, CHENIN BL, PINOT N, RIES.

Road 13 ★★★ Exciting Rhône-style SYRAH, VIOGNIER and treasured old-vine (1968) CHENIN BL (w/sp); premium Jackpot labels for Golden Mile Bench and Sim V.

Tantalus ★★★★ Electric old-vine (1968) RIES, silky PINOT N, CHARD and sparkling. Many of native artist Dempsey Bob's masks (seen on the labels) are displayed in the tasting room.

Nova Scotia

Benjamin Bridge ★★★ Traditional-method and pét-nat fizz. Excellent age-worthy Vintage and NV Brut from CHARD/PINOTS M/N, unique Nova 7.

The lake effect
Canadian wine country is never far from a lake or ocean; most vyds sit within 20 km (12 miles) of a large body of water. The so-called lake effect moderates the extreme temperatures that are now a climate-change part of life. Lakes also provide natural quality control, preventing the planting of vines on the lowest, most fertile land where they should never be grown.

South America

Abbreviations
used in the text:

CHILE

Aco	Aconcagua
Bío	Bío-Bío
Cach	Cachapoal
Casa	Casablanca
Cho	Choapa
Col	Colchagua
Coq	Coquimbo
Cur	Curicó
Elq	Elqui
Ita	Itata
Ley	Leyda
Lim	Limarí
Mai	Maipo
Mal	Malleco
Mau	Maule
Rap	Rapel
San A	San Antonio

ARGENTINA

Cata	Catamarca
La R	La Rioja
Luján	Luján de Cuyo
Men	Mendoza
Neu	Neuquén
Pat	Patagonia
Río N	Río Negro
Sal	Salta
San J	San Juan
Uco V	Uco Valley

CHILE

There are few countries quite as topographically thrilling as Chile. It has extreme landscapes that range from Andean peaks and plateaus to ice fields and glaciers, and a roaring Pacific coastline that runs the entirety of this long, skinny country. All this means a huge diversity of microclimates and a plethora of wine styles. From racy Chardonnay on Limarí's limestone coast, to rich Syrah from Apalta's sunny valley, and from fragrant Cinsault from Itata's granite hills, to world-class Cabernet Sauvignon from the Andean foothills, there's no lack of wines to discover in Chile. And each year Chile's winemakers are making them better and better. Could this be a golden age of Chilean wine?

Recent vintages

Chile is big, so vintages vary depending on region. But in general, 2021 a nice cool yr, 20 warmer, 19 drought yr with concentration, v.gd; 18 cool, dry, v.gd; 17 v. hot and dry.

Abolengo Cach ★★ Promising CARMENÈRE/SYRAH blends from Peumo.

CHILE

Aconcagua Large wine region n of Santiago, extending from mtns to coast. Wines range from rich reds to fresh coastal whites. SYRAH a particular star.

Almaviva Mai ★★★★ Iconic, sumptuous red blend from MAI, 1st New World wine sold on Place de B'x, B'x's internal market. It helps Mouton Rothschild family are co-owners, with CONCHA Y TORO.

Altaïr Wines Rap ★★★ Much-improved top blend by SAN PEDRO from high-altitude CACH. Complex, brooding.

Antiyal Mai ★★★ Chile's legendary bio winemaker Alvaro Espinoza, elegant wines. Worth hunting down.

Apaltagua ★★ Wide-ranging portfolio from all over Chile, gd for everyday.

Aquitania, Viña Mai ★★★ Family winery with top Lazuli CAB SAUV and mtn reds. In deep S MALLECO, excellent CHARD, PINOT N, SAUV BL and bubbles.

Arboleda, Viña Aco ★★ Costa focus and zippy wines from ERRÁZURIZ team. Try CHARD, PINOT N, SYRAH and red blend.

Bío-Bío Traditional region in s, coming back into fashion for its old vines (esp PAÍS) and new plantings of fresh RIES, SAUV BL, PINOT N.

Bouchon Mau ★★→★★★ Innovative family winery in MAU turning heads with old-vine PAÍS, excellent SEM, energetic red blends.

Caliboro Mau ★★→★★★ Boutique wines from Mau with organic focus. Classy portfolio by Count Cinzano (Italy).

Calyptra Cach ★★→★★★ Family winery in heights of Andes in CACH. Lively mtn reds and fresh SAUV BL.

Carmen, Viña Casa, Col, Mai ★★→★★★★ Wide and increasingly diverse portfolio from historic MAI winery. DO range is superb, incl flor-aged SEM AND Gold CAB SAUV a modern icon of Chile.

Casablanca Chile's 1st, and now largest, cool-climate Costa region. Known for its SAUV BL, PINOT N but increasingly SYRAH, CHARD, CAB FR too.

Casa Marín San A ★★★→★★★★ Top-notch coastal producer in Lo Abarca. Mother and son make powerful reds, racy whites.

Casas del Bosque Casa, Mai ★★ Prominent CASA producer with fab restaurant. Pequeñas Producciones is top range.

Casa Silva Col, S Regions ★★→★★★ Multi-generational family with penchant for wine and polo. Richer reds from COL, coastal whites from Lolol and racy cool-climate wines from Osorno.

Clos des Fous Cach, Casa, S Regions ★★→★★★ François Massoc makes eccentric lineup ranging from old vines in ITA to brand-new RIES in MAL.

Concha y Toro Cent V ★→★★★★ One of the world's biggest players, making everything from almost all regions. For everyday, try Casillero del Diablo, then move into Marquis and *Terruño*. Amelia, Gravas, MAYCAS DEL LIMARÍ where wines get v. serious indeed. Icon wine Don Melchor CAB SAUV is better than ever. *See also* ALMAVIVA, TRIVENTO (Argentina).

Cono Sur Bío, Casa, Col ★★→★★★ Focused on *Pinot N* but with gd range from all over. Other top picks: RIES from BÍO and CAB SAUV from MAI.

Cousiño Macul Mai ★★→★★★ Just outside Santiago, large range; v.gd Lota CAB SAUV.

80% of Chile's wines now adopt national wine Sustainability Code.

Elqui Narrow valley from coast to high Andes. Fresh coastal whites, powerful mtn reds. Top for SYRAH.

Emiliana Bío, Casa, Rap ★★ Noelia Orts is winemaker at impressive bio estate, one of largest in world. Great value with G and Coyam esp complex.

Errázuriz Aco, Casa ★★→★★★★ Large, historic estate with modern vision. Rich mtn reds as well as cooler coastal wines. Pizzaras CHARD esp gd. Don Maximiliano CAB SAUV a Chilean classic. *See also* SEÑA, VIÑA ARBOLEDA, VIÑEDO CHADWICK.

> **Slicing up by climate**
> Chile's regions are divided into regions from n to s, but also by climate
> from w to e. Costa = maritime influence. Entre Valles = warmer valley
> floor. Andes = Andes mtns.

Falernia, Viña Elq ★★ Giorgio Flessati is a pioneer of fine wine in ELQ. Range reflects diverse terroir from zesty coastal whites to spicy reds. *Appassimento*-style CARMENÈRE too.

Garcés Silva, Viña San A ★★→★★★ Pioneer in LEY with racy SAUV BL as well as opulent oaked style; gd PINOT N, SYRAH too.

Haras de Pirque Mai ★★→★★★ Italy's Antinori MAI outpost with glossy mtn reds and SAUV BL, CHARD from coast.

Itata Jewel of a region in deep s. Old vines, ungrafted, unirrigated. Top CINSAULT, MUSCAT, PAÍS. New CHARD also promising.

Koyle Col, Ita ★★→★★★ Bio winery with estates in COL and ITA. Juicy CINSAULT and hearty Cerro Basalto Med blend favourites.

Laberinto Mau ★★★→★★★★ Rafael Tirado is a maestro with wild, distinctive wines from hillside vyd by Colbún Lake. Don't miss SAUV BL.

Lapostolle Cach, Casa, Col ★★→★★★★ Andrea Leon is winemaker at noble bio estate owned by Grand Marnier family. Top Clos Apalta increasingly vibrant.

Leyda, Viña Col, Mai, San A ★★→★★★ Leading Ley estate with Viviana Navarrete in driving seat. Fresh coastal wines, notable SAUV BL, SYRAH, PINOT N.

Limarí Thrilling white and red (CHARD, SAUV BL, PINOT N, SYRAH) from cool-coast region with chalky soils.

Luis Felipe Edwards ★★ Large family winery with reliably gd value from all over Chile. Based in Col.

Maipo Home to some top sites for CAB SAUV and B'x blends, esp in higher subregions. Also home to Chile's most historic and glamorous estates.

Malleco Steadily growing region in s. Volcanic soils + mild temperatures = age-worthy PINOT N, CHARD, RIES.

Martino, De Cach, Casa, Elq, Ita, Mai, Mau ★★→★★★★ True explorers hunting vines from ITA to LIM. Covering all classics with a subtle style. Old-vine wines v.gd, esp CINSAULT, MALBEC.

Matetic Casa, San A ★★★ Stunning bio estate. Excellent SYRAH, v.gd reds and whites across all ranges.

Maule Old vines with a new rhythm; Mau coming back into fashion for juicy PAÍS, fragrant red blends and delicious CARIGNAN (*see* VIGNO).

Maycas del Limarí Lim ★★→★★★ Coastal outpost of CONCHA Y TORO with vibrant SAUV BL, CHARD, meaty SYRAH.

Montes Casa, Col, Cur, Ley ★★→★★★★ Innovative father and son run impressive winery in Apalta with vyds from Zapallar to Chiloe. *Folly Syrah* an icon, Outer Limits more experimental range.

MontGras Col, Ley, Mai ★★ Bold reds from Col and MAI, while Ley Amaral offers more electric whites.

Montsecano Casa ★★★ With focus on nervy PINOT N, bio estate is a gem.

Morandé Casa, Mai, Mau ★★→★★★ Under Ricardo Baettig's direction, makes wide range; Med red blends esp gd.

Neyen Col ★★★ Iconic boutique winery owned by VERAMONTE makes old-vine CAB SAUV/CARMENÈRE blend in Apalta.

Odfjell Cur, Mai, Mau ★→★★★ Horses take pride of place in vyd as do bio and organic preparations. Winery in MAI vyd, but MAU is where top CARIGNAN comes from.

Pérez Cruz, Viña Mai ★★→★★★ Handsome reds, incl varietal PETIT VERDOT. CAB SAUV another classic.

Pisco What keeps many Chileans and Peruvians merry through the weekend. Local brandy best served in a sour.

Polkura Col ★★→★★★ Sven Bruchfeld makes some mean SYRAH: meaty, delicious.

Quebrada de Macul, Viña Mai ★★→★★★ Old vines of CAB SAUV star of Alto MAI producer. Domus Aurea cellar-worthy.

Rapel Big, broad wine region covering Col and Cach.

Atacama Desert is world's driest: <20mm rain p.a. But can be cooler than Casa.

RE, Bodegas Casa ★★★ Family winery of winemaking legend Pablo Morandé Snr and innovative son, Jnr. Not afraid to break rules.

San Antonio Growing coastal wine region with zesty whites (CHARD, RIES, SAUV BL), perfumed reds (PINOT N, SYRAH). Star subregions incl LEY and Lo Abarca.

San Pedro Cur ★→★★★ One of Chile's largest with enormous range n to s. Cabo de Hornos CAB SAUV is icon, but 1865 offers some snazzy finds too, as does Tayu PINOT N from MAL. Everyday: 35 Sur, Castillo de Molina. (*See* ALTAÏR, TARAPACÁ.)

Santa Carolina, Viña Mai ★★→★★★ Historic winery, wide-reaching range. Try plush Herencia CARMENÈRE, elegant Luis Pereira CAB SAUV.

Santa Rita Mai ★★→★★★★ A stalwart but always innovative, esp in Floresta range. Classics incl Bougainville PETITE SIRAH, Triple C red blend and stunning *Casa Real Cab Sauv*. Everyday: 120, Medalla Real, Tres Medallas.

Seña Aco ★★★★ Ageable B'x blend from prime plots selected by ERRÁZURIZ team.

Tabalí Lim ★★→★★★★ Leading light making precise cool-coast wines from limestone terraces. Pai one of Chile's best PINOT N, and Talinay CHARD also second to none.

Tarapacá, Viña Casa, Ley, Mai ★★ Historic estate in a natural clos surrounded by hills. VSPT-owned with big portfolio. CAB SAUV increasingly juicy.

Torres, Miguel Cur ★★→★★★★ Latest release is Los Inquetos, "The Restless Ones", a gd name for this innovative and ever-evolving winery. Pioneers of Fairtrade, bubbly PAÍS and MAU PINOT N on daring schist slopes. Old vines and new aplenty. Cordillera is range to start your Torres exploration.

Undurraga Casa, Ley, Lim, Mai ★→★★★ Historic producer, enormous portfolio from all over. Best to explore Terroir Hunter range with pitch-perfect varietals from distinctive terroirs.

Valdivieso Cur, San A ★→★★★ Popular for bubbles in Chile but mainly exports still. Caballo Loco blends age-worthy, interesting.

Vascos, Los Rap ★★ Lafite Rothschild's outpost making rich, structured reds from Col and pithy whites from CASA.

Ventisquero, Viña Casa, Col, Mai ★→★★★ Col-based but with vyds running n up to Atacama. For coastal fresh, look to Kalfu or Grey; for opulent reds, try Enclave CAB SAUV, Pangea SYRAH. Natural-wine lovers should try Tara.

Veramonte Casa, Col ★★ Inland, with fresh, fruit-driven whites/reds. Best: Ritual range. Owned by González Byass (*see* Spain).

Vigno Mau 14 wineries make delicious, dry-farmed CARIGNAN with common cause to save the old vines and growers who cultivate them. Range of styles but all with fresh acid and deep concentration.

Latest novelty? Flor-aged white and rosé. Pink sort-of "Sherry". Hmmm.

Vik Cach ★★→★★★ Lovely winery set in 4000-ha estate. Main pour is iconic red B'x blend. Winemaker is experimenting with amphorae made with clay from vyd.

Villard Casa, Mai ★★ CASA winery, one of valley's 1st. Arganat CHARD and Tanagra SYRAH top pours.

Viñedo Chadwick Mai ★★★★ Pure Puente Alto elegance. Fine CAB SAUV by Chadwick/ERRÁZURIZ clan.

Viu Manent Casa, Col ★★ Horse-rides and family food make this an inviting Col winery. Hearty reds also, esp MALBEC.

Von Siebenthal, Viña Aco ★★→★★★ Mauro von S from Switzerland settled in ACO two decades ago to make full-bodied complex reds. Parcela 7 top value.

ARGENTINA

A land of mountains, gauchos and excellent steak is obviously going to love its red wines. Malbec rules in Argentina. With a whopping 44,000 ha and growing, it comes in all sorts of personalities – from floral with finesse, to rich plum fruit with breadth. But Argentina has also refined its Bonarda, Cabernet Franc and Cabernet Sauvignon. And in whites, there are some stellar Chardonnays today from both high altitudes in Mendoza and low latitudes in the south in Chubut. Torrontés has flamboyant floral and tropical notes, and Criolla is making juicy reds. Argentina's wine scene is ripe to explore, ideally on horseback.

Achaval Ferrer Men ★★→★★★ Famed estate that helped put premium MALBEC on map. Still luscious wines today.

Aleanna Men ★★→★★★★ One of Argentina's best winemakers, Alejandro Vigil, El Enemigo brand. CAB FR, MALBEC, CHARD top.

Alicia, Viña Men ★★★ Neat portfolio of lesser-known varieties, old vines, from NEBBIOLO to SAVAGNIN.

Alpasión Men ★★ Founded by a collective of wine-lovers in Uco V (2011). Mainly red, appetizing. Try PETIT VERDOT.

Alta Vista Men ★→★★★ Long-standing French-owned winery. Excellent reds, sparkling and crisp TORRONTÉS.

Altocedro Men ★★→★★★ Karim Mussi's focus on La Consulta terroir in Uco V. Classy, bio, incl v.gd TEMPRANILLO.

Altos las Hormigas Men ★★★→★★★★ Premium MALBEC pioneer, now also red blends and SEM. Excellent Appellation range showing Malbec by terroir. Also BONARDA under Colonia Las Liebres label.

Anita, Finca La Men ★★ Old-vine estate in Luján, notable for CAB SAUV, SYRAH.

Atamisque Men ★→★★★ Lovely Uco V estate, with range from v.gd-value Serbal to age-worthy Atamisque (now with icon PETIT VERDOT) and Cave Extrême fizz.

Benegas Men ★★→★★★ Historic estate: CAB FR, SANGIOVESE, old-vine MALBEC.

Bianchi, Bodegas Men ★→★★★ Two wineries, two terroirs, strong portfolio from San Rafael and Los Chacayes, Uco V. Leader in local sparkling.

Bosca, Luigi Men ★★→★★★ 120-year-old winery, impressive portfolio. White blend, MALBEC and old-vine CAB SAUV top picks.

Bressia Men ★★→★★★ Walter B and family make handsome portfolio. Sylvestra everyday, Ultima Hoja for special occasions.

Callia San J ★→★★ Weekday wines with focus on SYRAH.

Canale, Bodegas Humberto Río N ★→★★★ Some v. old vines: MALBEC, PINOT N, RIES.

Caro Men ★★★ Making opulent red blends since 2003, marriage of CATENA ZAPATA and (Lafite) Rothschild.

Casarena Men ★★→★★★ All about singular vyds in Luján; v. tempting CAB SAUV, MALBEC in top range.

Catena Zapata, Bodega Men ★★→★★★★ Renowned family and winery in Luján, but with major focus in Uco V. Top incl seductive CHARD, CAB FR, MALBEC from Gualtallary, but enormous range from everyday Alamos to La Marchigiana natural wines. (*See also* CARO.)

Chacra Río N ★★★→★★★★ Delicious PINOT N from old vines, CHARD too. Pedigree bio wines from Piero Incisa della Rocchetta of Sassicaia (*see* Italy).

Cheval des Andes Men ★★★★ One red, two vyds. Blend of Luján and Uco V terroirs made by folks from Cheval Blanc (*see* B'x).

Clos de los Siete Men ★★ One B'x blend made from fruit of estates (*see* BODEGA ROLLAND, CUVELIER LOS ANDES, DIAMANDES, MONTEVIEJO) in this 850-ha clos in Uco V. Blended by Michel Rolland.

Cobos, Viña Men ★★★→★★★★ Paul Hobbs's structured but fine CHARD, CAB SAUV, MALBEC from Uco V and Luján.

Colomé, Bodega Sal ★★→★★★ From tippy-tops of Calchaquí V (up to 3100m/10,171ft no less). Intense reds, mouthwatering whites.

Cruzat Men ★★★ Fabulous traditional-method fizz, now also pét-nat. Hunt down Finca La Dama and Millésime (10 yrs on lees).

Cuvelier Los Andes Men ★★→★★★ Structured but with poise, these B'x blends in Uco V are owned by Léoville Poyferré family (*see* B'x).

Decero, Finca Men ★★→★★★ Focused on reds in Luján, with MALBEC, CAB SAUV as flagships but also mini editions of PETIT VERDOT, CAB FR.

DiamAndes Men ★★ Bonnie family (Malartic Lagravière, *see* B'x) make classy B'x, inspired wines in Uco V, complete with diamond sculpture overlooking Andes.

Doña Paula Men ★★→★★★★ Large winery in Ugarteche with vyds in Uco V too. Martin Kaiser makes precise, varietally expressive wines from everyday Los Cardos to sumptuous Parcel range. Owned by SANTA RITA, Chile.

Durigutti Men ★★→★★★ Durigutti brothers' Las Compuertas line most elegant in otherwise classic range, with a wilder Cara Sucia Criolla line.

Esteco, El Sal ★★→★★★ High-altitude, fragrant, intense wines from Calchaquí V. Old Vines series v.gd; Don David, Ciclos gd value.

Etchart Sal ★→★★ Traditional Cafayate producer with 170 yrs' experience; v.gd value in MALBEC, TORRONTÉS.

Fabre Montmayou Men, Río N ★★→★★★ Classy MERLOT and red B'x blends.

Fin del Mundo, Bodega Del Neu ★→★★ Winery that put this new region on map. Mainly bold reds (try CAB FR, PINOT N) but also crisp whites. Postales, Ventus, Newen are everyday.

Flichman, Finca Men ★★ Set in historic corner of Maipú but with vyds elsewhere too. Reliable everyday wines, to Dedicado at top. Sogrape-owned (*see* Portugal).

Kaikén Men ★★→★★★ MONTES outpost (r/w/sp). Obertura is new icon CAB FR.

Manos Negras / Tinto Negro / TeHo / ZaHa Men ★★→★★★ "Colo" Sejanovich one of Argentina's top wine minds, making vibrant styles from all over country. Particularly gd MALBEC.

Masi Tupungato Men ★★→★★★ A taste of Italia in Uco V. *Ripasso*-style MALBEC/CORVINA blend worth the hunt.

Mendel Men ★★★ Roberto de la Motta is one of Argentina's finest winemakers. Lunta always stellar, Finca Remota top league, SEM and CHENIN BL age brilliantly.

Mendoza Heart of wine industry with lion's share of production. Three main regions: innovative Uco V, classic Luján and warmer, old-vine Maipú.

Moët-Hennessy Argentina Men ★→★★ Best known as Chandon, LVMH has

---

Cri-oy-ja!

Criolla (pronounced *cri-oy-ja*) is not one vine but several. The name means "Creole", and they are the ur-vines of S America – some grown from seed there, others imported as cuttings and known by other names in Europe. There's C Grande (indigenous), C Chica (the PAÍS of Chile) and its offspring (with MUSCAT of Alexandria), Cereza, MOSCATEL Amarillo, TORRONTÉS Mendocino (indigenous), Torrontés Riojano, Torrontés Sanjuanino. Torrontés is most famous, but the reds, regarded for yrs as hopelessly rustic, can respond to better winemaking.

been making fizz in MEN since 1959. Most is everyday party fizz; Baron B is traditional method and a step up. *See* TERRAZAS DE LOS ANDES.

Monteviejo Men ★★→★★★★ Family behind Le Gay in B'x makes solid portfolio in Uco V. MALBEC speciality, but B'x blends also classy.

Moras, Finca Las San J ★→★★ Top-value everyday wines, esp SYRAH. Gran Syrah is deliciously juicy. Owned by TRAPICHE.

Neuquén Pat Neighbouring historic RÍO N, Neu is still new kid in town, 1st planted in 2000s. Bold, fruit-driven thanks to sunny exposure.

Albariño, Garnacha, Mencía: just a few of the Spanish grapes gaining ground today.

Nieto Senetiner, Bodegas Men ★→★★★ Founded 1888, still evolving. Mainly everyday (Benjamin, Emilia), incl lots of fizz. Winemaker Santiago Mayorga takes charge of more innovative Cadus range.

Noemia Pat ★★★→★★★★ Hans Vinding-Diers's elegant, old-vine MALBEC and B'x blends in RÍO N.

Norton, Bodega Men ★→★★★ Founded 125+ yrs ago and still prominent players. Sound still and sparkling. Lot MALBEC top terroir range.

Passionate Wine Men ★★→★★★ Wilder side of Uco V specialist Matías Michelini. Laser-sharp acid, grip and mineral tension are de rigueur in white, red, orange.

Pelleriti, Marcelo Men ★★→★★★ Acclaimed winemaker's personal range of Uco V red, white. Often made in collaboration with Argentine rock stars.

Peñaflor Men ★→★★★ One of biggest wine companies in New World, incl EL ESTECO, FINCA LAS MORAS, Mascota, Navarro Correas, Santa Ana, Suter, TRAPICHE.

Piatelli Sal ★★ Two wineries, one in Cafayate, one in MEND. Both make fragrant whites and polished reds. MALBEC from n a particular treat.

Piedra Negra Men ★→★★★ François Lurton pioneer in Los Chacayes: bold fruit, structured tannin and freshness. White blend also a real gem.

Porvenir de Cafayate, El Sal ★★→★★★ Beautiful old estate in Cafayate, modern styles, often top value. Laborum MALBEC, TORRONTÉS highlights.

Pulenta Estate Men ★★→★★★ Prestigious family with 100 yrs+ wine history, but this estate is modern vision in Luján. Complex reds (excellent CAB FR, MALBEC) best.

Renacer Men ★★ Luján winery in Chilean hands. Known for gd CAB SAUV, memorable Amarone-style MALBEC blend.

Riccitelli, Matías Men ★★→★★★ Inspired range from Uco V, Luján and RÍO N. Tireless innovator.

Riglos Men ★★→★★★ In Gualtallary since 2002; single-vyd reds, CHARD.

Riojana, La La R ★→★★ Co-op in Chilecito, over 400 families. Biggest organic wine exporter in Argentina.

Río Negro Historic river region at gateway to Pat. Best known for old-vine MALBEC, PINOT N, SEM but Trousseau and RIES too.

Rolland, Bodega Men ★★★ Famed B'x winemaker Michel R (*see* France) has his upscale wine production in CLOS DE LOS 7 in Uco V.

Salentein, Bodegas Men ★★→★★★ Drop-dead-gorgeous estate with vyds up to San Pablo. Pepe Galante still makes wines: mouthwatering MALBEC, PINOT N, SAUV BL. El Portillo everyday.

Salta Main city and wine province of N Argentina with most vyds in Calchaquí V to w. High altitude, lots of sunshine and cool nights = intense, perfumed, fresh wines. Top for TORRONTÉS, TANNAT, MALBEC.

San Juan Baby brother to MEN, just to n; mtn climate and warm temperatures make it top for rich MALBEC, SYRAH, TORRONTÉS.

San Pedro de Yacochuya Sal ★★★ Notable Cafayate estate owned by new generation of ETCHART family; consultant Michel Rolland (*see* France). Intense reds and flamboyant TORRONTÉS.

Schroeder, Familia Neu ★★ Focused on PINOT N but with wide range. Biggest find was dino fossil in cellar when constructing.

Sophenia, Finca Men ★★→★★★ Structured whites (v.gd SAUV BL) and classy reds from one of early wineries in Gualtallary. Altosur brand top value.

Susana Balbo Wines Men ★★→★★★★ Excellence across board by family winery, 1st female winemaker in Argentina. Known as Queen of TORRONTÉS for her deft hand with white, but complex reds are real highlight.

Tapiz Men ★★→★★★ Founded in MEN but now with vyds in coastal RÍO N too, incl underwater cellar. MALBEC Black Tears, Notas MERLOT top pours from San Pablo.

Terrazas de los Andes Men ★★→★★★★ LVMH's fine-wine branch in Luján with vyds in Uco V too. Sumptuous reds, unctuous whites. Best for old-vine MALBEC.

Tikal / Alma Negra / Animal / Stella Crintina Men ★★→★★★ Natural wine and bio adventures of Ernesto CATENA in Uco V and Luján. Red, orange, white, pink.

Toso, Pascual Men ★★→★★★ Founded in 1890 and keeping traditional style of rich reds. Some v.gd old-vine wines.

There are 200 active volcanoes in the Andes.

Trapiche Men ★→★★★ Winery so big that it used to have a train running wine out of it. Today all moves by truckload, and quality is from everyday to super-premium. *Medalla* is renowned, Iscay MALBEC/CAB FR and SYRAH/VIOGNIER blends are age-worthy, Costa y Pampa range comes from coast. Part of PEÑAFLOR.

Trivento Men ★→★★★ Large winery, big portfolio, gd value from Chile's CONCHA Y TORO. Best: old-vine Eolo MALBEC.

Vines of Mendoza / Winemaker's Village Men ★★ Fancy hotel and private vyd estate with c.100 owners making own-labels. Also incl Winemaker's Village where Abremundos (*see* MARCELO PELLERITI), Corazon del Sol, Gimenez Riili, Super Uco live.

Zorzal Men ★★→★★★ Vibrant wines from Gualtallary in Uco V. Mainly everyday, but Eggo is v.gd, esp CAB FR, SAUV BL.

Zuccardi Men ★★→★★★★ Family winery, several generations. Focused on Uco V, with freshness paramount. Mineral MALBEC best in Alluvional, Concreto, Piedra Infinita. Everyday wines made at Santa Julia in Maipú.

BRAZIL

A sleeping giant, beginning to wake. Brazil has some 76,000 ha of vines, mostly hybrids for juice and jug wine. But fine wines are increasing, especially in Rio Grande do Sul in the south (best for sparkling, but also Merlot, Chardonnay and Portuguese reds), and increasingly in Serra da Mantiqueira near São Paulo (where mountain SYRAH is tops). More tropical-scented whites come from Bahia, where two harvests per year are common. A natural-wine scene is also growing in Brazil's cities.

Aurora ★→★★ Large co-op, one of biggest wineries in Brazil, founded 75 yrs+ ago in Serra Gaucha.

Casa Valduga ★→★★★ Leading premium winery, with v.gd fizz and more complex B'x-style reds.

Cave Geisse ★★★ Pinto Bandeira (sp) by Mario Geisse and family. One of Brazil's best bubblies.

Lidio Carraro ★★ Two terroirs: Vale dos Vinhedos and Serra do Sudeste in Rio Grande do Sul. Range from NEBBIOLO to sparkling MOSCATEL.

Miolo ★→★★★ Premium wines only from all over Brazil. Best: TOURIGA N blend, Millésime bubbles, single-vyd range.

Pizzato ★→★★★ One of top producers, led by tireless Flavio P. Excellent CHARD and bubbles, also v.gd MERLOT.

Salton ★→★★ Major producer in Serra Gaucha. Everyday wines v. popular in Brazil, but hunt for Gerações blend for something more complex.

URUGUAY

Small country with enormous potential. Hundreds of wine families, all with distinctive wines. Tannat rules – ripe, sometimes chunky, and with good acidity. But there's also Arneis, Sangiovese and Zinfandel… Albariño, Cabernet Franc and Marselan are rising stars. Montevideo and Canelones are main wine regions, with mild Atlantic climate and clay-lime soils. To the west, Colonia is a popular wine destination for neighbouring Argentinians, and to the east Maldonado is a hotspot for Brazilians. Maldonado is also building its own reputation for fresh wines from coast and granite soils. Uruguay is on an upward trajectory.

Uruguay is vintage-car heaven. So is Cuba, but Uruguay has better wine.

Alto de la Ballena ★→★★ The 1st to plant in Maldonado this century; juicy SYRAH, savoury TANNAT, fleshy VIOGNIER among others.

Bouza ★★→★★★ Excellent winery in Montevideo with vyds in Canelones and Maldonado. Pioneer of ALBARIÑO and top RIES, MERLOT, TANNAT.

Deicas, Familia ★→★★★ Innovative Santiago Deicas makes excellent single-vyd wines from extreme sites, TANNAT aged in amphora and even orange wines. Establecimiento Juanicó wines everyday.

Garzón, Bodega ★→★★★★ Few wine estates as impressive; 1500 ha of forest and hills, 240 ha of vines. Excellent ALBARIÑO, MARSELAN, TANNAT. Plus well-organized tourism.

Marichal ★→★★ Family winery in Canelones with solid classic PINOT N, TANNAT and now ALBARIÑO. Also fabulous Tannat vermouth.

Pablo Fallabrino Wines ★★→★★★ One of Uruguay's most adventurous. Pablo loves his Italian varieties as much as TANNAT. Check out ARNEIS, NEBBIOLO, and new wines infused with cannabis.

Pisano ★→★★★ Top wine family in Canelones, with v gd juicy PINOT N, firm TANNAT, steely SAUV BL.

OTHER SOUTH AMERICAN WINES

Bolivia Seriously high altitude and seriously wild, if you can get your hands on them. Bolivian wines are an experience. Seek out 300-yr-old MUSCAT from Cinti V made with grapes grown around peppercorn trees, or intense reds (PETIT VERDOT, TANNAT) from Tarija at over 2150m (7053ft), or fragrant wines from sub-tropical Santa Cruz V. Notable wineries: Campos de Solana, Jardin Oculto, Kohlberg, Kuhlmann, La Concepción, Vinos 1750.

Peru Making wine since 1540s, Peru shifted to mainly Pisco in 1600s, but a wine renaissance is under way esp with old-vine Criolla and inky reds. Try Bodega Murga, Intipalka, Mimo, Tacama, Vista Alegre.

> **Pét-nat bubble rising**
> The pét-nat bubble is growing in S America, with many producers using this traditional method of single-ferment sparkling, known in Spanish as *método ancestral*. Mainly organic too. Argentina: CRUZAT, Chakana, MATIAS RICCITELLI, Stella Crinita. Chile: Agricola Luyt, Cacique Maravilla, L'Entremetteuse. Uruguay: PABLO FALLABRINO WINES, Proyecto Nakkal.

Australia

Abbreviations used in the text:

Ad H	Adelaide Hills, SA
Bar V	Barossa Valley, SA
Beech	Beechworth, Vic
Can	Canberra, NSW
Coonw	Coonawarra, SA
Fra R	Frankland River, WA
Gra	Grampians, Vic
Hea	Heathcote, Vic
Hunt V	Hunter Valley, NSW
Kang I	Kangaroo Island, SA
Lang C	Langhorne Creek, SA
Mac	Macedon, Vic
N/S Tas	North/South Tasmania
Qld	Queensland
Ruth	Rutherglen, Vic
Marg R	Margaret River, WA
McL V	McLaren Vale, SA
Mor P	Mornington Penninsula, Vic

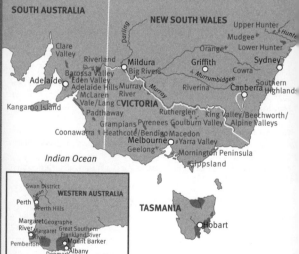

"Diverse" is one word to describe Australia's wines, with its producers best labelled as "resilient". The impact of a global pandemic aside, winemakers have endured a suite of difficult vintages – from small, drought-affected ones to bushfires in some regions in 2020 – and on top of that, a loss of more than $700m when China slammed its doors on Australian imports from 2020 onwards. Back to the land. A collective sigh of relief overall with vintage 2021; a blessing in South Australia and Victoria. It's timely to remember that Australian wine is not homogeneous. Its regions, subregions and a host of individual vineyards have distinct traits, many as complex and interesting as any in Europe, just different. Wine-lovers don't say, "I'm drinking French"; instead they say, "I'm really enjoying this Hermitage." Yet too often Australian wine is regarded as just one thing. About 4000 km (2485 miles) separates Margaret River in the west to the Hunter Valley on the east coast. The differences in styles and varieties are palpable. It doesn't take much to discover, hopefully with joy, the producers who excel in growing and making wines stamped with a sense of place. They're listed in the following pages. Now is the time to appreciate the distinctiveness of Yarra Valley Cabernet Sauvignon compared with Coonawarra or Mornington Peninsula Chardonnay or those from Tasmania. As always, the proof and reward are in the wine.

Recent vintages

New South Wales (NSW)

2021 Heavy rain, some disease pressure. Whites generally gd, choose wisely.
2020 Hot, fire-affected yr with all manner of challenges.
2019 Hot, with just enough rain for generous whites/reds.
2018 Big ripe reds for long haul. Hot/tough yr for whites.
2017 Hot summer followed a wet spring; whites/reds lapped it up, in general.
2016 Drink Hun V reds early, but gd mid-term ageing red/white elsewhere.

Victoria (Vic)

2021 Stark contrast to 20. Many calling it season of elegance, balance; gd yields.
2020 Fire-affected in ne and generally challenging. Yields down.
2019 Compressed season, but wines (r/w) look vibrant.
2018 Overshadowed by yr before. Slow evolving.
2017 Excellent yr. Reds/whites looked gd young and will stay.
2016 Warm, dry season produced many overripe reds. Tread carefully.

South Australia (SA)

2021 Said to be nr perfect. Bar V outstanding, Clare V exceptional (r/w).
2020 Hard/tragic yr, but Ries and reds look promising. Seriously low yields.
2019 Exceptionally low yields should produce concentrated wines.
2018 Gutsy reds with yrs up their sleeve. Whites gd but not in same class.
2017 High yield, high quality, highly drinkable young.
2016 Hopes are high for a special vintage for red/white.

Western Australia (WA)

2021 Challenging due to late-season rain but ultimately rewarding (esp w).
2020 Early, low-yield, high-concentration yr.
2019 Cool vintage, the kind to sort the wheat from the chaff.
2018 Reds will outlive most of us; whites will do medium-term in a canter.
2017 Tricky vintage. Medium-term wines.
2016 Humid, sultry vintage. Nothing wrong with wines; mid-termers.

Accolade Wines Name of once-mighty Constellation, HARDYS groups. BAY OF FIRES, Hardys, HOUSE OF ARRAS, PETALUMA, ST HALLETT.

Adams, Tim Clare V, SA ★★ Ever-reliable (in gd way) RIES, CAB SAUV/MALBEC blend, SHIRAZ and (full-bodied) TEMPRANILLO. FIANO to watch.

Adelaide Hills SA Cool 450m (1476ft) sites in Mt Lofty ranges. CHARD, SAUV BL, SHIRAZ outgun PINOT N. ASHTON HILLS, BASKET RANGE, HAHNDORF HILL, HENSCHKE, JERICHO, MIKE PRESS, SHAW & SMITH, TAPANAPPA all in excellent form.

Adelina Clare V, SA ★★★ Reds (SHIRAZ, GRENACHE, MATARO, NEBBIOLO) and subregional RIES rule. Stylish, detailed. Label designs of note too.

Alkoomi Mt Barker, WA (RIES) 17' 18' 19' 20 (CAB SAUV) 12' 16' 17' 18 Veteran maker of gd Ries; rustic reds but SHIRAZ making a stance, all priced keenly.

All Saints Estate Ruth, Vic ★★★ Rating for fortifieds. Serviceable table wines.

Alpine Valleys Vic In valleys of Victorian Alps. BILLY BUTTON, MAYFORD, Ringer Reef. TEMPRANILLO, SHIRAZ the headliners; aromatic whites for those in the know.

Angove's SA ★ MURRAY v family business. Cheapies (r/w) often standouts of a broad range. Mainstream face of organic grape-growing, both value and premium ends. Single-vyd MCL V GRENACHE, SHIRAZ can be excellent.

Arenberg, d' McL V, SA ★★ Wine names pretty wacky (eg. The Cenosilicaphobic Cat SAGRANTINO) but the styles are no-nonsense and the quality is generally high, incl sumptuous GRENACHE, SHIRAZ.

Ashton Hills Ad H, SA ★★ (PINOT N) 17' 18' 19' 20 Distinctive Pinot N from 30-yr-old+ vyds. Bought in 2015 by WIRRA WIRRA.

Baileys of Glenrowan Glenrowan, Vic, NE Vic ★ Sumptuous SHIRAZ, magnificent dessert MUSCAT (★★★★) and TOPAQUE; vyds all organic. CASELLA bought it in 2017.

Balgownie Estate Bendigo, Vic, Yarra V, Vic ★★ Capable of medium-bodied, well-balanced, minty CAB, SHIRAZ of finesse, character from its BENDIGO heartland. Separate YARRA V arm.

Balnaves of Coonawarra SA ★★ Family-owned COONW champion. Lusty CHARD; v.gd spicy, SHIRAZ, full-bodied Tally CAB SAUV flagship. "Joven-style" Cab gd.

Bannockburn Vic ★★★ (CHARD) 17' 18' 19 (PINOT N) 16' 17' 18' 19 Intense Chard; complex Pinot N. Put GEELONG region on map and continues to fly quality flag.

Barossa Valley SA Ground zero of Aussie red; v.-old-vine CAB SAUV, GRENACHE, MOURVÈDRE, SHIRAZ. Can produce bold, black, beautiful reds with its eyes closed, and has done just about forever. ELDERTON, GRANT BURGE, HENTLEY FARM, JOHN DUVAL, LANGMEIL, OCHOTA BARRELS, PETER LEHMANN, ROCKFORD, RUGGABELLUS, ST HALLETT, SALTRAM, SEPPELTSFIELD, SPINIFEX, TEUSNER, WOLF BLASS, YALUMBA.

Barry, Jim Clare V, SA ★★★ RIES rules thanks to top vyds and lots of them. New Ries LoosenBarry Wolta Wolta a collaboration with Mosel mogul Ernst Loosen (*see* Germany). Reds solid: age-worthy Armagh SHIRAZ. Also Oz's only ASSYRTIKO.

Basket Range Ad H, SA ★★ Lead player in new/avant garde of AD H producers. Pét-nat, CAB SAUV, PINOT N.

Bass Phillip Gippsland, Vic ★★★★ (PINOT N) 16' 17' 18' 19' 20 Tiny amounts of variable but mostly exceptional Pinot N.

Bay of Fires N Tas ★★★ PINOT N shouldn't be passed over but HOUSE OF ARRAS *super-cuvée sparklings* rightly dominate.

Beechworth Vic The rock-strewn highlands of NE Vic. Tough country. CHARD, SHIRAZ best-performing varieties, but NEBBIOLO fast rising from (winter) fog. A RODDA, CASTAGNA, DOMENICA, FIGHTING GULLY ROAD, GIACONDA, SAVATERRE, SCHMÖLZER & BROWN, SORRENBERG, TRAVIARTI essential producers.

Bekkers McL V, SA ★★★ Emmanuelle and Toby B, winemaker and viticulturalist respectively, crafting top GRENACHE, SYRAH with organic focus.

Bendigo Vic Hot central Vic region. BALGOWNIE ESTATE, PASSING CLOUDS, SUTTON GRANGE. Home of rich CAB SAUV, SHIRAZ.

Best's Great Western Gra, Vic ★★★ (SHIRAZ) 14' 15' 17' 18' 19 Shiraz master; plus

---

Italian connections

The CHALMERS family of growers/producers in Mildura and HEA (Vic) has introduced the most diverse range of Italian varieties into Australia, 46 and counting. Usual suspects, ie. FIANO, SANGIOVESE, but rarer stuff too. Pavana anyone?

v.gd mid-weight reds. Thomson Family Shiraz from 120-yr-old vines superb. Wines generally on plush side of elegant. Old-vine PINOT M insider's tip.

Billy Button Vic ★★ So many wines, such small quantities. Everything from SHIRAZ and CHARD to Verduzzo, VERMENTINO, SAPERAVI, SCHIOPPETTINO and more. Low oak in general; fruit-fresh wines.

Bindi Mac, Vic ★★★★ (PINOT N) 17′ 18′ 19′ 20 Meticulous maker of outstanding single-block Pinot N, super-fine CHARD. Small-production; high-density vyds (11,300 vines/ha) coming into play soon.

1st NFT: a barrel of Penfolds Magill Cellar 3 2021, US$130,000, via BlockBar.

Bleasdale Lang C, SA ★★★ Historic winery est 1850. Fabulous fortifieds, top CAB SAUV, MALBEC, SHIRAZ, plus gd-value range. Ever reliable.

Bondar McL V, SA ★★ GRENACHE, SHIRAZ stars. Dabbling in Italian varieties, plus v.gd AD H CHARD.

Bortoli, De Griffith, NSW, Yarra V, Vic ★★★ (Noble SEM) Both irrigation-area winery and leading producer. Excellent cool-climate PINOT N, SHIRAZ, CHARD, gd sweet, botrytized, Sauternes-style Noble Sem. YARRA V arm where high quality reigns.

Brash Higgins McL V, SA ★★ Brad Hickey is a smart cookie. He has degrees in English and botany but has also worked as a brewer, baker, sommelier and now makes radical expressions of MCL V (r/w).

Brave New Wine Marg R, WA ★★ Natural-wine producer capable of v. high highs, so to speak. CHARD, GEWURZ, SHIRAZ can all be excellent.

Bremerton Lang C, SA ★★ Silken CAB SAUV, SHIRAZ with mounds of flavour. Never mean, always generous. MALBEC can be v.gd too. All about the reds.

Brokenwood Hun V, NSW ★★★ (ILR Res SEM) 13′ 14′ 15 (Graveyard SHIRAZ) 13′ 17′ 18′ 19 A HUN V classic. *Cricket Pitch* Sem/SAUV BL gd value, quality high.

Brown Brothers King V, Vic ★ Wide range of crowd-pleasing styles, varieties. General emphasis on sweetness. Innocent Bystander (YARRA V) and Devil's Corner/Tamar Ridge (TAS) savvy acquisitions.

Burge, Grant Bar V, SA ★★ Smooth red/white from best grapes of Burge's large vyd holdings. Owned by ACCOLADE.

Campbells Ruth, Vic ★ Smooth ripe reds (esp Bobbie Burns SHIRAZ); extraordinary Merchant Prince Rare *Muscat*, Isabella Rare TOPAQUE (★★★★).

Canberra District NSW One of most significant cool-climate regions in Oz, for quality at least. CLONAKILLA the leader, but Collector, GUNDOG ESTATE, MOUNT MAJURA, Nick O'Leary, RAVENSWORTH all v.gd.

Cape Mentelle Marg R, WA ★★ A little wobbly recently and heavy use of oak, esp on CAB SAUV, ZIN; CHARD solid and SAUV BL/SEM popular. Part of LVMH.

Casella Riverina, NSW ★ Casella's Yellow Tail range of budget reds/whites has developed into an Australian wine empire. Now owner of heritage brands BAILEYS OF GLENROWAN, Brand's of Coonawarra, MORRIS, PETER LEHMANN.

Castagna Beech, Vic ★★★ (SYRAH) 12′ 14′ 15′ 16 Estate-grown bio SHIRAZ/VIOGNIER, SANGIOVESE/Shiraz excellent. Non-estate Adam's Rib range worth investigating.

Chalmers Hea, Vic ★ Grower, vine-nursery specialist with focus on Italian varieties. Range of diverse wines; FIANO, VERMENTINO gd.

Chambers Rosewood NE Vic Viewed with MORRIS as greatest maker of sticky TOPAQUE (★★★★), *Muscat*.

Chandon, Dom Yarra V, Vic ★★ Cool-climate fizz and table wine. Owned by LVMH. Known in UK as Green Point. NV cuvées in best-ever shape.

Chapel Hill McL V, SA ★★ High-profile MCL V producer. SHIRAZ, CAB SAUV the bread and butter, but TEMPRANILLO and esp GRENACHE on rise.

Charlotte Dalton Wines Ad H, SA ★★ Breath of fresh air in Oz wine scene. Classic varieties CHARD, SHIRAZ, but also v.gd Lang C FIANO.

Chatto Tas ★★★★ Young vyd, but already among Australia's best PINOT N producers. Fruit, spice and all things nice. Smoky savouriness abounds.

Clarendon Hills McL V, SA ★★ Full Monty reds (high alc, intense fruit) from grapes grown on hills above MCL V. Cigar wines.

Clare Valley SA Small, pretty, high-quality area 160 km (100 miles) n of Adelaide. Best toured by bike, some say. Australia's most prominent RIES region. Gumleaf-scented SHIRAZ; earthen, tannic CAB SAUV. ADELINA, GROSSET, KILIKANOON, KIRRIHILL, MOUNT HORROCKS, TIM ADAMS, WENDOUREE (esp) lead way.

Clonakilla Can, NSW ★★★★ (SHIRAZ) 10' 14' 15' 17' 18' 19 CAN superstar. Excellent RIES, VIOGNIER, Shiraz/Viognier famous but SYRAH making elegant statement.

Clyde Park Vic ★★ Single-vyd CHARD, PINOT N v.gd form. SHIRAZ turning heads.

Cobaw Ridge Mac, Vic ★★★ Certified bio producers of cool-climate, compelling CHARD, SYRAH. Super-fine PINOT N and finest LAGREIN in Oz.

Coldstream Hills Yarra V, Vic ★★★★ (CHARD) 17' 18' 19 (PINOT N) 13' 15' 17' 19' 20 Critic James Halliday est 1985. Delicious Pinot N to drink young, *Res and single-vyd to age*. Excellent Chard (esp Res). Part of TWE.

Coonawarra SA Home to some of Australia's best and most distinctive CAB SAUV, and certainly to its richest red soil (on limestone). WYNNS C ESTATE is the champion resident. BALNAVES, KATNOOK, LINDEMAN'S, MAJELLA, PARKER C ESTATE, YALUMBA too are all key. Cachet of region dimmed in recent decades.

Coriole McL V, SA ★★ (Lloyd Res SHIRAZ) 14' 15' 16' 18 Renowned producer of SANGIOVESE, old-vine Shiraz Lloyd Res. Compelling FIANO, NERO D'AVOLA.

Corymbia Marg R, WA, Swan V, WA ★★★ Producer of beautifully pitched CAB SAUV, CHENIN BL, among others. Star in making, if not already.

Craiglee Mac, Vic ★★★ (SHIRAZ) 14' 15' 16' 17 Salt-of-the-earth producer; N Rhône-inspired. Distinctive, fragrant, peppery Shiraz, age-worthy CHARD.

Crawford River Henty, Vic ★★★ Outstanding RIES producer. Cool, cold, scintillatingly dry but intense style, great for seafood. Highly age-worthy.

Crittenden Estate Mor P, Vic ★★★ Focus on v.gd CHARD, PINOT N, turned SAVAGNIN (once mistakenly labelled ALBARIÑO: error in vine import via Spain) into excellent Oz version of *vin jaune*-style, labelled Cri de Coeur Sous Voile.

Cullen Wines Marg R, WA ★★★★ (CHARD) 16' 17' 18' 19' 20 (CAB SAUV/MERLOT) 12' 13' 14' 15' 16' 17' 18' 19 Vanya C makes substantial but subtle SEM/SAUV BL, outstanding Chard, elegant, fine Cab/Merlot. Bio in all she does. Best of Oz.

Curly Flat Mac, Vic ★★★ (PINOT N) 14' 15' 16' 17' 18' 19' 20 Structured, perfumed Pinot N on two price/quality levels. Full-flavoured CHARD. Both age-worthy. Gradual turn towards elegance. Single-vyd offerings of note.

Dal Zotto King V, Vic ★★ "Prosecco" specialist across range of styles, the drier and funkier of which are v.gd. Col Fondo NEBBIOLO bit of a treat too.

Dappled Yarra V, Vic ★★★ Top-flight/value, elegant-but-complex CHARD, PINOT N.

Deep Woods Estate Marg R, WA ★★★ Compelling CHARD, CAB SAUV. Powerhouse wines, built to impress/last.

Devil's Lair Marg R, WA ★★ Opulent CHARD, CAB SAUV/MERLOT is this estate at its best. Solid performer. Owned and therefore largely hidden by TWE.

Domenica Beech, Vic ★★★ Exciting BEECH producer with est vyds. Exuberant, spicy SHIRAZ. Textural MARSANNE. But NEBBIOLO is "the one".

Dr Edge Tas ★★★ Peter Dredge makes wine for various TAS wineries; his own-brand CHARD and (esp) PINOT N more personality than clients would likely allow.

Eden Valley SA Closest neighbour of BAR V and cooler. Hilly region to e, home to Chris Ringland, HENSCHKE, PEWSEY VALE, Radford, TORZI MATTHEWS and others; textural RIES, (perfumed, bright) SHIRAZ, CAB SAUV of top quality.

Elderton Bar V, SA ★★★ Old vines; rich, oaked CAB SAUV, SHIRAZ. All bases covered. Some organics/bio. Rich reds in excellent form.

Eldorado Road Ruth, Vic ★★ Pet project of winemaker Paul Dahlenburg (BAILEYS OF GLENROWAN). DURIF, NERO D'AVOLA, SHIRAZ all show elegance and power not mutually exclusive. Always interesting.

Eldridge Estate Mor P, Vic ★★★ Winemaker David Lloyd is a fastidious experimenter. CHARD, PINOT N worth the fuss. Varietal GAMAY can be quite special.

Epis Mac, Vic ★★★ (PINOT N) Long-lived Pinot N; elegant CHARD. Cold climate. Powerful at release; complexity takes time. Pinot N 19 of particular note.

Faber Vineyards Swan V, WA ★★★ John Griffiths is a guru of WA winemaking. Home estate redefines what's possible for SWAN V SHIRAZ. Polished power.

Fighting Gully Road Beech, Vic ★★★ Touchstone producer of BEECH region. CHARD, AGLIANICO, TEMPRANILLO kicking goals, but SANGIOVESE is king.

Flametree Marg R, WA ★★★ CHARD and CAB SAUV of note, esp SRS (subregional series), Wallcliffe and Wilyabrup respectively; v.gd SHIRAZ surprise late entry.

Fogarty Group Increasingly important stable of wineries: Dalwhinnie, DEEP WOODS ESTATE, Evans & Tate, LAKE'S FOLLY, Smithbrook and Tasmanian Vintners.

Forest Hill Vineyard Gt Southern, WA ★★★ Kickstarting the region; est 1965. RIES the hero, Block 1 the star; v.gd MALBEC, SHIRAZ.

Frankland Estate Fra R, WA ★★★ (RIES) 17' 18' 19' 20' 21' Family-run; outstanding single-site Ries incl Isolation Ridge, the most remote vyd in WA, plus range of styles. Also medium-bodied CAB SAUV, SYRAH.

Fraser Gallop Estate Marg R, WA ★★ Concentrated CAB SAUV, CHARD, SEM/SAUV BL. Cab, Chard particularly strong in recent yrs.

Freycinet Tas ★★★ (PINOT N) 10' 17' 18 Pioneer family winery on TAS's e coast. Dense Pinot N, v.gd CHARD, excellent Radenti (sp).

Marg R's most widely planted Chard clone is the unique Gingin. A tonic.

Garagiste Mor P, Vic ★★★ CHARD, PINOT N of intensity, finesse. Quality always seems to be high or higher. Multi-vyd blends really take value cake.

Geelong Vic Region w of Melbourne. Maritime and dry climate. Best: BANNOCKBURN, By Farr, CLYDE PARK, LETHBRIDGE, PROVENANCE.

Gembrook Hill Yarra V, Vic ★★★ Cool site on upper reaches of Yarra River: fine-boned PINOT N, CHARD par excellence.

Gemtree Vineyards McL V, SA ★★ Warm-hearted SHIRAZ alongside TEMPRANILLO and other exotica, linked by quality. Largely bio.

Giaconda Beech, Vic ★★★★ (CHARD) 16' 17' 18' 19 (SHIRAZ) 15' 17' 18 In mid-80s Rick Kinzbrunner kickstarted BEECH region. Australian Chard royalty. Tiny production of powerhouse wines.

Giant Steps Yarra V. Vic ★★★ Top single-vyd CHARD, PINOT N, SHIRAZ: 15' 17' 18' 19' 20 exciting for three main varieties. Sold to Jackson Family Wines (*see* California, US) 2020.

Glaetzer-Dixon Tas ★★★ Nick G turned his family history on its head by setting up camp in cool TAS. Euro-style RIES, autumnal PINOT N, Rhôney SHIRAZ.

Goulburn Valley Vic Warm region in mid-Vic. Full-bodied, earthy table wines. MARSANNE, CAB SAUV, SHIRAZ the pick, MITCHELTON, TAHBILK perpetual flagbearers. Aka Nagambie Lakes.

Grampians Vic Temperate region in NW Vic previously known as Great Western. Spicy SHIRAZ, sparkling Shiraz, limey RIES. Home to SEPPELT (for now), BEST'S, Montara, MOUNT LANGI, The Story.

Granite Belt Qld High-altitude, (relatively) cool, improbable region just n of Qld/NSW border. Spicy SHIRAZ, rich SEM (Boireann, Golden Grove, Ridgemill Estate).

Great Southern WA Remote cool area at bottom left corner of Oz; Albany, Denmark, Frankland River, Mount Barker, Porongurup are official subregions; 1st-class RIES, SHIRAZ, CAB SAUV. *Style, quality, value here.*

Grosset Clare V, SA ★★★ (RIES) 17' 18' 19' 20' 21 (Gaia) 15' 16' 17' 18 Ries royalty with range of styles, incl high-end G110 from one-clone, organic site. *Gaia* CAB SAUV/MERLOT distinctive, FIANO appealing. Rich AD H CHARD, PINOT N interesting.

Gundog Estate Can, NSW ★★★ Highly aspirational SEM, SHIRAZ from CAN, HUN V. Shiraz starting to turn heads.

Hahndorf Hill Ad H, SA ★★★ Made GRÜNER v its own in Oz, consistently producing richly spiced, textured examples; interest and experimentation across range.

Hardys SA ★★ Historic company now part of ACCOLADE. CHARD, SHIRAZ can be excellent in fuller mode.

Heathcote Vic 500-million-yr-old Cambrian geology creates great potential for high-quality reds, esp SHIRAZ (ample body, spice). JASPER HILL, PAUL OSICKA, TAR & ROSES, Whistling Eagle, WILD DUCK CREEK.

Henschke Eden V, SA ★★★★ (SHIRAZ) 05' 06' 12' 13' 14' 15' 16 (CAB SAUV) 06' 10' 12' 15' 16 Pre-eminent 150-yr-old family business; delectable Hill of Grace (Shiraz), v.gd Cab Sauv, red blends, gd whites, scary prices. Wonder of modern world.

Hentley Farm Bar V, SA ★★★ Consistently produces SHIRAZ of immense power, concentration – wall-of-flavour territory – though importantly in a (generally) fresh, almost frisky context.

Hewitson SE Aus ★★ (*Old Garden Mourvèdre*) 14' 15' 16' 18 Dean H sources some of "oldest Mourvèdre vines on the planet"; v.gd SHIRAZ, various prices.

Hoddles Creek Estate Yarra V, Vic ★★★ Made name as value producer of cool-climate varieties but it's more than just value; quality is outstanding, full stop. Top single-block CHARD, PINOT N.

Houghton Swan V, WA ★★ (Jack Mann) 13' 15' 18' 17' 19 Once-legendary winery of SWAN V nr Perth. Part of ACCOLADE. Inexpensive white blend was long *a national classic*; v.gd CAB SAUV, SHIRAZ, etc. sourced from GREAT SOUTHERN, MARG R.

House of Arras Tas ★★★ Best-performing and most prestigious sparkling house in Oz. Serious ageing on lees. Serious sparkling. Part of ACCOLADE.

Howard Park WA ★★ Scented RIES, CHARD; earthy CAB. *MadFish* can be gd value.

Hunter Valley NSW Sub-tropical coal-mining area 160 km (100 miles) n of Sydney. Mid-weight, earthy SHIRAZ, racy SEM can live for 30 yrs. Arguably most terroir-driven styles of Oz: ANDREW THOMAS, BROKENWOOD, MOUNT PLEASANT, *Tyrrell's*.

Hutton Wines Marg R, WA ★★ CAB SAUV, SHIRAZ v.gd, but CHARD is where things tip into outstanding territory. Powerful palate, powerhouse finish.

Inkwell McL V, SA ★★ High polish, high opinion, high character. Full house of intriguing wines, mostly SHIRAZ-based.

Jacob's Creek Bar V, SA ★ Owned by Pernod Ricard. Almost totally focused on various tiers of uninspiring but reliable JC wines, covering all varieties, prices.

Jasper Hill Hea, Vic ★★ (SHIRAZ) 12' 17' 18' 19 Emily's Paddock Shiraz/CAB FR blend,

---

Marvellous Margaret River

Blessed with breathtaking beauty, MARG R sits 270 km (168 miles) s of WA's capital Perth. Thanks to the influence of the Indian Ocean, esp cooling breezes, the maritime climate creates high winter rain and warm, dry summers. An ancient granite bedrock runs through the region layered with limestone, gneiss and schist; those granite-based soils are the world's oldest, resulting in low-vigour vines and high-quality grapes. Reds are medium-bodied with fine tannins, and whites are flavoursome, led by natural acidity. CHARD and CAB SAUV the hero varieties. In 1967 VASSE FELIX planted 1st vines, and the rest is history. Other pioneers such as bio CULLEN WINES, which celebrated its 50th in 2021, MOSS WOOD, LEEUWIN ESTATE, WOODLANDS, XANADU WINES continue to be at top of their game. Newcomers are reshaping landscape incl LAS VINO and LS Merchant.

Georgia's Paddock Shiraz from dry-land estate are intense, burly, long-lived. NEBBIOLO to watch. Bio.

Jericho Ad H, McL V, SA ★★ Careful fruit selection and skilled winemaking produce modern, tasty wines, esp FIANO, GRENACHE, SHIRAZ, TEMPRANILLO.

John Duval Wines Bar V, SA ★★★ John D (former maker of PENFOLDS Grange) makes *delicious Rhône reds* of great intensity, character.

Kalleske ★★ Old family farm at Greenock, nw corner of BAR V, makes rather special single-vyd SHIRAZ among many other intensely flavoured things. Bio/organic.

Planted 1843, Langmeil Freedom vyd (Bar V) probably world's oldest Shiraz.

Katnook Estate Coonw, SA ★★ (Odyssey CAB SAUV) 14' 15' 18' 19 Pricey icons Odyssey, Prodigy SHIRAZ. Concentrated fruit, slathered in oak.

Kilikanoon Clare V, SA ★★★ RIES, SHIRAZ excellent performers. Luscious, generous, beautifully made. Owned by Chinese investment group (2017).

King Valley Vic Altitude range 155–860m (509–2821ft) has big impact stylistically. Noted for Italian varieties. Quality from BROWN BROTHERS, Chrismont, DAL ZOTTO, PIZZINI (esp).

Kirrihill Clare V, SA ★★ Makes v.gd CAB SAUV, SHIRAZ, RIES, often at excellent prices.

Knappstein Wines Clare V, SA ★ Reliable RIES, SHIRAZ, CAB SAUV; gd value.

Kooyong Mor P, Vic ★★★ PINOT N, *excellent Chard* of harmony, structure. PINOT GR of charm. High-quality single-vyd wines.

Lake Breeze Lang C, SA ★★★ Succulently smooth, gutsy, value CAB SAUV, SHIRAZ; few producers do mid-level wines so consistently well.

Lake's Folly Hun V, NSW ★★ (CHARD) 17' 18' 19 (CAB SAUV) 16' 17' 18' 19 Founded by surgeon Max Lake, pioneer of HUN V Cab Sauv. Chard often better than Cab blend. Idiosyncratic.

Lambert, Luke Yarra V, Vic ★ Off-beat producer of variable but at times v.gd (cool-climate, mostly) NEBBIOLO, PINOT N, SHIRAZ.

Langmeil Bar V, SA ★★ Holder of some of world's oldest SHIRAZ vines (planted mid-1800s), plus other old vyds, for full-throttle CAB SAUV, GRENACHE, Shiraz.

Larry Cherubino Wines Fra R, WA ★★★ Intense RIES, SAUV BL, *spicy Shiraz*, polished CAB SAUV. Ambitious label and justifiably so.

LAS Vino Marg R, WA ★★★ Wine in his blood. Nic Peterkin is grandson of MARG R pioneers late Kevin and Diana CULLEN and son of Pierro founder. Now he's a trailblazer, experimenter yet fastidious producer of outstanding CHARD, CHENIN BL, CAB SAUV. Remember the name.

Leeuwin Estate Marg R, WA ★★★★ (CHARD) 15' 16' 17' 18' 19 Iconic producer. All about the age-worthy Art Series Chard. SAUV BL, RIES less brilliant. *Cab Sauv* v.gd.

Leo Buring Bar V, SA ★★ 16' 17' 18' 20' 21 Part of TWE. Exclusively RIES; Leonay top label, *ages superbly*. Doesn't get much love but nothing wrong with wine quality.

Lethbridge Vic ★★★ Small, stylish producer of CHARD, RIES, SHIRAZ, PINOT N. Forever experimenting. Cool climate, but wines are meaty, substantial.

Limestone Coast Zone SA Important zone, incl Bordertown, COONW, Mt Benson, Mt Gambier, PADTHAWAY, Robe, WRATTONBULLY.

Lindeman's Owned by TWE. Low-price Bin range now focus, far cry from glory days.

Macedon and Sunbury Vic Adjacent regions: Macedon high elevation and cold, Sunbury nr Melbourne airport. Quality from BINDI, CRAIGLEE, CURLY FLAT, EPIS, PLACE OF CHANGING WINDS.

Mac Forbes Yarra V, Vic ★★★ Mover and shaker of YARRA V. Myriad (in both number, styles) single-vyd releases, mainly PINOT N, CHARD, RIES. Wines more about structure than brightness, unusual in Oz.

McHenry Hohnen Marg R, WA ★★ Solid producer of MARG R CHARD. Tends to focus more on breadth of flavour than mere depth.

McLaren Vale SA Beloved maritime region on s outskirts of Adelaide. Big-flavoured reds in general, but BRASH HIGGINS, CHAPEL HILL, CLARENDON HILLS, CORIOLE, D'ARENBERG, GEMTREE, INKWELL, JERICHO, Marius, PAXTON, SAMUEL'S GORGE, SC PANNELL, WIRRA WIRRA, YANGARRA can show elegance too. SHIRAZ the hero, but old-vine, dry-grown GRENACHE often outshines it.

McWilliam's SE Aus ★★★ Hanwood for value, *Mount Pleasant* for quality.

Main Ridge Estate Mor P, Vic ★★★ Pioneer of region, new owner (2015) respecting its history. Full-bodied, rich CHARD and distinctive PINOT N. Small quantities.

Majella Coonw, SA ★★ As reliable as the day is long. Opulent CAB SAUV, SHIRAZ. Essence of modern COONW.

Margaret River WA Maritime region s of Perth. Powerful CHARD, structured CAB SAUV. CAPE MENTELLE, CORYMBIA, CULLEN, DEEP WOODS ESTATE, DEVIL'S LAIR, FLAMETREE, FRASER GALLOP, HUTTON WINES, LAS VINO, LEEUWIN ESTATE, MOSS WOOD, PIERRO, TRIPE ISCARIOT, VASSE FELIX, VOYAGER ESTATE, WOODLANDS and more. Premium location of bushland, beaches and top surfing.

Mayford NE Vic, Vic ★★★ Tiny vyd in hidden valley. Put ALPINE v region on map. CHARD, SHIRAZ, silken TEMPRANILLO.

Meerea Park Hun V, NSW ★★ Brothers Garth and Rhys Eather create age-worthy SEM, SHIRAZ often as single-vyd expressions.

Mike Press Wines Ad H, SA ★★ Tiny production, tiny pricing. CAB SAUV, SHIRAZ. Crowd favourite of bargain-hunters.

Mitchelton Goulburn V, Vic ★★ Stalwart producer of CAB SAUV, SHIRAZ, RIES, plus speciality of *Marsanne*, ROUSSANNE. Top spot to visit; fancy new hotel set among those fabulous river red gums.

Montalto Mor P, Vic ★★★ A must-try of MOR P. Single-vyd PINOT N, CHARD releases can be eyebrow-raisingly gd.

Moorilla Estate Tas ★★ Pioneer nr Hobart on Derwent River; gd CHARD, RIES, PINOT N; v.gd restaurant, extraordinary art gallery. Owner also of nearby Dom A.

Moorooduc Estate Mor P, Vic ★★★ Long-term producer of complex CHARD, PINOT N.

Moppity Vineyards Hilltops, NSW ★ Affable SHIRAZ, CAB SAUV (Hilltops). Elegant CHARD (TUMBARUMBA). Best known for its value offerings.

Mornington Peninsula Vic Coastal area se of Melbourne. Cool climate. CHARD, PINOTS GR/N. Wine/surf/beach/food playground. ELDRIDGE ESTATE, GARAGISTE, KOOYONG, MAIN RIDGE ESTATE, MONTALTO, MOOROODUC ESTATE, PARINGA ESTATE, STONIER, TEN MINUTES BY TRACTOR, WILLOW CREEK, YABBY LAKE and more.

Morris NE Vic ★★★ RUTH producer of Oz's (the world's?) greatest dessert *Muscats*, TOPAQUES. Owned by CASELLA.

Moss Wood Marg R, WA ★★★ (CAB SAUV) 13' 14' 15' 16'17' 18 MARG R's most opulent (r) wines. CHARD, SEM, super-smooth *Cab Sauv*. Oak- and fruit-rich.

Mount Horrocks Clare V, SA ★★ Racy dry RIES, CAB SAUV, SHIRAZ, SEM in gd shape.

Mount Langi Ghiran Gra, Vic ★★★ (SHIRAZ) 13' 14' 15' 17' 18' 19 Peppery, *Rhône-like Shiraz*. Excellent Cliff Edge Shiraz. Estate-grown on a pretty special patch of dirt.

Chardonnay: cool rules

Burgundy has bragging rights for CHARD, but Australia can lay claim to crafting the finest outside it, at a fraction of Grand Cru prices. A bold statement? Not at all. Pick cooler climates for the exceptional: AD H, BEECH, MARG R, MOR P, TAS, YARRA V. The focus is on single-vyd or -block wines, thanks to a combination of better clones, better vyd management and more thoughtful winemaking, which has seen the rise of Chard to top-tier status. High-end producers: BINDI, CULLEN WINES, GIACONDA, GIANT STEPS, KOOYONG, OAKRIDGE, PENFOLDS Yattarna, SHAW & SMITH, TOLPUDDLE VYD, VASSE FELIX, XANADU WINES. Knock yourself out.

Mount Majura Can, NSW ★★ Leading TEMPRANILLO producer. CHARD, RIES, SHIRAZ all gd. Reds sturdy, spicy.

Mount Mary Yarra V, Vic ★★★★ (PINOT N) 16' 17' 18' 19' 20 (Quintet) 15' 16' 17' 18' 19' 20 YARRA V pioneer that goes from strength to strength. Best known for B'x blend Quintet, yet CHARD and Pinot N up there. Rhône varieties latest additions all made by Sam Middleton, grandson of founder, the late Dr. John Middleton.

Mount Pleasant Hun V, NSW ★★★★ Sold to Sydney-based investment company by MCWILLIAM'S (2021). Note single-vyd SEMS (esp *Lovedale*), SHIRAZ too. Historic vyds, great wine.

WA truffles (often labelled "Perigord" abroad) are perfect match for WA Cab Sauv.

Mudgee NSW Region nw of Sydney. Earthy reds, fine SEM, textural RIES, full CHARD. Gd quality but needs a hero.

Ngeringa Ad H, SA ★★ Perfumed NEBBIOLO and PINOT N. Rhôney SHIRAZ. Savoury rosé. Bio.

Oakridge Yarra V, Vic ★★★★ Leading producer of CHARD in Oz, more recently noteworthy producer of PINOT N, SHIRAZ. Multiple single-vyd releases.

Ochota Barrels Bar V, SA ★★★ Brilliant producer of old-vine GRENACHE, SHIRAZ from MCL V, BAR V.

O'Leary Walker Clare V, SA ★★ Low profile but excellent quality. CLARE V RIES, CAB SAUV standout. MCL V SHIRAZ oak-heavy but gd.

Orange NSW Cool-climate, high-elevation region. Lively SHIRAZ (when ripe), but best suited to (intense) aromatic whites and CHARD.

Osicka, Paul Hea, Vic ★★★ Vines dating back to 50s. Both character/flavour writ large. Small-scale, low-profile, high-impact and quality CAB SAUV, SHIRAZ.

Padthaway SA CAB SAUV, SHIRAZ, CHARD v.gd. Rarely mentioned these days but important region. Soil salinity on-going issue.

Pannell, SC McL V, SA ★★★ Trailblazer of the Oz wine scene. Intuitive winemaking, vyds of note. Midas touch with GRENACHE, SHIRAZ plus AGLIANICO. Dabbles with Spanish varieties. Always delicious drinking.

Paringa Estate Mor P, Vic ★★★ Complex PINOT N, SHIRAZ with impressive structure, defintion. CHARD too. Fleshy, fruity, bolshie, flashy styles.

Parker Coonawarra Estate Coonw, SA ★★ 1988 was 1st vintage. In gd yrs it produces full-bodied, age-worthy, tannic CAB SAUV of authority, distinction.

Passing Clouds Bendigo, Vic ★★ Down-to-earth folk producing reliable, robust reds esp SHIRAZ and top CAB SAUV often a show-stopper. Macedon CHARD has verve.

Paxton McL V, SA ★ Prominent organic/bio grower/producer: GRENACHE, SHIRAZ.

Pemberton WA Region between MARG R and GREAT SOUTHERN; initial enthusiasm for PINOT N replaced by CHARD, RIES, SHIRAZ.

Penfolds ★★★★ (Grange) 91' 96' 06' 08' 10' 12' 14' 15' 16' 17 (CAB SAUV Bin 707) 04' 08' 12' 14' 15' 18' 19 and of course *St Henri*, "simple" SHIRAZ. Originally Adelaide, now SA, Champagne and California. Oz's best warm-climate red-wine company. Superb *Yattarna* CHARD, Bin Chard now right up there with reds.

Petaluma Ad H, SA ★★ (RIES) 16' 17' 18' 20 (SHIRAZ AD H) 15' 17' 18 CAB SAUV COONW reliable. Has its moments but never quite been same since ex-owner/creator Brian Croser left the building.

Peter Lehmann Wines Bar V, SA ★★ Well-priced wines incl easy RIES. Luxurious, structured Stonewell SHIRAZ among many others (r/w). Now part of CASELLA (Yellow Tail).

Pewsey Vale Eden V, SA ★★ RIES v.gd: standard and (aged-release) The Contours, grown on lovely tiered vyd.

Pierro Marg R, WA ★★★ (CHARD) 14' 18' 19 Producer of full-bodied, often oaky B'x blend, tangy SEM/SAUV BL and full-throttle, utterly convincing Chard.

Pipers Brook Tas ★★ (RIES) 17' 18' 19' 20' 21 Cool-area pioneer; gd Ries, PINOT N variable, *restrained Chard and sparkling* from Tamar V. Second label: Ninth Island. Owned by Belgian Kreglinger family.

Pizzini King V, Vic ★★ (SANGIOVESE) 17' 18 Leader of Italian varieties in Oz, esp NEBBIOLO, Sangiovese (stepped up a gear). Dominant KING V producer.

Place of Changing Winds Mac, Vic ★★★ CHARD, PINOT N of great power, complexity, grown on high-density vyds. One of most exciting new wineries of recent yrs.

Pooley Tas ★★★ Historic, impressive property c.1832 in Coal River V, some of TAS's finest RIES, perfumed PINOT N, slinky CHARD. A star of the state.

Primo Estate SA ★★ Joe Grilli's many successes incl rich MCL V SHIRAZ, tangy COLOMBARD, potent Joseph CAB SAUV/MERLOT, (exceptionally) complex sparkling Shiraz, NEBBIOLO and more.

Provenance Geelong, Vic ★★★ Exciting CHARD from GEELONG and surrounding regions. PINOT N less reliable but can be super.

Punch Yarra V, Vic ★★ Lance family ran Diamond Valley for decades. Retained close-planted PINOT N vyd when they sold: can grow decisive, age-worthy wines.

Pyrenees Vic Central Vic region making rich, often minty reds. Blue Pyrenees, Dog Rock, Mitchell Harris, Summerfield, TALTARNI leading players, though it's also a happy hunting ground for assorted small producers.

Ravensworth Can, NSW ★★ Suddenly in hot demand for various wine experiments. SANGIOVESE best known but buzz over skin-contact whites and GAMAY Noir.

Rieslingfreak Clare V, SA ★★ Apt name as John Hughes only makes RIES. CLARE V and EDEN V ranges. Winemaker wife Belinda (ex-GRANT BURGE) joined 2021.

Riverina NSW Large-volume irrigated zone centred on Griffith.

Vespolina grape (r) arrives in Oz: 2 rows, 300 vines, Dalbosco family, Vic. Order now.

Robert Oatley Wines Mudgee, NSW ★★ Ambitious venture of ROSEMOUNT ESTATE creator Robert Oatley. Quality/price ratio usually well aligned.

Rochford Yarra V, Vic ★★ Main outdoor entertainment venue in YARRA V makes complex CHARD, PINOT N of note.

Rockford Bar V, SA ★★★ Sourced from various old, low-yielding vyds; reds best; iconic Basket Press SHIRAZ and noted *sparkling Black Shiraz*.

Rodda, A Beech, Vic ★★ Bright CHARD from est vyds; *Tempranillo* grown at high altitude can be a beauty.

Ruggabellus Bar V, SA ★★★ Causing a stir. More savoury, often lighter-framed version of BAR V. Old oak, min sulphur, wild yeast, whole bunches/stems. Blends of CINSAULT/GRENACHE/MATARO/SHIRAZ.

Rutherglen and Glenrowan Vic Two of four regions in warm NE Vic zone, justly famous for sturdy reds and esp magnificent fortifieds. ALL SAINTS, CAMPBELLS, ELDORADO ROAD, SCION, SIMAO & CO, TAMINICK CELLARS.

St Hallett Bar V, SA ★★ (Old Block) 15' 16' 17 Old Block SHIRAZ the star; rest of range is smooth, sound, stylish. ACCOLADE-owned.

Saltram Bar V, SA ★★ Value Mamre Brook (SHIRAZ, CAB SAUV) and (rarely sighted) No 1 Shiraz are leaders. Main claim to fame is ubiquitous Pepperjack Shiraz.

Samuel's Gorge McL V, SA ★★ Justin McNamee makes (at times) stunning GRENACHE, SHIRAZ, TEMPRANILLO of character and place.

Savaterre Beech, Vic ★★★ (CHARD) 17' 18' 19 (PINOT N) 16' 18 A v.gd producer of full-bodied Chard, meaty Pinot N, close-planted SAGRANTINO, SHIRAZ.

Schmolzer & Brown Beech, Vic ★★ CHARD, PINOT N and rosé of intense, spice-drenched interest. Textural RIES of note. One of highest vyds in BEECH.

Scion Ruth, Vic ★★ Fresh, vibrant approach to region's stalwart SHIRAZ and Durif.

Sentio Beech, Vic ★★★ Picks eyes out of various cool-climate regions to produce compelling CHARD, PINOT N, SHIRAZ.

Seppelt Gra, Vic ★★★ (St Peter's SHIRAZ) 14' 16' 17' 18' 19 Historic name owned by TWE. Impressive CHARD, RIES, (esp) peppery Shiraz.

Seppeltsfield Bar V, SA ★★★ National Trust Heritage Winery bought by Warren Randall (2013). Fortified wine stocks back to 1878.

Serrat Yarra V, Vic ★★★ Micro-vyd of noted winemaker Tom Carson (YABBY LAKE) and wife Nadège. Complex, powerful, precise SHIRAZ/VIOGNIER, PINOT N, CHARD.

Seville Estate Yarra V, Vic ★★★ (SHIRAZ) 15' 16' **17'** 18' 19 Excellent CHARD, spicy Shiraz, structured PINOT N. YARRA v pioneer with fresh lease on life.

Shaw & Smith Ad H, SA ★★★ Savvy outfit. Crisp *harmonious* SAUV BL, complex CHARD and, surpassing both, *Shiraz*. PINOT N v.gd.

Simao & Co Ruth, Vic ★★ Young Simon Killeen makes scrumptious TEMPRANILLO, UGNI BL, SHIRAZ and more. Personality+.

Sorrenberg Beech, Vic ★★★★ No fuss but highest quality. CHARD, SAUV BL/SEM, (Australia's best) GAMAY, B'x blend. Ultimate "in the know" winery of Oz.

Southern NSW Zone NSW Incl CAN, Gundagai, Hilltops, TUMBARUMBA. Savoury SHIRAZ, pure RIES, lengthy CHARD.

Spencer, Nick Can, NSW ★★ Former Eden Road winemaker. CHARD and red blend (SHIRAZ/TEMPRANILLO/TOURIGA N/CAB SAUV) of particular interest.

Spinifex Bar V, SA ★★★ Reds dominate, attention to detail in winemaking. Spicy GRENACHE, rich SHIRAZ and MATARO in the mix. Single vyds of note.

Stefano Lubiana S Tas ★★★ Beautiful bio vyds on banks of Derwent River, nr Hobart. Excellent sparkling, CHARD, MERLOT, PINOT N and, more recently, SHIRAZ. As homely as it is ambitious. Italian and Austrian varieties making a statement.

Stella Bella Marg R, WA ★★★ Solid performer across range of styles, prices. Top CHARD and CAB SAUV called Luminosa often illuminating.

Stoney Rise Tas ★★ Joe Holyman used to be a world-class wicketkeeper; he's an even better winemaker/grower, particularly with CHARD, PINOT N.

Stonier Wines Mor P, Vic ★★★ (CHARD) 18' 19' 20 (PINOT N) 15' **17'** 18' 19 Go-to producer to gauge best of MOR P. Entry level v.gd, but single-vyd and Res shine.

Sunbury Vic *See* MACEDON and SUNBURY.

Sutton Grange Bendigo, Vic ★★ (AGLIANICO) **17'** 18 (SYRAH) 17' 18' 19 Noteworthy Syrah, Rosé rocks; Aglianico, SANGIOVESE and FIANO more than a side-show.

Swan Valley WA Birthplace of wine in the w, 20 mins n of Perth. Hot climate makes strong, low acid wines. FADER VYDS leads way.

Swinney Fra R, WA ★★★ Grape-grower for decades; initial releases under own steam outstanding. RIES, GRENACHE, SHIRAZ all shine. Top-tier Farvie label excellent.

Tahbilk Goulburn V, Vic ★★★ (MARSANNE) 15' 17' 19' **20** (SHIRAZ) 13' 15' 16 Historic Purbrick family estate: long-ageing reds, also some of Oz's best old-vine *Marsanne*. Res CAB SAUV can be v.gd. Rare 1860 Vines Shiraz. Rustic styles.

Taltarni Pyrenees, Vic ★★ CAB SAUV, SHIRAZ in gd shape. Long-haul wines but jackhammer no longer required to remove tannin from your gums.

Taminick Cellars Glenrowan, Vic ★★ Booth family has farmed this tough, dry patch since 1914. Fair amount of character accumulated along way.

Tapanappa SA ★★★ WRATTONBULLY collaboration between Brian Croser, Bollinger and J-M Cazes of Pauillac (*see* France). Splendid CAB SAUV blend, MERLOT, SHIRAZ, CHARD. Surprising *Pinot N* from Fleurieu Peninsula.

Tar & Roses Hea, Vic ★★ SANGIOVESE, SHIRAZ, TEMPRANILLO of impeccable polish, presentation. Modern success story; 2017 death of co-founder Don Lewis a great loss but quality remains strong.

TarraWarra Estate Yarra V, Vic ★★ (Res CHARD) 16' **17'** 18' 19' (Res PINOT N) **17'** 18' 19 Always flavoursome yet more restraint in recent yrs with Chard and Pinot N. Dabbles in Rhône and Italian varieties: BARBERA a delight. A visit essential thanks to TarraWarra Museum of Art: Australian works, 50s onwards.

AUSTRALIA

Tasmania Cold island region with hot reputation. Outstanding sparkling, PINOT N, RIES; v.gd CHARD, PINOT GR, SAUV BL.

Taylors Wines Clare V, SA ★ Large-scale production led by RIES, CAB SAUV, SHIRAZ. Exports under Wakefield Wines brand.

Ten Minutes by Tractor Mor P, Vic ★★★ Top-notch wines, top-notch winery restaurant. Focus on single-vyd CHARD, PINOT N.

Teusner Bar V, SA ★★ Old vines, clever winemaking, pure fruit flavours. Leads a BAR V trend towards "more wood, no good".

Thomas, Andrew Hun V, NSW ★★ Old-vine SEM; silken SHIRAZ. Reds particularly gutsy in HUN V context.

Tolpuddle Tas ★★★ SHAW & SMITH own this outstanding 1988-planted vyd in Coal River V. Scintillating PINOT N, CHARD in lean, lengthy style.

Topaque Vic Replacement name for iconic RUTH sticky "Tokay", directive of EU; a decade later it's still hard to find anyone who likes the name.

Torbreck Bar V, SA ★★★ Dedicated to (often old-vine) Rhône varieties led by GRENACHE, SHIRAZ. Ultimate expression of rich, sweet, high-alc style.

Torzi Matthews Eden V, SA ★★★ Aromatic, stylish, big-hearted SHIRAZ. Value RIES, SANGIOVESE. Incredible consistency yr-on-yr.

Traviarti Beech, Vic ★★ Made name with TEMPRANILLO, but NEBBIOLO fast becoming force to be reckoned with.

Tripe Iscariot Marg R, WA ★★ Kooky name, compelling wines. Highlights greatness of CHENIN B yet excels in region's finest grapes, CHARD and CAB SAUV.

Tumbarumba NSW Cool-climate region tucked into Australian Alps. Sites 500–800m (1640–2625ft); CHARD the out-and-out star; PINOT N generally to avoid.

Turkey Flat Bar V, SA ★★★ Top producer of complex rosé, GRENACHE, SHIRAZ from 150-yr-old vyd. Controlled alc/oak. New single-vyd wines. Old but modern.

TWE (Treasury Wine Estates) Aussie wine behemoth. COLDSTREAM HILLS, DEVIL'S LAIR, LINDEMAN'S, PENFOLDS, ROSEMOUNT, SALTRAM, WOLF BLASS, WYNNS COONAWARRA ESTATE among them.

Two Hands Bar V, SA ★★★ Big reds and many of them. They've turned volume down a fraction lately; glory of fruit seems all the clearer.

Tyrrell's Hun V, NSW ★★★★ (SEM) 15' 16' 17' 18' 19' (Vat 47 CHARD) 15' 16' 17' 18' 19 Oz's greatest maker of Sem, Vat 1 now joined with series of individual vyd or subregional wines. *Vat 47*, Oz's 1st Chard, continues to defy climatic odds. Outstanding old-vine 4 Acres SHIRAZ, Vat 9 Shiraz. One of the true greats.

Vasse Felix Marg R, WA ★★★★ (CHARD) 17' 18' 19' 20 (CAB SAUV) 15' 16' 17 With CULLEN, pioneer of MARG R. Elegant Cab Sauv for mid-weight balance. Complex, funkified Chard. Estate grown-only now.

Voyager Estate Marg R, WA ★★★ Viticulture allowing range of styles/prices from value Girt by Sea to top-tier MJW CHARD, CAB SAUV.

Wanderer, The Yarra V, Vic ★★★ Upper YARRA V producer of exceptionally fine-boned PINOT N with equally fine CHARD.

Australia's new fine wines

Look for new varieties and blends, even from traditional producers. Think HENSCHKE: The Rose Grower NEBBIOLO, smells of roses. YANGARRA ESTATE: Blanc blend incl PICPOUL, BOURBOULENC, raised in 675-litre ceramic eggs and used barriques. LAS VINO (MARG R): thrilling wines incl Portuguese reds in The Pirate Blend. CRITTENDEN ESTATE: Cri de Coeur Sous Voile SAVAGNIN. STEFANO LUBIANA (TAS): fragrant BLAUFRÄNKISCH. HAHNDORF HILLS: GRÜNER V in many styles incl textural Gru. MAYFORD champions TEMPRANILLO, while SC PANNELL is master-blaster of Italian reds: AGLIANICO stunning.

Wantirna Estate Yarra V, Vic ★★★ Flies under radar for a regional pioneer. Worth seeking out. Lovely wines from B'x blend, PINOT N to CHARD, and cute labels via renowned Oz cartoonist and artist Michael Leunig.

Wendouree Clare V, SA ★★★★ The DRC of Australia. Old vines, commitment to quality and history. Beauty in a glass with long-lived reds now veering towards more refined, elegant styles of CAB SAUV, MALBEC, MATARO, SHIRAZ. Demand outstrips supply.

West Cape Howe WA ★ Affordable, flavoursome reds the speciality; gd RIES.

Westend Estate Riverina, NSW ★★ Thriving family producer of *tasty bargains*, esp Private Bin SHIRAZ/Durif. Recent cool-climate additions gd value.

Wild Duck Creek Hea, Vic ★★ Super-concentrated, high-octane reds using SHIRAZ (mostly), CAB SAUV, MALBEC. Vigour, freshness somehow kept intact.

Willow Creek Mor P, Vic ★★ Impressive producer of CHARD, PINOT N in particular. Power, poise.

Wirra Wirra McL V, SA ★★ High-quality, concentrated wines in flashy livery. The Angelus CAB SAUV named Dead Ringer outside Australia.

Wolf Blass Bar V, SA ★★ Owned by TWE. Not the shouty player it once was, but still churns through an enormous volume of clean, inoffensive wines.

Woodlands Marg R, WA ★★★ With 7 ha of 40-yr-old+ CAB SAUV among top vyds in region, plus younger but v.gd plantings of other B'x reds. Brooding impact.

Wrattonbully SA Important grape-growing region in LIMESTONE COAST ZONE; profile lifted by activity of TAPANAPPA, Terre à Terre, Peppertree.

Wynns Coonawarra Estate Coonw, SA ★★★ (SHIRAZ) 15' 16' 17' 18' 19 (CAB SAUV) 14' 15' 16' 17' 18' 19 Iconic and biggest producer in region; excellent vyds. TWE-owned. Fair CHARD, RIES, but reds star with Cab Sauv, Shiraz, esp Black Label and the pinnacle, *John Riddoch*, plus single-vyd.

Xanadu Wines Marg R, WA ★★★ Thoughtful winemaking across range with thrilling single-vyd and Res CAB SAUV and super-fine CHARD. Est 1977.

Yabby Lake Mor P, Vic ★★★ Made name with estate CHARD, PINOT N, boosted with single-site releases, now spice-shot SHIRAZ adds yet more to reputation.

Yalumba Bar V, SA, SA ★★★ 170-yrs+, family-owned. *Full spectrum of high-quality wines*, from budget to elite single vyd (eg. *The Caley*). Entry-level Y Series v.gd value.

Yangarra Estate McL V, SA ★★★★ Conventional in part, inventive in others. Whatever it takes to make great wine. Full box and dice here, across most price points. Varietal GRENACHE, SHIRAZ emphatically gd. Bio.

Yarra Valley Vic Thriving area just ne of Melbourne. Emphasis on CHARD, PINOT N, SHIRAZ, sparkling. Understated, elegant CAB SAUV. COLDSTREAM HILLS, DE BORTOLI, DOM CHANDON, GEMBROOK HILL, GIANT STEPS, HODDLES CREEK ESTATE, LUKE LAMBERT, MAC FORBES, MOUNT MARY, OAKRIDGE, PUNCH, ROCHFORD, SERRAT, SEVILLE ESTATE, TARRAWARRA, WANTIRNA ESTATE, YARRA YERING, YERINGBERG, YERING STATION is a formidable lineup.

Yarra Yering Yarra V, Vic ★★★★ (Dry Reds) 15' 16' 17' 18' **19'** 20 One-of-a-kind YARRA V pioneer. Powerful PINOT N; deep, refined CAB SAUV (Dry Red No 1); SHIRAZ (Dry Red No 2); CHARD up there. Absolute upper echelon (r/w).

Yellow Tail NSW *See* CASELLA.

Yeringberg Yarra V, Vic ★★★★ (MARSANNE/ROUSSANNE) 15' **16'** 17' 19 (CAB SAUV) 13' 14' 15' 16' 17' 18' 19 Historic estate still in hands of founding (1862) Swiss family, the de Purys. Small quantities of v. high-quality CHARD, Marsanne, Roussanne, Cab Sauv, PINOT N.

Yering Station / Yarrabank Yarra V, Vic ★★ On site of VIC's 1st vyd; replanted after 80-yr gap. Snazzy table wines (Res CHARD, PINOT N, SHIRAZ, VIOGNIER); Yarrabank (sparkling in joint venture with Champagne Devaux).

New Zealand

Abbreviations used
in the text:

Auck	Auckland
B of P	Bay of Plenty
Cant	Canterbury
Gis	Gisborne
Hawk	Hawke's Bay
Hend	Henderson
Marl	Marlborough
Mart	Martinborough
Nel	Nelson
N/C Ot	North/Central Otago
Waih	Waiheke Island
Waip	Waipara Valley
Wair	Wairarapa

Northland

Auckland
Auckland · Waiheke Island
Waikato
Bay of Plenty
Gisborne
Hawke's Bay
Wairarapa (incl Martinborough)
Nelson
Marlborough
Wellington
Blenheim
Waipara Valley
Canterbury
Christchurch

Tasman Sea

North Otago (incl Waitaki Valley)
Central Otago

Pacific Ocean

Dunedin

There is mounting competition here between artisan products and mass-produced, with large wineries getting larger and smaller ones struggling to stay in business. Those who can machine-harvest Sauvignon Blanc at night to make quantities of the pungent wines the world wants more of, while by day picking and hand-sorting more prestigious Pinot Noir and Chardonnay, manage to have a foot in both camps. 15- to 25-year-old vines are increasingly the norm, commitment to organic viticulture is expanding gradually, and many wine-growers have now been on the same site for decades. The good news is that the quality and diversity of NZ wine has never been greater.

Recent vintages

2021 Small, frost-affected crop in Marl: vigorous, flavour-packed.
    Hawk warmer and drier than usual, esp gd Chard, Syrah.
2020 Marl: crisp, intense. Hawk: outstanding reds, Chard.
2019 Weighty, textured Marl Sauv Bl. Hawk Chard and reds esp promising.
2018 Hottest-ever summer. Ripe, less herbaceous Marl Sauv Bl.

Akarua C Ot ★★★ Bought by Edmond de Rothschild Heritage, France, 2021. PINOT N: classy, complex Bannockburn, from mature vines, ages well. Top label The Siren: sturdy, lush. Lively fizz, incl Brut NV. Full-bodied, dry PINOT GR.

Allan Scott Marl ★★ Family-owned winery, controversial involvement in beer production (Moa). Fresh, easy-drinking. Top range with black labels.

Alpha Domus Hawk ★★→★★★ Family winery, Dutch background, aviation focus (nr airfield) in Bridge Pa Triangle. Mostly MERLOT-based B'x-style reds; classy CAB SAUV The Aviator. Fleshy VIOGNIER; The Skybolt CHARD: gd value.

Amisfield C Ot ★★★ Founded 1988, popular cellar door, bistro. Classy RIES (dr/medium); impressive SAUV BL. Rich PINOT N (RKV Res is Rolls-Royce model). Lake Hayes: drink young.

Astrolabe Marl ★★→★★★ Impressive range from vastly experienced Simon Waghorn, named after French explorer Dumont d'Urville's mighty ship. Punchy SAUV BL; refined CHARD; rich CHENIN BL. Scented PINOT N. Durvillea v.gd value.

Ata Rangi Mart ★★★→★★★★ Much-respected family affair, est 1980. PINOT N 14′ 15′ 16′ 17 18 19′: NZ classic. Crimson Pinot N: younger vines. Refined Craighall CHARD; dry, oak-aged Lismore PINOT GR; Kahu Botrytis RIES.

Auckland Largest city (n, warm, cloudy) in NZ; 1% vyd area (not expanding), but 14% of producers (incl head offices of big firms). Nearby districts with clay soils: W Auckland, incl Henderson, Kumeu, Huapai, Waimauku (long est); newer (since 80s): Matakana, Clevedon, WAIH (island vyds, popular with tourists). Savoury B'x blends in dry seasons, bold SYRAH rivals HAWK for quality; rich, ripe, underrated CHARD.

Auntsfield Marl ★★→★★★ Classy, characterful wines from site of region's 1st (1873) vyd (replanted 1999). See statue of founder David Herd, Scot who made sweet red wine, at Blenheim Airport. Weighty, oak-aged SAUV BL; citrus CHARD (esp deep Cob Cottage), powerful PINOT N.

Awatere Valley Marl Key subregion (pronounced "Awa-terry"), with silt loam soils, few wineries, but huge vyd area (more than HAWK). Major component in many regional blends of MARL SAUV BL. YEALANDS is key producer. Slightly cooler, drier, windier, less fertile than WAIRAU V, with racy ("tomato stalk") SAUV BL; vibrant RIES, PINOT GR; flavour-packed GRÜNER V; herbal PINOT N.

Pinot Gr often has a lemongrass note in NZ, a bit like a Sauv Bl crossover.

Babich Hend ★★→★★★ NZ's oldest family-owned winery, est 1916. Age-worthy single-vyd Irongate from GIMBLETT GRAVELS, refined CHARD, B'x-like CAB SAUV/ MERLOT/CAB FR. Graceful Winemakers' Res Marl PINOT N. Top red: The Patriarch (B'x-style, MALBEC-influenced).

Blackenbrook Nel ★★ Small, sloping, coastal vyd, Swiss-born owners. Impressive aromatic whites: Alsace-style GEWURZ, PINOT GR. Also MONTEPULCIANO; PINOT N.

Black Estate Cant ★★ Small organic WAIP producer with popular restaurant/cellar door and three sites; 1st vines 1994. Powerful Home CHARD, crisp rosé, refined CAB FR, perfumed Home PINOT N.

Blank Canvas Marl ★★★ Owned by husband-and-wife team Matt Thomson (ex-SAINT CLAIR) and consultant Sophie Parker-Thomson MW. Refined SAUV BL incl tropical single-vyd Holdaway. Seamless Reed CHARD; fragrant Escaroth PINOT N.

Borthwick Wair ★★★★ Stony vyd at Gladstone; Paddy Borthwick brand. Pungent SAUV BL; biscuity CHARD; citrus RIES; charming Pinot Rosé. Fragrant PINOT N.

Brancott Estate Marl ★→★★★ Owned by PERNOD RICARD NZ. Top wines: Letter Series (esp fleshy B SAUV BL, smoky O CHARD). Huge-selling crisp Sauv Bl. Living Land: organic. Flight: plain, low alc. Top-value bottle-fermented Brut Cuvée. Bargain-priced cherry-and-spice Res Awatere PINOT N.

Brightwater Nel ★★ Impressive whites from vyd on edge of river terrace, high-flavoured, low-alc Natural Light SAUV BL, supple PINOT N. Top: Lord Rutherford (deep Sauv Bl from mature vines, tight Chard).

Brookfields Hawk ★★→★★★ Smallish, with atmospheric winery from 1937. Typically great value. Full-flavoured Bergman CHARD, sturdy VIOGNIER. Firm Ohiti CAB SAUV, dense Sun-Dried MALBEC; deep Back Block SYRAH. Top: fleshy Marshall Bank Chard; rich Hillside Syrah; powerful Res Vintage Cab/MERLOT.

Burn Cottage C Ot ★★★ Organic, vyds at Pisa and Bannockburn. Moonlight Race PINOT N, gd young. Refined Burn Cottage Vyd Pinot N. Scented RIES/GRÜNER V.

Canterbury NZ's 4th-largest wine region, with 9% of all producers; most vyds in

> **The 85% rule**
> If an NZ label states that a wine is from a single grape variety, vintage or area of origin, it must be at least 85% that. Same in Europe. SAUV BL grapes, grown in NEL and N CANT, often trucked over hills to MARL. Yes, it happens in Europe too.

relatively warm WAIP district (increasingly called N Cant). Greatest success with aromatic RIES, savoury PINOT N. Strength in stylish CHARD, Alsace-style PINOT GR. SAUV BL heavily planted, but often minor component in other regions' wines.

Carrick C Ot ★★★ Bannockburn winery on n-facing slopes with gravel and sand glistening with silica. Organic focus. Classy RIES (gently sw Josephine); elegant CHARD, esp EBM. Powerful PINOT N, incl drink-young Unravelled.

Catalina Sounds Marl ★★→★★★ Australian-owned, gd regional blends, top estate-grown wines from large Sound of White vyd in upper Waihopai V.

Central Otago High-altitude, dry inland region in s of S Island, with 5% of total vyd, 18% of producers. Glacial outwash schist soils promote gd drainage. Sunny, hot days, v. cold nights. Most vines in Cromwell Basin. Crisp RIES, PINOT GR; fast-growing interest in vibrant CHARD; famous PINOT N has drink-young charm; older vines now more complex. Top Pinot N rosé and traditional-method fizz.

Chard Farm C Ot ★★ Pioneer in precipitous gorge, famous for bungee-jumping. Most vyds now in Cromwell Basin. Fleshy PINOT GR. Typically floral PINOT N.

Church Road Hawk ★★→★★★ PERNOD RICARD NZ winery with historic (1896) HAWK roots. Grand Res impressive, esp show-stopping CHARD. McDonald Series top quality, value. Prestige TOM selection. New single-vyd 1 range.

Churton Marl ★★ Elevated Waihopai V site with loess overlying clays, owned by Shropshire lad Sam Weaver and wife Mandy. Intense SAUV BL; refined PINOT N (esp The Abyss: oldest vines, greater depth). Honey-sweet PETIT MANSENG.

Clearview Hawk ★★→★★★ Coastal, shingly vyd at Te Awanga (also grapes from inland). Hedonistic Res CHARD (Beachhead Chard is jnr version); tight Three Rows Chard. Rich Enigma (MERLOT-based); complex Old Olive Block.

Clos Henri Marl ★★→★★★ Organic, est by Henri Bourgeois (Sancerre, France). Weighty SAUV BL from stony soils, one of NZ's best; fragrant PINOT N (on clay). Second label: Bel Echo. Third label: Petit Clos.

Cloudy Bay Marl ★★★ Large-volume, still-classy SAUV BL is NZ's most famous wine. Complex CHARD, savoury PINOT N. Stylish Pelorus NV (sp), esp Rosé. More involvement in C OT for Te Wahi Pinot N. Owned by LVMH.

Constellation New Zealand Auck ★→★★ Large producer owned by Constellation Brands. Strength in US market (KIM CRAWFORD MARL SAUV BL top-selling NZ wine). Strength mainly in solid, moderately priced (esp Sauv Bl) Kim Crawford, Monkey Bay, SELAKS brands.

Cooper's Creek Auck ★★→★★★ Innovative producer, Roman basilica-style hospitality centre, wide range, typically gd value. Top Res; SV (Select Vyd) range mid-tier. NZ's 1ST: ALBARIÑO, ARNEIS, GRÜNER V, MARSANNE.

Craggy Range Hawk ★★★→★★★★ High-profile; acclaimed Terroir restaurant. Large vyds in HAWK, MART. Stylish CHARD, PINOT N; excellent mid-range MERLOT, SYRAH, Te Kahu (B'x r blend) from GIMBLETT GRAVELS. Les Beaux Cailloux: age-worthy Chard (NZ$150). Showstopping Syrah Le Sol 14' 15' 16' 18 19'; sturdy The Quarry (CAB SAUV); dense Sophia (Merlot).

Deep Down Marl ★★→★★★ Organic, single-vyd wines by Clive Dougall (ex-SERESIN). Fragrant CHARD; complex SAUV BL; sulphur-free PINOT N.

Delegat Auck ★★ Large listed family company, est 1947, narrow range of popular, mid-priced wines. Hugely successful OYSTER BAY brand. Instantly likeable MERLOT under Delegat brand. Owns Barossa Valley Estates.

Delta Marl ★★→★★★ Owned by SAINT CLAIR; v.gd-value CHARD, PINOT GR, SAUV BL, PINOT N. Hatters Hill range: greater complexity.

Destiny Bay Waih ★★→★★★ Small, n-facing amphitheatre where expat Americans make pricey, brambly B'x-style reds. Flagship is age-worthy, mostly CAB SAUV Magna Praemia. Lush Mystae is mid-tier. Destinae: earlier drinking.

Deutz Auck ★★★ Champagne house gives name to great-value fizz from MARL by PERNOD RICARD NZ. Brut NV has min 2 yrs on lees. Much-awarded Vintage Blanc de Blancs. Vivacious Rosé NV; outstanding Prestige (disgorged after 3 yrs).

Dog Point Marl ★★★ Organic. Brand named after nearby hill, once home to pack of marauding dogs. Concentrated SAUV BL (Section 94); intense CHARD, fragrant PINOT N, all among region's finest. Larger volume Sauv Bl.

Domaine Thomson C Ot ★★→★★★ Small, organic PINOT N vyds in two hemispheres: Gevrey-Chambertin and Lowburn. Elevated, sloping site, deep gravels. Explorer: plummy, enjoyable young. Surveyor Thomson: complex.

Dry River Mart ★★★ Shingly, free-draining soils. Low-tech winemaking. Reputation for long-lived whites. Alsace-style PINOT GR (NZ's 1st outstanding example and still one of greatest). Classy CHARD, GEWURZ, RIES, sweet whites, PINOT N.

Elephant Hill Hawk ★★→★★★ Stylish winery within a stone's throw of the Pacific Ocean, also grapes from inland. Soils vary from shingle to clay. Powerful, rich wines. Outstanding Res range, incl CHARD, MERLOT blend, SYRAH. Top pair: Airavata Syrah, Hieronymus (r blend).

Escarpment Mart ★★★ Organic, owned by Torbreck (*see* Australia); vyd on alluvial loam over gravel. Citrus CHARD, savoury PINOT N. Top label: Kupe. Single-vyd, old-vine reds esp gd, MART Pinot N is regional blend. Lower tier: The Edge.

Esk Valley Hawk ★★→★★★★ Owned by VILLA MARIA. Impressive quality, value. MERLOT-based blends, classy Winemakers Res CHARD; vibrant Artisanal Collection CHENIN BL. Striking flagship red Heipipi The Terraces.

Felton Road C Ot ★★★★ Celebrated winery on n-facing slopes at Bannockburn, owned by Englishman Nigel Greening, former rock guitarist. Best known for PINOT N, but RIES, CHARD notably classy too. Notable Pinot N Block 3 + 5, both Elms Vyd; poised Bannockburn Pinot N, four-vyd blend.

Forrest Marl ★★ Big success with The Doctors' SAUV BL, low alc (9.5%), lively. Wide range of value whites; ALBARIÑO, racy GRÜNER V. Top: John Forrest Collection.

Framingham Marl ★★→★★★ Lovely walled gardens and subterranean cellar. Owned by Sogrape (*see* Portugal). Aromatic whites. Lush Noble RIES, silky PINOT N.

Fromm Marl ★★★ Focus on long-lived wines: intensity, structure. Distinguished PINOT N: hill-grown Clayvin Vyd. Fromm Vyd sturdier (powerful SYRAH). Savoury Cuvée H Pinot N. Racy RIES Spätlese. Refined 221 Brancott Road CHARD.

Gibbston Valley C Ot ★★→★★★ Pre-coronavirus was NZ's most-visited winery. Strong name for PINOT N, esp fragrant GV Collection; silky Le Maitre; powerful Res. Intense Red Shed RIES. Full-bodied GV PINOT GR (esp organic School House); classy CHARD (esp Chablis-like China Terrace).

With climate change, expect Marl Cab Sauv, Gisborne Zin. Could happen.

Giesen Cant ★★→★★★ Large winery, multiple vyds in MARL, family-owned; Marl SAUV BL v. successful in Australia. Single-vyd Gemstone Ries partly fermented in granite tanks. Generous Uncharted PINOT N. Memorable Clayvin SYRAH.

Gimblett Gravels Hawk Defined area (800 ha planted, mostly since early 80s) of old arid riverbed. Noted for rich B'x-style reds (mostly MERLOT-led, but stony soils also suit CAB SAUV – recent renewed interest). Super SYRAH. Best reds world-class 13' 14' 15' 18 19' 20'. Age-worthy CHARD from siltier soils.

Gisborne NZ's 5th-largest region, on e coast of N Island. Declining in planted area and producer numbers. Abundant sunshine but often rainy; highly fertile alluvial soils. Key is CHARD (often used for bubbly). Excellent CHENIN BL, GEWURZ, VIOGNIER; PINOT GR, MERLOT more variable. Interest in ALBARIÑO (rain-resistant). Top wines from MILLTON.

Gladstone Vineyard Wair ★★ Largest producer in n WAIR, est 1985 in old riverbed. Owned by Asian investment company. Generous PINOT N; single-vyd reds since 2018 best yet. 12,000 Miles lower-priced, early drinking.

Grasshopper Rock C Ot ★★→★★★ Estate-grown on n-facing slope by PINOT N specialist. One of Alexandra subregion's top reds: age-worthy, great value.

Greenhough Nel ★★→★★★ One of region's best boutiques; 1st vines 1979, deep river-stones over clay. Top label: Hope Vyd (refined Chard; PINOT BL is NZ's finest; mushroomy Pinot N).

Greystone Waip ★★★ Star organic producer (also owns Muddy Water). Planted on slopes and flats, with high limestone content. Top whites, esp honeyed RIES; fleshy CHARD. Thomas Brothers is distinguished PINOT N. Delicious dry rosé.

Greywacke Marl ★★★ Distinguished wines from Kevin Judd, ex-CLOUDY BAY. Named after NZ's most abundant bedrock (pronounced *greywacky*). Wild SAUV: barrel-fermented, full of personality.

Grove Mill Marl ★★ Attractive, gd-value whites with WAIRAU V subregional focus.

Haha Hawk, Marl ★★ Fast-growing producer, v.gd value. Peachy Hawk CHARD; generous Hawk MERLOT; gd drink-young Hawk SYRAH, excellent Brut Cuvée NV.

Hans Herzog Marl ★★★ Warm, stony, early ripening, organic vyd yields distinctive wines. Unexpectedly ripe (for S Island) MERLOT/CAB, delicious PINOT N Duc. Classy TEMPRANILLO, MONTEPULCIANO, dark LAGREIN. Hans brand in EU, US.

Hawke's Bay NZ's 2nd-largest region (14% producers). Founded 1850s; sunny, dryish climate, extreme soil diversity. Classy, B'x-like MERLOT and CAB SAUV-based reds in favourable yrs (19' 20'); spicy SYRAH; weighty CHARD; SAUV BL (tropical, suits oak) now most widely planted variety; NZ's best VIOGNIER. Promising PINOT N from cooler, elevated, inland districts. *See also* GIMBLETT GRAVELS.

Hunter's Marl ★★→★★★ Owned by Jane H, MD since 1987. Strength in whites (latest vintages best yet). Famous for SAUV BL. Excellent fizz Miru Miru NV, esp late-disgorged Res. GEWURZ, fleshy PINOT GR, RIES (off-dry) all rewarding, value.

Invivo Auck ★★ Young, entrepreneurial producer. Celebrity labels.

Johanneshof Marl ★★ Small winery with candle-lit tunnel 50m (164ft) into hillside, full of maturing wine. Exotic GEWURZ (one of NZ's finest).

Jules Taylor Wines Marl ★★→★★★ Stylish, gd value. Named after MARL-born winemaker, ex-Constellation NZ. Complex top OTQ ("On The Quiet").

Kim Crawford Wines Hawk ★→★★ Owned by CONSTELLATION NZ. Punchy MARL SAUV BL big seller in US, in bottles and cans. Top range: Small Parcels.

Kumeu River Auck ★★★★ Best known for weighty Estate CHARD. Top, single-vyd Mate's Vyd Chard (planted 1990); elegant single-vyd Hunting Hill Chard rising star. Lower-tier Village Chard great value. New Rays Road range from recently acquired vyd in HAWK.

Lawson's Dry Hills Marl ★★→★★★ World's 1st winery to use screwcaps for all (2002). Best known for exotic GEWURZ, lively SAUV BL. Top range: The Pioneer (gorgeous old-vine Gewurz). Mid-tier Res range.

Lindauer Auck ★→★★ Hugely popular (in NZ), low-priced fizz, esp bottle-fermented

---

Pink and popular

NZ's hot new style is rosé. It can be dry or slightly off-dry and is typically based on MERLOT, from N Island regions such as HAWK. Those from the South I, incl MARL and C OT, usually made from PINOT N ("Pinot Rosé" is often Pinots N/GR). Kiwis drink rosé as a fresh, chilled replacement for white wine, and it's perfect with NZ salmon.

Lindauer Brut Cuvée NV. Latest batches easy-drinking. Ever-expanding range: low-alc; single-variety; "strawberry-infused"; frothy, sweetish MOSCATO "perfect for lunching with the girls". Special Res range: more complexity, top value.

Mahi Marl ★★→★★★ Small producer, owner Brian Bicknell not after "fruit-bombs; we want texture, wines that give real palate satisfaction." CHARD, PINOT GR, SAUV BL, PINOT N.

Man O' War Auck ★★ Largest vyd on WAIH. Owned by Spencer family, one of NZ's richest. High, windy sites, loam-clay soils. Powerful Valhalla CHARD; complex Gravestone SAUV BL/SEM. Reds: supple SYRAH, sturdy Ironclad.

Aotearoa is increasingly popular name for NZ – gd for Scrabble?

Marisco Marl ★★ Large Waihopai V producer. Estate vyds on slopes and flats. Several brands: The Ned, The Kings Series, The Craft Series, Leefield Station. Best known for lively The Ned SAUV BL; more concentrated The King's Favour.

Marlborough NZ's predominant region (70% plantings) at top of S Island (land ideal for planting now in short supply); 1st modern vines 1973, SAUV BL 1975. Hot, sunny days, cold nights give aromatic, crisp whites and PINOT N-based rosés. Intense Sauv Bl. from green capsicum to ripe tropical fruit; some top wines oak-influenced. Fresh RIES (recent wave of sw, low-alc); some of NZ's best GEWURZ, PINOT GR; CHARD slightly leaner than HAWK, more vibrant, can age well. High-quality, gd-value fizz and classy botrytized Ries. Pinot N (45% of NZ's total in 2021) underrated, top examples (from n-facing clay hillsides) among NZ's finest. Interest stirring in ALBARIÑO, GRÜNER V. (*See also* AWATERE V, WAIRAU V.)

Martinborough Wair Small, prestigious but not expanding district in S WAIR (foot of N Island). Cold s winds reduce yields, warm summers, usually dry autumns, free-draining soils (esp on Martinborough Terrace). Success with several whites (CHARD, GEWURZ, RIES, PINOT GR, SAUV BL), but renowned for savoury PINOT N (higher % of mature vines than other regions).

Martinborough Vineyard Mart ★★★ Famous PINOT N since 1984 (perfumed Home Block). Weighty Home Block CHARD; intense Manu RIES; gd-value Te Tera range (PINOT GR, SAUV BL, Pinot N). Owned by American Bill Foley (2014).

Matawhero Gis ★★ Former star GEWURZ producer of 80s, now different ownership. Perfumed Gewurz; flavourful Church House CHARD; also PINOT GR, MERLOT.

Matua Auck ★→★★ Producer of NZ's 1st SAUV BL in 1974 (from AUCK grapes) long known as Matua V. Formerly an industry leader, but currently low profile in NZ. Owned by TWE. Most wines pleasant, easy-drinking.

Maude C Ot ★★→★★★ Mature vines on steep, terraced, n-facing Mt Maude Vyd, at Wanaka. Delicious floral rosé. Generous PINOT N; age-worthy Mt Maude Vyd.

Mills Reef B of P ★★→★★★ Easy-drinking Estate range (HAWK, MARL grapes). Top Elspeth range from Hawk (CHARD). Mid-tier Res range can be gd value (MERLOT).

Millton Gis ★★→★★★★ Region's top wines from NZ's 1st organic producer, despite warm, moist climate. Hill-grown, single-vyd Clos de Ste Anne range in favourable seasons. Long-lived CHENIN BL (honeyed in wetter vintages). Drink-young Crazy by Nature gd value.

Misha's Vineyard C Ot ★★→★★★ Large vyd at Bendigo, on gently sloping terraces and one steep face ("the ski slope"). Rich PINOT GR; vivacious RIES (Limelight, Lyric); perfumed GEWURZ. Charming dry rosé. PINOT N: graceful High Note, drink-young Impromptu.

Mission Hawk ★★→★★★ NZ's oldest producer, 1st vines 1851; still owned by Catholic Society of Mary. Large vyd in AWATERE V. Fine value regional varietals. Classy CHARD, SYRAH, CAB/MERLOT blends under Jewelstone and (esp) Huchet labels.

Mondillo C Ot ★★→★★★ Rising star at Bendigo; sandy, silty terraced vyd overlying gravels. Off-dry RIES; late-harvest Nina Ries; rosé; powerful PINOT N (Bella Res).

Mount Edward C Ot ★★→★★★ Small, respected, organic. Named after mtn looming over winery. CHARD, PINOT BL, racy RIES, delicious GAMAY, PINOT N. Ted by Mount Edward, drink young.

Mount Riley Marl ★★ Medium-sized family firm, named after peak on n side of WAIRAU V. Great value. Top range is Seventeen Valley.

Mt Beautiful Cant ★★ Large vyd at Cheviot, n of WAIP. Silt loams and light clay, overlying mudstone. PINOT GR, SAUV BL, ROSÉ, PINOT N.

Mt Difficulty C Ot ★★★ Sloping vyds (was "rabbit-infested, briar-covered wasteland") and popular restaurant with sweeping views, owned by US billionaire Bill Foley. Expanding range of single vyds, incl Chablis-like Packspur CHARD, elegant Packspur PINOT N. Classy whites, esp Dry RIES.

Mud House Cant ★★→★★★ Large, Australian-owned, MARL-based (also vyds in C OT, WAIP). Brands: Hay Maker (lower tier), Mud House, Waipara Hills. Gd-value regional blends. Excellent Single Vyd collection and gd Estate range.

Nautilus Marl ★★→★★★ Medium-sized, owned by S Smith & Sons (*see* Yalumba, Australia). Several vyds. Yeasty NV sparkler one of NZ's best (min 3 yrs on lees). Crisp SAUV BL. Excellent ALBARIÑO, CHARD, GRÜNER V, PINOT N.

Nelson Small region (3% NZ v'yd area) w of MARL; climate wetter but equally sunny. Clay soils of Upper Moutere hills (classy CHARD, PINOT N), and silty WAIMEA plains (strength in aromatic whites). SAUV BL most extensively planted, but also v.gd GEWURZ, PINOT GR, RIES.

Neudorf Nel ★★★→★★★★ Pronounced "Noy-dorf". Smallish; big reputation. Hillside, clay-gravel vyds ("Clays give depth of flavour," says co-founder Tim Finn). Refined, intense Home Block Moutere CHARD one of NZ's greatest; stylish Rosie's Block Chard. Savoury Home Block Moutere PINOT N (Tom's Block: gd 2nd-tier Pinot N), lightly oaked SAUV BL, off-dry PINOT GR.

No 1 Family Estate Marl ★★→★★★ Family-owned firm of regional pioneer Daniel Le Brun, ex-Champagne, dubbed "the mad Frenchman" by locals in early 80s. No longer controls Daniel Le Brun brand (owned by Lion). Specialist in v.gd fizz. Best known for CHARD-based NV, 2 yrs on lees. Res Blanc de Blancs toasty.

Nobilo Marl Popular brand, est 1943. Bought by E&J Gallo Winery in 2020 for US$130 million. Best known for herbal SAUV BL. Gallo's goal: make Nobilo the biggest-selling Sauv Bl in the US.

Oyster Bay Marl ★★ From DELEGAT. Named after a bay in MARL Sounds. Marketing triumph: huge sales in UK, US, Australia. Easy-drinking, mid-priced wines with touch of class from Marl and HAWK. Marl SAUV BL biggest seller.

Palliser Mart ★★→★★★ One of district's largest, named after s-most tip of N Island, Cape Palliser. Many shareholders. Classy CHARD, rich PINOT N, esp mature-vine Hua Nui, vivacious fizz. Lower tier: Pencarrow (gd value, majority of output).

Pegasus Bay Waip ★★★ Family firm, 1st vines 1986, n-facing terraces in "lean country": shallow loess over gravel. Superb range. Second label: Main Divide, top value. Aged Release range 10 yrs old.

Peregrine C Ot ★★ Architecturally inspired winery, elegant "blade of light" roof. Vibrant organic whites; 2nd tier, Saddleback, gd value.

Pernod Ricard NZ Auck ★→★★★ Paris-based, one of NZ's largest, originally

---

Top Syrahs

SYRAH was well known in NZ more than a century ago, but it needs hot sites, yielding its best results in HAWK and the upper N Island, esp WAIH, but also Northland. Top: BROOKFIELDS, CHURCH ROAD, CRAGGY RANGE, ELEPHANT HILL, ESK VALLEY, La Collina, MAN O' WAR, MISSION, Passage Rock, SACRED HILL, SMITH & SHETH CRU, STONECROFT, STONYRIDGE, TE MATA, TRINITY HILL, VIDAL, VILLA MARIA.

Montana. Wineries in HAWK, MARL. Extensive co-owned vyds for Marl whites, esp huge-selling BRANCOTT ESTATE SAUV BL. Major strength in fizz, esp big-selling DEUTZ Marl Cuvée (classy Mumm-branded Marl sp since 2021). Wonderful-value CHURCH ROAD reds and CHARD. Other key brands: STONELEIGH.

Prophet's Rock C Ot ★★→★★★ Small, top producer with vyds on opposite sides of Lake Dunstan, 40 mins by road. Alsace-style PINOT GR, Dry RIES. Savoury PINOT N. Top tier: authoritative Cuvée Aux Antipodes, bottle-aged Retrospect Pinot N; 2nd-tier Home Vyd; 3rd-tier floral Rocky Point.

"Puckerooed", satisfyingly expressive NZ term for "broken", from Maori "pakaru".

Puriri Hills Auck ★★★→★★★★ Distinguished, B'x-like MERLOT-based reds from small vyd on clay slopes at Clevedon. Estate: satisfying, gd complexity. Harmonie Du Soir (formerly Res) more new oak. Top label dense, silky Pope.

Pyramid Valley Cant ★★→★★★ High vyd on clay/limestone scarps at Waikari, n- and se-facing. Distinctive, rare, estate-grown wines: steely CHARD, delicate PINOT N. Regional range too.

Quartz Reef C Ot ★★★ Small, bio, on warm, n-facing slope with sandy clay soils. GRÜNER V, classy PINOT GR. Graceful PINOT N (Single Ferment). Stylish fizz, esp Vintage Blanc de Blancs.

Rapaura Springs Marl ★★→★★★ Skilfully crafted, gd value, esp Res whites. Impressive single-vyd Bull Paddock and Rohe (district) SAUV BL.

Rippon Vineyard C Ot ★★→★★★★ Pioneer vyd, 330m (1083ft) up, on gentle schist slope on shores of Lake Wanaka – arresting view and wines. Fragrant, savoury style. Mature Vine PINOT N (vines planted 1985–91) is "the farm voice". Majestic Tinker's Field Pinot N: oldest vines, age-worthy. Striking GAMAY. Wanaka Village: drink-young Pinot N.

Rockburn C Ot ★★ Originally "warm, fuzzy" hobby vyd planted 1991, uneconomic, now much extended. Eleven Barrels PINOT N; Devil's Staircase Pinot N, unoaked, drink-young charmer.

Sacred Hill Hawk ★★→★★★ Acclaimed Riflemans CHARD from mature vines on spectacular elevated site, low fertility. Rich Brokenstone MERLOT, Helmsman CAB/Merlot, Deerstalkers SYRAH from GIMBLETT GRAVELS. After financial collapse in 2021, now owned by VinLink (contract winemaking firm).

Saint Clair Marl ★★→★★★ Largest family-owned company in region. Much show success. Best known for pungent, lively SAUV BL. Classy PINOT N; James Sinclair (subregional focus), Origin (large-volume regional blends), Vicar's Choice (everyday). Also owns Delta, Lake Chalice.

Seifried Estate Nel ★★ Region's biggest winery, family owned. Extensive vyds. Medium-dry RIES, perfumed GEWURZ, punchy SAUV BL. Top wines: Winemakers Collection (esp honeyed Sweet Agnes Ries); 3rd tier, Old Coach Road, gd value.

Selaks Marl ★ Old producer of Croatian origin, now a supermarket brand of CONSTELLATION NZ. Solid, easy-drinking. The Taste Collection: Luscious HAWK PINOT GR (not really luscious); Silky Smooth Hawk MERLOT (actually fairly firm).

Seresin Marl ★★→★★★ Quality organic producer, owned by filmmaker Michael S. Sophisticated, partly oak-aged SAUV BL one of NZ's finest (Marama Sauv Bl even better); lively CHARD; single-vyd PINOT N; v.gd-value Momo range, esp dry rosé.

Sileni Hawk ★★ Large, owned by investment company. Estate vyds in warm Bridge Pa Triangle and on inland, cooler sites. Strong Grand Res range, incl complex Lodge CHARD, generous Triangle MERLOT. Attractive, gd-value Cellar Selection.

Smith & Sheth Cru Hawk ★★→★★★ Partnership of viticulturist Steve Smith MW (ex-CRAGGY RANGE) and US billionaire Brian Sheth. Classy single-vyd range, based on mature vines. Refined SYRAH, complex CHARD. (*See* PYRAMID V.)

Spy Valley Marl ★★→★★★ Waihopai V is used for monitoring satellite communications, hence name. High achievers, extensive vyds; gd-value whites, esp oak-aged SAUV BL. Generous PINOT N. Classy Envoy top selection.

Starborough Family Estates Marl ★★ Family-owned vyds in AWATERE V, WAIRAU V. Racy SAUV BL, full-flavoured CHARD, dryish PINOT GR, supple PINOT N.

Stonecroft Hawk, Marl ★★ Small organic winery with free-draining, gravelly home vyd. Excellent CHARD, GEWURZ, VIOGNIER. NZ's 1st serious SYRAH, still v.gd (dense Res). Delicious Undressed Syrah (no sulphur added).

NZ aims to lead world in low-alc wines of 9.5%. But Mosel Ries Kabinett c.8%.

Stoneleigh Marl ★★ Owned by PERNOD RICARD NZ. Based on relatively warm, stony Rapaura vyds. Popular MARL whites, esp SAUV BL. Top: Rapaura Series. Latitude, designed to be full-on.

Stonyridge Waih ★★★→★★★★ Boutique winery on n-facing, poor, free-draining clay soils. Known since mid-80s for exceptional CAB SAUV-based blend Larose, but profile lower than a decade ago. Dense, Rhône-style, SYRAH-based Pilgrim.

Te Kairanga Mart ★★→★★★ One of district's oldest, largest; troubled history, recently rejuvenated. Runholder is mid-tier (graceful PINOT N). Top tier is John Martin: complex CHARD; age-worthy Pinot N.

Te Mata Hawk ★★★→★★★★ Winery of high repute; 1st vintage 1895. CAB SAUV flourishes in Havelock North Hills, sheltered by Te Mata Peak. Coleraine (Cab Sauv/MERLOT/CAB FR) 10' 11 13' 14' 15' 16 17 18 19' 20' is B'x-like, breed, longevity. Much lower-priced Awatea Cabs/Merlot also classy, more forward. Bullnose SYRAH among NZ's finest. New powerful Alma HAWK PINOT N. Rounded Elston CHARD; gd-value Estate Vyds range for early drinking.

te Pa Marl ★★→★★★ Home vyd on Wairau Bar, archaeological site with ancient *pa* (fortified settlements). SAUV BL, CHARD both gd value. Top range: The Reserve Collection, incl weighty Chard, intense Seaside Sauv Bl.

Terra Sancta C Ot ★★→★★★ Bannockburn's 1st vyd, at end of Felton Road in deep, gravelly yellow earths. Fragrant Mysterious Diggings PINOT N; age-worthy mid-tier Bannockburn Pinot N; Slapjack Block Pinot N from district's oldest vines. Pinot N Rosé one of NZ's finest.

Te Whare Ra Marl ★★ Label: TWR. Small WAIRAU V producer, some of region's oldest vines, planted 1979. Estate-grown, organic. Known for perfumed GEWURZ; punchy RIES, SAUV BL.

Tiki Marl ★★ McKean family own extensive vyds in MARL, WAIP. Top tier: Koro. Mid-range: single vyd (esp generous N CANT SAUV BL). Second label: Maui.

Tohu Marl, Nel ★★ Maori-owned venture, vyds in MARL, NEL; v.gd Blanc de Blancs (sp). Incisive AWATERE V SAUV BL. Top intense Single Vyd Whenua Awa CHARD and savoury PINOT N.

Trinity Hill Hawk ★★→★★★★ Highly regarded, with vyds in GIMBLETT GRAVELS and winery built from tilt-slab concrete. Refined B'x-style blend The Gimblett. Stylish Gimblett Gravels CHARD. Outstanding single-vyd 125 Gimblett Chard. Prestigious Homage SYRAH 14' 15' 16 17 18' 19'. Impressive TEMPRANILLO. Lower-tier white-label range gd value.

Two Paddocks C Ot ★★ Actor Sam Neill makes several PINOT NS. Main label is vibrant, estate-grown, multi-site. Single-vyd Prop Res range: herbal First Paddock, riper Last Chance. Latest is earthy The Fusilier, grown at Bannockburn. Drink-young Picnic.

Two Rivers Marl ★★→★★★ David Clouston makes distinctive Convergence CHARD, SAUV BL, rosé; savoury Tributary PINOT N. Second label: Black Cottage v.gd value.

Valli C Ot ★★★ "Every winemaker wants to make a good red," says owner Grant Taylor. "Whites are just a warm-up." Superb range of single-vyd PINOT N, esp refined Bannockburn, graceful Bendigo.

Vavasour Marl ★★→★★★ Original pioneer of AWATERE V, founded by Peter V (whose ancestor was cup-bearer for William the Conqueror). Owned by Foley Family Wines. Rich CHARD, esp lovely Anna's Vyd, intense SAUV BL. Classy dry rosé; generous PINOT N.

Vidal Hawk ★★→★★★ Founded 1905. Owned by VILLA MARIA. Top Legacy and Soler ranges recently phased out, leaving standard and Res ranges.

Villa Maria Auck ★★→★★★ In receivership 2021, bought by MARL-based Indevin, best known for supermarkets' own-label. Also owns ESK V, VIDAL. Wine-show focus, with glowing success. Distinguished top ranges: Res (regional character) and Single Vyd (individual sites). Platinum Selection (some organic, much lees-ageing); 3rd-tier Cellar Selection (less oak) excellent, superb value (esp HAWK MERLOT); huge-volume, 4th-tier Private Bin can also be v.gd (Marlb SAUV BL). Small volumes of v.gd ALBARIÑO, GRENACHE, MALBEC. CAB SAUV-based, icon red Ngakirikiri (GIMBLETT GRAVELS) 13' 14' 18'.

Waiheke Island Helipad heaven at this lovely, sprawling, touristy island in AUCK's Hauraki Gulf (temperatures moderated by sea). Acclaim since 80s for stylish CAB SAUV/MERLOT blends, esp from warm Onetangi district; more recently for bold SYRAH. Sturdy CHARD. Largest producer MAN O' WAR.

Waimea Nel ★★ One of region's largest, best value; 2nd tier Spinyback.

Waipara Valley Cant Dominant CANT subregion, n of Christchurch. Gravelly soils on the flats and richer, clay-based on the e hills. Hot, dry nor'westers (wind) devigorate vines. High profile for weighty PINOT N, scented RIES (also heavy plantings of PINOT GR, SAUV BL). Increasingly calling itself "North Canterbury".

Wairarapa NZ's 6th-largest wine region (not to be confused with WAIP). *See* MART. Also incl Gladstone subregion in n (slightly higher, cooler, wetter). Driest, coolest region in N Island, but exposed to cold s winds (causing small crops); slight expansion in past 5 yrs. Strength in whites; SAUV BL esp widely planted, CHARD, PINOT GR, RIES. Famous for weighty, ripe, savoury PINOT N from relatively mature vines. Starting to promote itself as "Wellington Wine Country".

Wairau Valley Marl Largest MARL subregion (1st vyd 1873; modern era since 1973). Biggest town Blenheim gateway to 30+ cellar doors. Three important side valleys to s (Southern V): Brancott, Omaka, Waihopai. SAUV BL thrives on stony, silty plains (shingly soils speed ripening, giving riper, more tropical fruit notes); PINOT N on clay-based, n-facing slopes. Much recent planting in wetter, more frost-prone upper Wairau V.

Waitaki Valley C Ot Small subregion in N Ot (65 ha), limestone soils and cool, frost-prone climate, 0.03% NZ harvest in 2021. Handful of producers; v. promising PINOT N, can be leafy; scented PINOT GR, RIES superb in top vintages.

Whitehaven Marl ★★ Medium-sized family producer, best known in US, as part-owned by Gallo since 2005. Flavour-packed SAUV BL big seller in US. Rich CHARD. Generous PINOT N. Top range: Greg.

You like potato and I like potahto/You like tomayto and I like Te Mata...

Wither Hills Marl ★★ Big producer owned by Lion brewery. Popular, gd-value SAUV BL, plus CHARD, PINOT GR. Intense Single-Vyd Rarangi Sauv Bl from coastal site. Top, The Honourable PINOT N.

Yealands Marl ★★ NZ's biggest "single vyd", at coastal AWATERE V site, owned by utility firm Marlborough Lines. Partly estate-grown, mostly MARL. High profile for sustainability (incl NZ's largest solar panel array), but most not certified organic. Best known for green capsicum Marl SAUV BL. Top-value Res Awatere Sauv Bl. Other key brands: Babydoll, The Crossings.

Zephyr Marl ★★→★★★ Family vyd in lower WAIRAU V, bounded by Opawa River (whitebait with tangy, off-dry RIES). Impressive whites. Weighty MK III SAUV BL.

South Africa

Abbreviations used in the text:

Bre	Breedekloof	**Kl K**	Klein Karoo
C'dorp	Calitzdorp	**Oli R**	Olifants River
Cape SC	Cape South Coast	**Pie**	Piekenierskloof
Ced	Cederberg	**Rob**	Robertson
Coast	Coastal Region	**Sla**	Slanghoek
Const	Constantia	**Stell**	Stellenbosch
Ela	Elandskloof	**Swa**	Swartland
Elg	Elgin	**Tul**	Tulbagh
Fran	Franschhoek	**V Pa**	Voor Paardeberg
Hem	Hemel-en-Aarde	**Wlk B**	Walker Bay
Rdg/Up/V	Ridge/Upper/Valley	**Well**	Wellington

What a surge in dynamism and self-confidence there has been here. Covid, despite hitting producers hard with repeated prohibitions on local wine sales, has done little to stem their energy or the enthusiasm of their customers around the world. If anything, the pandemic has freed up time to reflect, prioritize and prepare, to work more intensively with the vines (and plant new ones) and push the boundaries in the cellar. This is abetting and accelerating trends like harnessing yeast species other than good old standard *Saccharomyces cerevisiae*, employing vessels of varying materials, shapes and sizes for ferments, and combining traditional wooden barrels with plastic, clay and concrete containers for ageing. The resultant wines are more nuanced, bright and fresh, with fruit (though not necessarily fruity) flavours front and centre, the role of oak diminished and even eliminated. Reds are lighter-textured and juicier, pinks paler and drier, whites tighter and zingier. Beyond the cellar, old and heirloom vines are prized and nurtured, established sites scientifically analysed and tweaked, suitable (higher and cooler) terroirs identified and groomed, and drought-/heat-resistant varieties planted. Tellingly, vineyard trials continue to be launched, inter alia among the maize fields of the remote Free State.

Recent vintages

2021 Late but bountiful, gd early rains, cool conditions. Much excitement (r/w).

2020 Post-drought vintage a humdinger: exceptional structure, intensity, verve, with moderate alc.

2019 Another arid yr, but milder temps balance concentrated fruit/freshness.

2018 Intense, flavourful wines, though probably not for long cellaring.

2017 Quality, character comparable to great 15. Accessible young, possibly peaking earlier too.

AA Badenhorst Family Wines Swa ★★→★★★★ Widely hailed portfolio reflecting every trend: focus on site, heirloom grapes, old vines, wild yeasts, orange wine. Centrepiece is suite of single-vyd CHENIN BL; gd-value range Secateurs is fun but serious. Brilliant Caperitif vermouth.

Alheit Vineyards W Cape ★★★★ Pinnacle of modern SA wine with fine, pure expressions of mostly old-vine CHENIN BL: multi-region Cartology, and site-specific bottlings eg. Broom Ridge from recently acquired SWA property. Also SEM Monument ex-FRAN and Vine Garden field blend from HEM. Former assistant Franco Lourens's own label, Lourens Family Wines, right up there.

Anthonij Rupert Wyne W Cape ★→★★★★ Extensive, impressive portfolio honours owner Johann R's late brother. From own vyds in DARLING, SWA, Overberg, elegant home farm L'Ormarins nr FRAN. Best ranges: flagship Anthonij Rupert, site-specific Cape of Gd Hope, premium Jean Roi (rosé).

Aristea Wines Coast ★★★ Recent UK/France/SA venture fronted by Matthew Krone, scion of famed local wine family. Mostly varietal bottlings of classic B'x/Burgundy grapes ex-STELL, ELG and HEM. Also MCC sparkling, a Krone speciality: his own bubblies named after daughters and released only in leap yrs.

Bartinney Private Cellar Stell ★★→★★★ Sustainability-focused boutique on steep Banhoek V sides: classic CAB SAUV, CHARD (regular, v. fine Res). Lifestyle sibling Noble Savage. High-energy Rose and Michael Jordaan now also own extensive Plaisir de Merle, former star Simonsberg Mtn estate in DISTELL portfolio.

Beaumont Family Wines Bot R ★★→★★★★ Excellence from C18 estate in Bot River. Rare solo-bottled MOURVÈDRE, CAPE BLEND Vitruvian, elegant, always superlative; Hope Marguerite CHENIN BL 12' 15' 16' 17' 18' 19' 20' 21 22 from old vines. Matriarch Jayne Beaumont's own PINOT N, CHARD Electrique worth a try

Beck, Graham W Cape ★★★ Front-rank MCC house nr ROB. Nine labels (Vintage/NV, Brut Nature to Demi-Sec) incl new Artisan Collection, debuting with Extended Lees Ageing 09, extraordinary 134 mths *sur lie*. Long-time maestro Pieter Ferreira's own-brand bubbles (CHARD Blanc de Blancs and PINOT N Rosé) arguably even finer.

BEE (Black Economic Empowerment) Initiative aimed at increasing wine-industry ownership and participation by previously disadvantaged groups.

Beeslaar Wines Stell ★★★★ KANONKOP winemaker's personal take on PINOTAGE 13' 14' 16' 17' 19 20 21. Refined and rather special.

Bellevue Estate Stellenbosch Stell ★→★★★ Local pioneers of PINOTAGE. Standout Res and single-vyd versions, latter from estate's original vines planted in 1953.

Bellingham W Cape ★★→★★★ Enduring DGB brand with low-volume, high-quality The Bernard Series, incl scarce monovarietal ROUSSANNE. Also gd-value Homestead Series.

Benguela Cove Lagoon Wine Estate Wlk B ★→★★★ Cellar, vyds, tourist destination on Bot River mouth, owned by Penny Streeter OBE. SEM Catalina and Vinography experimental range. Winemaker Johann Fourie also responsible for Streeter's UK wines (Leonardslee, Mannings Heath) and SA brand Brew Cru, owned by Fourie and friends, showcasing parcels (r/w) in cool S Cape.

Beyerskloof W Cape ★→★★★★ SA's PINOTAGE champion: ten versions (11, incl spirit to fortify Lagare Cape Vintage "Port"). Powerful varietal Diesel 13' 16' 17' 18' 19 20, clutch of CAPE BLENDS. Also classic CAB SAUV/MERLOT field blend.

Boekenhoutskloof Winery W Cape ★→★★★★ Top FRAN winery, exemplary quality, consistency: SWA SYRAH; Fran CAB SAUV (also newer STELL version); old-vines SEM; newer Patina CHENIN BL; simpler Porcupine Ridge, Wolftrap and Vinologist lines. Major development, Cap Maritime, under way in HEM; CHARD, PINOT N. *See* PORSELEINBERG.

Bon Courage Estate W Cape ★→★★★ Bruwer family in ROB with broad range. Stylish Brut MCC trio, aromatic desserts (RIES, MUSCAT), delightful COLOMBARD (DYA).

Boplaas Family Vineyards W Cape ★→★★★ Growers Carel Nel and daughter Margaux at C'DORP, known for Port styles, esp Cape Vintage Res 09' 12' 15' 16 17' 18' 19 20 and Tawny (mostly NV). Newer table wines of Portuguese grapes (r/w).

Boschendal Wines W Cape ★→★★★ Popular DGB brand on lovely C17 estate nr FRAN. Notable SHIRAZ, SAUV BL, CHARD, MCC in various tiers, some ex-ELG. B'x/ Rhône (r/w) blends (Black Angus, Nicolas, Suzanne) designed to impress.

Boschkloof Wines W Cape ★★→★★★★ Young gun Reenen Borman in STELL with eg. stellar SYRAH (varietals, blends), CHENIN BL under Boschkloof (home farm's name) and Kottabos labels. Also, with partners, notable Syrah, Chenin Bl, varietal COLOMBARD Doortjie in Patatsfontein range.

Botanica Wines W Cape ★★→★★★ American Ginny Povall on flower- and wine-farm Protea Heights nr STELL. Superlative CHENIN BL Mary Delany Collection from old w-coast bush vines; ALBARIÑO Flower Girl among 1st local solo bottlings of Iberian variety.

This Pinotage is made for walking: *veldskoen* with wine-red sole, by Beyerskloof.

Bouchard Finlayson Cape SC ★★→★★★★ HEM pioneer with fine versions of area specialities PINOT N (Galpin Peak, occasional barrel-selected Tête de Cuvée 13' 17' 19' 21) and CHARD (Missionvale, Sans Barrique; also Crocodile's Lair ex-Ela).

Breedekloof Large (c.12,700 ha) inland area known for bulk and entry-level wine. Fine winemaking in small/family cellars eg. Bergsig, Deetlefs, Le Belle Rebelle, OLIFANTSBERG, OPSTAL. But recent initiative, Breede Makers, prompts large/co-op ventures too, to step up by vinifying special/old parcels of mostly CHENIN BL. Some thrilling results.

Bruce Jack Wines W Cape ★★→★★★ Never a dull bottle at FLAGSTONE founder Bruce Jack's solo venture in cool S Cape. Endlessly creative blends, varietals, rosé and bubbly. Delightful labels by artist wife Penny.

Buitenverwachting W Cape ★★→★★★ Stylish family winery in CONST. Standout CHARD, Husseys Vlei SAUV BL, B'x reds Christine 09' 10 11 13 14 15 16 and Meifort. Labelled Bayten for export.

Calitzdorp DISTRICT (c.290 ha) in KLEIN KAROO climatically similar to the Douro (*see* Portugal), known for Port styles and latterly unfortified Port-grape (r/w) varietals and blends.

Cape Blend Usually red with significant PINOTAGE component; occasionally CHENIN BL blend, or simply wine with "Cape character".

Cape Chamonix Wine Farm Fran ★★→★★★ Excellent winemaker-run mtn property. Distinctive PINOT N, *ripasso*-style PINOTAGE, CHARD, SAUV BL, B'x blends (r/w), CAB FR, old-vine CHENIN BL, all worth keeping.

Cape Coast Umbrella appellation ("OVERARCHING REGION" in officialese) for COAST (w and central) and CAPE SC regions.

Capensis W Cape ★★★ SA/US venture, GRAHAM BECK's Antony Beck and Jackson Family's Barbara Banke, specializing in CHARD. Multi-region Capensis, single-region (STELL) Silene and single-vyd Fijnbosch. *See* JACKSON WINE ESTATES (SA).

Cape Point Vineyards Cape T ★→★★★★ Family winery nr tip of CAPE TOWN peninsula. Complex, age-worthy SAUV BL/SEM Isliedh, CHARD, Sauv Bl. Sibling venture Cape Town Wine Co.

Cape Rock Wines W Cape ★★→★★★ Leading boutique grower in OLI R. Characterful, strikingly packaged Rhône and Port-grape varietals, blends (r/w).

Cape South Coast Cool-climate REGION (c.2600 ha) comprising DISTRICTS of Cape Agulhas, ELG, Lower Duivenhoks River, Overberg, Plettenberg Bay, Swellendam, WLK B, plus standalone WARDS Herbertsdale, Napier, Stilbaai East. *See* CAPE COAST.

Cape Town Coastal DISTRICT (c.2700 ha) covering Cape Town city, its peninsula WARDS, CONST and Hout Bay, plus neighbour wards DUR and Philadelphia.

Cape West Coast Newer/1st subregion; incl Darling and Lutzville V DISTRICTS, and Bamboes Bay, Lamberts Bay and St Helena Bay WARDS.

Catherine Marshall Wines W Cape ★★★ Cool-climate (chiefly ELG) specialist Cathy M focuses mostly on PINOT N, CHENIN BL, SAUV BL; delightful dry, mineral RIES.

Cederberg Tiny (c.100 ha) high-altitude standalone WARD in Cederberg Mtns. Mostly SHIRAZ, CHENIN BL. Driehoek, CEDERBERG PRIVATE CELLAR main producers.

Cederberg Private Cellar Ced, Elim ★★→★★★★ Family cellar, among SA's highest (CED), most S (ELIM) vyds. Leashed power in CAB SAUV, PINOT N, rare Bukettraube, CHENIN BL, SAUV BL, SEM, MCC, SHIRAZ (incl exceptional CWG Teen die Hoog).

Central Orange River Previously an oversized WARD (c.8600 ha, almost two-thirds is Sultana for dried/table grape market), now a DISTRICT in N CAPE GU. Hot, dry, irrigated; traditionally white and fortified, but major producer ORANGE RIVER CELLARS pushing boundaries.

Certified Heritage Vineyard *See* OLD VINE PROJECT.

Charles Fox Cap Classique Wines Elg ★★★ Traditional-method bubbly house with French consultant. Six classic, delicious Bruts headed by Prestige Cuvée duo.

Coastal Largest REGION (c.45,100 ha); incl sea-influenced DISTRICTS of CAPE TOWN, DARLING, Lutzville V, STELL and SWA, plus WARDS Bamboes Bay, Lamberts Bay and St Helena Bay. Also incl non-maritime FRAN, PAARL, TUL, WELL DISTRICTS.

Colmant Cap Classique & Champagne W Cape ★★★→★★★★ Exceptional *méthode traditionnelle* house at FRAN, Belgian family-owned. Brut and Sec Res, Rosé, CHARD, Absolu Zero Dosage; all MCC, NV and excellent.

Constantia CAPE TOWN WARD (c.430 ha) on cool Constantiaberg slopes, SA's 1st and most historically famous wine-growing area, revitalized in recent yrs by GROOT CONST, KLEIN CONST et al.

Constantia Glen Const ★★★ Waibel family-owned gem on upper Constantiaberg. Trio of superb B'x blends (r/w), varietal SAUV BL.

Constantia Uitsig W Cape ★★★ Premium vyds and imposing glass-walled cellar producing mostly MCC and still white. Consistent, striking, individual SEM.

Creation Wines Cape SC ★★★ Family-owned portfolio of B'x, Burgundy, Rhône and newer Loire varietals, blends; bustling, awarded cellar door in HEM Rdg. Winemaker Gerhard Smith's solo project, Die Kat se Snor (The Cat's Whiskers), more than a joke.

Crystallum W Cape ★★★→★★★★ Star winemaker Peter-Allan Finlayson and brother Andrew, specializing in cool-climate PINOT N, CHARD. Eight bottlings, some single vyd, all superb, incl new Ferrum Chard. Vinified at GABRIËLSKLOOF.

CWG (Cape Winemakers Guild) Independent, invitation-only association of 43 top growers. Stages benchmarking annual auction of limited premium bottlings.

Darling DISTRICT (c.2730 ha) around this w-coast town. Best vyds in hilly Groeneekloof WARD. Cloof, Darling Cellars, Groote Post/Aurelia, Mount Pleasant, Ormonde, Withington bottle under own labels; most other fruit goes into 3rd-party brands, some spectacular.

David & Nadia Swa ★★★★ Sadie husband and wife follow natural-wine principles of SWA Independent Producers. Exquisite Rhône red Elpidios, GRENACHE Noir, CHENIN BL (varietals incl arresting, identically vinified single-vyd trio), PINOTAGE. Some from old vines. Former assistant, now solo André Bruyns's own collection City on a Hill also v. fine.

David Finlayson Wines W Cape ★★→★★★ David F, from esteemed Cape wine family, with portfolio previously known as Edgebaston; v.gd CAB SAUV GS, old-vine Camino Africana series, stylish early-drinkers, from own vyds nr STELL plus wider-sourced grapes. Stand-alone exploratory label, Sanniesrus, with resident winemaker, fine PINOTAGE, GRENACHES BL and N.

Bag-in-box now more popular than bottles. Corkscrews are so last century.

De Grendel Wines W Cape ★★★ Sir De Villiers Graaff's Table Mtn-facing venture in DUR draws on own and contracted vyds in far-flung areas: snowy Ceres Plateau for Op Die Berg PINOT N, CHARD; tip-of-Africa ELIM for brilliant newer SHIRAZ.

De Krans Wines W Cape ★→★★★ Nel family at C'DORP noted for Port styles (esp Vintage Res 10' 11' 12' 13' 16' 17' 18 19 20) and fortified MUSCAT. Also success with unfortified Portuguese grapes incl new VERDELHO.

Delaire Graff Estate W Cape ★★→★★★★ UK diamond merchant Laurence Graff's eyrie vyds, winery and visitor venue nr STELL. Glittering portfolio headed by age-worthy CAB SAUV Laurence Graff Res 09' 11 12' 13' 14 15 17' 18.

Delheim Wines Coast ★→★★★ Eco-minded family winery nr STELL. Vera Cruz SHIRAZ, PINOTAGE; cellar-worthy, best-yrs CAB SAUV Grand Res, scintillating botrytis RIES Edelspatz.

DeMorgenzon Stell ★→★★★★ Manicured property hits high notes with B'x, Rhône varietals/blends (r/w), CHARD, CHENIN BL. Occasional vine-selected Chenin Bl The Divas is spectacular. Classical music played in vyds/cellar 24/7.

De Toren Private Cellar Stell ★★★ Majority Swiss-owned, on ocean-facing Polkadraai Hills. Consistently flavourful B'x Fusion V, earlier-maturing MERLOT-based Z; light-styled Délicate (DYI).

De Trafford Stell ★★★→★★★★ Boutique grower David T with track record for bold yet harmonious wines: B'x/SHIRAZ Elevation 393, CAB SAUV, SYRAH Blueprint; CHENIN BL (dry, incl newer Skin Macerated, and *vin de paille*). *See* SIJNN.

DGB W Cape Long-est, WELL/FRAN-based producer/wholesaler, owner of high-end brands BOSCHENDAL, Fryer's Cove and The Bernard Series and, via subsidiary, newer Old Road Wine Company. Majority share in century-old family venture Backsberg; BELLINGHAM, Brampton, Douglas Green and other easy-drinkers.

Diemersdal Estate W Cape ★→★★★ DUR family farm excelling with various site-/row-/style-specific SAUV BL; incl ferments with frozen must and on skins, PINOTAGE, CHARD, SA's 1st commercial GRÜNER V.

Distell W Cape SA's biggest drinks company, in STELL. Owns or has interests in many wine brands, spanning styles/quality scales. *See* DURBANVILLE HILLS, FLEUR DU CAP, JC LE ROUX, NEDERBURG WINES.

District *See* GU.

Dorrance Wines W Cape ★→★★★★ French family-owned, with cellar in Cape Town city heritage building (more reason to visit). Fine-boned, consistently excellent SYRAH esp, also CHARD, CHENIN BL.

Durbanville Cool, hilly WARD (c.1440 ha) in CAPE TOWN DISTRICT, best known for pungent SAUV BL; also MERLOT, white blends. Corporate co-owned DURBANVILLE HILLS and many family ventures.

Durbanville Hills Dur ★→★★★ Owned by DISTELL, local growers and staff trust, with awarded PINOTAGE, CHARD, SAUV BL; v.gd B'x blend Tangram (r/w).

Eagles' Nest Coast ★→★★★ CONST family winery with reliably superior MERLOT, SHIRAZ, VIOGNIER. Also vibrant SAUV BL, cellar-door-only Little Eagle.

Edgebaston See DAVID FINLAYSON WINES.

Eikendal Vineyards W Cape ★★★ Swiss-owned high performer nr STELL; B'x red Classique, MERLOT, vintage-blend Charisma (r). Excellent wooded CHARD trio: multi-site, bush vine, single clone; unoaked version no slouch.

Elgin Cool-climate DISTRICT (c.740 ha) recognized for SAUV BL, CHARD, PINOT N; also exciting SHIRAZ, CHENIN BL, RIES, MCC. Mostly family boutiques, incl one of only two certified-bio wineries in SA, RADFORD DALE Organic, formerly known as Elg Ridge (other is REYNEKE in STELL).

Elim Windswept WARD (c.150 ha) in most s DISTRICT, Cape Agulhas, producing aromatic SAUV BL, white blends, SHIRAZ. Grape source for majors like DE GRENDEL and boutiques eg. TRIZANNE SIGNATURE WINES.

Ernie Els Wines W Cape ★→★★★★ Star golfer's wine venture with Baron Hans von Staff-Reitzenstein nr STELL; long-lived CAB SAUV under Signature, CWG, Major Series and Proprietor's labels; ready earlier Big Easy range. See STELLENZICHT.

Estate Wine Grown, made and bottled on "units registered for the production of estate wine". Not a quality designation.

Fable Mountain Vineyards W Cape ★→★★★ US-owned TUL grower with v.gd. SYRAH (varietal/blend) and rare MOURVÈDRE rosé, special-site/-vintage Small Batch Series, easy-drinking Raptor Post range. Winemaker Tremayne Smith moonlights with star viticulturist and college mate Jaco Engelbrecht as The Horsemen for special parcels; solo as The Blacksmith.

Fairview W Cape ★→★★★ Charles Back serves up a varietal, blended and single-vyd smorgasbord under Fairview, Goats do Roam, La Capra and Spice Route labels. Locally rare varieties eg. VERDELHO, qvevri vinification and innovative cellar-door experiences add piquancy.

FirstCape Vineyards W Cape ★→★★ DYA Huge export joint venture of four local co-ops and UK's Brand Phoenix. Mostly entry-level wines in over a dozen ranges, some sourced outside SA.

Flagstone Winery W Cape ★→★★★ Accolade Wines' high-end venture in former dynamite factory at Somerset W. Sources widely for impressive eg. PINOTAGE, SAUV BL, B'x white. Mid-tier Fish Hoek, entry level KUMALA are sibling brands.

Fleur du Cap W Cape ★→★★★ DISTELL premium label with headliner Laszlo (B'x r blend) and v.gd Series Privée Unfiltered (r/w).

Foundry, The Stell, V Pa ★★★→★★★★ Winemaker Chris Williams and partner James Reid among 1st to focus on (varietally bottled) Rhône grapes in 2000. Cellar in PAARL's V Pa; sensational GRENACHE BL; rare solo ROUSSANNE et al (STELL).

Franschhoek Huguenot-founded DISTRICT (c.1210 ha) known for CAB SAUV, CHARD, SEM, MCC. Home to some of SA's oldest farms and vines, eg. SEM planted in 1902 by great-grandfather of Eikehof owner-winemaker Francois Malherbe.

Free State Province and GU. Mile High Vyds sole producer (under The Bald Ibis label) in viticulturally challenging e highlands.

Visit Cape Town's Chenin Bl vine in Heritage Sq, planted 1771, still producing.

Gabriëlskloof W Cape ★→★★★ CWG member Peter-Allan Finlayson produces this lauded lineup in family cellar nr Bot River. Standard-bearing Landscape Series vies for attention/quality with Projects range, featuring whole-bunch, semi-carbonic SYRAH and amphora-vinified SAUV BL. See CRYSTALLUM.

Glenelly Estate ★★★→★★★★ Former Pichon Lalande (see B'x) owner May-Eliane de Lencquesaing's "retirement" venture in STELL. Top flagships Lady May (B'x r), Estate Res duo (B'x/SHIRAZ, CHARD). Superior-value Glass Collection.

Groot Constantia Estate Const ★★→★★★ Historic property, tourist mecca in SA's original fine-wine area. Suitably distinguished wines, esp MUSCAT de Frontignan Grand Constance 12 13 14′ 15′ 16 17 18 helping restore CONST dessert to C18 glory.

SOUTH AFRICA

GU (geographical unit) Largest of the wo demarcations: FREE STATE, KWAZULU-NATAL and LIMPOPO, plus E, N, W Cape – last three constitute the OVERARCHING GU known as Greater Cape. Other WO appellations (in descending size): overarching region, REGION, subregion, DISTRICT and WARD.

Hamilton Russell Vineyards W Cape ★★→★★★★ The 1st to succeed with Burgundy grapes in cool S Cape in early 80s at Hermanus. Elegant PINOT N, long-lived CHARD under HRV label. Same varieties in newer, stand-alone offshoot brand Tesselaarsdal, by long-time employee Berene Sauls. Super PINOTAGE, SAUV BL (varietals, blends) in Southern Right and Ashbourne ranges.

Hartenberg Estate W Cape ★★→★★★ Welcoming STELL family farm never disappoints with SHIRAZ (top-range Gravel Hill, several other varietals and blends); B'x red, CHARD, RIES (dr/s-sw/botrytis).

Hemel-en-Aarde Trio of cool-climate WARDS (Hem V, Up Hem, Hem Rdg) in WLK B DISTRICT, producing outstanding PINOT N, CHARD, SAUV BL.

Iona Vineyards Cape SC ★★→★★★★ Family winery co-owned by staff excels with ELG's signature grapes, CHARD, SAUV BL, PINOT N, some single vyd, from mostly exposed, high-lying sites. Also excellent SYRAH (blended One Man Band, varietal Brocha Solace). Cut-above lifestyle brand, Sophie, portion wider sourced.

Jackson Wine Estates (South Africa) Swa ★★★ California's Jackson Family Wines' newest SA venture is CHENIN BL from vines on Kalmoesfontein, SWA home farm of AA BADENHORST. *See* CAPENSIS.

JC le Roux, The House of W Cape ★→★★ SA's largest specialist bubbly producer nr STELL, DISTELL-owned. Best: Scintilla MCC. Offers on-trend canned and de-alcoholized fizz.

Joostenberg Wines W Cape ★★→★★★ Back-to-basics ethos in Joostenberg organic label from PAARL for SYRAH, CHENIN BL (incl botrytis dessert). Honours forebears via newer Myburgh Bros; also partner with STARK-CONDÉ in revitalized STELL estate Lievland in volume value brand MAN Family Wines.

Jordan Wine Estate W Cape ★→★★★★ Family venture nr STELL, admired for consistency, quality, value from entry Chameleon to immaculate CWG bottlings. Flagship CHARD Nine Yards, B'x reds Sophia and Cobblers Hill, fabulous Insiders CAB FR. Success with newer Chard MCC augurs well for foray into bubbly production in UK.

Kaapzicht Wine Estate Stell ★→★★★ Family winery; top range Steytler (best-yrs CAPE BLEND Vision, PINOTAGE, B'x red Pentagon, old-vines CHENIN BL The 1947). Angle of Skuinsberg CINSAULT site a tractor driver's nightmare.

South Africa's new fine wines

The notion of what a fine wine is has been moulded over centuries by a host of factors, such as custom, track record, perception, commercial interests, even politics, as well as site. Areas outside classic regions tended to rate poorly, if at all, on the fine-wine appreciation scale. SA, as producer of the fabled sweet CONST wines of C18/19, straddles old and new. Let's call the 1st great new wines Early Modern: HAMILTON RUSSELL PINOT N, CHARD; KANONKOP Paul Sauer (B'x r), PINOTAGE; MEERLUST Rubicon (B'x r); NEDERBURG Edelkeur (botrytis CHENIN BL), THELEMA CAB SAUV; Welgemeend Estate Res (B'x r). More recent claimants: SADIE FAMILY Columella, Old Vine Series, Palladius (SWA blends r/w). Even newer, stellar contenders: ALHEIT Cartology, Thistle & Weed Duwweltjie (Chenin Bl); Boschkloof Epilogue (SYRAH); LEEU PASSANT Cab Sauv; MVEMVE RAATS MR de Compostella (B'x r); PORSELEINBERG (Syrah); Restless River (Cab Sauv, Chard); RICHARD KERSHAW (ELG Chard); STORM WINES Ignis (Pinot N); VAN LOGGERENBERG Breton (CAB FR).

Kanonkop Estate Coast ★★→★★★★ Decades-long undisputed "First Growth" status, mainly with PINOTAGE (regular and old-vine Black Label), B'x red Paul Sauer and CAB SAUV. Insatiable demand for 2nd-tier Kadette.

Keermont Vineyards Stell ★★★ Low-key but high-performing family estate, neighbour and a grape supplier to top-ranked DE TRAFFORD, on steep STELL Mtn slopes. SHIRAZ (incl single-vyd pair) and CHENIN BL (varietal, blend).

Keet Wines Stell ★★★★ Owner-winemaker Chris K with a single, beautfully crafted, classic B'x blend First Verse.

Ken Forrester Wines W Cape ★→★★★ With international wine specialist AdVini as partner, STELL vintner/restaurateur Ken F sets a high bar for Med varieties and CHENIN BL (dr, off-dry, sp, botrytis). New one-of-a-kind rosé pét-nat Stained Glass. Gluggable budget lineup Petit.

Klein Constantia Estate W Cape ★★→★★★★ Iconic property focused on SAUV BL, with nine different varietal bottlings, and on luscious, cellar-worthy non-botrytis MUSCAT de Frontignan Vin de Constance, convincing re-creation of legendary C18 CONST. Sibling winery Anwilka (r) in STELL with SYRAH/CAB SAUV.

Kleine Zalze Wines W Cape ★→★★★★ STELL-based star with brilliant CAB SAUV, SHIRAZ, CHENIN BL, SAUV BL in Family Res and Vyd Selection lines. Exceptional value in Cellar Selection series; lab range Project Z features some spectacularly successful tinkerings.

Klein Karoo REGION (c.2000 ha), mostly semi-arid, known for fortified, esp Port style in C'DORP. Revived old vines beginning to feature in young-buck bottlings eg. COLOMBARD Patatsfontein (see BOSCHKLOOF).

Krone W Cape ★→★★★ Krone lineup now 100% MCC bubbly, vintage-dated, refined and classic, with avant-garde touches eg. some 1st ferments in amphorae. Made at revitalized Twee Jonge Gezellen estate in TUL. Recently revived TJG label features savoury GRENACHE Noir from Pie.

Kumala W Cape ★ DYA Major entry-level export label, and sibling to premium FLAGSTONE and mid-tier Fish Hoek. All owned by Accolade Wines.

KwaZulu-Natal Province and GU on e coast; summer rain; sub-tropical/tropical climate nr ocean; cooler in central Midlands plateau. Abingdon, Highgate wineries, and, further n, Cathedral Peak Estate in central Drakensberg Mtn area.

KWV W Cape ★→★★★ Formerly national wine co-op and controlling body, today one of SA's biggest producers and exporters, based in PAARL. More than a dozen labels, headed by serially decorated The Mentors; newer and rarely seen varietal CARMENÈRE.

Leeu Passant *See* MULLINEUX.

Le Lude Cap Classique W Cape ★★★ →★★★★ Celebrated MCC sparkling house in FRAN, family-owned. Innovative offering incl CHARD/PINOT N Agrafe, 1st locally to undergo 2nd ferment under cork.

Le Riche Wines Stell ★★★→★★★★ Fine, modern-classic boutique CAB SAUV (varietal, "heritage" blend with CINSAULT Richesse) by Christo le Riche and siblings.

Limpopo Most n province and GU in wo system.

Lowerland ★★★ Meaning "Verdant Land", part of family agribusiness in Prieska WARD beside Orange River; area's most exciting winery. Just 9 ha, vinified in W Cape by top names. Heirloom COLOMBARD re-imagined, TANNAT tamed.

MCC (méthode cap classique) EU-friendly name for bottle-fermented sparkling, one of SA's major success stories; c.370 labels and counting.

Meerlust Estate Stell ★★★→★★★★ Myburgh family-owned vyds, cellar since 1756. Elegance, restraint in flagship Rubicon 09' 10' 15' 16 17' 18 20, was among SA's 1st B'x reds; excellent CHARD, CAB SAUV, MERLOT, PINOT N.

Miles Mossop Wines Coast ★★★→★★★★ CWG member Miles M's sophisticated bottlings. CAB SAUV (r and w blend), botrytized CHENIN BL named after family members. New Cape Vintage honours late father and "Port" exponent Tony M. Mostly STELL, recently wider sourcing, same oustanding quality.

Morgenster Estate W Cape ★→★★★ Prime Italian-owned farm nr Somerset W, advised by Pierre Lurton (*see* Cheval Blanc, B'x). Elegant Morgenster Res and second label Lourens River Valley (both B'x r). Old-country varieties incl one of only four SA VERMENTINOS.

Sorry, flower children (and investors): cannabis banned as wine flavourant in SA.

Motte, La W Cape ★★→★★★ Graceful estate, winery and cellar-door at FRAN owned by Koegelenberg-Rupert family. Old-World-styled B'x/Rhône varietals, blends, CHARD, SAUV BL, VIOGNIER, MCC, *vin de paille*. Neighbour and sibling Leopard's Leap emphasizes food side of wine match.

Mulderbosch Vineyards W Cape ★★→★★★ Highly regarded, US-owned STELL winery, sibling to FABLE MTN VYDS, with single-block CHENIN BL, B'x-red Faithful Hound, huge-selling CAB SAUV rosé.

Mullineux & Leeu Passant Fran, Kl K, Stell, Swa, W Cape, Well ★★★→★★★★ Chris M and US-born wife Andrea with star viticulturist Rosa Kruger transform SWA SHIRAZ, CHENIN BL and handful of compatible varieties into ambrosial varietals and blends based on soil type (granite, quartz, schist), CWG bottlings and *vin de paille*. Wider-sourced Leeu Passant portfolio – incl old-vine CINSAULT pair – as sublime.

Mvemve Raats Stell ★★★★ Mzokhona Mvemve, SA's 1st qualified black winemaker, and Bruwer Raats (RAATS FAMILY): best-of-vintage B'x blend, MR de Compostella.

Nederburg Wines W Cape ★→★★★ Among SA's biggest (two million cases) and best-known brands, PAARL-based, DISTELL-owned. Excellent flag-bearer CAB SAUV Two Centuries; Heritage Heroes, Manor House ranges. Delicious, long-lived CHENIN BL botrytis Edelkeur. Many value quaffers.

Neil Ellis Wines W Cape ★★→★★★★ Pioneer STELL-based négociant sourcing mostly cooler-climate parcels for site expression. Masterly Site Specific range, esp Jonkershoek V CAB SAUV and Pie GRENACHE Noir. Newer No Added Sulphite lineup (r/w/rosé).

Newton Johnson Vineyards Cape SC ★→★★★★ Acclaimed family winery in UP HEM. Top Family Vyds PINOT N, CHARD, SYRAH/MOURVÈDRE Granum. SA's 1st commercial ALBARIÑO. Some fruit from partner vyds. Entry-level brand Felicité.

Northern Cape (N Cape) Largest province and GU in WO scheme. Semi-arid to arid, with temperature extremes. Giant ORANGE RIVER CELLARS and exciting boutique LOWERLAND in recent Prieska WARD. *See* SUTHERLAND-KAROO.

Oak Valley Estate Cape SC ★★★→★★★★ Extensive family agribusiness in ELG with stellar PINOT N, CHARD, RIES, SAUV BL from mtn vyds. Newer deconstructed range, Tabula Rasa, features fascinating single-clone bottlings (r/w).

Old Vine Project Recent groundbreaking initiative aided by businessman/vintner Johann Rupert (ANTHONIJ RUPERT) to locate, catalogue and preserve SA's old vyd blocks (35 yrs+). Certified Heritage Vyd seal on bottle shows planting date.

Olifantsberg Family Vineyards Bre ★★→★★★ Dutch-owned rising star in BRE, focused on Rhône grapes, new-wave PINOTAGE, CHENIN BL on mtn slopes. SHIRAZ-based Silhouette, polished Blanc from mostly ROUSSANNE/GRENACHE BL.

Olifants River REGION on W coast (c.6270 ha). Warm valley floors, conducive to organics; cooler, fine-wine-favouring sites in vaunted Citrusdal Mtn DISTRICT and its WARD, Pie.

Opstal Estate Bre ★→★★★ One of BRE's quality leaders, family-owned, in mtn amphitheatre; v. fine CAPE BLEND (r/w), old-vines CHENIN BL, SEM.

Orange River Cellars N Cape ★→★★★ Vast operation with 620 grower-owners, 2300 ha under vine and three cellars on Orange River banks. Increasingly impressive, interesting Res bottlings incl Lyra Demi-Sec Sparkling from Hungarian grape IRSAI OLIVÉR.

Overarching GU / region *See* GU.

Paarl DISTRICT (c.8700 ha) around historic namesake town with WARDS Agter Paarl, Simonsberg-Paarl, V Pa. Diverse styles, approaches; best results from Med vines (r/w), CAB SAUV, PINOTAGE, CHENIN BL.

Paul Cluver Estate Wines Elg ★★→★★★ Family-owned/run, Elg's pioneer; convincing PINOT N, elegant CHARD (new lightly oaked Village bottling), knockout RIES (partly *foudre*-fermented semi-dry and stylishly presented botrytis).

Porseleinberg Swa ★★★★ BOEKENHOUTSKLOOF's organically farmed vyds and cellar with superb SYRAH. Handcrafted, incl front label printed on-site by winemaker.

Raats Family Wines Stell ★★★→★★★★ Pure-fruited CAB FR and CHENIN BL (oaked and unwooded), esp Eden High Density single-vyd bottlings. Bruwer Raats and cousin Gavin Bruwer Slabbert also partners in B Vintners, unearthing vinous gems. *See* MVEMVE RAATS.

Radford Dale W Cape ★★→★★★ Thoughtful, dynamic STELL venture with Australian, French, SA and UK shareholders: Land of Hope, Radford Dale, Thirst, Winery of Good Hope brands. Also owns certified-bio winery, formerly known as ELG Ridge. Sources widely for creative, compatible blend of styles, influences, varieties, terroirs.

Rall Wines Coast ★★★→★★★★ Owner-winemaker and consultant Donovan R has phenomenal track record since debut 08. Original SWA blends (r/w) since joined by eg. Ava pair (SYRAH, CHENIN BL) showing signature flavour-filled understatement. Also vinifies Callender Peak boutique wines (currently all w) from vines (some ungrafted, v. rare) on high/cold Ceres Plateau.

Region *See* GU.

Reyneke Wines W Cape ★→★★★★ Leading producer nr STELL, certified bio with apt Twitter handle ZAVineHugger. Pure, subtle, beguiling varietal SHIRAZ, CHENIN BL, SAUV BL; characterful Organic range blends (r/w).

Richard Kershaw Wines W Cape ★★★→★★★★ UK-born MW Richard Kershaw's refined PINOT N, SYRAH, CHARD from ELG; consituent sites showcased in separate Deconstructed bottlings. Newer GPS and Smuggler's Boot ranges spotlight other areas and new techniques, respectively.

Robertson Valley Low-rainfall inland DISTRICT with record 14 WARDS; c.12,800 ha: lime soils; historically gd CHARD, desserts; more recently SAUV BL, CAB SAUV, SHIRAZ. Major cellars, eg. GRAHAM BECK, ROBERTSON WINERY, and many family boutiques, incl newcomer Prévoir Wines.

Robertson Winery Rob ★→★★ (Mostly DYA) Consistency, value throughout extended portfolio. Best: Constitution Rd (SHIRAZ, CHARD).

Rupert & Rothschild Vignerons W Cape ★★★ Top vyds and cellar nr PAARL owned by Rupert and Rothschild families. Red blends Baron Edmond and Classique, CHARD Baroness Nadine.

Rustenberg Wines W Cape ★→★★★★ Barlow family cellar and vyds nr STELL. Beautiful site. Flagship CAB SAUV Peter Barlow, outstanding red blend John X Merriman, distinctive single-vyd CHARD Five Soldiers.

Boomlet in ageing whites under flor, Sherry-style: complex, savoury, ageable.

Rust en Vrede Wine Estate W Cape ★→★★★★ Historic STELL property with acclaimed restaurant; powerful, polished reds. Also STELL Res (mostly single-variety wines, all wo Stell); Donkiesbaai (from Pie, incl v.gd CHENIN BL, dr, *vin de paille*); Afrikaans, with a trendy heritage blend (CAB SAUV/CINSAULT); revived Cirrus, now showcasing high/cold Ceres Plateau terroir; and Guardian Peak range, vyds and cellar, where all non-estate wines made.

Sadie Family Wines Oli R, Piek, Stell, Swa ★★★★ Traditionally made portfolio by Eben S. Signature Series: SYRAH blend Columella, multi-variety Palladius (w), both Cape benchmarks. Magnificent Old Vine Series a celebration of heritage, esp Mev Kirsten from SA's oldest CHENIN BL. Co-winemaker Paul Jordaan and partner Pauline make Chenin Bl Bosberaad, under brand name Paulus.

Saronsberg Cellar W Cape ★→★★★ TUL family estate; awarded B'x/Rhône blends/varieties (SHIRAZ), uncommon-in-SA varietal MOURVÈDRE, bracing CHARD MCC.

Savage Wines W Cape ★★★→★★★★ CWG member Duncan S ranges far and wide from Cape Town city base for thrilling, understated wines from mostly Med varieties, CHENIN BL. Talent for catchy names (Never Been Asked to Dance, Are We There Yet?). Consults to promising UK-owned Brookdale in PAARL.

Shannon Vineyards Elg ★★★→★★★★ Probably SA's top MERLOT, also cracking PINOT N, SAUV BL, SEM, newer B'x white Capall Bán, grown in ELG by Downes brothers, James and Stuart, vinified in HEM at/by NEWTON JOHNSON.

Sijnn Cape SC ★★★→★★★★ DE TRAFFORD co-owner David Trafford and partners' pioneer venture on CAPE SC. Pronounced "Sane". Stony soils, maritime climate, distinctive varietals and blends. Winemaker Charla Haasbroek's eponymous brand also compelling.

Silverthorn Wines W Cape ★★★→★★★★ MCC sparkling boutique in ROB owned-run by ex-STEENBERG John Loubser and wife Karen. Handcrafted Brut, mostly CHARD, some PINOT N, rosé ex-SHIRAZ, River Dragon (COLOMBARD, heritage grape long associated with area), aptly named CWG bottling Big Dog. For v. special occasions.

Simonsig Wine Estate W Cape ★→★★★ Malan family venture nr STELL admired for consistency and lofty standards. Pinnacle wine is powerful mtn CAB SAUV The Garland; SA's original MCC, Kaapse Vonkel, still a delicious celebrator; Rhône-inspired white blends under The Grapesmith label.

Spice Route *See* FAIRVIEW.

Spier W Cape ★→★★★★ Large winery and tourist magnet nr STELL. Frans K Smit flagship, Creative Block, 21 Gables and Seaward ranges, all certified vegan.

Spioenkop Wines Elg, Stell ★★★ Belgian Koen Roose on ELG estate with Second Boer War-themed ranges (Spioenkop, "1900"); v.gd and individual PINOT N, PINOTAGE, CHENIN BL, RIES.

Stark-Condé Wines Stell ★★★→★★★★ US-born boutique vigneron José Conde in STELL's Jonkershoek. Exceptional CAB SAUV, SYRAH and Field Blend (w) in High Altitude and STELL ranges. Kara-Tara, winemaker Rüdger van Wyk's young brand (PINOT N, CHARD), looking gd too. Also vinifies neighbours' Syrah blend, Lingen, dubbed "BBQ wine for billionaires".

Steenberg Vineyards W Cape ★★→★★★★ Top CONST winery, vyds and chic cellar door, GRAHAM BECK-owned; SAUV BL/SEM blend Magna Carta, Sauv Bl, MCC, polished reds incl rare varietal NEBBIOLO.

Swartland 2.0

For a decade or more, SWA has had a monopoly on charisma and interest. Now its pre-eminence is contested by bottlings from all over the map. Like refined, savoury COLOMBARD Patatsfontein by Reenen Borman (BOSCHKLOOF) from vines in semi-arid Montagu; lithe, light-styled PINOTAGE Féniks by Bernhard Bredell (Scions of Sinai) ex-old block in STELL's Helderberg area; aromatic CHARD Onderduivenhoksrivier by Trizanne Barnard (TRIZANNE SIGNATURE WINES) featuring grapes from nr S Cape resort Vermaaklikheid; deliciously austere CHENIN BL Revenge of the Crayfish by w-coast-based newcomer Sakkie Mouton (SM Family Wines); and debutantes Hanneke Krüger and Pauline Roux (Vino pH) with textured, part skin-fermented version of locally rare PALOMINO.

Stellenbosch University town, demarcated wine DISTRICT (c.12,350 ha) and heart of wine industry – the Napa of SA. Many top estates, esp for reds, tucked into postcard mtn valleys and foothills. All tourist facilities.

Stellenbosch Vineyards Coast, Stell ★→★★★ Big volume, with impressive Flagship range. Limited Release lineup has only solo bottling of locally developed white grape Therona. Budget bottlings under Welmoed, named after C17 home farm.

Ageing wine in spirit barrels now legal in SA. Single-malt Sauv Bl, anyone?

Stellenrust Coast ★→★★★★ Family winery with extensive portfolio from three prime STELL terroirs. Headliner is magnificent Barrel Fermented CHENIN BL from 50-yr-old+ vines. Envelope-pushing, partly wider-sourced ArtiSons lineup also has top-flight Chenin Bl The Mothership, plus surprises eg. new Amarone-style CAB SAUV La Dolce Vita.

Stellenzicht Wines Coast ★★★ Famous Helderberg Mtn cellar and vyds rejuvened by Baron Hans von Staff-Reitzenstein, also owner of neighbour Alto Estate and long-time partner in nearby ERNIE ELS. CAB SAUV, SYRAH (solo, blended), powerful but not blockbusting; new CINSAULT, pair of CHARD.

Storm Wines Hem Rdg, Hem V, U Hem ★★★→★★★★ PINOT N, CHARD specialist Hannes Storm expresses favoured HEM sites with precision and sensitivity. Side project: textured SAUV BL Wild Air.

Sutherland-Karoo DISTRICT in challenging N CAPE, not to be confused with separate, distant KLEIN KAROO. Only 5 ha under vine, chiefly PINOT N, SHIRAZ, CHARD nr SA's coldest town, Sutherland. Scintillating SYRAH by Super Single Vyds, vinified offsite in STELL. TEMPRANILLO also v. fine.

Swartland Coastal DISTRICT, popular abroad; c.9600 ha of mostly shy-bearing, unirrigated bush vines produce concentrated, distinctive, fresh wines. Home to heavies like MULLINEUX, RALL, SADIE FAMILY, source for lengthening list of others.

Testalonga Swa ★★→★★★ Range name El Bandito says it all: Craig Hawkins's natural vinifications of organic/bio-grown Med/heritage varieties and fizz defy convention; much respected, loved and Instagrammed nonetheless. Earlier/easier approachable Baby Bandito label for iconoclasts-in-training.

Thelema Mountain Vineyards W Cape ★→★★★★ STELL pioneer of SA's modern wine revival, still beacon of quality, consistency (and touch of hedonism); CAB SAUV, MERLOT Res, red blends Rabelais and The Abbey (B'x, Rhône). Extensive Sutherland vyds in ELG set local benchmark for varietal PETIT VERDOT.

Thorne & Daughters Wines W Cape ★★★→★★★★ John Thorne Seccombe and wife Tasha's wines, some from v. old vines, marvels of purity and refinement. White blend (with eg. CLAIRETTE Blanche) Rocking Horse epitomizes Bot River vintners' cerebral-yet-sensual style and love of heirloom varieties. Hen's-teeth-rare gd-value PINOT N in Copper Pot range.

Tokara W Cape ★★→★★★★ Wine, food, art showcase nr STELL; vyds also in ELG. Gorgeous Director's Res blends (r/w); elegant CHARD, SAUV BL. Newer CAB SAUV Res, Chard MCC. Restyled XO estate brandy from CHENIN BL delectable.

Trizanne Signature Wines W Cape ★→★★★★ Number of women in SA's cellar teams rising rapidly, but female soloists like Trizanne Barnard still v. unusual. Avid surfer sources from mostly COAST vyds for seriously gd boutique SYRAH, varietal/blended SAUV BL, new CHARD from fashionable Lower Duivenhoks River DISTRICT. Dawn Patrol one of several mostly exported brands. Sister lone rangers: Christa von La Chevallerie (Huis van Chevallerie), Jocelyn Hogan Wilson (Hogan Wines), Lucinda Heyns (Illimis Wines), Marelise Niemann (Momento Wines), newcomer Jolette Steyn (The Vineyard Party).

Tulbagh Inland DISTRICT (c.980 ha) historically associated with white wines and MCC, latterly also red, esp PINOTAGE, SHIRAZ, some sweet styles.

Uva Mira Mountain Vineyards Stell ★★★ Helderberg Mtn eyrie vyds and cellar owned by Toby Venter, CEO of Porsche, Bentley, Lamborghini SA. Toned lineup incl newer SYRAH plus long-time performers CHARD, SAUV BL.

Van Loggerenberg Wines W Cape ★★→★★★★ PAARL-based Lukas v L recent entrant (2016), but already star of light-styled, new-wave scene. Top CAB FR Breton, pair of CHENIN BL. Savoury pink from CINSAULT, Break a Leg, spun off into affordable range. Vinifies also-excellent Carinus Family portfolio of mostly Chenin Bl.

Vergelegen Wines Stell ★★★→★★★★ Historic mansion and gardens owned by resources company Anglo-American plc. Immaculate vyds, wines; stylish cellar door at Somerset W. Powerful CAB SAUV "V"; sumptuous, perfumed B'x blends (r/w) named GVB. Deserved WWF-SA Conservation Champion.

Vilafonté Paarl ★★★ California's Zelma Long (ex-Simi) and Phil Freese (ex-Mondavi viticulturist) partnering ex-WARWICK Mike Ratcliffe. Trio of deep-flavoured B'x blends given distinctiveness by CAB SAUV, MERLOT or MALBEC predominance.

Villiera Wines Hem V, Stell ★★→★★★ Grier family nr STELL with exceptional quality/value ratio, esp Brut MCC bubblies incl new magnum-only Shooting Star for CWG Auction.

Vondeling V Pa ★→★★★ UK-owned, sustainability-focused estate in PAARL's V Pa. Eclectic offering incl new glee-inducing carbonated Little Sparkle (CHARD).

Walker Bay Highly regarded maritime DISTRICT (c.1000 ha); WARDS HEM, Bot River, Sunday's Glen, Stanford Foothills, Springfontein Rim. PINOT N, SHIRAZ, CHARD, SAUV BL standout.

Ward *See* GU.

Warwick Estate W Cape ★→★★★★ Tourist drawcard on STELL outskirts, prime vyds extended, redeveloped under recent US owners; v. fine full-flavoured CAB SAUV, CAB FR, CHARD, mostly B'x Pitch Black and old-vine CHENIN BL.

Waterford Estate W Cape ★→★★★ Ord and Arnold families' winery nr STELL with savoury Kevin A SHIRAZ, elegant CAB SAUV, intricate Cab Sauv-based flagship The Jem, latter with own tasting lounge at stylish quadrangular cellar door.

Waterkloof W Cape ★→★★★ British wine merchant Paul Boutinot's organic-farmed vyds, winery and glass-curtained cellar door nr Somerset W. Top tiers: Astraeus MCC, Circle of Life, Seriously Cool, Waterkloof. Quality easy-drinking False Bay, Peacock Wild Ferment lines.

Wellington Warm-climate DISTRICT (c.3820 ha) bordering PAARL and SWA. Growing reputation for PINOTAGE, SHIRAZ, red blends, CHENIN BL.

Western Cape (W Cape) Most s province and most important GU in WO system, with 114 of 136 official appellations.

Worcester Sibling DISTRICT (c.6600 ha) to ROB, BRE in Breede River basin. Mostly bulk produce for export. Alvi's Drift, Arendskloof, Conradie, Leipzig, Stettyn, Survivor (by Overhex), Tanzanite taste-worthy, mostly family-made exceptions.

WO (Wine of Origin) SA's "AOC" but without French restrictions. Certifies vintage, variety, area of origin. Opt-in sustainability certification additionally aims to guarantee eco-sensitive production. *See* GU.

Tweaking *cap classique*

Traditional-method sparkling celebrated its 50th birthday in SA in 2021 by introducing new rules – how else would one celebrate? A more differentiated approach sees the category divided into four classes: **Bottle-fermented** fermented in bottle, aged min 3 mths on lees; **traditional method** fermented in bottle in which it's sold, aged min 9 mths on lees; MCC fermented in bottle in which it's sold, aged min 12 months, must be certified; **extended lees ageing** new class of MCC, made from CHARD, PINOT N and/or M only, min 36 mths on lees.

How wine ages (and why)

Why do some wines age, and others not? Some wines will improve for 20 years or more. Others die after five. All follow the same basic trajectory – let's call it youth, adolescence and old age.

Most everyday wines pack a lifetime into just a few years. They might be aged for a few months in the producer's cellar to settle down after fermentation, but then they are bottled and sent into the world with the expectation that people will buy them and drink them immediately. Leave them too long, and their fruit will fade, their freshness dim. Others are sold with the tacit message that they should not be touched for some years, that to open them too young is to miss half the pleasure they promise. We bypass their youth and (more gratefully) their adolescence to enjoy their maturity, when it arrives.

Once upon a time, ageing was a remedy – for dry, obstreperous tannins, or for tooth-dissolving acidity. We still talk of wines "coming round", like patients after surgery. If wines were sufficiently challenging in youth, then the solution was to leave them be for as many years as they needed for the tannins to soften, the acidity to calm. Not surprisingly, a degree of awkwardness in youth became an implicit suggestion of great quality to come.

Summers are warmer now, and tannin management is better. These days it's quite normal to taste a just-bottled red from a great property and think, I could drink this now. The winemaker might urge us to put it away for a few years – it will improve, they will say; it will soften and open and develop more complexity. And they are right, it will. But ageing is no longer a solution to a problem, because the problem no longer exists. Ageing wine is now mostly a matter of choice and taste. Age still carries kudos. Aged wines have more prestige than young ones. But should they? And what is the difference between improving with age and merely surviving?

These are some of the questions I will be addressing in this supplement. Not everything is understood about the process of ageing, and I frequently taste wines that prove every theory wrong. But that is why it is interesting, no?

Must wine be ageable to be considered fine?

Traditionally, wines that aged well had a claim to be considered fine; those that didn't were everyday fodder and not suitable for connoisseurs.

Why is this? At first glance it seems curious. Cars are not considered less fine because they do not improve with age; the MacBook Air on which I am writing this is showing its age after only a couple of years, but most people would consider it a decent laptop. But the ability not just to survive for some years but to evolve for the better is one reason why wine is so fascinating. It is what lifts it above commodity level. The ability to improve with age is built into the narrative of fine wine.

What if we challenge this? You could construct a case against it – perhaps citing Condrieu (usually best drunk young but most often regarded as "fine"). Or even white burgundy (people probably drink it earlier these days, though still not immediately); the premox (premature oxidation) problem revealed in the late 90s is no longer an issue, but the memory remains. The ageing camp says that wines that will age turn out better, more complex and more interesting than those that will not. And even white burgundy of Village level and above needs some years of age to show its full depth.

But then there are what are often called "the new fine wines". An old-vine Chenin Blanc from South Africa's Swartland, or a Pinot Noir from Hemel-en-Aarde, perhaps. You can drink these within two or three years, and they're brilliant – poised, complex, tense. What about Grenache from McLaren Vale or Spain's Gredos Garnacha? Or a minimum-intervention wine made from native grape varieties in Portugal's Douro Valley or Alentejo? The late Denis Dubourdieu, who transformed the white wines of Bordeaux, reckoned that to be considered fine, a wine had to be able to age and improve for at least ten years. But that was then, and this is now. Will these new fine wines age for ten years or more? Some will; perhaps most. All we know for certain now is that they don't need to. But they are undeniably "fine", if by "fine" we mean "of outstanding quality".

Wine styles are evolving; our judgement must evolve with them. Even traditional fine wines are made to be drinkable earlier than of yore: warmer summers, better tannin management, that sort of thing. Jean-Philippe Delmas, cellarmaster at Château Haut-Brion, believes that some of his 2020s will be drinkable at about five years old; but they will live as long as ever. Frank Schönleber of Nahe estate Emrich-Schönleber says that in the past you had to leave (dry) Grosses Gewächs Rieslings ten years before drinking them, and they wouldn't last that long. Now, because of lower yields, they are ready much earlier, and they last 20 years and more.

One reason that traditional fine wines need to be able to promise ageability is the secondary market. The high price of young red

Bordeaux, for example, depends on those wines maintaining and increasing their value over their lifetimes. Eventually, when they become rarities, they are expensive because they're rare; and so far they have moved seamlessly from expensive-because-old-and-delicious to expensive-because-rare.

What if that changed? What if they ceased to live as long? That would eventually affect their resale value, which would in turn affect their release price. Ageability is a necessity for these wines. For the new fine wines, it's less important. Which gives us, as consumers, a great many more options.

Must wine age for as long as this to be considered "fine"?

What does "too young" mean?

"Too young for what?" you might ask. And the only possible answer is, too young for perfection. And what is perfection? Well, quite.

We're back to that trajectory of youth, maturity and decline. In a perfect world, we would have the option of drinking everything at the point of perfect maturity – but we'd still have to define what that point is. Wines straight from the fermentation vat are bursting with life, with sweet fruit, with juiciness. They're plump with puppy fat. They're exuberant and fun, in the way that a puppy is fun. Even Vintage Port is like that; even great, long-lived Cabernet is like that. But these wines have a core of tannin, a firm scaffolding that you can clearly perceive under the puppy fat. If that structure isn't there, then they're for drinking young. Some should be drunk as young as possible. They will have nothing to offer except youth and freshness. The producer will probably age them for a few months in steel or concrete to enable them to settle down, and then they'll be out on the shelves. From the drinker's point of view, there's no such thing as too young here. Keep them too long, and the fruit will fade and the structure will fall apart. The trajectory from youth to old age is quick.

Other wines can be delicious in youth, and then they close up. They become tight and inexpressive; the usual simile is of a monosyllabic adolescent. You know they're bright, and they used to be charming, but right now... For a novice drinker, this is the trickiest stage to encounter a wine. You have to look hard to find signs of promise – and usually those signs will be in the finish. When you taste a wine, if the palate isn't saying much but the finish – after you've swallowed or spat – just keeps expanding in your mouth, showing signs of the layers of flavour it refused to reveal earlier, then it is simply too young. It will be good, but it needs time.

Bouncy, plump, irresistible youth – but dogs and wine both need time to grow up.

How much time? That depends on the wine. Wines that close up comprehensively will need some years to emerge. Others may go through a slightly sulky phase but open again in a year or even a few months. It's never completely predictable. Winemakers will say, tasting a wine, "This was tasting wonderful/terrible six months ago, and look at it now." A line on a graph tracking such a wine's progress might look quite wobbly. Typically, it's reds with plenty of tannin that go through a really closed phase. Médoc is the classic example. All that pretty plumpness will have been shed, and you'll just get a mouthful of chewy tannin with the fruit buried well beneath. Whites can close up as well but never quite so sulkily. That tannic cosh is missing, so whites tend just to be inexpressive; think of being stuck at dinner with a stranger who has nothing to say. If the wine has been aged in new oak, it might be that only the oak is showing: perfect maturity will be when that oak has been absorbed into the wine, and the rest of the wine has opened up to envelop the oak. These days we don't want mature wines that taste of oak.

Sometimes it can be the acidity that needs to calm down. Sparkling wine made to be drunk young will taste different, at the same early stage, from sparkling wine made to be drunk at ten years old or more. It will be open and approachable, whereas the wine meant for ageing will have more structure, more weight and more power – and again, it will be more closed in youth and have, apparently, less to say. The 2012 Champagnes, from a great year, are an example. They were largely released at around eight years old or more, an age at which their NV siblings would already have been drunk; and they were drinkable, but tight. They needed time, and they're tasting lovely now. But once a wine has started to emerge again, how do you know when it's at its peak? The only honest answer to that, unfortunately, is, "when it has passed it." And we'll come to that next.

If it's still fermenting, it's definitely too young.

What does maturity look like?

Historically, British drinkers liked their (traditional) wines more mature than French drinkers; Americans were somewhere in between.

There was a particularly British taste for old Champagne – so old it had turned mushroomy in flavour. Spanish drinkers, too, adored very old Rioja, so ancient that the reds had turned pale and tawny, and the whites had darkened to, well, tawny, and the flavours were of beeswax furniture polish and were fragile and ethereal. For these drinkers, any hint of youthful vivacity would have been a warning light. Those are extreme examples. For most people, that would be what old age looks like: faded, graceful, fragile, with flavours of mushrooms, undergrowth and beeswax. The next stage is decline, when the wine thins out, the fruit goes and the finish breaks up. My definition of perfect maturity would still have some youthful vigour: wine should be singing and dancing, still light on its feet.

You could say that a wine is most truly itself when it is at its peak. Youthful wines have all the ingredients but some aspects are still hidden: exuberance can hide subtlety. Perfect maturity is when everything has come together in perfect balance. Take top Australian Cabernets, for example. You can drink them young; they're expressive at two or three years old. However, there's still that sensation of a closed door, or a dancer on pointe but with her face turned slightly away. The tannins are not an obstacle, because they're ripe and tucked in, but you'll notice them halfway through; they can still soften a bit more.

At its peak? Still young? Or an interesting relic?

What happens inside the bottle is easy to describe but harder to explain. Tannins polymerize, which means that the molecule chains get longer, and longer chains taste softer. Some tannins also drop out, forming a deposit in the bottom of the bottle. The colour, because some of these tannins are anthocyanins, lightens. A young top Cabernet will be dark blackish blue-red in the glass, and the rim, when you tilt the glass, will be dark as well. With time, that rim gets paler as the colour fades. Oak flavours – which might be noticeable in youth, even in these days of less oak influence – will (or should) blend into the wine and become unnoticeable. Puppy fat disappears, leaving the wine perhaps lean in adolescence, until greater roundness reappears with maturity. The first juicy fruit aromas, the primary aromas of fermentation and grape variety, are replaced by secondary and eventually tertiary aromas. Primary aromas are such notes as blackcurrant in Cabernet Sauvignon or grass in Sauvignon Blanc. Secondary aromas often derive from the *élevage*: a bit of toastiness from oak and a bit of creamy brioche from lees contact would count as secondary aromas.

Tertiary aromas come from age. Think of the savouriness that replaces primary fruit in a good Cabernet and those classic smells of tobacco and cigar box; think of the honey of mature Riesling.

The Champenois have helpfully codified the different flavours of different stages of Champagne, and there are lots (find them at www.champagne.fr in tasting-and-appreciation/champagne-tasting-experience/champagne-aromas). But to summarize: young Champagne tastes of flowers and fresh fruit; Champagne with some age tastes of brioche, honey, dried fruit and nuts; fully mature Champagne tastes of gingerbread, mushrooms, perhaps cocoa and coffee.

How long each stage lasts depends on grape quality, terroir, winemaking style – umpteen variables. But all wines, in the end, arrive at a similar point: red wines lighten, white wines darken, sweet wines dry out. Some sweet and fortified wines will last for a century or more: 1921 Sauternes from a great château can be incredibly good now. If you want to taste even older wines than that, then look at fortifieds. You can find Ports from the 19th century that are still aromatic, vigorous and much drier than they would have been in youth. (We'll look at the difference between wood ageing and bottle ageing on p.332.) Or there's Madeira, which seems indestructible.

Bottle variation, which most probably comes from the difference between one individual cork and another, increases with age. So very old wines are risky. Two identical bottles of the same wine can be very different: one on the way out or dead; one still alive, even frisky. There's no way of telling without sampling the wine.

Next we'll look at the various elements that help wine to age, or not.

The role of acidity

Some of the longest-ageing whites have high acidity. Think of German Riesling, Champagne, Loire Chenin Blanc, Sauternes, Tokaji, Hunter Semillon, Chablis. Should one then infer that acidity is necessary for ageing?

Yes, usually. Reds have tannins to act as a preservative while they age; whites have acidity. Whites with low acidity cannot be expected to live as long. Except that sometimes – tryingly, for those who like tidy rules – they do. White Rioja can last a human lifetime. Even low-acid grapes like Marsanne and Roussanne can confound expectations.

One might additionally point out that some of those long-ageing whites also have a good slug of (residual) sugar in them; in the case of Sauternes and Tokaji, a very good slug of sugar. Sugar helps wines to age. And those long-ageing white Riojas are aged for a long time in old oak, which gives them a bigger structure. You can also find Australian Chardonnays that were aged for a long time in oak and are still alive after 40 years – but like old white Riojas, a lot depends on the year, how much sulphur was injected at bottling and whether the cork has held up. Ageing capacity depends on various factors. The main one, which we will come back to with relentless regularity, is balance.

Wines that age have good balance from the start. Very acidic whites need to have that acidity balanced with something else, usually residual sugar or dry extract – which is one of the things, along with alcohol, that give a sensation of weight in the mouth. A very acidic, very dry, very low-alcohol wine will taste skeletal. German Riesling Kabinett of 8.5% abv relies on residual sugar for that balance. Trocken Rieslings need nearer 13.5% abv. Will the latter live as long? The same question can be asked of zero-dosage Champagne, without the added grams of sugar that will keep even good NVs going for some years. They are possible partly because of warmer summers giving riper grapes, but the jury is still out on their longevity, and it will be out for a few more decades yet. In any case, it was never every wine that would last 50 years.

Yes to acidity for ageing, but not only, and not always.

The role of tannins

Stem tannins are optional, but need to be ripe to be useful.

Tannins in the plural, you notice. Tannins are not just one thing.

They encompass anthocyanins, the blue-black pigments that come from grapeskins and give reds their colour; they include myriad different compounds and can come from skins, pips, stems or indeed oak. All these tannins can feel different in the mouth. When winemakers talk about "tannin management", it's partly about getting ripe tannins as opposed to unripe ones; it's also about deciding which tannins to include and which to exclude or minimize.

Tannins are brilliant – we need them. But we need them to be (magic word) balanced. Reds with massive extraction feel wearyingly solid to drink and, perhaps counterintuitively, fall apart early, particularly if that massive extraction goes with overripeness and a lot of oak, which it often does – or used to.

Tannins, like acidity, can act as a preservative while wines age. And the tannins in a wine intended to age are not the same as the tannins in a wine made to be drunk young.

They need to be balanced with ripe fruit, alcohol and acidity – not too much of any one element, and not too little. Some grapes arrive in the winery with the concentration to stand up to more extraction; some don't. (We'll discuss concentration on p.330.) For new-wave Grenache the buzzword now is often "infusion" – extraction so delicate that it does not impinge on the perfumed, floral fruit, instead giving it a structure, a frame. But it's the frame of an 18th-century watercolour, not the big black frame of a 17th-century Dutch portrait. Tannins also need to be ripe – and they ripen, on the vine, with time, whereas sugar levels rise with heat. Ideally, both kinds of ripeness occur together: it's part of the tricky art of matching grape variety to site. Get it wrong, and you'll have high sugar levels and green tannins. And that won't mature well either, no matter how long you leave it.

The role of concentration

For every wine, you could argue, there is an ideal point of concentration.

Lets use a bottle of squash as an example. Undiluted, it's undrinkable, but too much water makes it insipid. You can add water bit by bit from the start, but you have to judge when to stop; you can't take it away. Winemakers, at least in the EU, cannot add water. They have to judge the ideal concentration by the crop level on the vines; they can remove grapes early to increase concentration in the rest, but they can't put them back later. For the last few decades, until recently, the word was that a smaller yield was better. There was a mine-is-smaller-than-yours one-upmanship among growers.

The results, until the whole thing became silly, were good. Smaller crops ripened better and had better balance and more concentration: the wines were nicer young, and they aged more gracefully. It was one of the great improvements in wine, producing better quality across the board. Nowadays, though, many growers are quietly increasing their yields – with the aim, again, being better balance. What has changed?

Climate is one change; fashion is another. Concentration, taken to extremes, produces wines that do well in blind tastings but are undrinkable. Too much alcohol, too much new oak, too much tannin, too much everything. One way of dealing with warmer summers and the high sugar levels they produce is to delay sugar ripeness with a larger crop. Another is to recognize that pushing concentration to the point where Grenache and Shiraz become indistinguishable is to push it too far.

We're in a sweet spot now. Winemakers are focused on drinkability and freshness, but the great thing is that the lesson – that concentration is good and improves quality – is understood. There is no danger of returning to green insipidity – unless fashion suddenly discovers it, of course. There is no arguing with fashion.

Skin thickness and berry size affect concentration.

Achieving balance

Adjusting for perfect balance needs a good winemaker.

Balance is everything when it comes to wine ageing. Poor balance makes wines, and people, fall over. And sometimes – often – balance needs to be adjusted in the winery. Nature isn't perfect.

There is chaptalization, the adding of sugar during fermentation to increase alcohol levels. Acidification is adding acidity. You're not supposed to do both. Does either process actually improve balance? Yes, if done in small amounts. Can they be avoided with better viticulture? To some extent, but most winemakers like to have the option. Vines are plants, not automatons. You can't completely control what they do, and you can't control the weather. You can add tannin and structure with oak ageing; and you can choose your cooper and your barrel according to what it can do for your wine. Some barrels will add aroma; some will add structure at the end of the palate, some in the mid-palate. That's why barrels are so expensive; and some of the most expensive of all are those that are undetectable on the palate, so subtle are they. Then there is malolactic fermentation, making acidity taste softer without actually reducing it. Winemakers have a whole bag of tricks at their disposal, to tilt their wines to greater ageability, or not.

And then comes blending. This is how final balance is achieved in wine. A winemaker might blend different parts of a vineyard or blend different grape varieties. Cabernet Sauvignon makes really good varietal wines in only a few parts of the world, and the Médoc is not one of them. Napa can be; so can Margaret River, though even in these places winemakers might add a bit of Merlot, a bit of Cabernet Franc. It's how you balance structure with richness with aroma with acidity. Time does the rest. If you want to taste perfect balance without much intervention in the winery, try David & Nadia from Swartland. Or Schloss Lieser from the Mosel. Or Ceretto from Piedmont.

What happens in a barrel compared to a bottle?

Wines are bottled at different stages of their development – when the winemaker reckons their balance is just right.

That can be "just right" for drinking now, or in six months, or after ten years' ageing; we've seen that the balance will be slightly different in all those cases. But what is going on in the meantime? Why are some wines aged in oak, some in steel and some in concrete or even clay amphorae before being bottled? And why are some wines aged in oak for a decade or even longer?

Wines need to settle down after fermentation; even wines for immediate drinking can't be bottled the moment they stop bubbling. A steel tank might do the job, if you want crispness; concrete is fashionable again, for its neutrality. But an oak barrel will allow some slow ingress of oxygen through the pores of the wood, enough to

Pipes of Port ageing at Taylor's.

soften the wine over a few months, and soften it more over a year or two. Large old oak barrels are popular once more, for softening acidity and tannin without adding oak flavour.

Italy was a battleground when it came to small new oak barrels versus large old ones. The issue divided families; it's hard to imagine now that it could ever have mattered. But for a long time the traditional way of ageing, say, Barolo was in large old *botti*, during which the wine softened, became graceful and absorbed its sometimes-ferocious tannins. By the time it was bottled, it was entirely civilized. Then came the new idea of ageing it for less time in new oak barriques: the new oak did the job of softening the grape tannins, while adding some more of its own; the colour was darker, the style more bumptious, sometimes more vanilla-scented. Which aged better in the end? There's no clear answer, and it depended, as it always does, on balance. But they aged differently. The small-oak wines could end up graceful; not all the big-oak wines retained much vigour.

The difference is even clearer in Port. Tawny Port is aged in wood and bottled when it's ready to drink; Vintage Port is bottled young, for future drinking. Both, from good vineyards in good years, will age astonishingly well – for a century or even more. But they develop differently. Wood-aged Port loses more colour and becomes Tawny, eventually yellowish; with age it loses sweetness and keeps acidity. Old Vintage Port will have more colour, more sweetness, more fruit.

Once bottled, wine continues to age, but differently and more slowly. Very slowly indeed, if the bottle is screwcapped: wines bottled this way keep their youth for much longer. Providing the seal is good, of course.

And what if you age wine in bottle but throw a spanner in the works by making it re-ferment in its bottle first? We are talking, of course, about Champagne and other sparklers made by the traditional method. And the answer is, it's complicated.

These wines have two kinds of ageing – or three, if you count the ageing before they're bottled. They age on lees in the bottle, and here it's the breakdown of the lees over several years that has extraordinary effects on the wine, releasing minute quantities of substances that change its flavour. But it will keep all its freshness in this time: it gains complexity but tastes tight and young. Then, when it is disgorged, another phase of ageing begins, in which the wine softens and opens. Three bottles of the same wine, aged for 20 years in all, but one bottle-aged for three years on the lees, one aged for ten years on the lees and one aged for 19 years on the lees will taste like three different wines. The last one will taste the youngest.

Complicated? Certainly – very complicated indeed. And intriguing, don't you think?

Food for youth and food for age

This is not intended to lead anyone into ever-greater refinements of food-and-wine matching. As you can see from the chapter on p.25, there are always plenty of options and not many hard and fast rules.

Probably the only rule that applies here is that the older the wine, the more delicate the flavours and the more you need to make the food stand back. You would probably not put an old red burgundy with chilli con carne. Roast a chicken instead. Generally, the more subtle the food flavours, the more subtle the wine should be, and vice versa. Texture is important too. With casseroles, I would tend to go for older wines, silky in texture; obvious tannic grip is better with roast meat. But while this applies to Bordeaux or burgundy and other Cabernets and Pinots, it rather falls down with Barolo and Barbaresco, which go so well with risotto and sauced pasta in spite of their considerable grip.

In whites, old Riesling or Chardonnay will be happy with some good, firm-textured fish with, perhaps, beurre blanc. Old Champagne, full of umami flavours, is remarkably versatile and resilient, and it goes very well with a wide range of flavours. Young Champagne is trickier with food; keep it for the apéritif. And the same applies to top English wine and Riesling Sekt. Old still Riesling is a useful match for all sorts of complicated meals that mixes fermented foods, spices, citrus and whatever else.

Sweet wines? A very old Sauternes or Tokaji, dark with age but still bursting with life, offers few problems. You might want to showcase it with some perfect cheese, perhaps, rather than smother it in dessert. Leave the younger generation to do battle with pastry and cream. I'd say the same for ancient Port and other fortifieds: keep it simple with great cheese. The cheese will thank you too.

Probably the most wine-friendly dish of all time: roast chicken.

Storage

If you want to store wines that will age well, you must store them properly. All the usual rules apply: cool conditions without big temperature swings; darkness; not too dry and not too damp.

Vibration is also supposed to be bad, though nobody seems to know exactly why. But the worst things are heat and bright light. The latter can cause light strike, which you won't know about until you open the bottle and find the wine smells of drains. Wines in clear glass bottles are the most vulnerable, which is why you see so few ageable wines in clear glass. Sauternes and Tokaji are commonly sold this way, but the sugar seems to give a degree of protection. Even so, it's better not to risk it. Sparkling wine is very much at risk. And light strike can happen quickly, in minutes. Heat? It's like sunbathing. You know that it will age your skin, but you go on doing it because for a long time it doesn't show – and then it does. Wines exposed to heat won't be ruined immediately (unless it's so hot the corks have pushed out), but they will age faster; and again, you won't know until you open them.

Producers these days are alert to the dangers, and they ensure that when wine is shipped halfway across the world, it is sent in refrigerated containers – a major consideration for anyone buying at the top level. In fact, anyone buying mature wines on the secondary market should enquire where they've been stored. Wine that has been in the same reputable storage facility since it was shipped, even if it has changed owners several times, will have been cared for. If it has moved around a lot, there is a risk. The older the wine, the greater the risk. It's why old wines released from the producer's cellar carry a premium; the next page suggests producers who sell wines this way.

However one of the greatest dangers, if you plan to age wine for a long time, is forgetting about it. I had some lovely German Riesling that I forgot, then found – but too late. The corks, never very good, had become spongy, and the wines were brown and maderized.

And yes, I did curse.

Buying maturity

Auctions are good sources of mature classic wines, and they don't have to be expensive, especially if you don't go for blue-chip names or famous vintages.

Tip: lesser vintages from good producers are likely to be better than top years from lesser producers. But always ask how they've been stored in the past. www.wine-searcher.com is a good tool for discovering which merchants have stocks of mature vintages. You can also use it to search for single bottles of particular vintages – if you're looking for a 40th birthday present, or a 50th, or a 60th for example.

Some producers make a practice – regular or occasional – of releasing old vintages from their cellars. These may be available retail, or only to a mailing list, or they may only be available in person from the cellars. Here are some to look for:

AUSTRALIA: Barossa Seppeltsfield has century-old fortified – the world's only unbroken lineage of single-vintage wines; Pewsey Vale Museum Reserve. **Hunter** Semillon from Brokenwood and Tyrrell's (Vat 1). **Rutherglen** old releases from All Saints, Campbells Isabella, Campbells The Merchant Prince.

AUSTRIA: Domäne Wachau, Kracher, Nikolaihof. Also look at maywines.com and zarbach-weinhandel.shop, both based in Vienna.

FRANCE: Champagne regular releases of late-disgorged wines from Bollinger, Deutz, Dom Pérignon, Gosset, Jacquesson, Roederer, Veuve Clicquot; Boizel vintages back to the 70s. **Jura** Bourdy has decades-old vintages of *vin jaune*. **Loire** Lamé Delisle Boucard (Bourgueil) vintages currently back to 2005. Also Famille Bourgeois (Sancerre), Dom de Beauséjour (Chinon).

GERMANY: Wegeler (Mosel, Rheingau) Vintage Collection.

ITALY: Cantina Terlano (Alto Adige) top Rarity wines, 10 years old.

NZ: Pegasus Bay's Aged Release range, 10 years old.

PORTUGAL: Alentejo Herdade do Mouchão. **Lisboa** Quinta de Pancas 10-year-old cellar releases. **Madeira** Henriques & Henriques, Blandy's. **Port** Graham's, Poças, Taylor's, Quinta do Vallado very very very old Tawnies. Andresen will bottle century-old Colheitas on demand, though maybe not for a single bottle. *See* Very Old Tawny (Portugal).

SLOVENIA: Puklavec Family Wines for archive wines from the 70s.

SOUTH AFRICA: Boekenhoutskloof, Kanonkop, Klein Constantia and Nederburg.

SPAIN: López de Heredia (Rioja) as old as it gets.

SWITZERLAND: Louis Bovard (Vaud).

THE US: California Ridge, Chappellet for wine club members. **Oregon** Eyrie.

A little learning...

A few technical words

Winemaking terms inevitably creep into any discussion of wine styles and changing fashions. Here are the ones we use most in the book.

Acidity is both fixed and volatile. **Fixed** is mostly tartaric, malic and citric, all from the grape, and lactic and succinic, from fermentation. Acidity may be natural or (in warm climates) added. **Volatile (VA)**, or acetic acid, is formed by bacteria in the presence of oxygen. A touch of VA is common and can add complexity. Too much = vinegar. Total acidity is fixed + VA combined.

Alcohol content (mainly ethyl alcohol) is expressed as per cent (%) by volume of the total liquid. (Also known as "degrees".) Table wines are usually 12.5–14.5%. Controlling alcohol levels is a big challenge of modern viticulture.

Amphora the fermentation vessel of the moment, and the last 7000 years. Remove lid, throw in grapes, replace lid, return in six months. Risky.

Barrique small (225-litre) oak barrel, for fermentation and/or ageing. The newer the barrel, the stronger the smell and taste of oak; French oak is more subtle than American. Oak use is now far more restrained across most of the globe.

Biodynamic (bio) viticulture uses herbal, mineral and organic preparations in homeopathic quantities, in accordance with the phases of the moon and the movements of the planets. Now mainstream. Note "bio" in French means organic as well, but in this book it means biodynamic.

Carbonic maceration whole (red) berries go into a closed vat; fermentation starts within each berry. Gives a juicy, light style.

Concrete egg the fermentation vessel of the moment; keeps lees moving. Concrete generally has returned to fashion as part of the move away from oak.

Field blend different varieties planted together, picked and fermented together. Ultra-trendy.

Malolactic fermentation occurs after the alcoholic fermentation, and changes tart malic acid into softer lactic acid. Can add complexity to red and white alike. Often avoided in hot climates, where natural acidity is low and precious.

Micro-oxygenation is a widely used bubbling technique that allows controlled contact with oxygen during the wine's maturation. Softens flavours and helps to stabilize wine.

Minerality a tasting term to be used with caution: fine as a descriptor of chalky/stony flavours; often wrongly used to imply transference of minerals from soil to wine, which is impossible.

Natural wines are undefined but start by being organic or bio, involve minimal intervention in the winery and minimal sulphur or none. At best, wonderful; shouldn't be an excuse for faults. Often made in amphorae or concrete eggs.

Old vines give deeper flavours. No legal definition: some "vieilles vignes" turn out to be c.30 years. Should be 50+ to be taken seriously.

Orange wines are tannic whites fermented on skins, perhaps in amphorae. Like natural wines, some good, some not. Excellent with food.

Organic viticulture prohibits most chemical products in the vineyard; organic wine prohibits added sulphur and must be made from organically grown grapes.

Pét-nat (pétillant naturel) bottled before the end of fermentation, which continues in bottle. Slight residual sugar, quite low alcohol. Dead trendy.

pH is indication of acidity: lower pH indicates more acidity. Wine is normally 2.8–3.8. High pH can be a problem in hot climates. Lower pH gives better colour, helps stop bacterial spoilage and allows more of the sulphur dioxide to be free and active as a preservative. So low is good, in general.

Residual sugar is that which is left after fermentation has ended or been stopped, measured in grams per litre (g/l). A dry wine has almost none.

Sulphur dioxide (SO₂) added to prevent oxidation and other accidents in winemaking. Some combines with sugars etc., and is "bound". Only "free" SO_2 is effective as a preservative. Trend worldwide is to use less. To use none is brave.

Tannins a vital component of most reds, all orange, and some whites. Derived from